ELECTROMAGNETIC FIELDS AND WAVES

ELECTROMAGNETIC FIELDS AND WAVES

R.L. YADAVA
Professor
Department of Electronics and Communication Engineering
Galgotias College of Engineering and Technology
Greater Noida, Uttar Pradesh (*India*).

KHANNA BOOK PUBLISHING CO. (P) LTD.
Publisher of Engineering and Computer Books
4C/4344, Ansari Road, Darya Ganj, New Delhi-110002
Phone : 011-23244447-48 **Mobile:** +91-9910909320
E-mail : contact@khannabooks.com
Website : www.khannabooks.com

Dear Readers,

To prevent the piracy, this book is secured with HIGH SECURITY HOLOGRAM on the front title cover. In case you don't find the hologram on the front cover title, please write us to at *contact@khannabooks.com* or whatsapp us at +91-99109 09320 and avail special gift voucher for yourself.

Specimen of Hologram on front Cover title:

Moreover, there is a SPECIAL DISCOUNT COUPON for you with EVERY HOLOGRAM.

How to avail this SPECIAL DISCOUNT:

Step 1: Scratch the hologram
Step 2: Under the scratch area, your "coupon code" is available
Step 3: Logon to *www.khannabooks.com*
Step 4: Use your "coupon code" in the shopping cart and get your copy at a special discount
Step 5: Enjoy your reading!

Thanks,
Khanna Publishing House

Price: ₹ 425.00

ELECTROMAGNETIC FIELDS AND WAVES

R.L. Yadava

ISBN: 978-93-86173-54-6

First Edition: 2018
Reprint: 2021

Published by:

Khanna Book Publishing Co. (P) Ltd.
4C/4344, Ansari Road, Darya Ganj, New Delhi-110 002
Phone: 011-23244447-48 **Mobile:** +91-9910909320
E-mail: contact@khannabooks.com

Printed by: SPS Printers

Dedicated

To

MY PARENTS AND CHILDREN

PREFACE

This is the first edition of the book named Electromagnetic Fields and Waves. In this edition authors have tried to cover all the basic fundamental of the electromagnetic theory and waves. This book intended for use as text materials and is designed for B.E /B.Tech, M.E/M.Tech AMIE, IETE, M.Sc. Polytechnic students in Electronics and Communication, Telecommunication and in Electronics and Electrical Engineering and Information Technology, as well as for the student appearing in competitive examination of all public sectors and government sector. Having taught the subject for more than a decade along with valuable experience gained in industry, it was felt that a comprehensive book covering all the aspect as per Indian Universities was not available to the students. It inspired to author to write such comprehensive book.

The contents of the each chapter are described in lucid manner and tried to cover all the related topics of the electromagnetic. The planning of the each chapter is done in such a way that even beginner can get maximum benefits from the book. The book is recognized into eleven chapters.

Chapter 1. Briefs about historical views of the electromagnetic theory and their properties, applications and also the modern trends in Electromagnetic.

Chapter 2 discussed the basics of the scalar and vector algebra and the rectangular, circular and cylindrical coordinate systems and their transformations from one coordinate system to another.

Chapter 3 treats the static electric fields. In which Columb's law, Electric Field Intensity, Fields due to Different Charge Distributions, Electric Flux Density, Gauss Law and its Applications, Electric Potentials, Relation between E & V, Electric Dipole & Flux lines, Properties of the Materials, Propagation in dielectrics, Poisson's & Laplace Equations, Continuity Equations, Isotropic & Homogeneous Dielectrics, Resistance Capacitances, Methods of Images, Fields Mapping & concept of Fields cells have been discussed. Chapters 4 dedicated to the static magnetic fields in the same way as static electric fields while Chapter 5 focuses the Maxwell's equation in differential and integral forms, as well as cover Uniform plan waves.

Chapter 6 deals with the electromagnetic wave and power flow in different mediums, Poynting theorem in general and complex vector forms. It also deals with power flow in coaxial cable, surface resistance as well as wave impedance. Chapter 7 deals with introduction of the transmission lines, transmission line equation, different parameters of the transmission lines, impedance of the transmission line, application of smith chat in transmission line parameter calculations along with the application of the transmission lines itself.

Chapter 8 describes the impedance matching in antennas. Impedance Matching Techniques, Matching Section Networks, Balun transformer matching, and RF Balun Transformers have dealt in depth. In Chapter 9 author has focused on the basics of the waveguide, rectangular as well as circular wave guides, different modes in the wave guide field configuration, different parameters likes cutoff wavelength & cut off frequency, phase as well as group velocity, wave impedance also optical wave guides.

Chapter 10 deals with radiation and antennas in which authors have concentrated on basics of antennas and radiation from the antennas. Chapter 11 deals with the EMI/EMC effect and method of prevention, testing methods of EMI/EMC etc.

The arrangement of the topics is flexible; an instructor may conveniently select or order the several topics to suit either in one semester or two semesters. Several solved examples for each chapter establish the reader further understands of the subject discussed. The each chapter is finished with objective question bank as well as set of unsolved questions for the exercise.

ACKNOWLEDGEMENT

I thank Almighty for providing me this opportunity to be the author of such a comprehensive book. It gives immense pleasure in offering my heartiest and humble thanks to the Management of Galgoitas Educational Institutions (GEI), without which it has not been possible to give shape to this book. The positive support of senior professors, HODs, Registrar and Director of Galgotias College of Engineering and Technology deserves special thanks.

The constant motivation and help of colleagues and technical staff of College are sincerely acknowledged. Also, it is a great pleasure and honor for me to be associated with KHANNA BOOK PUBLISHING COMPANY, LTD New Delhi. I express my sincere gratitude to the entire team, including the production department for maintaining a high degree of precision and accuracy in the production of the book. I indebted to all reviewers for their valuable suggestions and comments, which have helped us in adding more student-friendly material in the book. I am grateful to my parents who have inculcated in me good values and culture. My sincere thanks go to my wife Om Lalitha, and children; Om Vikash, Saichethana and Om Prakash as without their wholehearted cooperation and sacrifices, this book would not have come in the present shape.

Last but not the least, I wish to avail myself of this opportunity to express a sense of gratitude and love to my beloved research scholars and students for their helpful support during the preparation of this book. Though every effort have been made to produce an error-free text. However, we shall feel much obliged to readers who can provide valuable suggestions/feedback to enhance the quality of the book.

CONTENTS

Chapter 1

INTRODUCTION

Electromagnetic wave may be regarded as the study of the interactions of electric charge particles in rest as well as in motion. Electromagnetic theory can be defined as a branch of physics or engineering which deals with electric and magnetic fields and their interaction with charge particles. Since ancient times people used to investigate the things like :

(*a*) Why do small paper clips attracted to an amber rod?
(*b*) What causing lightening?
(*c*) Why does a magnetic needle deflect when it kept close to current carrying conductors.?
(*d*) Why does stars twinkle but star does not?
(*e*) How does light reach the earth from the sun?
(*f*) Why does light separated into different colors when it passed through the prisms?

In modern days also people have questions like :

(*a*) How do we receive T.V. radio and other communication signals?
(*b*) How do the transmitted signal affect by atmosphere when it is propagated?
(*c*) Why we cannot receive communication signal inside the railway coach?
(*d*) Why mobiles receive the signal at one place whereas other places it does not?
(*e*) Why are T.V. antenna mounted at the top of the tower, whereas mobile antenna inside the mobile?
(*f*) Why do the things heated into the microwave oven.?
(*g*) Why radio signal and T.V. signal get disturbed, but not mobile signals when motorcvcle are started.?

The answers of above said questions can be given only from a well known threat called electromangetisms.

Earlier it was thought that electricity and magnetism are two independent phenomenonas. BM as time progressed the relationship between electricity and magnetism became evidence and ward electromasgnetism emerged. The man who relates the electricity and magnetics is Maxwell.

He has given four basic relationship between them as:

$$\nabla . D = \rho_v \quad ...(1.1a)$$

$$\nabla . B = 0 \quad ...(1.1b)$$

$$\nabla \times E = -\frac{\partial B}{\partial t} \quad ...(1.1c)$$

$$\nabla \times H = J + \frac{\partial D}{\partial t} \quad ...(1.1d)$$

where, ∇ = differential vector operator

D = electric flux density

B = magnetic flux density

E = electric field intensity

H = magnetic field intensity

ρ_v = volume charge density

and J = current density.

From these four relationship it is clear, that both electric and magnetic fields are such that if one will change it causes the effect on the others. Here *E, D, H,* Band *J* are vector quantities, hence for better understanding one should be familiar with vector quantity also.

Today, the applications of electromagnetics can be broadly divided into two categories :

(*i*) The low frequency and high power.

(*ii*) The high frequency and low power.

In addition there are few applications which have high frequency and high power also. For low frequency applications; The analysis of static fields like electrostatics and magnetos tales is adequate to investigate the electromagnetic phenomenon. However as the frequency increases electric and magnetic fields get more strongly coupled and we always observe a composite phenomenon of electric and magnetic fields. One can divide the subject of electromangetics into two parts; state electromagnetic and dynamic electromagnetic. The lime varying electric and magnetic fields always constitutes a wave phenomenon.

The electromagnetic theory is the generalization of the circuit theory, as the circuit theory is rather special case of the electromagnetic theory. The circuit theory deals with quantities like voltage, current, resistance, inductance and capacitance which are scalar in nature, while the electromagnetic fields deal with quantities which are vector.

Historical Views

Electromagnetic field theory is really the result of the union of three distinct sciences. The oldest of these is electrostatics, which was first studied by the Greeks. They discovered that if one tubbed certain substance, it becomes waveable and attracts lighter object, and one of these substances was amber, whose Greek name is electron. This is the name where we get the word electric. In 1785, a French physiciest, Charles Augustin de Coulomb (1736-1806), shows that electrically charged materials sometimes attract and sometimes repel each others. This was first indication that there are two types of charges, positive and negative.

In 1700s two Italians were working on the new science of current electricity. One, Luigi Galvani (1737-1798), was a physiologist and physician who thought that animal tissues generates electricity. Although he was later proved wrong but his experiments stimulated count Alessan dro Volta (1745-1827) to invent the first electric battery in 1800. Most of the early experiments in current electricity were performed on Frog's legs, this was a result of Galvani's work.

Later, a favourite party trick was to get a group of people to hold hands and then connected them to a voltaic battery. The battery produced quite a large voltage, which then caused current to flow through quests. This made them to jump uncontrollably! It was not until 1833, that battery Michel Faraday (1791-1867), shows that the current electricity of Volta and Galvani was the same as the electrostatic electricity of coulombs.

Although the ancient Greeks also know about the magnetism in the form of lodestone, the chinese incrented the magnetic compass and in 1600, William Gilbert of Gloucestor laid down some fundamentals. It was Oessted who demonstrated to a group of student that a current, carrying conductor produce the magnetic fields. This was first sign that electricity and magnetism could be interrelated. This link was strengthed in 1831 by the work of Faraday who showed that a changing magnetic field could induce a current into a wire. It was the French physicist Andre Ampere who first formulated the idea that the fields of permanent magnet could be due to currents in the materials.

Atomic Structure Circuits and Fields

When we learn to operate a mobile system, we do not necessarily need to know exactly how the mobile works. However, if we do understand how the different part (Antenna, RF section, Mixer display etc.) works and why the mobile used, it can help us to be better mobile user.

A similar situation occurs with electricity and magnetism, when we use electricity and magnetism we seldom have to worry about exactly how the effects are produced. However, it can make better engineers if we have an adequate model of electricity and magnetism. This is where we study the structure of atom and circuit equivalent of the parameters to analyze in better ways.

The simplest model of electricity can be given by hydrogen atom. This atom has one electron that orbits the nucleus containing a single proton. The charge on the electron is equal and opposite to the charge on the proton and has the value.

$$q = 1.6 \times 10^{-19} \text{ C}$$

where C is unit called coulombs.

We owe our present view of 'field theory' to Faraday who performed experiments on electricity and magnetism. After Faraday preferred to work without mathematics, he did introduce the idea of fields in free space. This greatly influenced later workers and it was in the mid 1800s that the British physicist James Clerk Maxwell (1831-1879) formalized Faraday's results using mathematics. Among other things Maxwell was able to predict the existance of electromagnetic waves. This work inspired others in the fields, such as Oliver Heaviside (1850-1925) who worked in the first transalantic telegraph cable, as well as predicted the existance of ionosphere.

The rest, as thing say, is history, due to the work of the German physicist Heinrich Rudolf Hertz (1851-1844) and the Italian engineer Guglictino Markionol (1874-1937), we are able to communicate even to great distance. We can also use electrical machinery to make our lives more comfortable. The complex materials, such as amber for instant have many atoms held in a crystalline structure. If we rub amber, the friction removes electrons, so, leaving the materials positivity charged. This is the basis of electrostatic electricity. In some materials the electrons are wery lightly bound to the nucleus and considerable energy, must be expanded to remove these electrons. These are insulators.

In a metal atom, the electron in the outermost orbit is not tightly bound to the nucleus. When a number of metal atoms are closed to each other, they form a crystalline structure in which these outermost electrons are free to more around metals, are electrically neutral with the source of electrons exactly balancing the number of protons. If we connect a source of electrons to the metal, injected electron will travel through lattice. As the charges repels, these electrons force the free electrons away from them. The net effect produces a disturbance that travels down the metal.

The rate of flow of charge is the current, *i.e.*,

$$i = \left(\frac{dq}{dt}\right) \quad ...(1.2)$$

We know that mass of electrons are 1837 times lighter than proton. So the conduction ground occurs from electron, not from proton.

Let us now turn our attention to magnetism. A flowing current generates magnetic fields. In a antenna we can regard, the motion of electronics around a nucleus constitute as current. After knowing these basics, we can conclude that it is necessary to know the atomic structure before going into details of electromagnetic.

Since, electric field E and magnetic field H are high frequency version of voltage and current respectively, then we can replace the whole thing to circuit model for better analysis of electromagnetic theory.

Properties of Electromagnetic Wave

An electromasgnetic wave originates from a point in free space and spreads out uniformly in ail the directions and it forms a spherical wave. However, at large distance from the source one is able to observe only the small part of the wave in his immediate vicinity and it appears to one as plane wave just as the oceans appear flats to a person who can only see a few km around him. Thus, at large distance from the source the wave has similar properties to the plane wave in the strip line and so by analogy of strip line, the properties of plane wave in free space may be as follows :

(*i*) At every point in free space, the electric vector field £ and the magnetic field vector H are perpendicular to each other and to the direction of propagation as given in below Fig. 1.1.

(*ii*) The directions of $\overline{E}$ and $\overline{H}$ are perpendicular everywhere in a plane wave. A uniform plane wave is one in which E and H lies in a plane and have the same value everywhere in that plane. At any fixed instant, the E and H fields are uniform over plane surface,

(*iii*) If wave propagation is in z-direction, then fields are not dependence on x and y, axes *i.e.*

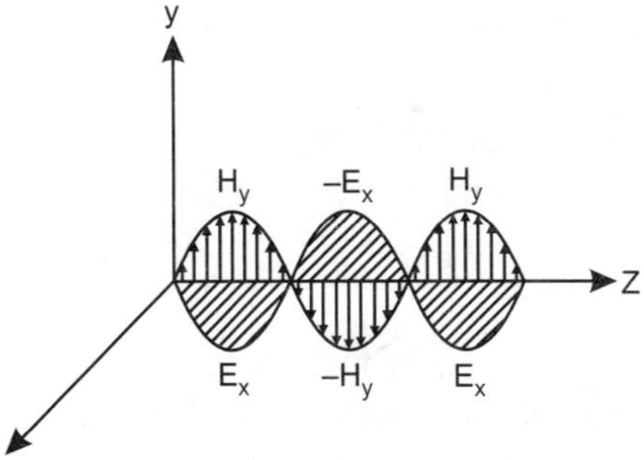

Fig. 1.1

$$\frac{\partial}{\partial x} = 0 \text{ and } \frac{\partial}{\partial y} = 0$$

i.e., variation of electromagnetic wave in x and y directions are zero.

(*iv*) The velocity of propagation in free space is,

$$c = \frac{1}{\sqrt{\mu\varepsilon}} = \frac{1}{\sqrt{\mu_r \mu_0 \varepsilon_r \varepsilon_0}} = \frac{1}{\sqrt{\mu_0 \varepsilon_0}} = 3 \times 10^8 \text{ m/sec}$$

(*v*) In free space charge density (and current density) is zero everywhere.

(*vi*) E and H field oscillates in phase and ratio of their amplitude is constant being equal to 120 π or 377 Ω;

$$\frac{E}{H} = \sqrt{\frac{\mu}{\varepsilon}} = \sqrt{\frac{\mu_r \mu_0}{\varepsilon_r \varepsilon_0}} = \sqrt{\frac{\mu_0}{\varepsilon_0}} = \sqrt{\frac{4\pi \times 10^{-7}}{\frac{1}{36\pi} \times 10^{-9}}} = 120\,\pi\Omega = 377\ \Omega$$

(*vii*) Whatever the frequency, electromagnetic wave travels in free space with velocity of light,

(*viii*) A plane wave is transverse, *i.e.*, E and H are transverse (90°) in the direction of propagation, so, they are also called TEM wave.

Applications of Electromagnetic Waves

Now-a-days all electrical and electronic devices somehow based on electromagnetic waves. Since of the applications can be given in the following sections :

1. **Transmission line :** At two frequency an electrical circuit is completely characterised by the electrical parameters like resistance inductance etc. and the physical size of the electrical components play no role in the circuit analysis. Simple KCL or KVL are adequate for analyzing a circuit. However, as frequency increases the size of component becomes significant and space starts playing a role in the performance of the circuit. The voltages and currents exist in the form of waves even a small change in the length of a simple connecting wire may alter the behaviour of the circuit. The circuit approach then re-investigated with inclusion of the space in the analysis. This approach then called transmission line. Two conductor transmission line media like the coaxial cable, flat ribbon cable etc. are example of transmission lines. Although the primary objective of a transmission line is to carry electromagnetic energy efficiently from one location to another location however they also find wide applications in high frequency circuit analysis.

2. **High frequency microwave circuits :** As the frequency increases any discontinuity in the circuit path leads to electromagnetic radiations. Also transit time effect can not be ignored at high frequency applications, where data rate are approaching to a GB/sec. This phenomena is also related with electromagnetic waves; like the bit distortion and signal reflection. Impedance matching also play important role in high speed data communication networks.

Antennas : An antenna is a device which can launch and receive electromagnetic waves efficiently. An antenna which appears as passive looking device in a communication system, is one of the most important devices.

Fiber optic communication : Fiber optic communication is the most modern form of guided wave communication. Electromagnetic theory is used to investigate light inside the optical fibers. The phenomenon of dispersion which has a direct bearing on the speed of communications, is due to the difference in speed of light for different mode of optical fibers. The model phenomenon of propagation inside an optical fiber is a direct consequences of the wave nature of the light. EM theory is is also important to analysis the optical fiber communication devices like, laser and photodiode.

Mobile communication. Knowledge of EM wave propagation plays a vital role in understanding the radio environment. Repeatation on the variation of the signal strength as a function of distance difference and frequency reuse schemes are employed in a cellular system. Fading is one of the important aspect of the mobile communication.

EMI and EMC: An electrical circuit which especially carrying heavy current, tends to give the EM wave radiation. This radiation interferes the other nearby electrical and electronics system and this affects the overall operation of the system. Switch mode power supply and high speed of digital circuit creates a substantial amount of EMI. To protect the system from EMI shielding technique are employed.

Radio Astronomy : This is the branch of astronomy in which observation of sky are carrial out at radio frequency. The radio signals received from the sky are very weak in nature. Studs of art communication receivers (are antennas) are required to detect these signals. This is one of the fields where a full understanding of electromagnetic wave is necessary.

• "The function of education is to teach one to think intensively and to think critically.

Martin Luther King, Jr.

Chapter 2

VECTOR ANALYSIS AND CO-ORDINATE SYSTEMS

2.1 INTRODUCTION

As such vector analysis is a mathematical tool which is used to express the concept of electromagnetism conveniently, and electromagnetism (EM) mainly as the study of the interaction between electric charges at rest as well as in motion. As well as electromagnetic fields and wave propagation is concern the study of co-ordinate systems, vector analysis and electromagnetic theory is compulsory. In EM theory we often deal with those fields which gave both electric and magnetic fields. The study of electromagnetic theory needs the knowledge and usage of certain mathematical concepts as tools for better understanding. The vectors and their transformation as well as elements of differential and integral calculus are essential for describing the basic concept of electromagnetism. The convenience of employing vector concept allows us to visualize problem with or without the specification of a co-ordinate system. After choosing co-ordinate system that will most accurately described the field, the field is then specified with the component determined with regard to the co-ordinate system. The co-ordinate system that is considered in the following section are; Cartesian, cylindrical and spherical co-ordinate systems. There are number of orthogonal systems which will also be considered. The term orthogonal means; every point in that section can be defined as the intersection of three orthogonal surfaces.

A quantity can be either scalar or vector, suppose a car is traveling with a velocity 'v' from a location 'A' to another location 'B' *i.e.*, car have certain velocity $\hat{v} = |v|$ in a prescribe direction.

In this case the speed is the magnitude of the velocity. Basically there are two classes of physical quantities, each with characteristic properties and an appropriate algebra; Vector and Scalar.

Vector: Vector is a quantity which has both magnitude as well as direction. Vector quantities includes; velocity, force, acceleration and displacement. For the complete specifications of vector quantity, it is necessary that;

(*i*) A unit of same kind must be used.

(*ii*) A number giving the magnitude of the quantity in terms of this unit, should be considered.

(*iii*) The direction of vector must be clear.

Scalar: Scalar is a quantity which has only magnitude and no direction. Scalar quantity can be specified by a single (positive or negative) scalar quantity such as; mass, density, temperature electric potential, voltage and power.

The product of a vector and a scalar is defined as a vector whose magnitude is equal to the numerical product of the magnitude of the original vector and the magnitude of the scalar and whose direction coincide with that of the original or is opposite to it depending on whether the scalar is positive or negative. Thus a vector × (–1) is equivalent of reversing the direction of the vector (A) *i.e.*,

$$A \times (-1) = -A$$

There are two types of vectors;

(*a*) **Axial Vector :** The vector in which rotary action of same kind takes place about an axis. For example angular velocity and angular acceleration is called axial vectors. Note that the vectors cannot represent the finite rotation of a body about a fixed axis.

(*b*) **Polar vector :** Force, displacement and linear velocity in which the vector is drawn in the direction of the quantity concerned. That is linear action in a particular direction such vector is called polar vector.

In almost all cases we concern with scalar and vector fields. A field is function that specifies a particular quantity everywhere in a region. The field is scalar/vector that depends on the quantity. That is there are two fields; *Scalar field and Vector field.* The temperature distribution in a building, sound distribution in a theater, electric potential in a region and refractive index of a stratified medium are the examples of scalar fields. Whereas the gravitational force on a body in the space and the velocity of raindrops in the atmosphere are examples of vector fields.

Unit vector : A unit vector 'a_A' along a vector A, is a vector whose magnitude is unity *i.e.*, 1 and its direction is along A, that is

$$a_A = \frac{\overline{A}}{|A|} = \frac{\overline{A}}{A} \qquad ...(2.1)$$

As $|a_A| = 1$, we can write $\overline{A}$ as $\overline{A} = Aa_A$ which describes a vector $\overline{A}$ in terms of its magnitude A and its direction a_A.

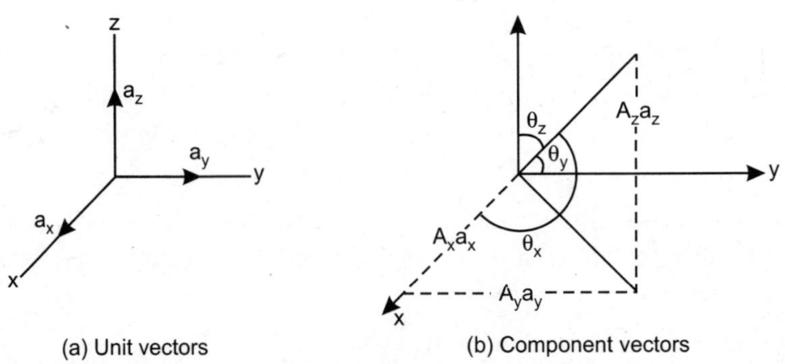

(a) Unit vectors (b) Component vectors

Fig. 2.1

In general, a vector $\overline{A}$ can be expressed in terms of its unit vector in different co-ordinate systems as follows;

$$(A_x, A_y, A_z) \text{ or } \overline{A}_x a_x + \overline{A}_y a_y + \overline{A}_z a_z$$
$$(A_\rho, A_\phi, A_z) \text{ or } \overline{A}_\rho a_r + \overline{A}_\phi a_\phi + \overline{A}_z a_z$$
$$(A_r, A_\theta, A_\phi) \text{ or } \overline{A}_r a_r + \overline{A}_\theta a_\theta + A\phi a_\phi$$

where (A_x, A_y, A_z), $(A_\rho\ A_\phi, A_z)$ and (A_r, A_θ, A_ϕ) are the components of a vector A in (x,y,z), (ρ, θ, z) and (ρ, θ, ϕ) co-ordinates respectively. The magnitude of a vector A is given by

$$A = \sqrt{A_x^2 + A_y^2 + A_z^2}$$

Hence, an unit vector along $\overline{A}$ is given by

$$a_A = \frac{\overline{A}_x a_x + \overline{A}_y a_y + \overline{A}_z a_z}{\sqrt{A_x^2 + A_y^2 + A_z^2}} \quad ...(2.2)$$

In Fig. 2.1 a_x, a_y and a_z are the unit vectors of the unit magnitude in the direction of increase *i.e.*, x, y, z axes respectively. From the Fig. 2.1 (*b*) we get;

$$A_x = A \cos \theta_x, A_y = A \cos \theta_y \text{ and } A_z = A \cos\theta_z \quad ...(2.3)$$

2.2 SCALAR AND VECTOR ALGEBRA

Addition and Subtraction of the Vectors

The addition of two vectors give another vector and follows the parallelogram rule *i.e.*,

$$\overline{A} + \overline{B} = \overline{C} \text{ and } \overline{C} = \sqrt{(A^2 + B^2 + 2AB \cos \theta)} \quad ...(2.4)$$
$$\overline{D} = \overline{A} - \overline{B} = \overline{A} - (+\overline{B}) \quad ...(2.5)$$

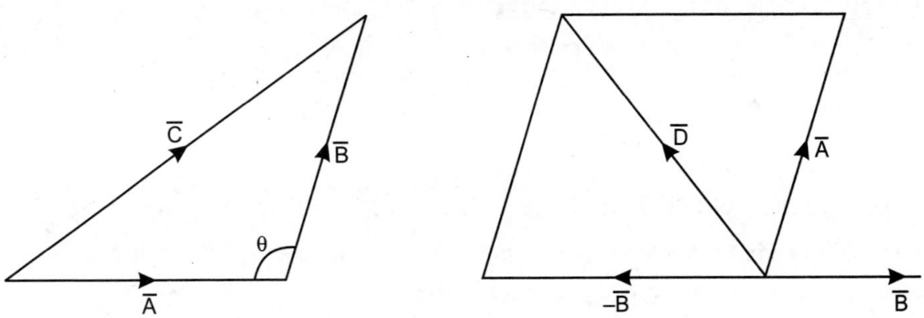

Fig. 2.2

If the vectors are expressed in terms of their components *i.e.*, $A = (A_x, A_y, A_z)$ and $\overline{B} = (B_x, B_y, B_z)$

Then

$$\overline{C} = (A_x + B_x)\, a_x + (A_y + B_y)a_y + (A_z + B_z)a_z$$
$$\overline{D} = (A_x - B_x)a_x + (A_y - B_y)a_y + (A_z - B_z)a_z$$

If both vectors are equal in magnitude and in the direction, then

$$\overline{C} = 2\overline{A} \quad ...(2.6)$$

Hence, by reversing the process of addition, any vector can be expressed as sum of *n*-vectors of which $(n-1)$ are arbitrary and the last one closes the polygon.

There are three basic laws associated with addition of vectors and they are;

Commutative law;

$$\overline{A} + \overline{B} = \overline{B} + \overline{A}, K\overline{A} = \overline{A}K$$

Associative law;

$$(\overline{A} + \overline{B}) + \overline{C} = \overline{A} + (\overline{B} + \overline{C})$$

Distributive law;

$$K(\overline{A} + \overline{B}) = K\overline{A} + K\overline{B}$$

$$\text{K}(l\overline{A}) = (Kl)\overline{A}$$

where '*K*' and '*l*' are scalars.

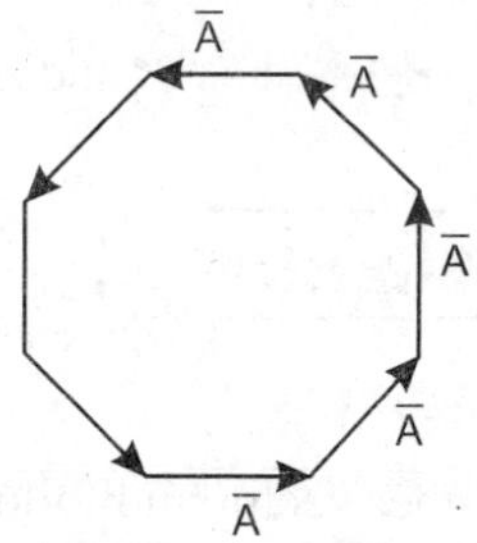

Fig. 2.3 Polygon formed by vectors.

Multification (Product) of Vectors

In the case of multification of two vectors, the result is either a scalar or vector that depends on nature of multification. However, there are two types of vector multification;

(*i*) Scalar or dot product; $\overline{A}\,.\,\overline{B}$

(*ii*) Vector or cross product; $\overline{A} \times \overline{B}$

The scalar product of two vectors $\overline{A}$ and $\overline{B}$ is defined as the product of magnitudes of both the vectors and the cosine of the angle between them *i.e.*,

$$\overline{A}\,.\,\overline{B} = |A||B| \cos\theta \qquad ...(2.7)$$

The scalar (dot) product yields the projection of the $\overline{A}$ and the vector $\overline{B}$, since $\cos\theta = \cos(-\theta)$, this can also be interpreted as the projection of the vector $\overline{B}$ on $\overline{A}$. The physical meaning of dot product is that it represents the work done by a force $\overline{A}$ on moving an object through $\overline{B}$ along a straight line.

$$W = F.d \cos\theta \qquad ...(2.8)$$

i.e., Work done = Magnitude of displacement × the component of the force along displacement.

If both the vectors are in their component form, then

$$\overline{A}\,.\,\overline{B} = A_x.\,B_x + A_y.\,B_y + A_z.\,B_z \qquad ...(2.9)$$

The dot-product of unit-vector yields;

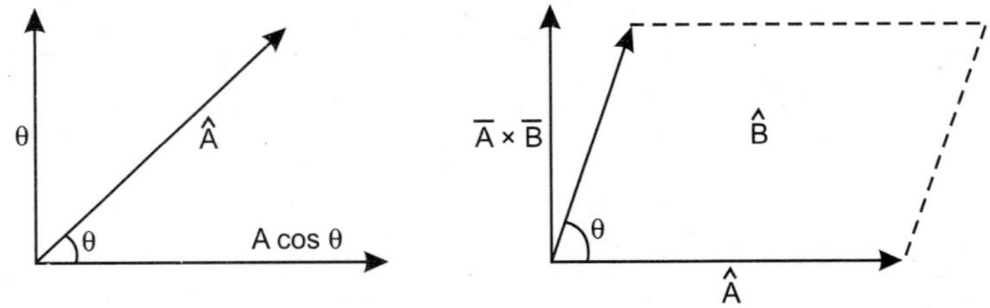

Fig. 2.4 Vector representations of 'dot' and 'cross' products of $\hat{A}$ and $\hat{B}$.

$$a_x.a_x = a_y.a_y = a_z.a_z = 1 \quad ...(2.10)$$

$$a_x.\, a_y = a_y.a_z = a_z.a_x = 0 \quad ...(2.11)$$

The vector product of two vectors is a new vector quantity whose magnitude is the area of the parallelepiped formed by $\overline{A}$ and $\overline{B}$ and perpendicular to place formed by $\overline{A}$ and $\overline{B}$ *i.e.*,

$$\overline{A} \times \overline{B} = \text{AB}\sin\theta.\hat{a}_n$$

where 'a_n' is unit vector normal to the plane containing $\overline{A}$ and $\overline{B}$.

In its components form, vector product of the two-vector is given by

$$\overline{A} \times \overline{B} = \begin{vmatrix} a_x, & a_y, & a_z \\ A_x, & A_y, & A_z \\ B_x, & B_y, & B_z \end{vmatrix}$$

Apply to unit vectors, cross-product gives;

$$a_x.\, a_x = a_y.\, a_y = a_z.\, a_z = 0 \quad ...(2.12)$$

$$a_x \times a_y = a_z\, a_y \times a_z = a_x,\, a_z \times a_x = a_y \quad ...(2.13)$$

Position and Distance Vectors

In order to define these vectors, let us consider, a point (P) located in Cartesian co-ordinate system by $P\,(x, y, z)$

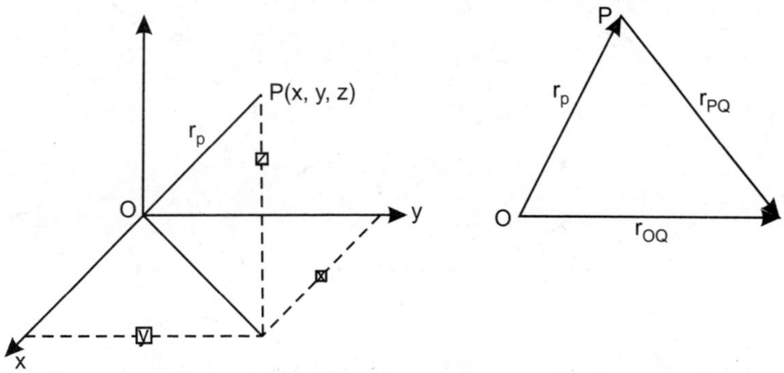

Fig. 2.5 Location and position of point 'P'.

The position vector (r_p) of the point 'P' is defined as the directional distance from the origin 'O' to 'P'. *i.e.*,

$$r_p = OP = x.a_x + y.a_y = z.a_z \quad ...(2.14)$$

However, displacement vector is defined as the displacement from one to another point. *i.e.*,

$$r_{PQ} = r_Q - r_p$$

On the basis of space-variables *x,y* and *z*, the vectors are classified into; uniform vector and non-uniform vector. Uniform vector does not depend upon space variables (*x, y, z*) *e.g.*,

$$A = 2a_x + 3a_y + 4a_z$$

However, uniform vector means, vector varies with space variables; (*x, y, z*);

$$A = 2xa_x + 3ya_y + 4za_z$$

2.3 VECTOR CALCULUS

Vector calculus means differentiation and integration of the vectors. It provides means for expressing certain ideas in antenna theory. Differential elements; length (*dl*), area (*ds*) and volume (*dV*) are useful in vector calculus. They are expressed in Cartesian, Cylindrical and Spherical co-ordinate systems as follows;

In Cartesian Co-ordinate System

$$dl = dx\,.\overline{a_x} + dy\,.\overline{a_y} + dz\,.\overline{a_z}$$

$$ds = dydz\,.\overline{a_x} + dxdz\,.\overline{a_y} + dx\,.dy\,.\overline{a_z}$$

$$dv = dx\,.dy\,.dz$$

Where *dx*, *dy* and *dz* are the elementary length along *x*, *y*, *z* axes and *dl*, *ds* and *dv* are the differential displacement, area and volume respectively (Fig 2.6).

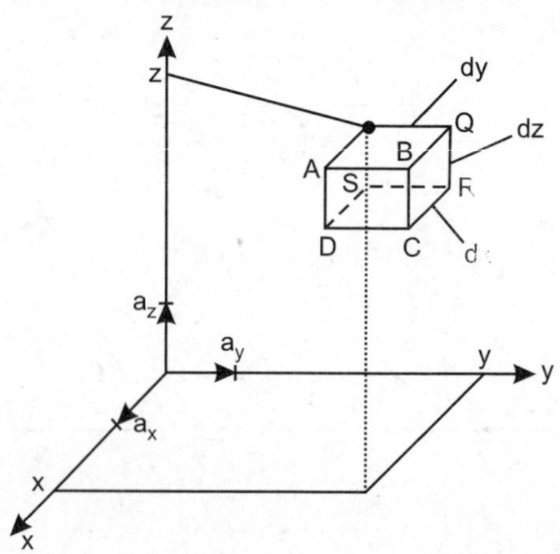

Fig. 2.6 Differential elements in Cartesian co-ordinates.

in cylindrical co-ordinate system

$$dl = d\rho\,\overline{a_\rho} + \rho.\,d\phi.\,\overline{a_\phi} + dz.\overline{a_z}$$

$$ds = \rho\, d\phi.\, dz.\, \overline{a_\rho} + d\rho\, dz\, \overline{a_\phi} + \rho\, d\rho\, d\phi.\, \overline{a_z}$$

$$dv = \rho\, d\rho.\, d\phi.\, dz$$

Where $d\rho$, $\rho\, d\phi$ and dz are the elementary lengths along x, y and z directions respectively.

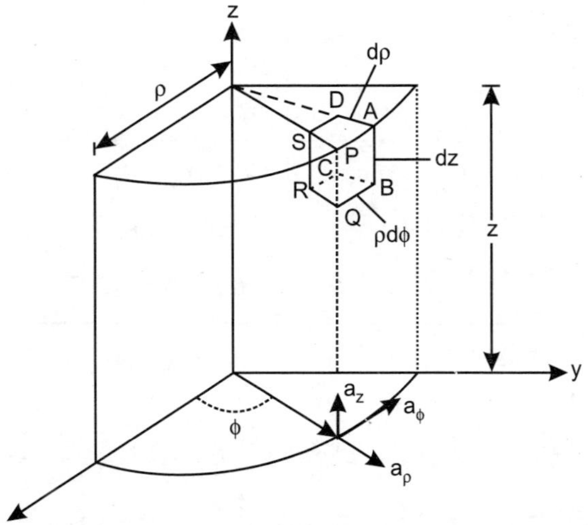

Fig. 2.7 Differential elements in cylindrical co-ordinates.

In spherical co-ordinate system

$$dl = dr.\, \overline{a_r} + r\, d\theta.\, \overline{a_\theta} + r \sin\theta.\, d\phi.\, \overline{a_\phi}$$

$$d_s = r^2 \sin\theta\, d\theta\, d\phi.\, \overline{a_r} + r \sin\theta\, dr\, d\phi.\, \overline{a_\theta} + r\, dr\, d\theta.\, a_\phi$$

$$dv = r^2 \sin\theta.\, dr.\, d\theta.\, d\phi.\, \overline{a_r}$$

Where dr, $(r\, d\theta)$ and $(r \sin\theta\, d\phi)$ are the elementary lengths along r, θ and ϕ directions respectively (Fig. 2.8). Table 2.1 shows the details of vectors and related parameters.

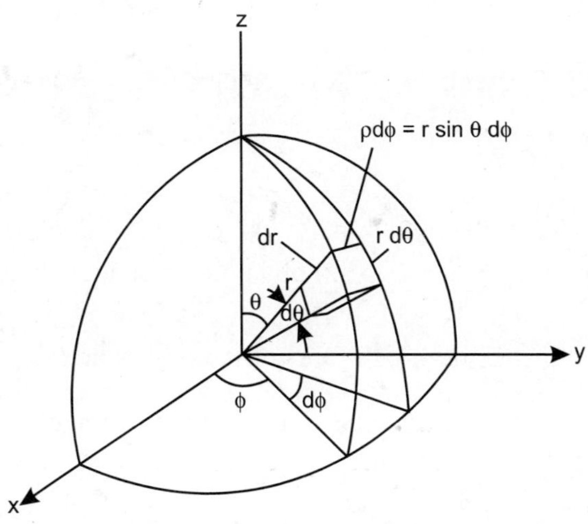

Fig. 2.8 Differential elements in spherical co-ordinates.

Table 2.1

Co-ordinate systems	Cartesian (x, y, z)	Spherical (r, θ, ϕ)	Cylindrical (ρ, ϕ, z)
Unit Vectors	a_x, a_y, a_z	a_r, a_θ, a_z	a_ρ, a_ϕ, a_z
Differential	$dx.\, a_x$ $dy.\, a_y$ $dz.\, a_z$	$dr.\, a_r$ $r \sin\theta.\, d\theta.\, a_\theta$ $r\, d\phi.\, a_\phi$	$d\rho.\, a_\rho$ $r\, d\phi.\, a_\phi$ $dz.\, a_z$
Differential surface Area	$dy.\, dz.\, a_z$ $dx.\, dz.\, a_y$ $dx.\, dy.\, a_z$	$r_2 \sin\theta\, d\theta\, d\phi\, \overline{a_r}$ $r \sin\theta\, dr\, d\phi\, \overline{a_\theta}$ $r\, dr\, d\theta\, \overline{a_\phi}$	$\rho d\phi.\, dz.\, \overline{a_\rho}$ $d\rho.\, dz.\, \overline{a_\phi}$ $\rho\, d\rho.\, d\phi.\, \overline{a_z}$
Differential Volume	$dx.\, dy.\, dz$	$r^2 \sin\theta.\, dr\, d\theta\, d\phi\, \overline{a_r}$	$\rho\, d\rho.\, d\theta\, dz$

Differentiation of Vectors

To describe the concept of differentiation of vector, let a vector A is vector function of the scalar variable, *i.e.*, when 't' changes from 't' to '$t + \delta t$' the vector 'A' becomes '$A + \delta A$'. Therefore the average rate of change of 'A' with respect to 't' as $\delta t \to 0$ is

$$\frac{dA}{dt} = \lim_{\delta t \to 0} \left(\frac{\delta v}{\delta t}\right) \qquad ...(2.15)$$

Since in cartesian co-ordinate

$$A = A_x.\, a_x + A_y a_y + A_z a_z$$

The

$$\frac{dA}{dt} = a_x\left(\frac{dA_x}{dt}\right) + a_y\left(\frac{dA_y}{dt}\right) + a_z\left(\frac{dA_z}{dt}\right)$$

For example, if A is displacement, then $\frac{dA}{dt}$ is velocity and $\frac{d^2A}{dt^2}$ is acceleration.

Partial-Differentiation

This is the case where a vector is function of more than one scalar independent variable. In case a vector is function of co-ordinates (x, y, z) of a point in space. Then $\frac{\partial A}{\partial x}$ represents the rate of change of A with respect to x, while y and z are constant. Similarly $\frac{\partial A}{\partial y}$ and $\frac{\partial A}{\partial z}$ are the rate of change of vector A *w. r.* to y and z respectively, while other two variable would be constant. If x, y and z changes simultaneouly by the elementary increment, then total differential dA will be

$$dA = \left(\frac{\partial A}{\partial x}\right)dx + \left(\frac{\partial A}{\partial y}\right)dy + \left(\frac{\partial A}{\partial z}\right)dz$$

or

$$dA = \left[\left(\frac{\partial}{\partial x}\right)dx + \left(\frac{\partial}{\partial y}\right)dy + \left(\frac{\partial}{\partial z}\right)dz\right]A$$

As the radius vector

$$r = a_x . x + a_y . y + a_z . z$$

$$dr = a_x . dx + a_y . dy + a_z . dz$$

Therefore $$dV = (\nabla . dr) . V$$

where $$\nabla = del = a_x \left(\frac{\partial}{\partial x}\right) + a_y \left(\frac{\partial}{\partial y}\right) + a_z \left(\frac{\partial}{\partial z}\right)$$

In next section, let extends the concept of integration to the case where the integral involves a vector. Basically there are three types of integrals; line integral, curve or surface integral and volume integral. The line integral $\int A\ dl$, is the integral of the tangential component of A along curve L_1 to L_2. Thus for a given vector function A over a path AO, the line integral will be;

$$\int_Q^P \vec{A} . dl = \int_Q^P A . \cos\theta . dl$$

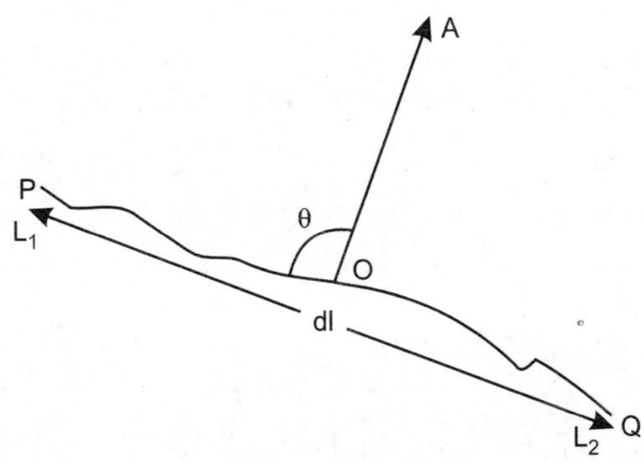

Fig. 2.9 Vector representation of partial differentiation.

When the path of integration is a closed curve, this integral is written $\oint A . dl$, which is called the circulation of A around L. Surface integral is similar to the line integral only if P and Q joins forming a closed surface (S), then surface $\int A\, dS$ is defined the flux of A through S and given by

$$\psi = \int_S A . \cos . ds = \int_S A . a_n . ds$$

where at any point on S, the a_n is the unit vector normal to S. For a closed surface (similar to line)

$$\psi = \oint_S A . ds$$

Which is referred to as the net outward flux of A from S. Similarly volume integral is defined as

$$\int_V f(x, y, z)\ \mathrm{dv}$$

where f (x, y, z) is a scalar function. The physical significance of line, surface and volume integrals depend on the nature of physical quantity say A or f in above cases.

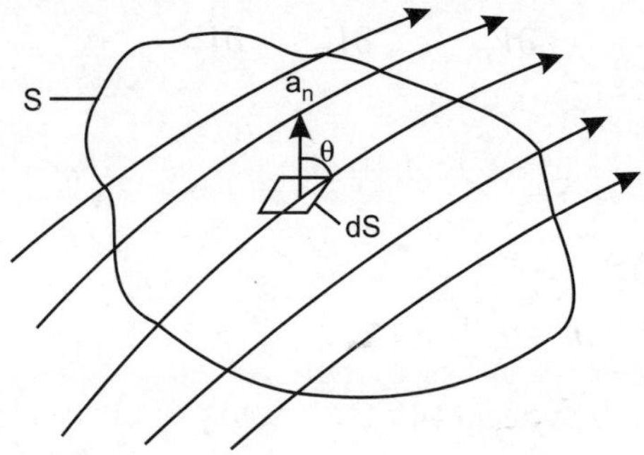

Fig. 2.10 Field lines across a surface.

2.4 VECTOR OPERATORS AND FUNCTIONS

Del Operator- The del operator was first introduced by Sir, W.R. Hamilton and developed by P.G. Tail. Originally it was named as '*nabla*' but later it was renamed as '*del*' by J.W. Gibbs. It is a vector differential operator and denoted by symbol ∇. The ∇ is also know as the gradient operator, and it is a scalar function. In various co-ordinate systems, it is defined as follows;

- in Cartesian co-ordinates;

$$\nabla = \frac{\partial}{\partial x}a_x + \frac{\partial}{\partial y}a_y + \frac{\partial}{\partial z}a_z \qquad ...(2.16a)$$

- in Cylindrical co-ordinates;

$$\nabla = \frac{\partial}{\partial \rho}a_\rho + \frac{1}{\rho}\frac{\partial}{\rho\partial \phi}a_\phi + \frac{\partial}{\partial z}a_z \qquad ...(2.16b)$$

- in Spherical co-ordinates.

$$\nabla = \frac{\partial}{\partial r}.a_r + \frac{1}{r}\frac{\partial}{\partial \theta}.a_\theta + \frac{1}{r\sin\theta}\frac{\partial}{\partial \phi}.a_\phi \qquad ...(2.16c)$$

All these can be applied as a directive differentiator to scalar as well as vector functions in the space. There are three useful operations with ∇, which are defined as follows;

(*i*) $\nabla V \rightarrow$ The gradient of a scalar V.

(*ii*) $\nabla . A \rightarrow$ The divergence of a vector A.

(*iii*) $\nabla \times A \rightarrow$ The curl of a vector A.

The gradient of a scalar V

The gradient of a scalar pertains specially to scalar function of space, It is a vector with both magnitude and direction of the maximum space rate of increase of a scalar field V. It is of significant importance in the study of electric potential. If V_T represent temperature, then ∇V_T will represent temperature gradient of the rate of change of temperature with distance i.e., temperature at various location in space. Though, temperature is a scalar quantity, the temperature gradient is a vector quantity having both magnitude and direction in the direction of maximum rate of change of temperature. The gradient vectors ∇V_T are expressed in direction of co-ordinate axes x, y and z, in terms of its components. The resultant temperature gradient can be expressed as

$$\nabla V_T = a_x \frac{\partial V_T}{\partial x} + a_y \frac{\partial V_T}{\partial y} \, a_z \frac{\partial V_T}{\partial z} \quad ...(2.17)$$

Above description reveals that, it is possible to connect the locations where the temperatures are same. The graphical plots of these equi-temperature contours show significant advantage in intersecting various effects. Like determining the magnitude and the directions where the most rapid changes occur or ascertaining the direction and the amount of heat that will flow. The gradient of scalar quantity (here temp.) help us to compute the magnitude and required direction that we should follow to attain the desired conditions.

If a scalar function ϕ represents electrical potential or gravitation, the 'grad ϕ' will represent electric potential or gradient of gravitation of respectively. The gradient V (*i.e.*,$\hat{\nabla}$V) can also be written in Cartesian, Cylindrical and Spherical co-ordinate systems using Equation 2.16. (*a, b, c*), just by inducing 'V' on both sides.

Properties of grad V

(*i*) If $A = \nabla V$, V is said to be the scalar potential of A.

(*ii*) The magnitude of ∇V equals maximum rate of changes in 'V' per unit distance.

(*iii*) ∇V points in the direction of the maximum rate of change in 'V'.

(*iv*) ∇ V at any point is perpendicular to the constant 'V' surface that passes through that point.

(*v*) The project of ∇V in the direction of a unit vector 'a' is (∇V. a) and it is called the directional derivative of ' V' along V.

(*vi*) When a vector field is expressed as the gradient of a scalar field, the line integral of the vector between two points is independent of the path followed and equals the difference of the values of the scalar at its ends. Also for such a vector field the line integral along any closed path in zero. A vector field so derived from the gradient of a scalar is sometimes called a scalar potential field, and corresponding vector is known as 'lamellar'

Divergence of Vector

In previous section, we operated Del (∇) on a vector field, which will produce a scalar function. Here we will operate Del (∇) on a vector field, which will produce a scalar function. The divergence of a vector/divergence operation is denoted by ∇. A. The divergence of a vector 'A' at a given point 'P' is defined as the outward flux per unit volume as the volume shrink about 'P'. The $\nabla.A$. that applies at any point is defined from the expression

$$div\, A = \nabla.A = \lim_{\nabla V \to 0} \frac{\oint A.ds}{\nabla V} \quad ...(2.18)$$

Where ∇V is the volume enclosed by the closed surface 'S' in which point 'P' is located.

The divergence of a vector can be viewed by the limit of the fields source strength per unit volume *i.e*, source density- It is positive at a source point in the field (divergence) and negative at a sink point (convergence) or zero where is neither source nor sinks. These three cases are presented in Fig 2.11 as follows.

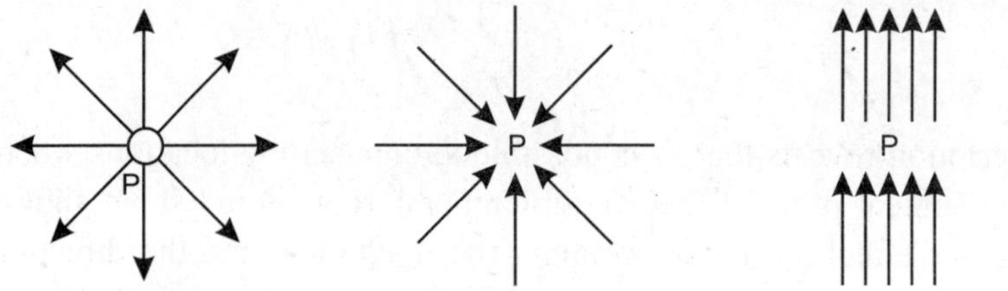

(a) Positive divergence (b) Negative divergence (c) Zero divergence

Fig. 2.11

The divergence of a vector '*A*' in different coordinate systems are expressed as follows;

$$\nabla A = \frac{\partial A_x}{\partial x} + \frac{\partial A_y}{\partial y} + \frac{\partial A_z}{\partial z}$$

$$\nabla A = \frac{1}{\rho}\frac{\partial(\rho A_\rho)}{\partial \rho} + \frac{1}{\rho}\frac{\partial(A_\phi)}{\partial \phi} + \frac{\partial A_z}{\partial z}$$

$$\nabla A = \frac{1}{r^2}\frac{\partial(r^2 A_r)}{\partial r} + \frac{1}{r\sin\theta}\frac{\partial(A_0 \sin\theta)}{\partial\theta} + \frac{1}{r\sin\theta}\frac{\partial(A_\phi)}{\partial\phi}$$

Properties of Divergence

(*i*) $\nabla.(A + B) = \nabla.\mathrm{A} + \nabla.\mathrm{B}$

i.e., Divergence of sum of two vectors is equal to sum of divergence of individual vectors.

(*ii*) $\nabla.\,(VA) = V\,\nabla.A + A.\nabla V = V.\,(Div.\,A) + A.$ (grad V) and

(*iii*) The divergence of a vector is scalar and has significant values, whereas divergence of scalar makes no sense (meaningless).

Physical Interpretation of ∇ . A

If the divergence of A is zero, means there is no source to create the vector A (field) and no sinks to absorbed the vector A at that location. Also, the divergence of a vector zero means, everything that enters into volume would come out without loss/gain. In context of electromagnetic field positive and negative divergence determines where positive or negative electric charges exist within a volume and the act that the magnetic charges of magnetic monopole do not exist in nature.

Curl and Curl of a Vector

The curl is a vector operation that is used to sense whether there here is a rotation, associated with a vector field or not. The curl of vector operation determines both the direction and magnitude of the rotation. In German, this vector also called '*Rotation Vector*'. It is second application of 'del' operator an a vector, and results another vector. The curl *A* is an axial vector, whose magnitude is the maximum circulation of A per unit area as the area tends to zero and whose direction is the normal direction of the area, when the area is oriented so as to make the circulation maximum. That is

$$\text{Curl A} = \nabla \times A = \left(\lim_{\Delta s \to 0} \frac{\oint A.dl}{\Delta s}\right)_{max} .a_n \qquad ...(2.19)$$

where Δs is the area bounded by the curve 'L' and 'a_n' is unit vector normal to the surface Δs, and is determined by the right-hand screw rule. The curl of a vector (curl A) in different co-ordinate systems is defined as follows.

$$\nabla \times A = \begin{vmatrix} a_x & a_y & a_z \\ \frac{\partial}{\partial x} & \frac{\partial}{\partial y} & \frac{\partial}{\partial z} \\ A_x & A_y & A_z \end{vmatrix}, \nabla \times A = \begin{vmatrix} a_\rho & \rho a_\phi & a_z \\ \frac{\partial}{\partial \rho} & \frac{\partial}{\partial \phi} & \frac{\partial}{\partial z} \\ A_\rho & \rho A_\phi & A_z \end{vmatrix}$$

and

$$\nabla \times A = \frac{1}{r^2 \sin^2 \theta} \begin{vmatrix} a_r & r a_\theta & r \sin \theta a_\phi \\ \frac{\partial}{\partial r} & \frac{\partial}{\partial \theta} & \frac{\partial}{\partial \phi} \\ A_r & r A_\theta & r \sin \theta A_\phi \end{vmatrix}$$

Properties of Curl A

(*i*) The curl of a vector is another vector.

(*ii*) The curl of a scalar is meaningless.

(*iii*) The curl of the gradient of a scalar field vanishes *i.e.*, $\nabla \times (\nabla V) = 0$

(*iv*) The divergence of the curl of a vector field vanishes *i.e.*, $\nabla \times (\nabla A) = 0$

(*v*) $\nabla \times (VA) = V \nabla \times A + \nabla V \times A$

Physical significance of curl A

The curl A provides the maximum value of the circulation of the field per unit area (*i.e.*, Circulation density) and indicates the direction along which this maximum values occurs. The curl A at point P in space regarded as how much the field curls around the point P. The curl of A in different situations is illustrated in Fig 2.12

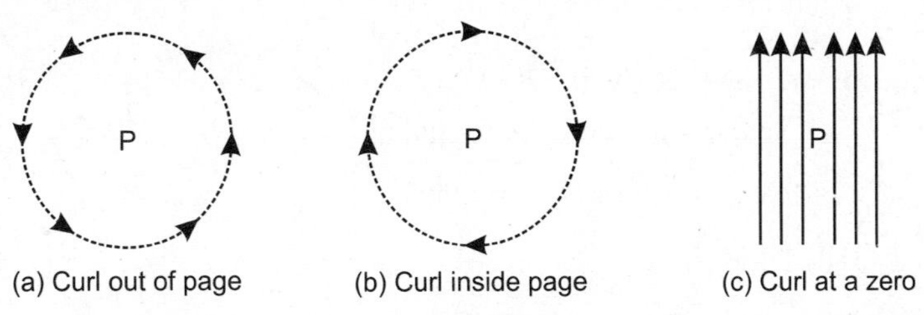

Fig. 2.12

2.5 CLASSIFICATION OF VECTOR FIELDS

As we have seen a vector field is uniquely characterized by its divergence and curl. In fact, neither divergence nor curl of a vector field is completely describes the field. All the fields can be classified on the basis of their vanishing/non-vanishing properties of divergence and curl as follows;

A. Solenoidal and Irrotational Fields

A vector field A is said to be solenoid (divergence less), if $\nabla. A = 0$. Examples of such field are; Magnetic field and conduction current density under steady state condition. A vector field A is said to be irrotational or conservative, if $\nabla \times A = 0$. Examples are electrostatic and gravitational fields.

Since curl $A = 0$, *i.e.* $A =$ grad V, where V is scalar potential. Therefore div $(\nabla V_\rho) = \nabla^2 V_\rho = 0$. This is called the Laplace's fields. This is reason A is also called potential field and V is scalar potential of A.

B. Irrotational but not Solenoid Fields

That is curl $A = 0$ but div $A \neq 0$. Hence, $A =$ grad V, but div $(\nabla V_s) = \nabla^2 V_s \neq 0$. This equation is called the Poission's equation and such fields are called Poissonian. For example; Electrostatic field in a charge medium, electrons in a thermionic tube and gravitational force inside the mass.

C. Solenoidal but Irrotational Field

That is $\nabla. A = 0$, but curl $A \neq 0$.

Since Curl $A \neq 0 \Rightarrow A =$ Curl V_V Where V_V is the vector potential.

$$\text{div } (V) = \text{div curl } V_V = 0$$

and

$$\text{Curl } A = \text{Curl curl } V_V \neq 0$$

$$= \text{grad div } (Vv) - \nabla^2 V_V. \neq 0$$

If however, div $(V_V) = 0$, then above equation reduces to $\nabla^2 V_V \neq 0$, which is similar to Poission's equation, but in turns a vector potential (V_V). For example;

(*i*) Magnetic field within conductor carrying a steady current.

(*ii*) Time-varying E.M field in a charge free and current free-region.

D. Neither Irrotational nor Solenoidal

In this case, $\quad$ curl $A \neq 0$ and $\nabla.A \neq 0$

Here $\quad A = grad\ V_s + \text{curl } V_v$

Where V_s and $V_v \rightarrow$ Scalar and vector potentials.

$$\text{div } A = (\text{div}\nabla V_s + \text{div curl } V_v) \neq 0$$

But div curl $V_v = 0$, Therefore

$$\text{div } (\nabla V_s) \neq 0$$

Which is the Poisson's equation to determine V_s.

Also

$$\text{curl A} = (\text{curl } \nabla V_s + \text{curl curl } V_v) \neq 0$$

but

$$\text{curl } \nabla V_s = 0$$

Hence curl curl $V_V \neq 0$, that is;

$$\nabla \times \nabla \times V_v \neq 0$$

Which can be reduced to the form of $\nabla^2 V_v \neq 0$ by using div $V_V = 0$ and then solving for the V_v. This is most generalized type of field. This is basic of Helmtroltz theorem. Examples of such fields are rotational notion of compressible fluids.

Laplacian Operator

Laplacian is a single operator and composite of gradient and divergence operators. It is denoted by ∇^2. The Laplacian of a scalar field *V*, is the divergence of the gradient of *V*. *i.e.*

$$\text{Laplacian V} = \nabla.(\nabla V) = \nabla^2 V$$

That is Laplacian of a scalar field is another scalar field. Laplacian operator may be scalar or vector Laplacian, depending upon whether operated on a scalar function or a vector respectivelv. In fact ($\nabla^2 A$) does not define as the divergence of the gradient of *A* as well as does not bring any sense. Rather ($\nabla^2 A$) is defined as the gradient of the divergence of A minus the curl of the curl of *A*.

i.e.,
$$\nabla^2 A = \nabla(\nabla.A) - \nabla \times \nabla \times A$$

In the Cartesian co-ordinate systems, its becomes;

$$\nabla^2 A = \nabla^2 A_x.\ \overline{a_x}\ \nabla^2 A_y.\ \overline{a_y} + \nabla^2\ \overline{a_z}$$

Laplacian operator in different co-ordinate systems is defined as.

$$\nabla^2 V = \frac{\partial^2 V}{\partial x^2} + \frac{\partial^2 V}{\partial y^2} + \frac{\partial^2 V}{\partial z^2}$$

$$\nabla^2 V = \frac{1}{\rho}\frac{\partial}{\partial \rho}\left(\rho \frac{\partial V}{\partial \rho}\right) + \frac{1}{\rho^2}\frac{\partial^2 V}{\partial \phi^2} + \frac{\partial V}{\partial z^2}$$

$$\nabla^2 V = \frac{1}{r^2}\frac{\partial}{\partial r}\left(r^2 \frac{\partial V}{\partial r}\right) + \frac{1}{r^2 \sin\theta}\frac{\partial}{\partial \theta}\left(\sin\theta \frac{\partial V}{\partial \theta}\right) + \frac{1}{r^2 \sin^2\theta}\frac{\partial^2 V}{\partial \phi}$$

A scalar field '*V*' is said to be harmonic in a given region, provided its Lappacian vanishes in that region. That is

$$\nabla^2 V = 0$$

Which is called Laplace's equation.

2.6 CO-ORDINATE SYSTEMS

We have seen that any physical quantity; vector or scalar is mainlv defined in three different co-ordinate systems; Cartesian (x, y, z), Cylindrical (ρ, ϕ, z) and Spherical (r, ϕ, θ) co-ordinate systems. In order to describe the spatial variation of these quantities completely, we must define all the concern point uniquely in space in a proper manner which may be done by using an appropriate reference co-ordinate system. In three dimensional space, a co-ordinate system can be specified by the intersection of three surfaces. Each surface could be described by A_i=constant, where i = 1,2,3 and define i^{th} axis of the orthogonal co-ordinated systems. An orthogonal system is one in which the surfaces are mutually perpendicular at a point. These surfaces may be planer of curved. Examples of orthogonal co-ordinates systems are Cartesian, Cylindrical, Spherical, Elliptisoidal and Prolate spheroidal. In Cartesian co-ordinate system, all of the surfaces are planar and they are specified by each of independent variable *x,y* and *z* separately being constants. In cylindrical co-ordinate system two surfaces are plane and third one is cylinder. However, in spherical co-ordinate system the surfaces are; a plane, a sphere and a cone.

Cartesian Co-ordinate System (*x, y, z*)

In practice, there are two basic ways to describe this co-ordinate system (Fig 2.13): *Right handed and left handed systems*:

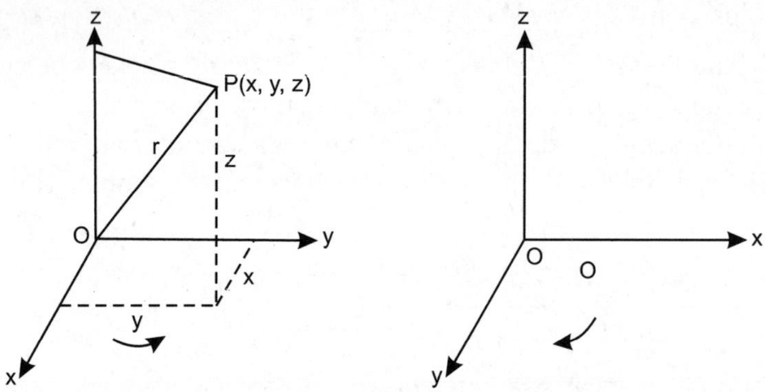

Fig. 2.13 Right and left handed cartesian co-ordinate systems.

A point *P* can be represented by *P* (*x, y, z*), where *x, y* and *z* indicate the position of the particle corresponding distances. The range of the co-ordinate *x, y, z* lies;

$$-\infty < x < \infty$$

$$-\infty < y < \infty$$

$$-\infty < z < \infty$$

A vector $\overline{A}$ in Cartesian co-ordinate system is represented by

$$(A_x, A_y, A_z) \text{ or } A_x\, a_x + A_y\,.a_y + A_y\,.a_z$$

and $$|A| = \sqrt{A_x^2 + A_y^2 + A_z^2}$$

where $\overline{a_x}$, $\overline{a_y}$ and a_z are unit vectors and A_x, A_y and A_z are the components of vector $\overline{A}$ along *x, y* and *z* directions. The radial distance '*r*' of the paints is given as follows;

$$r = \sqrt{x^2 + y^2 + z^2}$$

Cylindrical Co-ordinates System (ρ, ϕ, *z*)

This co-ordinates system is obtained by taking radius vector in *x-y* plane and rotating it about the z-axis. *i.e.*, translation of the *x-y* plane normal to the z-axis. This system is very convenient whenever we deal with case having cylindrical symmetry.

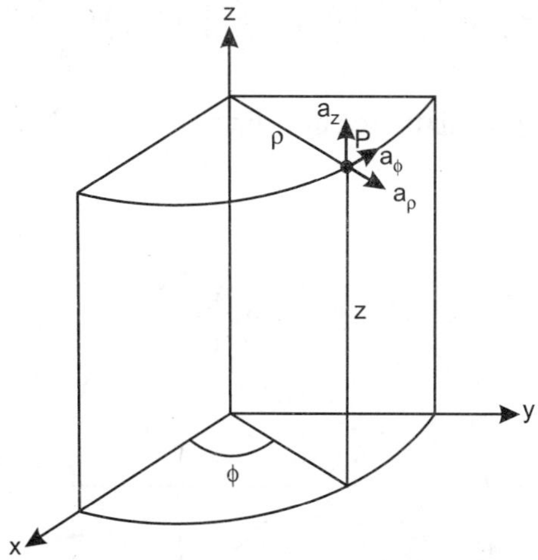

Fig. 2.14 Cylindrical co-ordinate system.

A point in cylindrical co-ordinate is represented by (ρ, ϕ, z) as shown in Fig 2.14, where ρ is the radial distance from the z- axis, ϕ is the azimuth angle and is measured from the x- axis in x-y plane, and z is remains same as in previous one. The range of cylindrical variables lies;

$$0 \le \rho \le \infty$$

$$0 \le \phi \le 2\pi$$

$$-\infty \le z \le \infty$$

Any vector in this co-ordinate system is represented by

$$(A_\rho, A_\phi, A_z) \text{ or } A_\rho a_\rho + A_\phi a_\phi + A_z a_z$$

and $$|A| = \sqrt{A_\rho^2 + A_\phi^2 + A_z^2}$$

where a_ρ, a_ϕ and a_z are unit vectors along ρ, ϕ and z-directions. Note that a_ϕ is not in degree and it is assumed the unit vector of A.

If any force = 20 a_ϕ . N acting along ϕ in circular motion, it has a_ϕ in Newton. The relation between Cylindrical and Cartesian co-ordinates are

$$\rho = \sqrt{x^2 + y^2},$$

$$x = \rho \cos \phi,$$

$$y = \rho \sin \phi, z = z$$

and $$\tan \phi = \left(\frac{y}{x}\right)$$

Spherical Co-ordinates System

Spherical co-ordinate system is obtained by rotating the radius about the origin (Fig. 2.15). A point P in this co-ordinates system is represented by P (r, ϕ, θ).

A vector **A** from the origin of the co-ordinate system to the point is determined by the intersection of the sphere whose radius 'r' is the plane that makes an angle 'ϕ' with respect to the axis and the cone that makes an angle θ w.r. to the z-axis.

The range of variable (r, ϕ, θ) in these co-ordinates lies:

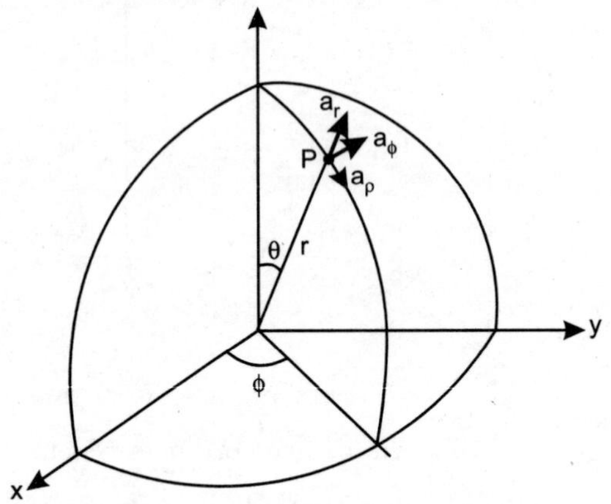

Fig. 2.15 A point 'P' in spherical co-ordinate system.

$$0 \leq r \leq \infty$$

$$0 \leq \phi \leq 2\pi$$

$$0 \leq \phi \leq \pi$$

Any vector A in spherical co-ordinate system is given by

$$(A_r, A_\phi, A_\theta) \text{ or } A_r a_r + A_\phi a_\phi + A_\theta a_\theta$$

and $$|A| = \sqrt{A_r^2 + A_\phi^2 + A_\theta^2}$$

Relation between Spherical and Cartesian co-ordinates is

$$\theta = \tan^{-1} \frac{\sqrt{x^2 + y^2}}{z}, \phi = \tan^{-1}\left(\frac{y}{x}\right)$$

$$x = r \sin\theta. \cos\phi,$$

$$y = r \cos\theta \sin\phi \text{ and } z = r\cos\theta$$

However, relation between Spherical and Cylindrical co-ordinate system is

$$r = \sqrt{\rho^2 + z^2},$$

$$\theta = \tan^{-1}\left(\frac{\rho}{z}\right)$$

$$\rho = r\sin\theta,$$

$$\phi = \phi \text{ and } z = r\cos\theta$$

2.7 CO-ORDINATES TRANSFORMATIONS

Cartesian to Cylindrical ($x, y, z \to \rho, \phi, z$)

We know that $x = \rho \cos \phi$, $y = \rho \sin \phi$ and $z = z$, which represents the transformation of a point (ρ, ϕ, z) to (x, y, z), Fig. 2.16. The vice-versa of the same could be represented by

$$\rho = \sqrt{x^2 + y^2},\ \phi = \tan^{-1}\left(\frac{x}{z}\right)$$

and
$$z = z$$

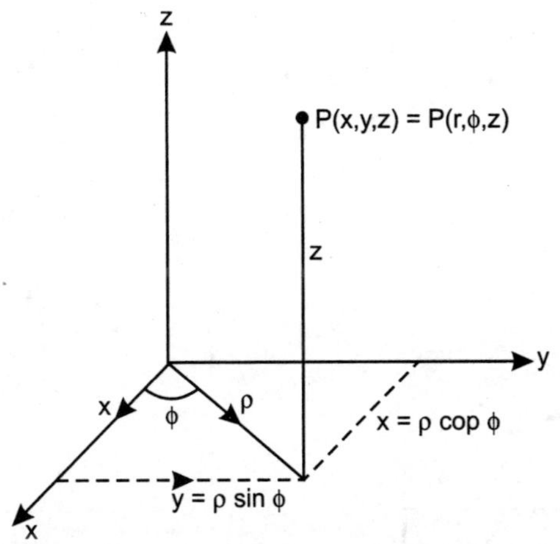

Fig. 2.16 Relation between (x, y, z) and (ρ, ϕ, z).

To derive relation between the components of a vector $\vec{A}$. Let refer Fig. 2.17, which components are acting in (x, y, z) and (ρ, ϕ, z) directions respectively, are A_x , A_y and A_z and A_r, A_ϕ and A_θ. However the decomposition of vectors 'a_x' and 'a_y' are shown in Fig. 2.18 (*a* and *b*).

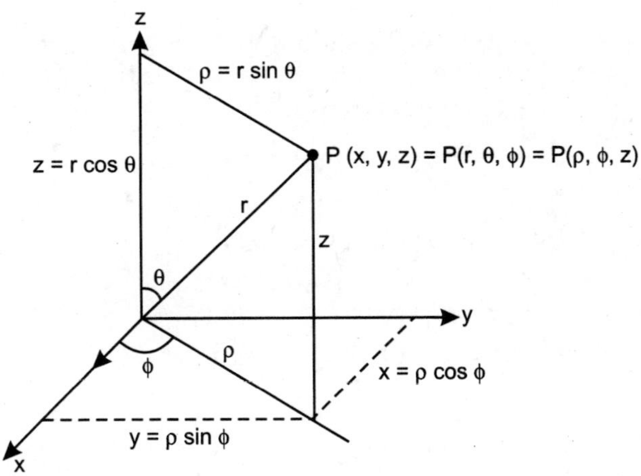

Fig. 2.17 Relation between (x, y, z) and (r, θ, ϕ).

From Fig. (2.18*a*), it is clear that

$$\text{Line } (A, O) = -a_\phi \sin\phi \;(\Delta\, \text{AOB})$$

and $$\text{Line } (OC) = a_\rho \cos\phi \;(\Delta\, COB)$$

Therefore, $$\text{Line } AC = AO + OC$$

$$a_x = a_\rho \cos\phi - a_\phi \sin\phi$$

Again for the cylindrical components of a_y, from Fig. (2.18 *b*), it is clear that

$$\text{Line } (AO') = a_\phi \cos\phi \;(\Delta AO'\, B')$$

$$\text{Line } (O'\, C') = a_\rho \sin\phi \;(AB'\, O'\, C')$$

Hence, $$\text{Line } A'C' = AO' + O'C'$$

$$a_y = a_\rho \sin\phi + a_\phi \cos\phi$$

$$a_z = a_z$$

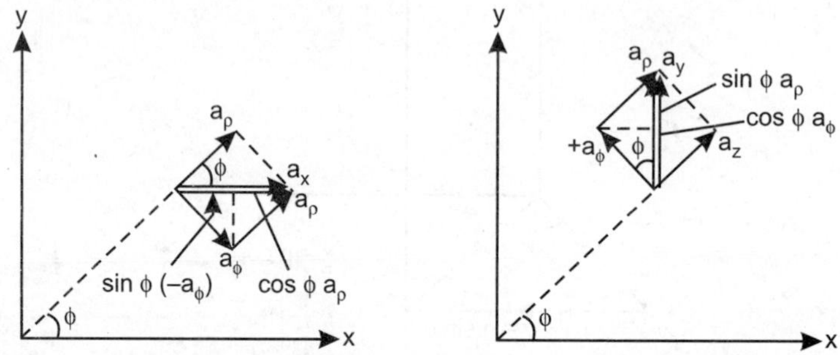

(a) Cylindrical components of a_x (b) Cylindrical components of a_y

Fig. 2.18 Decomposition of vectors.

Above equations for a particular vector A, can be written in the matrix form as follows:

$$\begin{bmatrix} A_x \\ A_y \\ A_z \end{bmatrix} = \begin{bmatrix} \cos\phi & -\sin\phi & 0 \\ \sin\phi & \cos\phi & 0 \\ 0 & 0 & 1 \end{bmatrix} \begin{bmatrix} A_p \\ A_\phi \\ A_z \end{bmatrix}$$

$$\begin{bmatrix} A_p \\ A_\phi \\ A_z \end{bmatrix} = \begin{bmatrix} \cos\phi & \sin\phi & 0 \\ -\sin\phi & \cos\phi & 0 \\ 0 & 0 & 1 \end{bmatrix} \begin{bmatrix} A_x \\ A_y \\ A_z \end{bmatrix}$$

These equations represent the transformation of $(\rho, \phi, z) \rightarrow (x, y, z)$ and $(x, y, z) \rightarrow (\rho, \phi, z)$ respectively.

Spherical to Cartesian, *i.e..*, $(r, \theta, \phi) \rightarrow (x, y, z)$

The relations between variables of these co-ordinates systems with reference to a point are shown in Fig. 2.17. Where

$$r = \sqrt{x^2 + y^2 + z^2}$$

$$\theta = \tan^{-1} \frac{\sqrt{x^2 + y^2}}{z}$$

$$\phi = \tan^{-1}\left(\frac{y}{x}\right)$$

or,
$$x = r \sin\theta \cos\phi$$
$$y = r \sin\phi \sin\phi,$$
$$z = r \cos\theta$$

Any vector in Cartesian and spherical co-ordinates systems can be written as;

$$A = A_x a_x + A_y a_y + A_z a_z = A_r a_r + A_\phi a_\phi + A_\theta a_\theta$$

By using the concept of component of vector by taking the dot-product of unit vectors, we find

$$A_r = \text{A}.a_x,$$
$$\text{A}_\theta = \text{A}.a_\theta,$$
$$A_\phi = A.a_\phi,$$

Therefore,

$$A_r = [A_x a_x + A_y a_y + A_z a_z].\ a_r = A_x a_x.\ a_r, + A_y a_y.\ a_r + A_z a_z.\ a_r$$

Referring to Fig. 2.17, the dot-product of Cartesian unit vectors a_x a_y, and a_z require first projection of the spherical unit-vector on (xy), (yx) and (xz) planes and then find projection onto the desired axes. Taking the projection of a_r, a_θ and a_ϕ on (x y), (y z) and (x z) planes respectively and using the Cartesian unit vectors a_x . a_y and a_z, from Fig. 2.17, we get;

$$a_r.\ a_x = \sin\theta \cos\phi$$
$$a_r.\ a_y = \sin\theta \sin\phi \text{ and } a_r.\ a_z = \cos\theta$$
$$a_\theta.\ a_x = \cos\theta.\ \cos\phi,$$
$$a_\theta.\ a_y = \cos\theta.\ \sin\phi \text{ and } a_r.$$
$$a_z = -\sin\theta$$

and
$$a_\phi.a_x = -\sin\phi,$$
$$a_\phi.\ a_y = \cos\phi \text{ and } a_\phi.a_z = 0$$

Hence
$$A_r = A_x \sin\theta \cos\phi + A_y \sin\theta.\ \sin\phi + A_z \cos\theta$$
$$= A_x\, a_r.\ a_x + A_y a_r.\ a_y + A_z\, a_r.\ a_z$$
$$A_\phi = A_x a_\phi.\ a_x + A_y a_\phi.\ a_y + A_z a_\phi\ .a_z$$

or
$$A_\phi = -A_x \sin\phi + A_y \cos\phi$$
$$A_\theta = A_x a_\theta.\ a_x, + A_y a_\theta\ .a_y + A_z a_\theta\ .a_z,$$

or
$$= A_x .\cos\theta \cos\phi + A_y .\cos\theta \sin\phi - A_z \sin\theta$$

Therefore in matrix form;

$$\begin{bmatrix} A_r \\ A_\theta \\ A_\phi \end{bmatrix} = \begin{bmatrix} \sin\theta.\cos\phi, & \sin\theta.\sin\phi & \cos\theta \\ \cos\theta.\cos\phi & \cos\theta.\sin\phi & \sin\theta \\ -\sin\phi, & \cos\phi & 0 \end{bmatrix} \begin{bmatrix} A_x \\ A_y \\ A_z \end{bmatrix}$$

or

$$\begin{bmatrix} A_x \\ A_y \\ A_z \end{bmatrix} = \begin{bmatrix} \sin\theta.\cos\phi, & \cos\theta.\cos\phi & -\sin\phi \\ \sin\theta.\sin\phi, & \cos\theta.\sin\phi & \cos\phi \\ \cos\theta & -\sin\theta & 0 \end{bmatrix}$$

SOLVED EXAMPLES

Example 1. Show that the cross-product of two vectors that are in a plane can be interpreted in terms of the area included the two vectors.

Solution : We can use the magnitude of the vector product $|A \times B|$ to the area of a parallelogram whose sides are specified by the vectors $\overline{A}$ and $\overline{B}$ as shown in Fig. 2.19. It is clear that the area of parallelogram is given AB sin θ. This is equal to the area of the rectangle which area is given by the product of the horizontal distance $\overline{A}$ and the vertical distance $B \sin\theta$. From the definition of products of two vectors, *i.e.*,

$$A \times B = AB \sin\theta.a_n$$

and

$$|A \times B| = AB \sin\theta$$

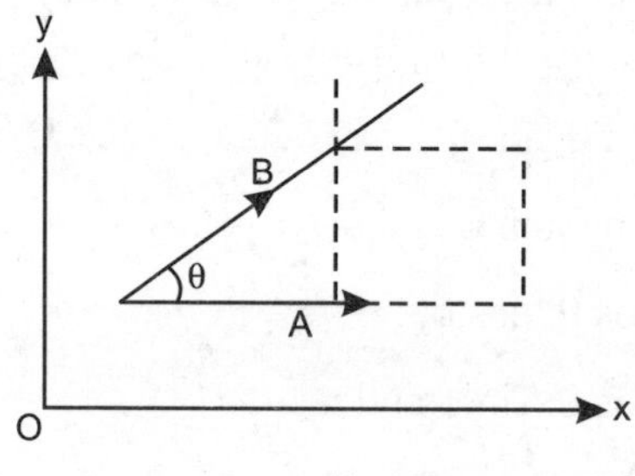

Fig. 2.19 For example (1).

Example 2. Derive the expression for the following in Cartesian co-ordinates.

(*i*) grad. div (A) (*ii*) div grad V

(*iii*) curl grad (V) (*iv*) div curl A

Solution : grad div = $V(\nabla A)$

$$= \left(\overline{a}_x \frac{\partial}{\partial x} + \overline{x}_y \frac{\partial}{\partial y} + \overline{a}_z \frac{\partial}{\partial z}\right)\left(\frac{\partial A_x}{\partial x} + \frac{\partial A_y}{\partial y} + \frac{\partial A_2}{\partial z}\right)$$

$$= \frac{\partial}{\partial x}\left(\frac{\partial A_x}{\partial x} + \frac{\partial A_y}{\partial y} + \frac{\partial A_z}{\partial z}\right)\overline{a}_x + \frac{\overline{\partial}}{\partial y}\left(\frac{\partial A_x}{\partial x} + \frac{\partial A_y}{\partial y} + \frac{\partial A_z}{\partial z}\right)\overline{a}_y$$

$$+ \frac{\partial}{\partial z}\left(\frac{\partial A_x}{\partial x} + \frac{\partial A_y}{\partial y} + \frac{\partial A_z}{\partial z}\right)\overline{a}_z$$

$$= \overline{a_x}\left(\frac{\partial^2 A_x}{\partial x^2} + \frac{\partial^2 A_y}{\partial x \partial y} + \frac{\partial A_z}{\partial x\, \partial z}\right)\left(\frac{\partial^2 A_x}{\partial y\, \partial x} + \frac{\partial^2 A_y}{\partial y^2} + \frac{\partial^2 A_z}{\partial x\, \partial z}\right)\overline{a_y}$$

$$+ \overline{a_z}\left(\frac{\partial^2 A_x}{\partial z \partial x} + \frac{\partial^2 A_y}{\partial z\, \partial y} + \frac{\partial^2 A_2}{\partial^2 z^2}\right)$$

$$= \left(\overline{a_x}\frac{\partial^2 A_x}{\partial x^2} + \overline{a_y}\frac{\partial^2 A_y}{\partial y^2} + \overline{a_z}\frac{\partial^2 A_z}{\partial z^2}\right) + \overline{a_x}\left(\frac{\partial^2 A_y}{\partial x\, \partial y} + \frac{\partial^2 A_z}{\partial x\, \partial z}\right)$$

$$+ \overline{a_y}\left(\frac{\partial^2 A_x}{\partial y\, \partial x} + \frac{\partial^2 A_z}{\partial x\, \partial z}\right) + \overline{a_x}\left(\frac{\partial^2 A_x}{\partial z\, \partial x} + \frac{\partial^2 A_x}{\partial z\, \partial x} + \frac{\partial^2 A_y}{\partial z\, \partial y}\right)$$

(*ii*) Div grad $V = \nabla .(\nabla V)$

$$= \left(\overline{a_x}\frac{\partial}{\partial x} + \overline{a_y}\frac{\partial}{\partial y} + \overline{a_z}\frac{\partial}{\partial z}\right)\left(\overline{a_x}\frac{\partial V}{\partial x} + \overline{a_y}\frac{\partial V}{\partial y} + \overline{a_z}\frac{\partial V}{\partial z}\right)$$

$$= \left(\frac{\partial^2 V}{\partial x^2} + \frac{\partial^2 V}{\partial y^2} + \frac{\partial^2 V}{\partial z^2}\right)$$

Using the properties of unit vectors.

(*iii*) Curl grad $V = \nabla V = \nabla \times \left(\overline{a_x}\frac{\partial V}{\partial x} + \overline{a_y}\frac{\partial V}{\partial y} + \overline{a_z}\frac{\partial V}{\partial z}\right)$

$$= \begin{vmatrix} a_x & a_y & a_z \\ \frac{\partial}{\partial x} & \frac{\partial}{\partial y} & \frac{\partial}{\partial z} \\ \frac{\partial V}{\partial x} & \frac{\partial V}{\partial y} & \frac{\partial V}{\partial z} \end{vmatrix}$$

$$= a_x\left(\frac{\partial^2 V}{\partial y\, \partial z} - \frac{\partial^2 V}{\partial x\, \partial y}\right) + a_y\left(\frac{\partial^2 V}{\partial x\, \partial z} - \frac{\partial^2 V}{\partial z\, \partial x}\right) + a_z\left(\frac{\partial^2 y}{\partial x\, \partial y} - \frac{\partial^2 V}{\partial y\, \partial x}\right)$$

$$= a_x\left(\frac{\partial^2 V}{\partial y\, \partial z} - \frac{\partial^2 V}{\partial x\, \partial y}\right) + a_y\left(\frac{\partial^2 V}{\partial x\, \partial z} - \frac{\partial^2 V}{\partial z\, \partial x}\right) + a_z\left(\frac{\partial^2 V}{\partial x\, \partial y} - \frac{\partial^2 V}{\partial y\, \partial x}\right)$$

$$= 0 + 0 + 0 = 0$$

Since $\qquad \dfrac{\partial^2 V}{\partial x\, \partial y} = \dfrac{\partial^2 V}{\partial y\, \partial x}$ etc.

(*iv*) Div curt A = ∇.∇ × A

$$= \left(\overline{a_x}\frac{\partial}{\partial x} + \overline{a_y}\frac{\partial}{\partial y} + \overline{a_z}\frac{\partial}{\partial z}\right)\begin{bmatrix} a_x & a_y & a_z \\ \frac{\partial}{\partial x} & \frac{\partial}{\partial y} & \frac{\partial}{\partial z} \\ \frac{\partial A}{\partial x} & \frac{\partial A}{\partial y} & \frac{\partial A}{\partial z} \end{bmatrix}$$

$$= \left(\overline{a_x}\frac{\partial}{\partial x} + \overline{a_y}\frac{\partial}{\partial y} + \overline{a_z}\frac{\partial}{\partial z}\right).\left[\frac{\partial A_z}{\partial y} - \frac{\partial A_y}{\partial z}\right.$$

$$\left. + \overline{a_y}\left(\frac{\partial A_x}{\partial z} - \frac{\partial A_z}{\partial x}\right) + \overline{a_z}\left(\frac{\partial A_y}{\partial x} - \frac{\partial A_x}{\partial y}\right)\right]$$

$$= \frac{\partial}{\partial x}\left(\frac{\partial A_z}{\partial y} - \frac{\partial A_y}{\partial z}\right) + \frac{\partial}{\partial y}\left(\frac{\partial A}{\partial z} - \frac{\partial A_z}{\partial x}\right) + \frac{\partial}{\partial z}\left(\frac{\partial A_y}{\partial x} - \frac{\partial A_x}{\partial y}\right)$$

$$= \frac{\partial^2 A}{\partial x\,\partial y} - \frac{\partial^2 A_y}{\partial x\,\partial z} + \frac{\partial^2 A_x}{\partial y\,\partial z} - \frac{\partial^2 A_z}{\partial y\,\partial x} + \frac{\partial^2 A_y}{\partial z \partial x} - \frac{\partial^2 A_x}{\partial z\,\partial y} = 0$$

Example 3. If r is a position vector defined by

$$r = \bar{a}_x . x + \bar{a}_y . y + \bar{a}_z .z$$

Show that $\nabla . r = 3$

Solution:

$$\nabla . r = \left(\bar{a}_x \frac{\partial}{\partial x} + \bar{a}_y \frac{\partial}{\partial y} + \bar{a}_z \frac{\partial}{\partial z}\right)\left(\bar{a}_x . x + \bar{a}_y . y + \bar{a}_z .z\right)$$

$$= \left(\bar{a}_x .a_x \frac{\partial x}{\partial x} + \bar{a}_y \frac{\partial y}{\partial y} + \bar{a}_x .a_z \frac{\partial z}{\partial z}\right) = 1.1 + 1.1 + 1.1 = 1 + 1 + 1 = 3$$

Using properties of unit vector.

Example 4. Find the volume of a sphere of radius 'a' using the differential volume.

Solution: We know that for spherical co-ordinates;

$$dv = r^2 \sin\theta .\ d\theta .\ d\phi .\ dr$$

Hence

$$V = \int dv = \int_{r=0}^{a} \int_{\theta=0}^{\pi} \int_{\phi=0}^{2\pi} r^2 \sin\theta \, dr\, d\theta\, d\phi$$

$$= \int_0^a (r^2\, dr) \int_0^{\pi} \sin\theta . d\theta \int_0^{2\pi} d\phi$$

$$= \left(\frac{r^3}{3}\right)_0^a (-\cos\theta)_0^{\pi} (\phi)_0^{2\pi}$$

$$= \frac{a^3}{3} \times 2 \times 2\pi = \frac{4\pi\, a^2}{3}$$

Example 5. Show that $(r_n \bar{r})$ is an irrotational vector for any value of (n) but it is solenoidal only for $(n - 3)$, where $\bar{r}$ is a position vector of magnitude r.

Solution: As given $r_n \bar{r}$ is irrotational and Solenoidal;

$$\nabla \times (r_n \bar{r}) = 0$$

$$\nabla . (r_n \bar{r}) = 0$$

(A)

$$\nabla \times r_n \bar{r}$$

$$= \left(\bar{a}_x \frac{\partial}{\partial x} + \bar{a}_y \frac{\partial}{\partial y} + \bar{a}_z \frac{\partial}{\partial z}\right) \times r_n \left(\bar{a}_x x + \bar{a}_y y + \bar{a}_z z\right)$$

$$= \left(\bar{a}_x \times \bar{a}_y\right)\frac{\partial}{\partial x} r^n y + \left(\bar{a}_x \times \bar{a}_z\right)\frac{\partial}{\partial x} r^n z + \left(\bar{a}_y \times \bar{a}_z\right)\frac{\partial}{\partial x} r^n y$$

$$+ \left(\bar{a}_z \times \bar{a}_z\right)\frac{\partial}{\partial z} r^n y + \left(\bar{a}_z \times \bar{a}_y\right)\frac{\partial}{\partial x} r^n z + \left(\bar{a}_y \times \bar{a}_z\right)\frac{\partial}{\partial y} r^n z$$

$$= \overline{a_z}\frac{\partial}{\partial_x} r^n y - \overline{a_y}\,\frac{\partial}{\partial x}\, r^n z\; \overline{a_z}\frac{\partial}{\partial y} r^n x + \overline{a_x} + \overline{a_x}.\frac{\partial}{\partial y}\, r^n z$$

$$+ \overline{a_y}\frac{\partial}{\partial z} r^n x - \overline{a_x}\frac{\partial}{\partial z} r^n y$$

$$= \overline{a_x}\left(\frac{\partial}{\partial y} r^n z - \frac{\partial}{\partial z} r^n y\right) + \overline{a_y}\left(\frac{\partial}{\partial z} r^n x - \frac{\partial}{\partial x} r^n z\right) + \overline{a_z}\left(\frac{\partial}{\partial x} r^n y - \frac{\partial}{\partial x} r^n z\right)$$

(I) (II) (III)

Let consider

$$I = \overline{a_x}\left[\frac{\partial}{\partial y}(x^2 + y^2 + z)^{\frac{n}{2}}.z - \frac{\partial}{\partial z}(x^2 + y^2 + z^2)^{\frac{n}{2}}.y\right]$$

$$= a_x\left[\frac{n}{2}(x^2 + y^2 + z^2)^{\frac{n}{2}}.2yz - (x^2 + y^2 + z^2)^{\frac{n}{2}-1}.2yz\right]$$

$$= a_x\left[\frac{n}{2}\, r^{n-1}.\, 2yz - r^{n-1}.\, 2xz\right] = 0$$

Similarly,

(II) $$a_y\left[\frac{n}{2} r^{n-1}.\, 2xz - r^{n-1}.\, 2xz\right] = 0$$

and (III) $$a_z.\left[\frac{n}{2} r^{n-1}.\, 2xy - r^{n-1}.2xy\right] = 0$$

$$= 0 + 0 + 0 = 0,$$

and independent from the value of n.

(B) $\nabla.(r^n r) = 0$

$$\left(a_x\frac{\partial}{\partial x} + a_y\frac{\partial}{\partial y} + a_z\frac{\partial}{\partial z}\right).\, r_n\,(z_x x + a_y y + z_z z) = 0$$

As $a_x\,.a_x = 1 = a_y\,.a_y. = a_z a_z$

$$= \frac{\partial}{\partial x}\,(r^n x) + \frac{\partial}{\partial y}\,(r^n y) + \frac{\partial}{\partial z}\,(r^n z) = 0$$

(*i*) $$\frac{\partial}{\partial x}\,(r^n x) = \frac{\partial}{\partial x}\,(x^2 + y^2 + z^2)^{\frac{n}{2}}.\text{x}$$

$$= \frac{n}{2}(x^2 + y^2 + z^2)^{\frac{n}{2}-1}.2x^2 + (x^2 + y^2 + z^2)^{\frac{n}{2}}.1$$

Let $$r_1 = (x^2 + y^2 + z^2)^{\frac{n}{2}} - 1$$

$$= \frac{n}{2}\,(r_1.\, 2x^2) + r^2 = n.r_1 x^2 + r^n$$

Similarly

(*ii*) $$n(r_1\,.\,y^2) + r^n$$

(*iii*) $$n(r_1 z^2) + r^n$$

Therefore from (*i*), (*ii*) and (*iii*), we get,

$$n(r_1)(x^2+y^2+z^2)+3r^n = 0$$

$$= n(x^2+y^2+z^2)^{\frac{n}{2}-1}(x^2+y^2z^2)+3r^n=0$$

$$= n.r^n+3r^n=0$$

$\Rightarrow$ $$r^n(n+3) = 0$$

Hence, $$\nabla.r^n.r = r^n.(n+3)=0$$

i.e., $$n = -3.$$

Example 6. Let $A = 3a_y + 4a_z$ and $B = 4a_x - 10a_y + 5a_z$

(*i*) Find the component of A along *B*.

(*ii*) Determine a unit vector perpendicular to both *A* and *B*.

Solution: (*i*) $A_B = (A.\ a_B) = \dfrac{(A.B)B}{|B|^2} = \dfrac{-10(4,-10.5)}{|4|}$

$$= [-0.284a_x + 0.709a_y - 0.355a_z]$$

(*ii*) $A \times B = \begin{vmatrix} a_x & a_y & a_z \\ 0 & 3 & 4 \\ 4 & -10 & 5 \end{vmatrix} = (55, 16, -12)$

$$a_{A.B} = (0.939, 0.2734, -0.205)$$

Example 7. Determine the (*i*) grad of

$$V = e^{-z}\sin 3x + \cos 2hy$$

Solution: $$\nabla V = \overline{a_x}\frac{\partial V}{\partial x} + \overline{a_y}\frac{\partial V}{\partial y} + \overline{a_z}\frac{\partial V}{\partial z}$$

$$= 3.\cos 3x.\ \text{cosh}y.\ \overline{a_x} + e^{-z}\sin 3x\sin hy.\ \overline{a_y} - e^{-2}\sin 3x\ \text{cosh}y\ \overline{a_z}$$

(*ii*) div of

Solution $$A = \rho\sin\phi.a_\rho + \rho^2z^2.\ a_\phi + \cos\phi.\ a_z$$

$$\nabla.A = \frac{1}{\rho}\frac{\partial}{\partial\rho}(\rho A_\rho) + \frac{1}{\rho}\frac{\partial}{\partial\phi}A_\phi + \frac{\partial}{\partial z}Az$$

$$= \frac{1}{\rho}\frac{\partial}{\partial\rho}(\rho^2\sin\phi) + \frac{1}{\rho}\frac{\partial}{\partial\phi}(\rho^2z^2) + \frac{\partial}{\partial z}(z\cos\phi)$$

$$= \frac{1}{\rho}\times 2\rho\sin\phi + 0 + \cos\phi$$

$$= 2\sin\phi + \cos\phi$$

(*iii*) Curl of $A = a_x + ya_y + (4y - z)$az

Solution $$\nabla\times A = \begin{vmatrix} a_x & a_a & a_z \\ \dfrac{\partial}{\partial x} & \dfrac{\partial}{\partial y} & \dfrac{\partial}{\partial z} \\ 1 & y^2 & (4y-z) \end{vmatrix}$$

$$= \overline{a_x}\,(4-0) - \overline{a_y}\,(0-0) + \overline{a_z}\,(0-0)$$

$$= 4.\,\overline{a_x}$$

Example 8. Find the (*i*) Laplacian of

$$V = 10r^2 \sin^2\theta\,.\,\cos\phi$$

Solution:

$$\nabla^2 V = \frac{1}{r^2}\frac{\partial}{\partial r}\left(r^2\frac{\partial V}{\partial r}\right) + \frac{1}{r^2\sin\theta}\frac{\partial}{\partial\theta}\left(\sin\theta\frac{\partial V}{\partial\theta}\right) + \frac{1}{r^2\sin^2\theta}\frac{\partial^2 V}{\partial\phi}$$

$$= \frac{1}{r^2}\frac{\partial}{\partial r}\,(10r^3\sin^2\theta\cos\phi)$$

$$+ \frac{1}{r^2\sin\theta}\frac{\partial}{\partial\theta}\left(10r^2\sin 2\theta.\sin\theta.\cos\phi - \frac{10r^2\sin\theta.\cos\phi}{r^2\sin^2\theta}\right)$$

$$= 30\sin^2\theta.\cos\phi + 20\cos 2\theta.\cos\phi + \frac{10r2\sin 2\theta.\cos\theta.\phi}{\sin\theta}$$

$$- 10\cos\phi$$

$$= 10.\cos\phi.(3\sin^2\theta + 2\cos\theta + 2\cos^2\theta - 1)$$

$$= 10\cos\phi.(\sin^2\theta + 2 + 2\cos 2\theta - 1)$$

$$= 10\cos\phi.(\sin^2\phi + 2\cos 2\theta + 1)$$

(*ii*) Given vectors

$$A = \alpha\,.a_x + a_y + 4.a_z$$

$$B = 3.a_x + \beta.a_y - 6.a_z$$

$$C = 5.a_x - 2.a_x + \gamma.a_z$$

Calculate the value of α, β and γ such that the *x* vectors are mutual *y* orthogonal.

Solution: The vectors are mutual *y* perpendicular *i.e*,.

$$A.B = AB\cos 90° = 0$$

$\therefore$ $$A.B = 0,$$

$$A.C = 0$$

and $$B.C = 0, \text{ which gives;}$$

$$3\alpha + \beta - 23 = 1 \qquad ...(i)$$

$$5\alpha - 2 + 4y = 0 \qquad ...(ii)$$

$$15\alpha + 2\beta - 6\gamma = 0 \qquad ...(iii)$$

$$3\alpha + \beta + 0 = 24$$

or $$5\alpha + 0 + 4\gamma = 2$$

$$0 - 2\beta - 6\gamma = 15$$

$$\begin{bmatrix} 3 & 1 & 0 \\ 5 & 0 & 4 \\ 0 & -2 & -6 \end{bmatrix} = \begin{bmatrix} \alpha \\ \beta \\ \gamma \end{bmatrix} = \begin{bmatrix} 24 \\ 2 \\ 15 \end{bmatrix}$$

$$\Delta = \begin{vmatrix} 3 & 1 & 0 \\ 5 & 0 & 4 \\ 0 & -2 & -6 \end{vmatrix} = 3(0-8) - 1(30-0) + 0(10-0) - 24 - 30) = -54$$

$$\Delta_\alpha = \begin{vmatrix} 24 & 1 & 0 \\ 2 & 0 & 4 \\ 15 & 2 & 6 \end{vmatrix} = -144$$

$$\Delta_\beta = \begin{vmatrix} 3 & 24 & 0 \\ 5 & 2 & 4 \\ 0 & 15 & 24 \end{vmatrix} = -864$$

$$\Delta_\gamma = \begin{vmatrix} 3 & 1 & 24 \\ 5 & 0 & 2 \\ 0 & 2 & 15 \end{vmatrix} = 153$$

Hence,
$$\alpha = \frac{\Delta_\alpha}{\Delta} + \frac{-144}{-54} = 2.667$$

$$\beta = \frac{\Delta_\beta}{\Delta} = \frac{-864}{-54} = 16$$

$$\gamma = \frac{\Delta_\gamma}{\Delta} = \frac{153}{-54} = -2.833$$

Example 9. $A = x^2 a_x - 2xz\ a_y - y^2 a_z$, calculate the circulation of A around the closed path shown in Fig. 2.20.

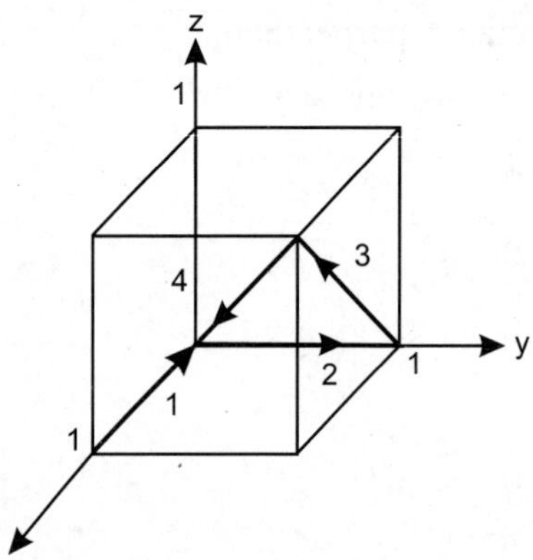

Fig. 2.20

Solution : The circulation of A around path L can be given by

$$\oint_L A.dl = \left(\int_1 + \int_2 + \int_3 + \int_4\right) A\, dl$$

where integral $\int 1$, $\int 2$, $\int 3$ and $\int 4$, represents broken segments of path.

For segment (1), $y = 0 = z,$

Hence $A = x^2\overline{a_x}, dl = dx.\ \overline{a_x}$

Since dl is always taken along $\pm\ a_x$, the direction on segment (1) is taken by the limits of integration. Therefore

$$\int A\ dl = \int_0^1 x^2\ dx = \left.\frac{x^3}{3}\right|_1^0 = -\frac{1}{3}$$

For segment (2), $x = 0, A = -y^2\ a_z, dl = dy.\ ay$

Hence, $A.\ dl = 0, i.e., \int_2 A.al = 0$

For segment (3) $y = 1$

$$A = x^2 a_x - 2xza_y - a_z$$

and $dl = dx.\ a_x + dz\ .a_z$

So, $\int_3 A.\ dl = \int (x^2.\ dx - 2dz)$

But on (3), $z = x$, that is $dx = dz$,

Hence $\int_3 A.\ dl = \int_0^1 (x^2 - 2).\ dx = \frac{x^3}{3} - 2x|_0^1 = \frac{1}{3} - 2 = \frac{-5}{3}$

For segment (4), $x = 1$, so

$$A = a_x - 2za_y - y^2 az$$

$$dl = dy.ay + 2dz.a_z$$

Hence, $\int_4 A.dl = \int (-2z\ dy - y^2\ dz)$

But on (4),

$$z = y, i.e., dz = dy$$

So $\int_4 A.dl = \int_1^0 (-2y - y^2)dy = -y^2 - \left.\frac{-y^3}{3}\right|_1^0 = -1 - \frac{1}{3} = +\frac{4}{3}$

Hence $\oint_L A.dl = -\frac{1}{3} + 0 - \frac{5}{3} + \left(+\frac{4}{3}\right) = \frac{2}{3}$

Example 10. (*i*) Given the vector field

$$A = (10xy - z)\overline{a_x} + 5x^2\overline{a_y} - x\overline{a_z}$$

Find the nature of A. Is a conservative?

$$\nabla \times A \begin{vmatrix} \overline{a_x} & \overline{a_y} & \overline{a_z} \\ \frac{\partial}{\partial x} & \frac{\partial}{\partial y} & \frac{\partial}{\partial z} \\ (10xy - z) & 5x^2 & -x \end{vmatrix} = \overline{a_x}\ (0) - \overline{a_y}\ (-1 + 1) + \overline{a_z}\ (10x - 10x) = 0$$

That is A is conservative

(*ii*) If a vector $A = (\alpha x + \beta z^2)\,\overline{a_x} + (3x^2 - yz)\,\overline{a_y} + (3xz^2 - y)\overline{a_z}$

is irrotational, determine α, β and γ and also find ∇. *A* at *A* (2, – 1,0).

$$\nabla \times A = \begin{vmatrix} i & j & k \\ \frac{\partial}{\partial x} & \frac{\partial}{\partial y} & \frac{\partial}{\partial z} \\ (\alpha x + \beta z^2) & (3x^2 - yz) & (3xz^2 - y^2) \end{vmatrix}$$

$$= (-1 + \gamma)\overline{a_x} + (3\beta z^2 - 3z^2)\overline{a_y} + (6x - \alpha x)\overline{a_z}$$

Since *A* is irrotational:

$$\nabla \times A = 0$$

$$\Rightarrow \qquad \alpha = 6, \beta = \gamma = 1$$

$$\nabla . A = \frac{\partial A_x}{\partial x} + \frac{\partial A_y}{\partial y} + \frac{\partial A_z}{\partial z}$$

$$= \alpha y + 0 + 6xz$$

$$= \alpha y + 6xz$$

At (2, – 1, 0)

$$\nabla . A = -1 + 0 = -1$$

Example 11. Transform a vector $A = y\overline{a_x} - x\overline{a_y} + z\overline{a_z}$ into cylindrical and spherical co-ordinate systems.

Solution: We know that

$$A_\rho = A_x \cos\phi + A_y \sin\phi$$

$$A_\phi = -A_x \sin\phi + A_y \cos\phi$$

$$A_z = A_z$$

And

$$x = \rho\cos\phi$$

$$y = \rho\sin\phi$$

From given vector $A_x = y, A_y = -x, A_z = z$

$$A_\rho = y\cos\phi - x\sin\phi$$

$$= \rho\sin\phi\cos\phi - \rho\cos\phi\sin\phi = 0$$

$$A_\phi = -y\sin\phi + (-x)\cos\phi$$

$$= [\rho\sin^2\phi + \rho\cos^2\phi] = -\rho$$

$$\therefore \qquad \overline{A}\,(\delta.\,\phi, z) = -\rho\, a_\phi + z\, a_z$$

and since

$$\begin{bmatrix} A_r \\ A_\theta \\ A_\phi \end{bmatrix} = \begin{bmatrix} \sin\theta\cos\phi & \sin\theta\sin\phi & \cos\theta \\ \cos\theta\cos\phi & \cos\theta\sin\phi & -\sin\theta \\ -\sin\phi & \cos\phi & 0 \end{bmatrix} \begin{bmatrix} A_x \\ A_y \\ A_z \end{bmatrix}$$

$$A_r = A_x.\sin\theta\cos\phi + A_y.\sin\theta\sin\phi + A_z\cos\theta$$

$$= y.\sin\theta\cos\phi - x\sin\theta\sin\phi + z\cos\theta$$

$$= r.\sin^2\theta\cos\phi - r\sin^2\theta\cos\phi + r\cos^2\theta$$

$$A_r = r.\cos^2\theta$$

$$A_\theta = A_x\cos\theta\cos\phi + A_y\cos\theta\sin\phi - A_z\sin\theta$$

$$= y\cos\theta\cos\phi - x\cos\theta\sin\phi - z\sin\theta$$

$$= r\sin\theta\cos\theta\cos\phi - r\sin\theta\cos\theta\sin\phi + r\sin\theta\cos\theta$$

$$A_\theta = r\sin\theta\cos\theta$$

$$A_\phi = -A_x\sin\phi + A_y\cos\phi$$

$$= -y\sin\phi - x\cos\phi$$

$$= -r\sin\theta\sin^2\phi - r\sin\theta\cos^2\phi$$

$$A_\phi = -r\sin\theta$$

$$\therefore \quad \overline{A}\,(r, \theta, \phi) = A_r\,(r\cos^2\theta) + A_\theta\,(r\sin\theta\cos\theta) - A_\phi\,(r\sin\theta)$$

Example 12. Prove that

$$a_x.\,a_\rho = \cos\phi \text{ and } a_x a_r = \sin\theta\cos\phi$$

Solution : We know that

$$a_x = \cos\phi\,a_\rho - \sin\phi.a_\phi$$

$$a_x.\,a_\rho = (\cos\phi a_\rho - \sin\phi a_\phi).\,a_\rho$$

$$= \cos\phi.\,1 - 0 = \cos\phi$$

and

$$a_x.\,a_r = (\sin\phi\cos\phi\,\overline{a_r} + \cos\theta\cos\phi\,a_\theta - \sin\phi\,a_\phi).\,a_r$$

$$= \sin\phi\cos\phi(\overline{a_r}.\,\overline{a_r}) + \cos\theta\cos\phi(a_\theta\,.a_r) - \sin\phi(a_\phi\,.a_r)$$

$$= \sin\theta\cos\phi 1 + 0 - 0$$

$$= \sin\theta\cos\phi$$

Similarly we can also prove that

$$a_x\,.a_\phi = -\sin\phi,$$

$$a_y\,.a_\phi = \cos\phi$$

$$a_x\,.a_\theta = \cos\theta\cos\phi$$

Example 13. (*i*) Transform a vector $A = z\sin\phi a_\rho - \rho\cos\phi a_\phi + 2\rho z\,a_z$ into Cartesian co-codinate system and calculate its magnitude at point (3, 4, 0).

Solution: We know

$$\begin{bmatrix} A_x \\ A_y \\ A_z \end{bmatrix} = \begin{bmatrix} \cos\phi & -\sin\phi & 0 \\ \sin\phi & \cos\phi & 0 \\ 0 & 0 & 1 \end{bmatrix} = \begin{bmatrix} z\sin\phi \\ -\rho\cos\phi \\ 2\rho z \end{bmatrix}$$

$$A_x = z\sin\phi\cos\phi + \rho\sin\phi\cos\phi$$

$$= \frac{xyz}{x^2+y^2} + \frac{xy\sqrt{x^2+y^2}}{x^2+y^2}$$

$$= 0 + \frac{3\times4\times5}{25} = \frac{12}{5} = 2.4$$

$$A_y = z\sin^2\phi - \rho\cos^2\phi = \frac{y^2 z}{x^2+y^2} - \frac{x^2\sqrt{x^2+y^2}}{x^2+y^2}$$

$$= 0 + \frac{9\times5}{25} = 1.8$$

$$A_z = 2.\rho.z = 2z\sqrt{x^2+y^2} = 0$$

$$\therefore \quad A = 2.4\,\overline{a_x} - 1.8\,\overline{a_y}$$

(*ii*) If $D = \rho\cos\theta.a_\rho + \rho z^2\sin\phi.a_z$

Transform D into spherical co-ordinates system and find its magnitude at (3, 4 and 0) point.

Solution: Since D is given in Cylindrical co-ordinate, better to transform it in Cartesian co-ordinate and then into spherical. Hence

$$A_x = \rho\cos^2\phi = \frac{x^2}{\sqrt{x^2+y^2}} = \frac{x^2}{\rho}$$

$$A_y = \rho.\sin\phi\cos\phi = \frac{xy}{\sqrt{x^2+y^2}} = \frac{xy}{\rho}$$

$$A_z = \rho z^2\sin\phi = \frac{z^2 y}{\sqrt{x^2+y^2}} = \frac{yz^2}{\rho}$$

$$\begin{bmatrix} A_r \\ A_\theta \\ A_\phi \end{bmatrix} = \begin{bmatrix} \sin\theta\cos\phi & \sin\theta\sin\phi & \cos\theta \\ \cos\phi\cos\phi & \cos\theta\sin\phi & -\sin\theta \\ -\sin\phi & \cos\phi & 0 \end{bmatrix} \begin{bmatrix} \frac{x^2}{\rho} \\ \frac{xy}{\rho} \\ \frac{yz^2}{\rho} \end{bmatrix}$$

where $x = r\sin\theta\cos\phi,\ y = r\sin\theta\sin\phi$

$$z = r\cos\theta, \text{ and } \rho = r\sin\theta$$

$$A_r = \frac{r^2 \sin^2 \theta \cos^2 \phi}{r \sin \theta} \sin \theta \cos \phi + \frac{r^2 \sin^2 \theta \sin \phi \cos \phi}{r \sin \theta} \sin \theta \sin \phi$$

$$+ \frac{r^3 \sin \theta \cos^2 \phi}{r \sin \theta} \sin \phi \cos \theta$$

$$A_r = r \sin^2 \theta \cos \phi + r^2 \cos^2 \theta \sin \theta$$

At point (3, 4, 0)

$$r = \sqrt{x^2 + y^2 + z^2} = \sqrt{9+16} = \sqrt{25} = 5$$

$$\theta = \frac{\pi}{2}, \text{ and } \phi = 306.83$$

$$\cos \phi = \frac{3}{5} \text{ and } \sin \phi = \frac{4}{5}$$

$$A_r = 5.1^2 \times \frac{3}{5} + r^2 \times 0 = 3$$

$A_\phi = r \sin \theta \,.\, \cos^2 \phi \cos \theta \cos \phi + r \sin \theta \cos \phi .\, \sin \phi \cos \theta \sin \phi$
$- r^2 \cos^2 \theta \sin \phi \sin \theta$

$$= r \sin \theta \sin \theta \,[\cos \phi - r \cos \theta \sin \phi]$$

and $\qquad = 0$ At (3, 4, and 0)

$$A_\phi = -\sin \theta .\, \cos^2 \phi \sin^2 \phi + r \sin \theta \cos \phi \sin \phi .\, \cos \phi = 0$$

$\therefore \qquad |\overline{A}| = 3$

Example 14. The surface $r = 2$ and 4, $\theta = 30°$, and $\phi = 20°$ and $60°$ identify a closed surface.

(*a*) Find the closed volume.

(*b*) Find the total area of the enclosing surface.

(*c*) Find the total length of the twelve edges of the surface.

Solution: (*a*) $\quad$ Volume $= \int_{20°}^{60°} \int_{30°}^{50°} \int_2^4 r^2 \sin \theta \, dr \, d\theta \,.\, d\phi$

$$= \int_{20}^{60} \int_{30}^{50} \left[\frac{r^3}{3}\right]_2^4 \sin \theta \, d\theta .\, d\phi$$

$$= 2.87 \int_{\phi = 20}^{60} \int_{\theta = 20}^{50} \sin \theta \, d\theta .\, d\phi$$

$$= 2.87 \times 40° \int_{\theta = 20}^{50} \sin \theta \, d\theta.$$

$= 2.91$ (After changing the radius)

(*b*) Area $= \int_{20°}^{60°} \int_{30}^{50°} (4^2 + 2^2) \sin \theta d\theta \, d\phi + \int_2^4 \int_{30}^{60} r \, (\sin 30° \sin 50°) \, dr .\, d\phi +$

$$2 \int_{30°}^{50°} \int_2^4 r dr .\, d\theta = 12.61$$

(*c*) The total length of the twelve edges of the surface.

$$L = 4\int_2^4 dr + 2\int_{30°}^{50°}(4+2)d\theta) + \int_{20}^{60}(4\sin 50° + 4\sin 30° + 2\sin 50° + 2\sin 30°)\, d\phi$$

$$= 8 + 12 \times 20° + \int_{20}^{60}(3.064 + 2 + 1 + 2.532)\, d\phi$$

$$= 17.49$$

Example 15. (*i*) Given a point *P*(–2, 6, 3) and a vector $\vec{A} = ya_x + (x+z)y$, express *P* and $\overline{A}$ in cylindrical and spherical coordinate system. Evaluate *A* at point *P* in the cartesian, cylindrical and spherical system.

Solution. At point *P* (–2, 6, 3), $x = -2, y = 6, z = 3$

$$\rho = \sqrt{x^2+y^2} = \sqrt{4+36} = \sqrt{40} = 6.32$$

$$z = 3$$

Since

$$x = \rho\cos\phi$$

$$y = \rho\sin\phi$$

So

$$\phi = \tan^{-1}\left(\frac{y}{x}\right) = \tan^{-1}\left(\frac{6}{-2}\right) = 71.56°$$

$$r = \sqrt{x^2+y^2+z^2} = \sqrt{4+36+9=7}$$

$$\phi = \tan^{-1}\frac{\sqrt{x^2+y^2}}{z} = \tan^{-1}\frac{\sqrt{40}}{3} = 64.62°$$

So *P* in cylindrical coordinate $(\rho, \phi, z) = (6.32, -71.56, 3)$

P in spherical coordinate system $(r\ \theta\ \phi) = (7, 64.62° - 71.56)$

In the cartesian coordinate system $A = ya_x + (z+x)\, a_y$

$$= 6ax + \text{ay}$$

$$\begin{bmatrix} A_\rho \\ A_\phi \\ A_z \end{bmatrix} = \begin{bmatrix} \cos\phi & \sin\phi & 0 \\ -\sin\phi & \cos\phi & 0 \\ 0 & 0 & 1 \end{bmatrix}\begin{bmatrix} A_x \\ A_y \\ A_z \end{bmatrix}$$

$$\begin{bmatrix} A_\rho \\ A_\phi \\ A_z \end{bmatrix} = \begin{bmatrix} \cos\phi & \sin\phi & 0 \\ -\sin\phi & \cos\phi & 0 \\ 0 & 0 & 1 \end{bmatrix}\begin{bmatrix} 6 \\ 1 \\ 0 \end{bmatrix}$$

$$\begin{bmatrix} A_\rho \\ A_\phi \\ A_z \end{bmatrix} = \begin{bmatrix} 6\cos\phi + \sin\phi + 0 \\ -6\sin\phi + \cos\phi + 0 \\ 0+0+0 \end{bmatrix}\begin{bmatrix} A_\rho = 6\cos\phi + \sin\phi \\ A_\phi = -6\sin\phi + \cos\phi \\ A_z = 0 \end{bmatrix}$$

$$x = \rho\cos\phi$$

$$y = \rho\sin\phi$$

$$r = \sqrt{\tan\phi} = \left(\frac{y}{x}\right)$$

$$\theta = \tan^{-1}\left(\frac{6}{-2}\right) = -71.56$$

$$A = A_\rho a_\rho + A_\phi a_\phi + A_z a_z$$

$$= (6 \cos \phi + \sin \phi)\, a_\rho + (-6 \sin \phi + \cos \phi)\, a_\phi$$

$$(1.897 + (-0.948)a_\rho + (5.697 + 0.3162)a_\phi$$

$$= 0.9483 a_\rho + 6.600 a_\phi$$

Similarly, for spherical coordinates system.

$$\begin{bmatrix} A_r \\ A_\theta \\ A_\phi \end{bmatrix} = \begin{bmatrix} \sin\theta\cos\phi & \cos\theta\cos\phi & \cos\phi \\ \sin\theta\sin\phi & \cos\theta\sin\phi & -\sin\phi \\ \sin\phi & -\cos\phi & 0 \end{bmatrix} \begin{bmatrix} A_x \\ A_y \\ A_z \end{bmatrix}$$

$$\begin{bmatrix} A_r \\ A_\theta \\ A_\phi \end{bmatrix} = \begin{bmatrix} \sin\phi\cos\phi & \sin\theta\sin\phi & \cos\theta \\ \cos\theta\cos\phi & \cos\theta\sin\phi & -\sin\theta \\ -\sin\phi & \cos\phi & 0 \end{bmatrix} \begin{bmatrix} A_x \\ A_y \\ A_z \end{bmatrix}$$

$$A_\rho = 0.8571$$

$$A_\theta = 0.4066$$

$$A_\phi = 6.00$$

$$\overline{A} = 0.8571 a_\rho + 0.4066 a_\theta + 6008 a_\phi$$

(*ii*) Express the vector, $B = \frac{10}{r} a_r + r \cos \theta a_\theta + a_\phi$ in cartesian cylindrical coordinate system.

Solution: We know that

$$A_x = \overline{B}.\text{ax}$$

$$= \left(\frac{10}{r} a_r + r \cos \theta a_\theta + a_\phi\right) a_x$$

$$= \left(\frac{10}{r} .a_r .a_x + r \cos \theta\, a_\theta .a_x + a_\phi a_x\right)$$

$$= \left(\frac{10}{r} \sin \theta . \cos \phi + r \cos \theta \cos \theta \cos \phi + - \sin \phi\right)$$

$$= \left[\frac{10}{r} \sin \theta \cos \phi + r \cos^2 .\theta . \cos \phi - \sin \phi\right]$$

$$A_y = B.a_y$$

$$= \left(\frac{10}{r} a_r + r \cos \theta\, a_\theta + a_\phi\right) a_y$$

$$= \frac{10}{r} a_r .a_y + r \cos \theta .\, a_y + a_\phi .a_y$$

$$= \left[\frac{10}{r} \sin \theta \cos \phi + r \cos^2 \theta \sin \phi + \cos \phi\right]$$

$$A_z = \overline{B}.a_z$$

$$= \left(\frac{10}{r} a_r + r \cos \theta \, a_\theta + a_\phi\right) a_z$$

$$= \frac{10}{r} a_x a_z + r \cos \theta . \, a_\theta az + a_\phi a_z$$

$$= \left[\frac{10}{r} \cos \theta - r \cos \theta \sin \theta\right]$$

$$r = \sqrt{x^2 + y^2 + z^2},$$

$$\phi = \tan^{-1}\left(\frac{y}{x}\right)$$

$$x = \rho \cos \phi,$$

$$y = \rho \sin \phi$$

$$\rho = \sqrt{x^2 + y^2}$$

$$x = r \cos \theta . \cos \phi$$

$$y = r \sin \theta \cos \phi$$

$$\cos \phi = \frac{x}{\rho} = \frac{x}{\sqrt{x^2 + y^2}}$$

$$\sin \phi = \frac{y}{\rho} = \frac{y}{\sqrt{x^2 + y^2}}$$

EXERCISES

1. Two vectors A and B are given by

$$A = 3a_x + 2a_y + a_z$$

$$B = 2a_x + 2a_y + 3a_x$$

(*i*) Find the angle between them

(*ii*) Dot and cross product of the vector.

2. Find the gradient of a scalar function F, when

$$F(x; y, z) = xy^2 + 3e^z$$

Also, find the magnitude sand direction of the vector at point $P(2, 4, -2)$.

3. Find $\nabla \left(\frac{1}{A}\right)$, if $A = 2xa_x + ya_y + 2za_z$

4. Find the expressions and compute the magnitude of $\nabla \times$ A and ∇. A at the origin.

If $$A = 3ya_x + (x^2 + 3z)a_y + 3a_z$$

5. Find the divergence and curl for the vector field.

$$F = a_x \left(\frac{2x^2}{r}\right), \text{ where r} = ax_x + 2ya_y + za_z$$

6. Find $\nabla^2 A$, if $A = (x + y)a_x + (2x^2 + y^2)a_y$.

7. Two vectors having equal magnitudes A, make an angle 'θ' with each-other. Show that the magnitude (R) and direction (α) of the resultant vector is

$$R = 2A\cos\left(\frac{\theta}{2}\right) \text{ and } \alpha = \frac{\theta}{2}$$

8. If $P(3, 5, -2)$ is a point in a Cartesian co-ordinate. Express its position in cylindrical and spherical co-ordinates.

9. If a vector $A = 3a_x + 4a_y + 5a_z$. Convert it into spherical co-ordinate.

10. Show that the vector fields

$$A = a_r\left(\frac{\sin 2\theta}{r^2}\right) + 2a_\theta\left(\frac{\sin\theta}{r^2}\right)$$

$B = a_r\,(r\cos\theta) + a_\theta\,(r)$ are every where parallel to each-other.

11. Given points $P(r = 0.8.5, \theta = 40°, \phi = 50°)$

and $$E = \frac{1}{r^2}\left[\cos\phi . a_r + \frac{\sin\phi}{\sin\theta}a_\theta\right]$$

Find a unit-vector in the direction of E at point P.

- Have the courage to follow your heart and intuition.

Steve Jobs

Chapter 3

ELECTRO STATIC FIELDS

3.1 INTRODUCTION

Static electric fields and '*dc*' electric fields are also known as **Electrostatic.** This types of field produces due the **static charge** or **electricity at rest.** Since like charges repel to each other and **unlike attract** to each other, so there **exist a force** between them called electrostatic force.

The source of electrostatic fields are electric charges, and strength of the fields at any point depends upon the magnitude, position, velocity and accelerations.

Electrostatic divides the materials into two groups.

(*i*) Conductor Bodies permitting to pass the charges.

(*ii*) Insulator, Non-conductor, and dielectric, Bodies don't permitting to pass charges.

Colomb's Law of Electrostatic Forces

It is also called colomb's Inverse Squre law, which gives, the force existing between two electric charges. It states that, the force (F) between to charges (Q_1 and. Q_2) varies directly as the product of the charges and inversely as the square of the distance between them (Fig. 3.1)

Mathematically,, $$F \propto Q_1 Q_2 \quad \text{...(3.1}a\text{)}$$

$$F \propto \frac{1}{r^2} \quad \text{...(3.1}b\text{)}$$

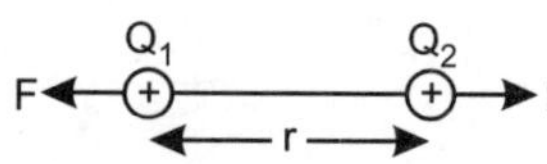

Fig. 3.1 Two Changes at finite distance

Combing 3.1 (*a*) and 3.1 (*b*), gives;

$$F \propto \frac{Q_1 Q_2}{r^2} \quad \text{...(3.1}c\text{)}$$

or $$F = K\frac{Q_1 Q_2}{r^2} \qquad \text{...(3.1}d\text{)}$$

where, $$K = \text{proportionality constant} = \frac{1}{4\pi\varepsilon}$$

So, $$F = \frac{1}{4\pi\varepsilon}\frac{Q_1 Q_2}{r^2} \qquad \text{...(3.2}a\text{)}$$

and ε = permitivity of the medium or dielectric constant of the medium in which the charges are placed.

Permitivity of the medium can be expressed as,

$$\varepsilon = \varepsilon_r \varepsilon_0$$

where ε_r = relative permitivity of the medium w.r.t. free space and

$$\varepsilon_0 = \frac{1}{36\pi} \times 10^{-9} \text{ F/m}$$

$$= 8.854 \times 10^{-12} \text{ F/m}, \qquad \text{(for free space or air, } \varepsilon_r = 1\text{)}$$

So, $$F = \frac{1}{4\pi\varepsilon_0}\frac{Q_1 Q_2}{r^2} \qquad \text{...(3.2}b\text{)}$$

The factor 4π appears in colombs law, so that it could not appear in Maxwell's equation and simplify the electromagnetic theory.

Since force is a vector quantity and has magnitude as well as direction hence equation (3.2 *b*) can be written as:

$$\vec{F} = \frac{1}{4\pi\varepsilon_0}\frac{Q_1 Q_2}{r^2}\overline{a_r} \qquad \text{...(3.3)}$$

where $\overline{a_r}$ = unit vector. In general, the vector force can be given as, shown in Fig. 3.2,

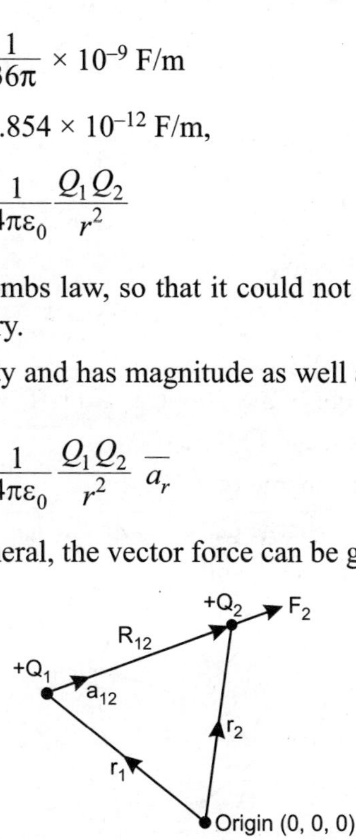

Fig. 3.2 Vector force F_2 at the charges Q_1

Let the vector r_1 locates Q_1 while vector r_2 locates Q_2. Then

$$\overline{R_{12}} = \overline{r_2} - \overline{r_1}$$

$$|\overline{R_{12}}| = \frac{R_{12}}{|\overline{r_2} - \overline{r_1}|}$$

The force $\overline{F_2}$ on charge Q_2 due to Q_1.

$$\overline{F_2} = \frac{1}{4\pi\varepsilon}\frac{Q_1 Q_2}{R_{12}^2} a_{12}$$

where, $$a_{12} = \frac{\overline{R_{12}}}{|\overline{R_{12}}|} = \frac{\overline{R_{12}}}{|R_{12}|} = \frac{\overline{r_2} - \overline{r_1}}{|(r_2 - r_1)|}$$

Similarly the force F_1 on charge Q_1 due to Q_2 can be expressed as shown in Fig. 3.3

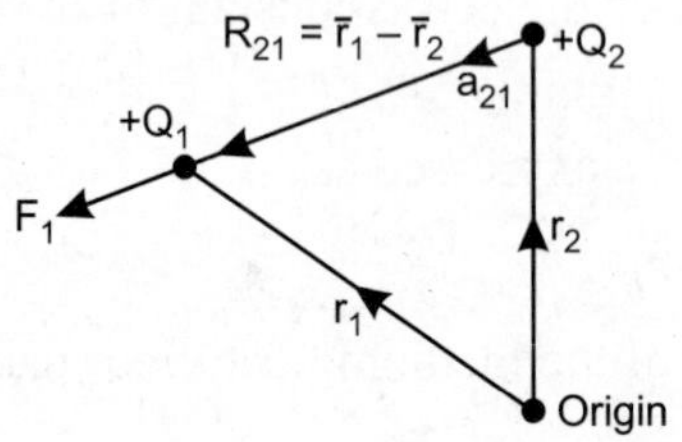

Fig. 3.3 Vector force F_1 at the charge Q_2

$$a_{21} = \frac{\overline{R_{21}}}{|\overline{R_{21}}|} = \frac{\overline{R_{21}}}{R_{21}} = \frac{\overline{r_1} - \overline{r_2}}{|r_1 - r_2|}$$

$$\overline{F_1} = \frac{1}{4\pi\varepsilon}\frac{Q_1 Q_2}{R_{21}^2} \cdot \overline{a_{21}} \qquad \text{...(3.4)}$$

SOLVED EXAMPLES

Example 1. Considering a charge of 3×10^{-4} C at P (1, 2, 3) and a charge of -10^{-4} C at Q (2, 0, 5) in vaccum.

(*a*) Find the force on Q_2 due to charge Q_1.

(*b*) Find the force on Q_1 due to charge Q_2.

(*c*) Give the conclusion from force F_2 and F_1.

Solution: (*a*)

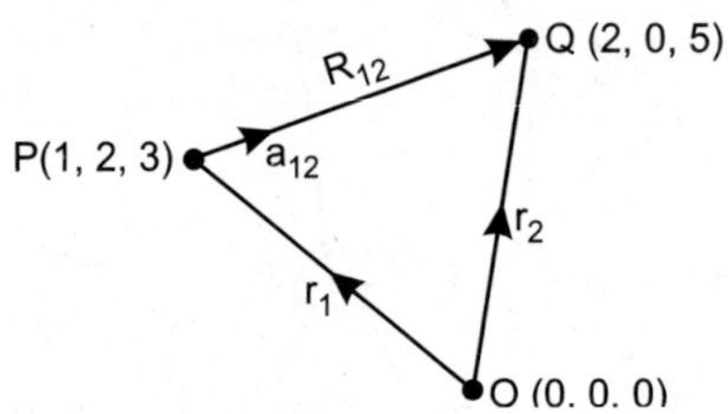

Fig. 3.4 Geometry for example 1(*a*).

$$r_1 = 1a_x + 2a_y + 3a_z$$

$$r_2 = 2a_x + 5a_z$$

So $$\overline{R_{12}} = \overline{r_2} - \overline{r_1}$$

$$= (2a_x - 5a_z) - (a_x + 2a_y + 3a_z)$$

$$= a_x - 2a_y + 2a_z$$

$$a_{12} = \frac{\overline{R_{12}}}{|R_{12}|} = \frac{(a_x - 2a_y + 2a_z)}{\sqrt{1^2 + 2^2 + 2^2}} = \frac{(a_x - 2a_y + 2a_z)}{\sqrt{9}}$$

$$= \frac{(a_x - 2a_y + 2a_z)}{3}$$

$$F_2 = \frac{1}{4\pi\varepsilon_0} \cdot \frac{Q_1 Q_2}{R_{12}^2} a_{12}$$

$$= \frac{1}{4\pi \times \frac{1}{36\pi} \times 10^{-9}} \times \frac{3 \times 10^{-4} \times -1 \times 10^{-4}}{9} \frac{(a_x - 2a_y + 2a_z)}{3}$$

$$= -10\,(a_x - 2a_y + 2a_z)\,N$$

$$\overline{F_2} = (-10a_x + 20a_y - 20a_z)\,N$$

(*b*) Now the force on charge at Q_1 at P with Q_2 can be given as

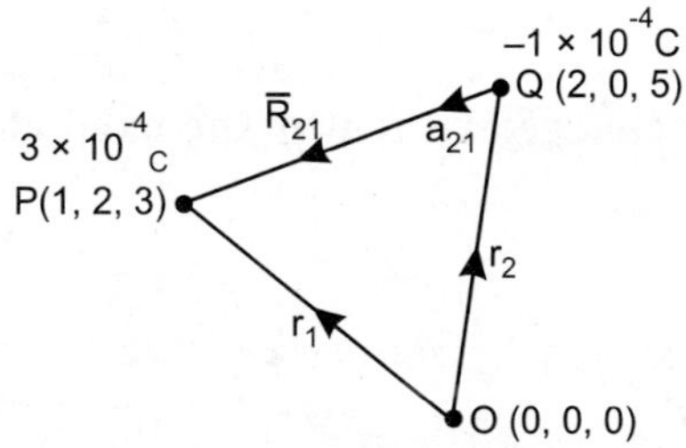

Fig. 3.5 Geometry for example 1(b).

$$\overline{r_1} = a_x + 2a_y + 3a_z,\ \overline{r_2} = 5\overline{a_y}$$

$$\overline{R_{21}} = \overline{r_1} - \overline{r_2}$$

$$= (-a_x + 2a_y - 2a_z)$$

$$R_{12} = |R_{12}| = \sqrt{-1^2 + 2^2 + (-2)^2}$$

$$= \sqrt{1 + 4 + 4}$$

$$= \sqrt{9} = 3$$

$$a_{21} = \frac{\overline{R_{12}}}{|R_{12}|} = \frac{r_1 - r_2}{|r_1 - r_2|} = \frac{(-a_x + 2a_y - 2a_z)}{3}$$

$$\overline{F_1} = \frac{1}{4\pi\varepsilon_0} \frac{Q_1 Q_2}{R_{12}^2} a_{21}$$

$$= \frac{1}{4\pi \times \frac{1}{36\pi} \times 10^{-9}} \times \frac{3 \times 10^{-4} \times -1 \times 10^{-4}}{9} \times \frac{(-a_x + 2a_y - 2a_z)}{3}$$

$$\overline{F_1} = -10(-a_x + 2a_y - 2a_z) = (10a_x - 20a_y + 20a_z)$$

From force equation F_1 and F_2, it is cleared that forces are equal in magnitude but out of phase or in opposite direction,

So $$F_1 = F_2 = \frac{Q_1 Q_2}{4\pi\varepsilon_0 R_{12}^2} a_{12} = -\frac{Q_1 Q_2}{4\pi\varepsilon R_{12}^2} a_{21}$$

(*c*) Columb's law is linear means if we multiply the charge Q_1 by a factor *n*, the force on Q_2 is also get multified by the same factor *n*, also if there are several charges the force on a given charge is the vector sum of forces due to individual charges acting alone.

Example 2. A 2 mC positive charge is located in vacuum at point P_1 (3, –2, –4) and 5,μC negative charge is at P_2 (1, -4,. 2),

(*a*) Find the vector force on negative charge.

(*b*) What is the magmtudue on the negative charge at P_1?

[Ans. (*a*) $0.616a_x + 0.616a_y - 1.848\ a_z$ (*b*) 2.04 N.

Electric Field Intensity : Electric field, or electric field intensity is denoted by *E*. If a test charge Q_t is placed at any point near a fixed charge Q_1 the test charge experience a force due to Q_1 as shown in Fig 3.6

"The force per unit on test charge exerted by the fixed charge is called electric field or electric field intensity."

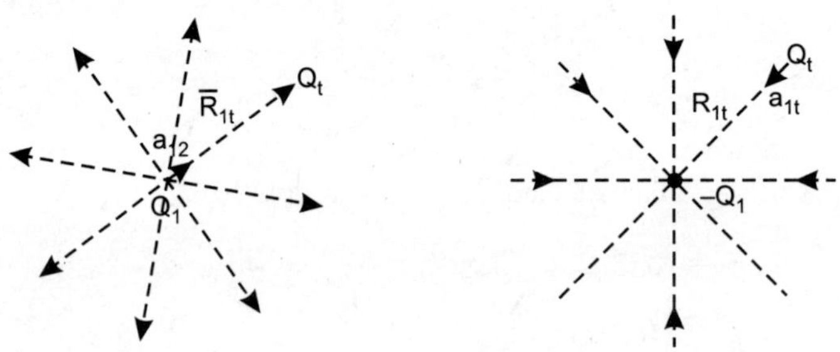

Fig. 3.6 Flux lines due to charges (+ Q_1) and (- Q_1)

$$\Downarrow \qquad\qquad \Downarrow$$

$$F_t = \frac{Q_1 Q_t}{4\pi\varepsilon R_{1t}^2} a_{1t} \qquad F_t = -\frac{Q_1 Q_t}{4\pi\varepsilon R_{1t}^2} a_{1t}$$

$$\frac{F_t}{Q_t} = E = \frac{Q_1}{4\pi\varepsilon R_{1t}^2} \qquad \frac{F_t}{Q_t} = E = -\frac{Q_1}{\pi\varepsilon R_{1t}^2} a_{1t}$$

It is also called electric field strength.

In the simple from the electric field intensity in free-space can be given as,

$$E = \frac{Q}{4\pi\varepsilon_0 R^2} a_R \qquad ...(3.5)$$

where *R* is the magnitude of $\overline{R}$ vector and a_R is the unit vector in the direction *R*.

Let Q_1 is at the centre of the spherical coordinate system Figs (3.7 and 3.8) then;

$$E = \frac{Q_1}{4\pi\varepsilon_0 r^2} \underline{a_r}$$

If cartesian coordinate system is taken then

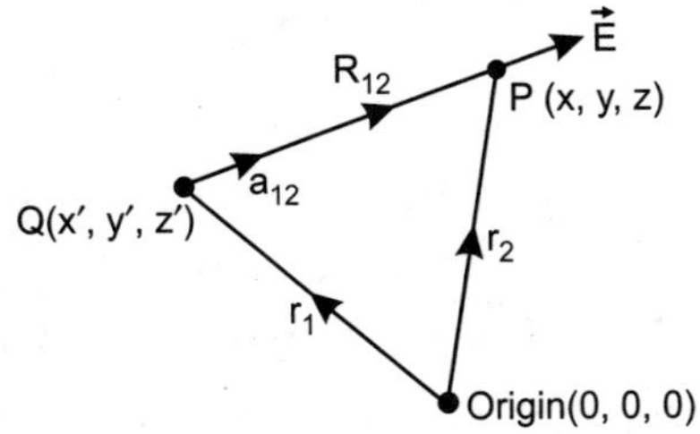

Fig. 3.7 E at a charge Q

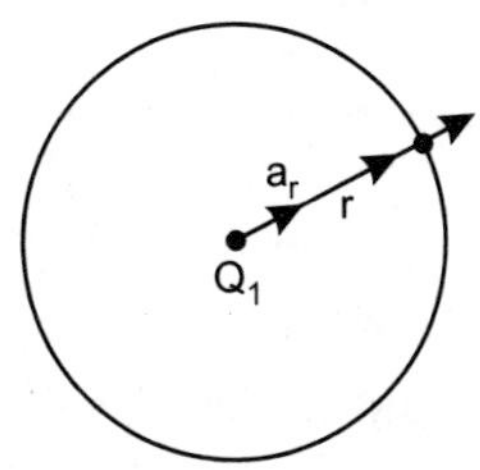

Fig. 3.8 Charge Q_1 at the centre of sphere

$$\overline{r_1} = x'a_x + y'a_y + z'a_z.$$

$$\overline{r_2} = xa_x + ya_y + za_z$$

$$\overline{R_{12}} = \overline{r_2} - \overline{r_1}$$

$$= (xa_x + ya_y + za_z) - (x'a_x + y'a_y + z'a_z)$$

$$= (x - x')a_x + (y - y')a_y + (z - z')a_z$$

$$\overline{E} = \frac{Q}{4\pi\varepsilon_0 R_{12}^2} a_{12}$$

$$a_{12} = \frac{\overline{R_{12}}}{|R_{12}|} = \frac{(x - x')a_x + (y - y')a_y + (z - z')a_z}{\sqrt{(x - x')^2 + (y - y_1)^2 + (z - z')^2}}$$

$$E = \frac{Q}{4\pi\varepsilon_0[(x - x')^2 + (y - y')^2 + (z - z')^2]} \times \frac{(x - x')a_x + (y - y')a_y + (z - z')a_z}{[(x - x')^2 + (y - y')^2 + (z - z')^2]^{1/2}}$$

$$E = \frac{Q}{4\pi\varepsilon_0[(x - x')^2 - (y - y')^2 + (z - z')^2]^{3/2}} [(x - x')a_x + (y - y')a_y + (z - z')a_z] \quad ...(3.6)$$

Since Colomb's force is linear, so the electric field is also linear.

If there are two point charges Q_1 at r_1 and Q_2 at r_2 then electric field intensity at the test charge will be

$$E = E_1 + E_2$$

$$E = \frac{Q_1}{4\pi\varepsilon_0 r_1^2} a_{r1} + \frac{Q_2}{4\pi\varepsilon_0 r_2^2} a_{r2} \quad ...(3.7)$$

If there are n point charges; $Q_1, Q_2, Q_3, ...Q_n$ be located at a distance $r_1, r_2, r_3,...,r_n$ from the origin (0, 0, 0) then the electric field intensity at point P is given by

$$E = \frac{Q_1}{4\pi\varepsilon_0 r_1^2} a_{r1} + \frac{Q_2}{4\pi\varepsilon_0 r_2^2} a_{r2} + \frac{Q_3}{4\pi\varepsilon_0 r_3^2} a_{r3} + ... + \frac{Q_n}{4\pi\varepsilon_0 r_n^2} a_{rn}$$

$$= \sum_{i=1}^{n} \frac{Q_i}{4\pi\varepsilon_0 r_i^2} a_{ri}$$

$$E = \frac{1}{4\pi\varepsilon_0} \sum_{i=1}^{n} \frac{Q_i}{r_i^2} a_{ri} \qquad ...(3.8)$$

Example 3. Find the electric field intensity E at point P (1, 1, 1) caused by four identical $3nC$ point charges located at P_1 (1, 1, 0), P_2 (–1, – 1, 0), P_3 (–1, – 1, 0) and P_4 (1, – 1, 0)

Solution : From Fig. 3.9

$$r_1 = a_x + a_y$$

$$r_2 = -a_x + a_y$$

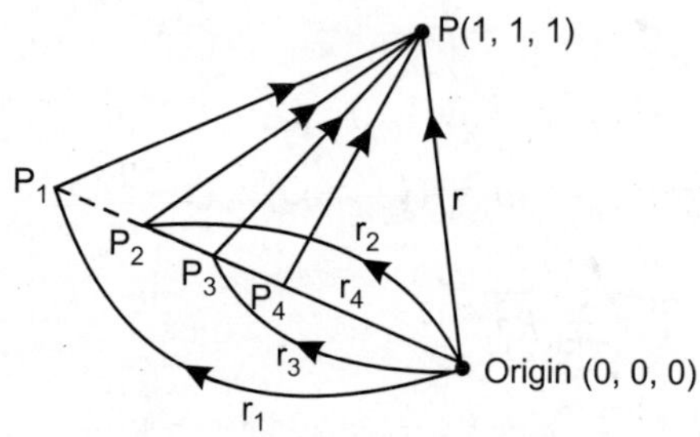

Fig. 3.9 Geometry for example 3.

$$r_3 = -a_x - a_y$$

$$r_4 = a_x - a_y$$

$$r = a_x + a_y + a_z$$

From the definition $\frac{Q_1}{4\pi\varepsilon_0 |r - r_1|^2} a_1 + \frac{Q_2}{4\pi\varepsilon_0 |r - r_2|^2} a_2 + \frac{Q_3}{4\pi\varepsilon_0 |r - r_3|^2} + \frac{Q_4}{4\pi\varepsilon_0 |r - r_4|^2} a_4$

$$r - r_1 = a_z \Rightarrow |r - r_1| = 1$$

$$a_1 = \frac{r - r_1}{|r - r_1|} = \frac{a_z}{1} = a_z$$

$$r - r_2 = a_x + a_y + a_z + a_x - a_y$$

$$\Rightarrow \qquad |r - r_2| = \sqrt{5}$$

$$= 2a_x + a_z$$

$$a_z = \frac{r - r_2}{|r - r_2|} = \frac{2a_x + a_z}{\sqrt{5}}$$

$$r - r_3 = a_x + a_y + a_z + a_x + a_y$$

$$2a_x + 2a_y + a_z \Rightarrow |r - r_3| = \sqrt{9}$$

$$a_3 = \frac{r - r_3}{|r - r_3|} = \frac{(2a_x + 2a_y + a_z)}{\sqrt{9}}$$

$$r - r_4 = a_x + a_y + a_z - a_x + a_y$$

$$= 2a_y + a_z$$

$$\Rightarrow \quad |r - r_4| = \sqrt{4+1} = \sqrt{5}$$

So,

$$E = \frac{Q_1}{4\pi\varepsilon_0 |r - r_1|^2} a_1 + \frac{Q_2}{4\pi\varepsilon_0 |r - r_1|^2} a_2 + \frac{Q_3}{4\pi\varepsilon_0 |r - r_3|^2}$$

$$a_3 + \frac{Q_4}{4\pi\varepsilon_0 |r - r_4|^2} \quad Q_1 = Q_2 = Q_3 = Q_4 = Q = 3 \times 10^{-9} \text{ C}$$

$$\Rightarrow \quad E = \frac{Q}{4\pi\varepsilon_0}\left[\frac{a_1}{|r - r_1|^2} + \frac{a_2}{|r - r_2|^2} + \frac{a_3}{|r - r_3|^2} + \frac{a_4}{|r - r_4|^2}\right]$$

$$= \frac{3 \times 10^{-9}}{4\pi \times \frac{1}{36\pi} \times 10^{-9}}\left[\frac{a_z}{1^2} + \frac{(2a_x + a_z)}{(\sqrt{5})^2.\sqrt{5}} + \frac{(2a_x + 2a_y + a_z)}{(\sqrt{9})^2\sqrt{9}} + \frac{(2a_y + a_z)}{(\sqrt{5})\sqrt{5}}\right]$$

$$= 27\left[a_z + \frac{(2a_x + a_z)}{(5)^{3/2}} + \frac{(2a_x + 2a_y + a_z)}{(9)^{3/2}} + \frac{(2a_y + a_z)}{(5)^{3/2}}\right]$$

$$= 27\left[\frac{a_2}{1} + \frac{2a_x + 2a_y + 2a_z}{(5)^{3/2}} + \frac{2a_x + 2a_y + a_z}{(9)^{3/2}}\right]$$

$$= 27\left(\frac{45a_z + (18a_x + 18a_y + 18a_z) + 10a_x + 10a_y + 5a_z}{45}\right)$$

$$= 27\left(\frac{(45 + 18 + 5)a_z + (18 + 10)a_x + (18 + 10)a_y}{45}\right)$$

$$= 27\left[\left(\frac{28}{45}\right)a_x + \left(\frac{28}{45}\right)a_y + \left(\frac{68}{45}\right)a_z\right]$$

$$= \left(\frac{27 \times 28}{45}\right)a_x + \left(\frac{27 \times 28}{45}\right)a_y + \left(\frac{27 \times 68}{45}\right)a_z$$

$$= \left(\frac{84}{5}\right)a_x + \left(\frac{84}{5}\right)a_y + \left(\frac{204}{5}\right)a_z$$

$$E = (6.82a_x + 6.82a_y + 32.8a_z) \text{ V/m}$$ Ans.

Example 4. Calculate electric field at point P (3, – 4, 2) in free space caused by

(*a*) Charge $Q = 2\mu\ C$ at P_1 (0, 0, 0).

(*b*) A charge $Q_2 = 3\mu C$ at P_2 (–1, 2, 3)

(*c*) A charge $Q_1 = 2\mu C$ at (0, 0) & $Q_2 = 3\mu C$ at P_2 (–1, 2, 3)

[Ans. (*a*) $(345a_x - 460a_y + 230a_z)$, (*b*) $(280a_x - 419a_y - 69.9a_z)$ (c) $629a_x - 880a_y + 160.3a_z$]

Electric Field Due to Charge Distribution

Before discribing the electric field due to charge distribution, let us first discuss the charge density. Charge distribution may be in the form of volume, sheet or line, so there are three types of charge density.

Volume charge density (ρ_v): The electric charge density ρ_v is the ratio of total charge Q in a given volume to volume V, *i.e.* the total change for unit V.

$$\rho = \frac{Q}{V} \qquad \text{...(3.9)}$$

If electric field is uniformely distributed throughout the region, then charge ($\Delta\rho$) in a region at a point in volume ΔV can be given as

$$\rho_v = \lim_{\Delta v \to 0} \frac{\Delta Q}{\Delta v} \text{ C/m}^3 \qquad \text{...(3.10)}$$

it is called volume charge density.

Surface charge density : If the charge is distributed uniformly over a surface, then surface charge density is used. This is defined as charge per unit area and denoted by ρ_s.

$$\rho_S = \lim_{\Delta S \to 0} \frac{\Delta Q}{\Delta S} \text{ C/m}^2$$

Line charge density : Now the last is line charge density, the charge per unit length is called line charge density, and it is given as,

$$\rho_L = \lim_{\Delta L \to 0} \frac{\Delta Q}{\Delta L} \text{ C/m}$$

Now total charge can be calculated as

$$\rho_v = \frac{\Delta Q}{\Delta v} = \frac{dQ}{dv}$$

$$dQ = \rho_v dv$$

$$Q = \int_{Vol} dQ = \int \rho_v dv \qquad \text{...(3.11a)}$$

Similarly, for surface and line distributions:

$$Q = \int_S dQ = \int_S \rho_s ds \qquad \text{...(3.11b)}$$

$$Q = \int_L dl = \int_L \rho_L dL \qquad \text{...(3.11c)}$$

Field Due to Line Charge Distribution : Let us consider a filament like distribution of volume charge, such as very fine sharp beam in CRT (cathode ray tube) or a charged conductor of very small diameter. Also let straight line charge extended along Z-axis in a cylindrical coordinate system (r, ϕ, z)

The charges are extended along the z-axis from $-\infty$ to ∞ (Fig. 3.10). Before calculating electric field, we must consider the symmetry. There are there cases of symmetry.

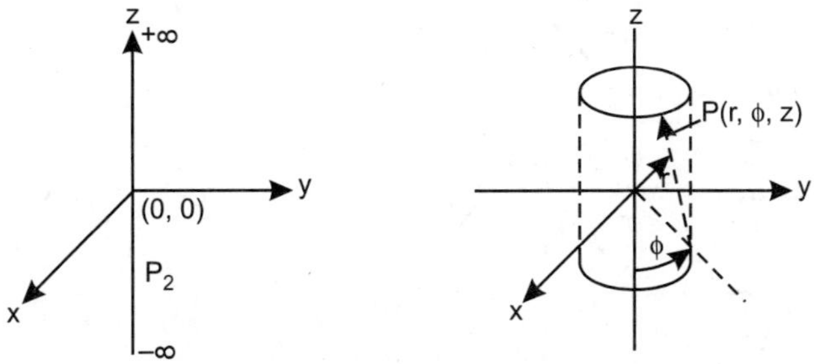

Fig. 3.10 Charge extended along z-axis

(*i*) If we vary Q while keeping r and z constant, the line charge appears the same from every angle,*i.e.*, azimuthal symmetry present and no field component vary with ϕ

(*ii*) Again if we maintain r and ϕ constant and moving along z axis, it shows symmetry such that no field changed.

(*iii*) Third if we make ϕ and z constant, but vary r, the problem get change, there is no symmetry so electric field will be obtained in this direction. (Fig. 3.11)

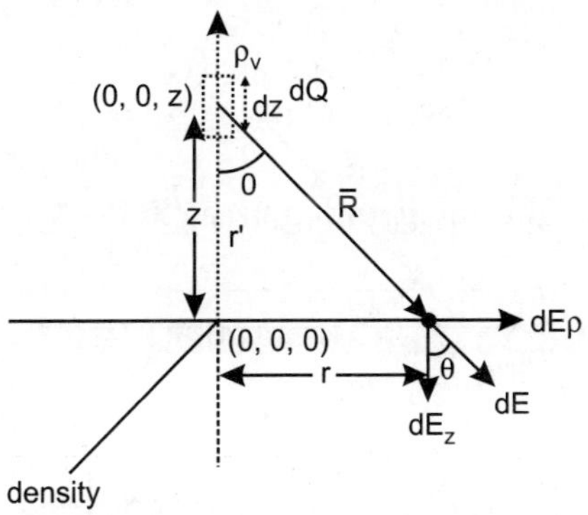

Fig. 3.11 Field due to a line charge

$$E = \frac{Q}{4\pi\varepsilon_0 R^2} a_r \qquad \text{...(3.11d)}$$

$$\overline{R} = r - r'$$

$$a_r = \frac{r - r'}{|r - r'|}$$

hence,

$$dE = \frac{dQ}{4\pi\varepsilon_0 |r - r'|^2} \frac{r - r'}{|r - r'|}$$

$$= \frac{dQ}{4\pi\varepsilon_0} \cdot \frac{(r - r')}{(r - r')^3}$$

$$r' = z a_z$$

$$dQ = \rho_L d_z$$

$$r = \rho a_p$$

$$dE = \frac{(\rho_L dz)}{4\pi\varepsilon_0}\frac{(za_z - pa_p)}{(z^2 + p^2)^{3.2}} \quad \text{...(312)}$$

Since only ρ component is present then

$$dE = \frac{\rho_L \, dz \, \rho a_\rho}{4\pi\varepsilon_0(\rho^2 + z^2)^{3/2}}$$

$$\int dE = E = \int_{-\infty}^{\infty} \frac{\rho_L dz \rho a_\rho}{4\pi\varepsilon_0(\rho^2 + z^2)^{3/2}}$$

Let $z = \rho \cot\theta$

$$c = \rho \operatorname{cosec}^2\theta \, d\theta$$

$$\int_0^{\pi/2} \rho_L \quad \frac{\rho \operatorname{cosec}^2 \theta d\theta}{4\pi\varepsilon(\rho^2 + \rho^2 \cot^2\theta)^{3/2}}$$

$$\int_0^{\pi/2} \rho_L \quad \frac{\rho^2 \operatorname{cosec}^2 \theta d\theta}{\rho^3(1 + \cot^2\theta)^{3/2}}$$

$$\int_0^{\pi/2} \rho_L \quad \frac{\rho^2}{\rho^3} \cdot \frac{\operatorname{cosec}^2\theta}{\operatorname{cosec}^3\theta} \, d\theta$$

$$= \frac{1}{4\pi\varepsilon_0}\frac{\rho_L}{\rho}\int_0^{\pi/2} \frac{1}{\operatorname{cosec}\theta} \, d\theta$$

$$= \frac{1}{4\pi\varepsilon_0}\frac{\rho_L}{\rho}\int_0^{\pi/2} \sin\theta \, d\theta$$

$$= \frac{1}{4\pi\varepsilon_0}\frac{\rho_L}{\rho} - [\cos\theta]_\pi^0 = \frac{1}{4\pi\varepsilon_0}[1+1]\frac{\rho_L}{\rho}$$

$$\Rightarrow \quad E = \frac{\rho_L}{2\pi\varepsilon_0\rho} \quad \text{...(3.13)}$$

In vector form

$$\vec{E_\rho} = \frac{\rho_L}{2\pi\varepsilon_0\rho}\vec{a_\rho} \quad \text{...(3.14)}$$

That is 'E' vary inversely with distance in line charge.

Example 5. A infinite line charge shect is placed parallel to the z axis at $x = 6, y = 8$ (Fig. 3.12). Find electric field 'E' at a general point $P(x, y, z)$.

$$r = 6a_x + 8a_y + za_z$$

$$r' = xa_x + ya_y + za_z$$

$$\overline{R} = r - r' = (x - 6)\, a_x + (y - 8)\, a_y$$

$$a_R = \frac{(x-6)a_x + (y-8)a_y}{[(x-6)^2 + (y-8)^2]^{1/2}}$$

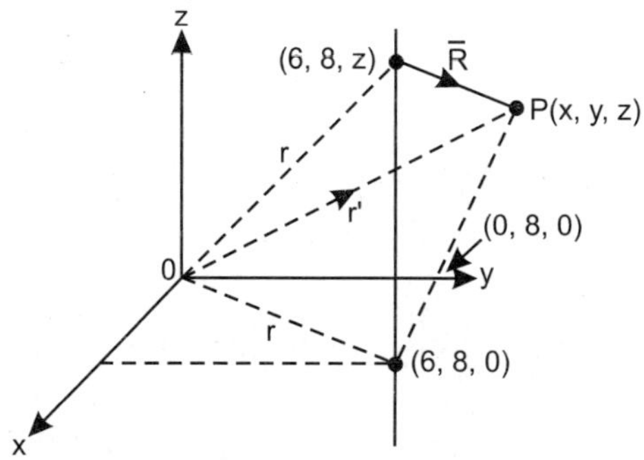

Fig. 3.12. Geometry for example 5.

$$E = \frac{\rho_L}{2\pi\varepsilon_0\rho} a_R = \frac{\rho_L}{2\pi\varepsilon_0} \frac{(x-6)a_x + (y-8)a_y}{[(x-6)^2 + (y-8)^2]} \quad \text{Ans.}$$

The Electric Dipole and Dipole Moment

The combination of two equal point charges of opposite nature and separated by a small distance is called an electric dipole or simply dipole and product of charge Q and distance is called dipole moment *i.e.*

$$m = Ql$$

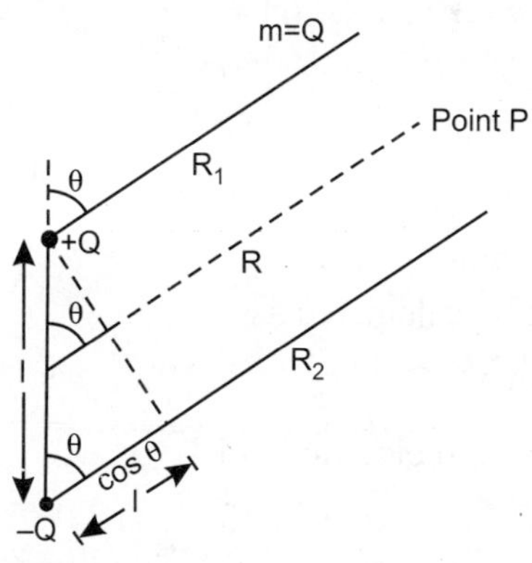

Fig. 3.13. Sample diagram of a dipde.

The spacing between two charges l is considered as vector $\vec{l}$ directed from negative charge (Fig. 3.13). Potentials due to charge + Q

$$V_1 = \frac{Q}{4\pi\varepsilon R_1}$$

and due to –Q

$$V_2 = -\frac{Q}{4\pi\varepsilon R_2}$$

Total electric Potential at point P will be:

$$V = V_1 + V_2 = \frac{Q}{4\pi\varepsilon R_1} - \frac{Q}{4\pi\varepsilon R_2}$$

$$= \frac{Q}{4\pi\varepsilon} \times \left(\frac{1}{R_1} - \frac{1}{R_2}\right)$$

$$V = \frac{Q}{4\pi\varepsilon}\left[\frac{R_2 - R_1}{R_1 \,.\, R_2}\right]$$

$$= \frac{Q}{4\pi\varepsilon}\frac{l\cos\theta}{R^2} \text{ Assuming (As } R_1 = R_2 = R \;\&\; R_1 . R_2 = R^2\text{)}$$

$$V = \frac{Q}{4\pi\varepsilon}\frac{l\cos\theta}{R^2}$$

where $R_1 = R - \frac{l}{2}\cos\theta$

$$R_2 = R + \frac{l}{2}\cos\theta$$

$$\text{V} = \left(\frac{1}{4\pi\varepsilon}\right)\left(\frac{1}{R^2}\right)(\text{Ql})\cos\theta$$

Angle law $R_1 . R_2 = \left(R - \frac{1}{2}\cos\theta\right) \times \left(R + \frac{l}{2}\cos\theta\right)$

$$= \left(R^2 - \frac{l^2}{2^2}\cos^2\theta\right) \approx R^2 \qquad \text{(since } R >> l\text{)}$$

$$R = \sqrt{R_1 . R_2}$$

Therefore, $V = \dfrac{m\cos\theta}{4\pi\varepsilon R^2}$ As $m = Q.l$

The electric field intensity can be calculated as

$$\text{E} = -\nabla\text{V}$$

i.e., Negative gradient of V equal to electric field

$$\nabla = \frac{\partial}{\partial x}a_x + \frac{\partial}{\partial y}a_y + \frac{\partial}{\partial z}a_z \qquad \text{in cartesian coordinate system}$$

$$\nabla = \frac{\partial}{\partial r}a_r + \frac{1}{r}\frac{\partial}{\partial\theta}a_\theta + \frac{1}{r\sin\theta\partial\phi}a_\theta \qquad \text{in spherical cordinate system}$$

Again in spherical coordinate system;

$$R_1 = R - \frac{l}{2}\cos\theta$$

and $R_2 = R + \frac{l}{2}\cos\theta$

So, $\overline{E} = -\nabla.V$

$$= -\left[\frac{\partial}{\partial r}a_r + \frac{1}{r}\frac{\partial}{\partial \theta}a_\theta + \frac{1}{r\sin\theta}\frac{\partial}{\partial \phi}a_\phi\right]V$$

$$= -\left[\frac{\partial V}{\partial r}a_r + \frac{1}{r}\frac{\partial V}{\partial Q}a_\theta + \frac{1}{r\sin\theta}\cdot\frac{\partial V}{\partial \phi}a_\phi\right]$$

Since the electric field also has three component (E_v, E_θ E_ϕ), then

$$E_r a_r + E_\theta a_\theta + E_\phi a_\phi = -\left[\frac{\partial V}{\partial r}a_r + \frac{1}{r}\frac{\partial V}{\partial \theta}a_\theta + \frac{1}{r\sin\theta}\frac{\partial V}{\partial \phi}a_\phi\right] \quad ...(3.15)$$

i.e. $E_r = -\left(\frac{\partial V}{\partial r}\right)$

$$E_\theta = \frac{1}{r}\frac{\partial V}{\partial \theta}$$

$$E_\phi = \frac{1}{r\sin\theta}\frac{\partial V}{\partial \phi}$$

Again, $E_r = -\frac{\partial}{\partial r}\frac{1}{4\pi\varepsilon_0}\left(\frac{m\cos\theta}{r^2}\right) = -\frac{1}{4\pi\varepsilon_0}\text{ m}\cos\theta\,\frac{\partial}{\partial r}\left(\frac{1}{r^2}\right)$

(from the value of V.)

$$= -\frac{1}{4\pi\varepsilon_0}m\cos\theta\left(-\frac{1}{r^3}\right)$$

$$E_r = \frac{1}{4\pi\varepsilon_0}\frac{m\cos\theta}{r^3} \quad ...(3.16\ a)$$

Similarly

$$E_\theta = \frac{1}{r}\frac{d}{d\theta}\frac{m\cos\theta}{4\pi\varepsilon r^2} = \frac{1}{4\pi\varepsilon_0 r^3}m\sin\theta \quad ...(3.16\ b)$$

and $E_\phi = 0$ ...(3.16 *c*)

Since V does not depend on ϕ.

From above equations it is cleared that electric field increase as $\frac{1}{r^3}$ where as electric potential as $\frac{1}{r^2}$.

3.2 ELECTRIC FIELD DUE TO A SHEET OF CHARGE

There are another charge distribution known as sheet charge distribution. Example of such distribution is, charge on the strip line, and parallel plate capacitor etc. In these examples the charges are located at the surface of the conductor not interior of the conductor. It is also called surface charge distribution having surface charge density $\rho_s\left(\frac{C}{m^2}\right)$.

Let us consider charge at a sheet situated in the *yz* plane as shown in Fig. 3.14. Now we first see that the field can not vary with *y* or *z* and hence E_y and E_z components of electric field will not exist. That is only E_x component will be found.

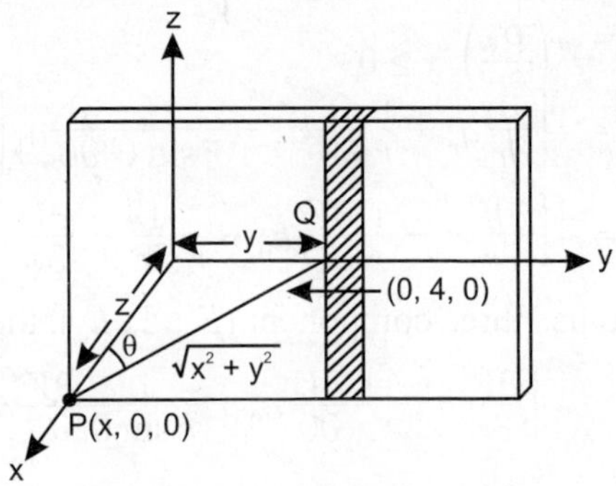

Fig. 3.14. Charge distribution on a sheet

Now charge in differential *dy* strip can be given as:

$$\rho_L = \rho_s dy$$

The electric field intensity at point *Q* will be

$$dE_x = \left(\frac{\rho_s dy}{2\pi\varepsilon_0\sqrt{x^2+y^2}}\right)\cos\theta, \text{ in } x \text{ direction}$$

$$= \frac{\rho_s dy}{2\pi\varepsilon_0\sqrt{x^2+y^2}}\left(\frac{x}{\sqrt{x^2+y^2}}\right) = \frac{\rho_s dy}{2\pi\varepsilon_0}\frac{x}{(x^2+y^2)^{3/2}}$$

Let

$$y = x\tan\theta$$

$$dy = x\sec^2\theta.\, d\theta$$

$$= \frac{\rho_s x^2\sec^2\theta d\theta}{2\pi\varepsilon_0(x^2+x^2\tan^2\theta)}$$

$$= \frac{\rho_s x^2\sec^2\theta d\theta}{2\pi\varepsilon_0 x^2(1+\tan^2\theta)}$$

$$= \frac{\rho_s x^2\sec^2\theta d\theta}{2\pi\varepsilon_0 x^2(\sec^2\theta)}$$

$$\int dE_x = \int\frac{\rho_s}{2\pi\varepsilon_0}d\theta$$

$$E_x = \frac{\rho_s}{2\pi\varepsilon_0}\theta$$

$$= \frac{\rho_s}{2\pi\varepsilon}\left[\tan^{-1}\frac{y}{x}\right]_{-\infty}^{\infty}$$

$$= \frac{\rho_s}{2\pi\varepsilon_0}\left[\tan^{-1}\frac{\infty}{x} - \tan^{-1}-\frac{\infty}{x}\right]$$

$$= 2\pi\times\frac{\rho_s}{2\pi\varepsilon_0} = \frac{\rho_s}{2\varepsilon}\; x \geq 0$$

$$\Rightarrow \qquad E_x = \left(\frac{\rho_s}{2\varepsilon}\right), x \geq 0 \qquad \text{...(3.17 a)}$$

Since 'E' is a vector quantity and it is always away from positive charge So a unit vector can be taken as normal to sheet and away from positive charge. *i.e.*

$$E = \frac{\rho_s}{2\varepsilon_0} = \frac{\rho_s}{2\varepsilon_0} a_x \qquad \text{...(3.17 b)}$$

Note: If a second infinite sheet of charge having a negative charge density ($-\rho_s$) is located at the plane $x = a$, we may find total field by adding the contribution of each sheet in the REGION, $x > a$.

$$E_+ = \frac{\rho_s}{2\varepsilon_0} a_x,$$

$$E_- = \frac{-\rho_s}{2\varepsilon_0} a_x$$

$$E = E_+ + E_- = 0$$

For $x < 0$

$$E_+ = \frac{\rho_s}{2\varepsilon_0} a_x, E_- = \frac{\rho_s}{2\varepsilon_0} a_x$$

$$E = E_+ + E_- = \frac{\rho_s}{\varepsilon_0} \Rightarrow E = \frac{\rho_s}{\varepsilon_0} \qquad \text{...(3.18)}$$

Fields Due to Volume Charge Distribution

If a region of space filed with an extremely large number of charges separated by a minute distances, then this distribution of very small particles can be replaced with a smooth continous disribution, called volume charge distribution. Mathematically we know that

$$\frac{\Delta Q}{\Delta v} = \rho$$

$$\Delta Q = \rho \Delta_v \qquad \text{...(3.19)}$$

$$\rho_v = \lim_{\Delta v \to 0} \frac{\Delta Q}{\Delta v} \qquad \text{...(3.20)}$$

From (3.19)

$$dQ = \rho dv$$

$$Q = \int_V dQ = \int_V \rho_v dv$$

The incremental electric field

$$\Delta E = \frac{\Delta Q}{4\pi\varepsilon_0 r^2} a_r = \frac{\rho dv}{4\pi\varepsilon_0 r^2} a_r$$

Total field

$$E = \sum_{i=1}^{n} \frac{\rho_i \Delta v}{4\pi\varepsilon_0 r^2} a_{ri}$$

When Δv shrink to zero, the summation is converted into integration:

$$E = \int_V \frac{\rho dv}{4\pi\varepsilon_0 r^2} a_r \qquad ...(3.21)$$

Electric Flux and Density: Electric flux and electric flux density are another two important parameters of field, and they were invented by Faraday. Faraday did an experiment for this taking a sphere as an example (Fig. 3.15).

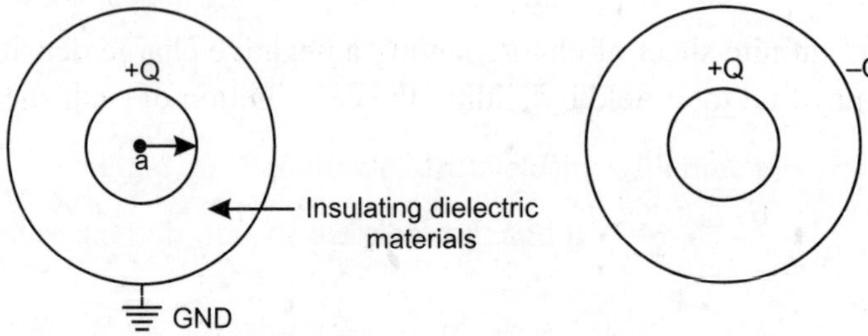

Fig. 3.15. Faraday experiment

Here linear sphere is given a known positive charge and the outer sphere is discharged through a connecting wire from the ground. It is separated carefully, and negative charge induced on outer sphere, is measured. It was seen that the total charge on outer sphere is equal in magnitude to original charge on the inner sphere and it is traced regardless of the dielectric material. It is due to some short of displacement from inner sphere to outer sphere, and called as displacement flux or electric flux. Large charge in inner sphere induced larger negative charge on the outer sphere and there is direct relationship between flux and charge;

$$\psi = Q$$

If inner sphere has radius 'a' then area '$4\pi a^2$' and flux density, will be

$$\frac{Q}{4\pi a^2} = \frac{\psi}{4\pi a^2} = D \qquad ...(3.22)$$

We know that electric field intensity due to point charge is

$$E = \frac{Q}{4\pi\varepsilon_0 r^2} a_r$$

$$\varepsilon E = \frac{Q}{4\pi r^2} a_r \qquad ...(3.23)$$

Now comparing (3.22) and (2.33), gives

$$D = \varepsilon E = \frac{Q}{4\pi r^2} a_r \qquad ...(3.24)$$

called electric flux density. Sometimes electric flux density is also called as displacement vector and flux associated with it is called displacement flux. It has unit of C/m^2. It has same direction as electric field intensity. The above formula is for point charge now if charge distribution is volume of radius 'R', then

$$D = \frac{\int_v \rho dv}{4\pi R^2} a_r \qquad ...(3.25)$$

The flux density does not depends on the proparty of the medium (ε_0).

Electrostatic Energy

When a capacitor is charged, a potential difference V exist between the plates and electrostatic energy is stored. The amount of electrostatic energy stored can be found by calculating the work done in charging the capacitor and potential is defined as the work done per unit charge.

i.e., Work done is $Q.V$. If we take small amount of charge dQ

Work done $\Delta W = V\Delta Q$

$$\Delta W = \frac{Q}{C}\Delta Q$$

As, $\frac{Q}{C} = V$

$$dW = \frac{Q}{C}\,dQ$$

Total work done, $W = \int dW = \int \frac{Q}{C}\,dQ = \frac{1}{C}\int Q\,dQ$

$$= \frac{1}{C}\int Q\,dQ$$

$$= \frac{1}{C}\left[\frac{Q^2}{2}\right]_0^Q = \frac{Q^2}{2C}$$

$$W = \frac{Q^2}{2C} \quad \text{...(3.26)}$$

or $W = \frac{Q^2}{2C} = \left(\frac{Q}{C}\right).\frac{Q}{2}$ (Since $Q = CV$)

$$= \frac{VQ}{2} = \frac{QV}{2} = (CV)\,\frac{V}{2}$$

$$W = \frac{CV^2}{2}$$

$$W = \frac{Q^2}{2C} = \frac{VQ}{2} = \frac{CV^2}{2} \quad \text{...(3.27)}$$

where, $C = \frac{\varepsilon A}{d}$

and represented in Fig. 3.16 as follows:

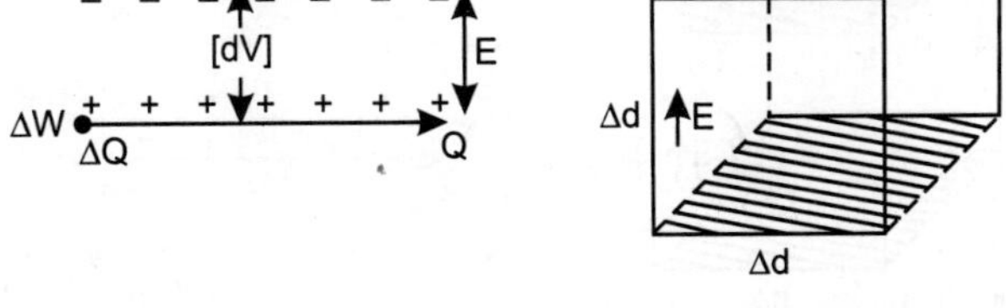

Fig. 3.16. Charge distributions on a capacitor

The small volume Δv has the value on capacitance ΔC;

$$\Delta C = \frac{\varepsilon(\Delta d)^2}{\Delta d}$$

$$\Delta C = \varepsilon \Delta d$$

The potential difference ΔV on thin sheet is

$$\Delta V = E\Delta d$$

or $$\frac{E}{\Delta d} = \Delta V$$

The stored energy in the volume Δv is;

$$\Delta W = \frac{1}{2} CV^2$$

$$= \frac{1}{2} \Delta C (\Delta V)^2$$

$$= \frac{1}{2} . \varepsilon\Delta d. (E\Delta d)^2$$

$$= \frac{1}{2} \varepsilon\Delta d. E^2 (\Delta d)^2$$

$$\Delta W = \frac{1}{2} \varepsilon E^2 (\Delta d)^3$$

$$= \frac{1}{2} \varepsilon E^2 \text{ (volume)}$$

$$= \frac{1}{2} \varepsilon E^2 (\Delta v)$$

Per unit volume, $$\frac{\Delta W}{\Delta v} = \frac{1}{2} \varepsilon E^2$$

That is the stored energy/unit volume is $\frac{1}{2} \varepsilon E^2$.

Now taking Δv as tends to zero.

$$W = \lim_{\Delta v \to 0} \frac{\Delta W}{\Delta v} = \frac{1}{2} \varepsilon E^2 \text{ J}/m^3$$

The total energy can be calculated as;

$$W = \int_V W\,dV = \frac{1}{2}\int_V \varepsilon E^2\,dV$$

$$= \frac{1}{2}\int_V \varepsilon E.EdV \qquad \text{As } \varepsilon E = D$$

$$W = \frac{1}{2}\int_V DEdV \qquad ...(3.28)$$

Since, $E = -\nabla.V$

$$= \frac{1}{2}\int_V D.\nabla\,dV$$

So, total stored energy is

$$W = \frac{1}{2}\int_V \varepsilon E^2 dV = \frac{1}{2}\int_V \varepsilon E^2 dV = \frac{1}{2}\int_V D.\nabla V dV \quad ...(3.29)$$

Methods of Electrostatic Images

It is a method used to convert an electrostatic fields into another equivalent field which is easier to calculate. It is possible to replace the conductors by one or more point charges in such a way that conductor surfaces can be treated as equipotential surfaces at the same potential (Fig. 3.17)

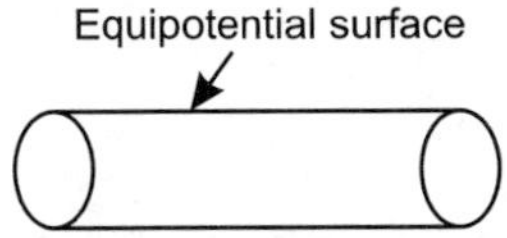

Fig. 3.17. Conductors as equipotential surfaces

Image methods permit the determination of electric fields behavior in dielectric without knowning the actual charge distribution on the plane. Since charge density can easily be determined, thousand of images may be applicable to spherical charge distribution.

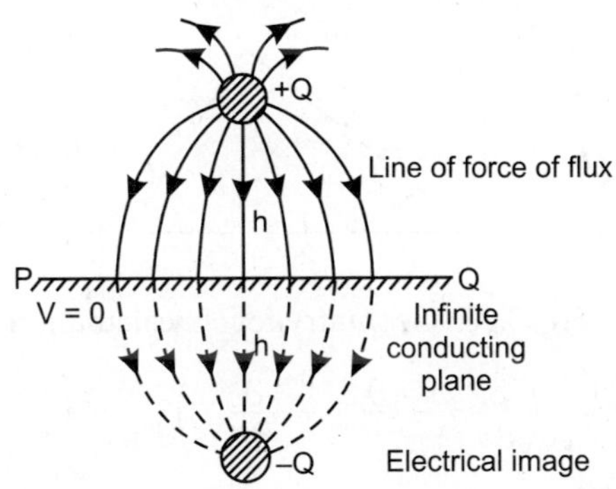

Fig. 3.18. Field lines in image method

Where (–Q), is fictitious charge and called electrostatic electrical image similar to electrical image. Now let ground plane removed. The upper portions of the two field will be identical and equivalent configuration can be used to determine all field quantities in upper plane half space.

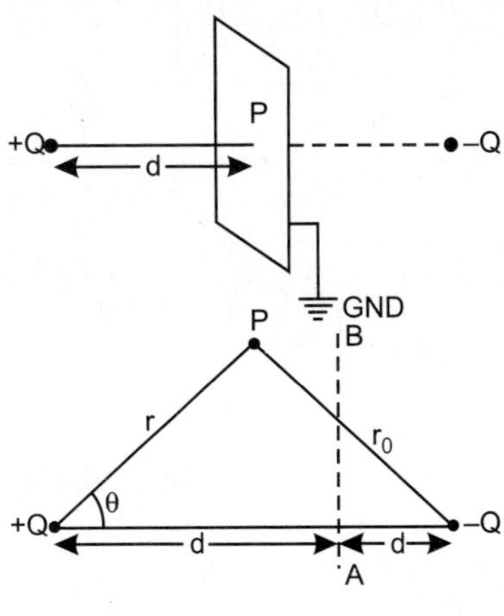

Fig. 3.19. Equivalent geometry of Fig. 3.17.

Total potential

$$V = V_1 + V_2$$

$$V_1 = \frac{Q}{4\pi\varepsilon_0 r}$$

$$V_2 = \frac{-Q}{4\pi\varepsilon_0 r}$$

$$V = \frac{1}{4\pi\varepsilon_0}\left[\frac{Q}{r} - \frac{Q}{r_0}\right]$$

$$r_0^2 = r^2 + (2d)^2 + 2.2rd\cos\theta$$

or
$$r_0 = \sqrt{r^2 + (2d)^2 + 4rd\cos\theta}$$

Now electric field intensity

$$E = -\nabla V$$

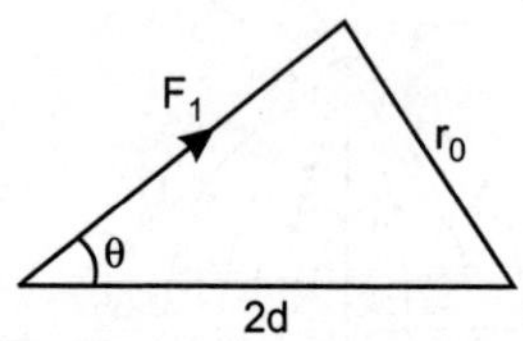

Fig. 3.20. Geometry for 'E' calculation

$$= -\left(\frac{\partial}{\partial x}a_x + \frac{\partial}{y}a_y + \frac{\partial}{\partial z}a_z\right)\mathrm{V}$$

or in cylindrical coordinate

$$\left(\underset{E_r}{\frac{\partial V}{\partial r}a_r} + \underset{E_\theta}{\frac{1}{r}\cdot\frac{\partial V}{\partial \theta}a_\theta} + \underset{E_z}{\frac{\partial V}{\partial z}a_z}\right)$$

$$E_r = -\frac{\partial}{\partial r}\left[\frac{Q}{4\pi\varepsilon_0}\left(\frac{1}{r} - \frac{1}{r_0}\right)\right]$$

$$= \frac{\partial}{\partial r}\cdot\frac{Q}{4\pi\varepsilon_0}\left[\frac{1}{r} - \frac{1}{\sqrt{r^2 + 4d^2 + 4rd\cos\theta}}\right]$$

$$= \frac{Q}{4\pi\varepsilon_0}\left[-\frac{1}{r^2} - \left(-\frac{1}{2}\right)(r^2 + 4d^2 + 4rd\cos\theta)^{3/2}(2r + 4d\cos\theta)\right]$$

$$E_r = \frac{Q}{4\pi\varepsilon_0}\left[-\frac{1}{r^2} + \frac{1}{(r^2 + 4d^2 + 4rd\cos\theta)^{3/2}}.(r + 2rd\cos\theta)\right]$$

Now
$$E_\theta = \frac{1}{r}\frac{dV}{d\theta}$$

$$= \frac{1}{r}\frac{dV}{d\theta}\left[\frac{1}{4\pi\varepsilon_0}\left(\frac{Q}{r} - \frac{Q}{\sqrt{r^2 + 4d^2 + 4rd\cos\theta}}\right)\right]$$

$$= \frac{Q}{4\pi\varepsilon_0 r}\left[\frac{d}{d\theta}\left(\frac{1}{r} - (r^2 + 4d^2 + 4rd\cos\theta)^{-1/2}\right)\right]$$

$$= \frac{Q}{4\pi\varepsilon_0 r}\left[-\left(-\frac{1}{2}\right)(r^2 + 4d^2 + 4ar + \cos\theta)^{-3/2}(-4dr\sin\theta)\right]$$

$$= \frac{Q}{4\pi\varepsilon_0 r}\left[-\frac{2dr\cos\theta}{(r^2 + 4d^2 + 4dr\cos\theta)^{3/2}}\right]$$

$$= \frac{Q}{4\pi\varepsilon r}\frac{dr\sin\theta}{r^3}$$

$$E_\theta = \frac{2Qd}{4\pi\varepsilon_0 r_0^3}\sin\theta$$

$$\Rightarrow \quad E_\theta = \frac{Q.d}{2\pi E_0 r_0^3}, \quad \text{and} \quad E_z = \frac{\partial V}{\partial z} = 0$$

$$E_n = \frac{\rho_s}{e_0}$$

$$\Rightarrow \quad \rho_s = E_n\varepsilon_0. \text{ charged density}$$

Then the electric flux D;

$$\varepsilon_0 E_n = D = \rho_s$$

The Divergence Theorem

The divergence of a vector flux density flux density A, is the out from small closed surface per unit volume as volume shrinks to zero. The divergence indicates the spreading or diverging of a quantity from a point and it is applicable for a vector quantity.

Mathematically it is indicated as, ∇.A.

Divergence of a vector $A = \nabla.A$.

$$= \text{div } A = \frac{\partial A_x}{\partial x} + \frac{\partial A_y}{\partial y} + \frac{\partial A_z}{\partial z}$$

The divergence of a vector quantity indicates the net flow of a quantity like, gas, fluid, vapour electric magnetic flux lines so on. In other word it measure the difference of inward and otuward flows.

1. The divergence is positive if net flow is outward.
2. And negative if the net flow is inward.
3. If $\nabla.A = 0$, then the condition is incompressible and net flow is zero.

Example:

1. Leaking air from ballon yield +ve divergence.
2. The reshing the air into the dram under the carriage of a train yields negative divergence.

The divergence of a air, oil is almost zero since they are almost uncompressable.

Divergence is dot product of one vector to another as follows:

$$\text{div A} = \nabla.\text{A} = \left(\frac{\partial}{\partial x}a_x + \frac{\partial}{\partial y}a_y + \frac{\partial}{\partial z}a_z\right).\ (A_xa_x + A_ya_y + A_za_z)$$

$$= \frac{\partial}{\partial x}a_x.\ A_xa_x + \frac{\partial}{\partial x}a_x.A_ya_y + \frac{\partial}{\partial x}a_x.A_za_z$$

$$+ \frac{\partial}{\partial y}a_y.A_xa_x + \frac{\partial}{\partial y}a_y.\ A_ya_y + \frac{\partial}{\partial y}a_y.A_za_z$$

$$+ \frac{\partial}{\partial z}a_zA_xa_x + \frac{\partial}{\partial z}a_zA_ya_y + \frac{\partial}{\partial z}a_z.A_za_z$$

$$= \frac{\partial A_x}{\partial x}(a_x.a_x) + \frac{\partial A_y}{\partial x}\ (a_x\ .a_y) + \frac{\partial A_z}{\partial x}\ (a_x.a_z)$$

$$+ \frac{\partial A_x}{\partial y}(a_y.a_x) + \frac{\partial A_y}{\partial y}\ (a_y.a_y) + \frac{\partial A_z}{\partial y}\ (a_z.z_x)$$

$$+ \frac{\partial A_x}{\partial z}\ (a_z.a_x) + \frac{\partial A_y}{\partial y}\ (a_z.a_y) + \frac{\partial A_z}{\partial z}\ (a_z\ .a_x)$$

Since, $a_x\ .a_x = 1$ $a_y\ a_y = 1$ $a_z\ .a_z = 1$

$a_x\ .a_y = 0$ $a_y\ .a_z = 0$ $a_z\ .a_x = 0$

$a_x\ .a_z = 0$ $a_y\ .a_x = 0$ $a_z\ .a_y = 0$

So, $$\text{div } A = \nabla.A = \left(\frac{\partial A_x}{\partial x} + \frac{\partial A_y}{\partial y} + \frac{\partial A_z}{\partial z}\right) \text{ scalar quantity.}$$

On the mathmatical basis, it can be defined as the limit of the net outward flow flux, $\oint_S Ads$ per unit volume as the volume dv enclosed by the surface S approaching toward zero. Symbollically

$$\nabla.A = \text{Div } A = \lim_{dv \to 0} \frac{\oint_S Ads}{dv} \qquad \text{(flux lines per cubic meters)}$$

$$= \frac{\partial A_x}{\partial x} + \frac{\partial A_y}{\partial y} + \frac{\partial A_z}{\partial z}$$

A positive divergence for any quantity indicates source of that vector at that point. Similarly, a negative divergence indicates sink.

Differential volume in differential co-ordinates systems:

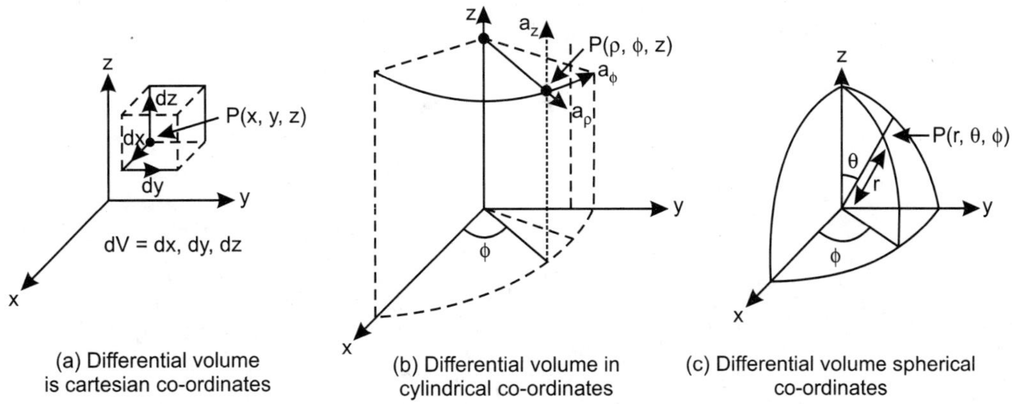

Fig. 3.21. dV in different co-ordinate system

Now divergence in other coordinate systems.

In cylindrical co-ordinates:

$$\nabla . A = \frac{1}{\rho}\frac{\partial(\rho A_\rho)}{d\rho} + \frac{1}{\rho}\frac{dA_\phi}{\partial\phi} + \frac{\partial A_z}{\partial z}$$

In spherical coordinate system

$$\nabla . \text{A} = \frac{1}{r^2}\frac{\partial}{\partial r}(r^2 A_r) + \frac{1}{r \sin\theta}\frac{\partial}{\partial\theta}(\sin\theta \, \text{A}\theta) + \frac{1}{r\sin\theta}\frac{\partial}{\partial\phi}\text{A}\phi$$

Example 6: If $D = e^{-x} \sin y a_x - e^{-x} \cos y a_y + 2z_a$ then find $\nabla . D$, *i.e.*, divergence of D.

Solution. We know that

$$\nabla . D = \frac{\partial D_x}{\partial x} + \frac{\partial D_y}{\partial y} + \frac{\partial D_z}{\partial z}$$

$$= \frac{\partial}{\partial x}(e^{-x} \sin y) + \frac{\partial}{\partial y}(-e^{-x} \cos y) + \frac{\partial}{\partial z}(2z)$$

$$= \sin y \frac{\partial}{x} e^{-x} - e^{-x}\frac{\partial}{\partial y} \cos y + \frac{2\partial z}{\partial z}$$

$$= -e^{-x} \sin y + e^{-x} \sin y + 2 = 0 + 2$$

$$\nabla . D = 2$$ Ans.

Gauss's Law

Gauss low states that electric flux passing through any closed surface is equal to the total charge enclosed by that surface The explanation and prove is given as follows :

Fig. 3.22 shows point charges, surrounded by a closed surface of any shape. The closed surface may be the surface of the some real materials. If total charge is Q, then Q columbs flux will pass through the enclosed surface. At every point on the surface, the electric flux density vector D will have same value D_S.

Now let an incremental surface ΔS and it is vector quantity as well as has direction normal to that plane which is tangent, at any point *P*.

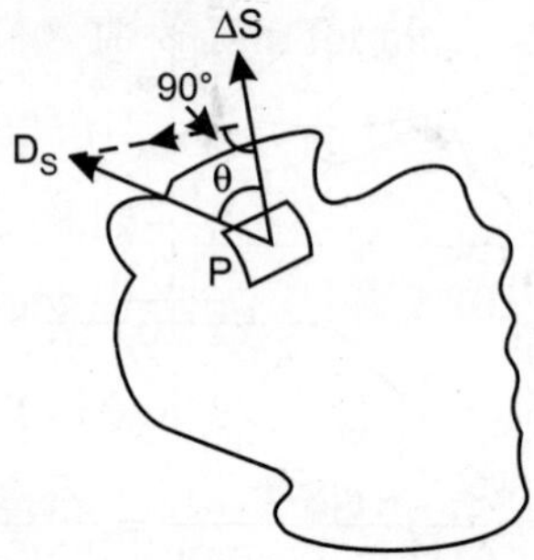

Fig. 3.22. Point charge on a closed surface

Let the D_S makes an angle θ with ΔS then flux crossing ΔS can be given as

$$\Delta\psi = \text{flux crossing through, } \Delta S = D_{snorm} \times \Delta S$$

$$= D_S \cos\theta . \Delta S = \overline{D_S}\ \Delta\overline{S}$$

The total flux passing through surface

$$\int \Delta\psi = \psi = \int_{closed} \overline{D_S}.\bar{d}\,S \qquad ...(3.30)$$

$$\psi = \oint_S D_S.dS = \text{charge enclosed} = Q$$

If charge distribution is volume then

$$Q = \int_V \rho_v d_v \qquad ...(3.31)$$

Now from eqn. (3.30) and (3.31),

$$\oint_S \overline{D_S}\,\bar{d}\,s = \int_V P_v dv \qquad ...(3.32)$$

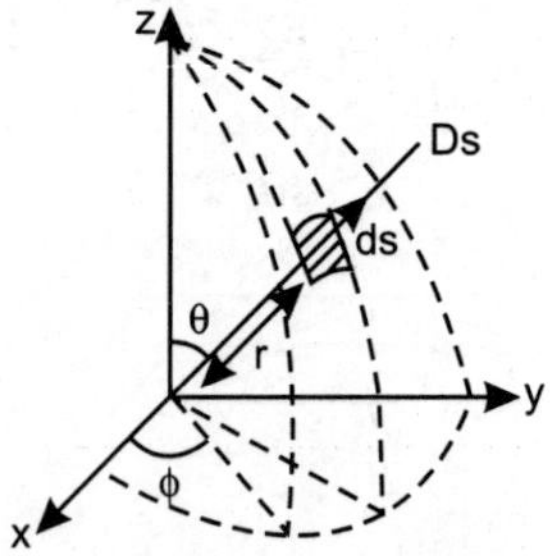

Fig. 3.23. Ds in cartision coordinate system

Now we can check the Gauss law as:

$$E = \frac{Q}{4\pi\varepsilon r^2} a_r$$

$$\varepsilon E = \frac{Q}{4\pi r^2} a_r$$

$$\overline{D_S} = \frac{Q}{4\pi r^2} a_r$$

As, $$\overline{d} s = r^2 \sin\theta \, d\theta \, d\phi a_r$$

$$\overline{D_S} . \overline{d} s = \frac{Q}{4\pi r^2} a_r (r^2 \sin\theta \, d\theta \, d\phi \, a_r)$$

$$= (Q \sin\theta d\theta \, d\phi a_r . a_r). \frac{1}{4\pi} \qquad \text{(as } a_r . a_r = 1\text{)}$$

$$\oint_S D_S ds = \oint_S \frac{Q}{4\pi} \sin\theta \, d\theta \, d\theta$$

$$= \frac{Q}{4\pi} \int_{\phi=0}^{2\pi} d\phi \sin\theta \, d\theta . \int_{\phi=0}^{2\pi} d\phi$$

$$= \frac{Q}{4\pi} - [\cos\theta]_0^{\pi} . [2\pi]$$

$$= \frac{Q}{\pi} - [\cos\pi - \cos\theta] \, 2\pi$$

$$= \frac{Q}{4\pi} [2] \times 2\pi = Q,$$

$$\oint D_s \overline{d_s} = Q = \text{flux passing through the closed surface.}$$

Application of Gauss Law (Maxwell's First Equation)

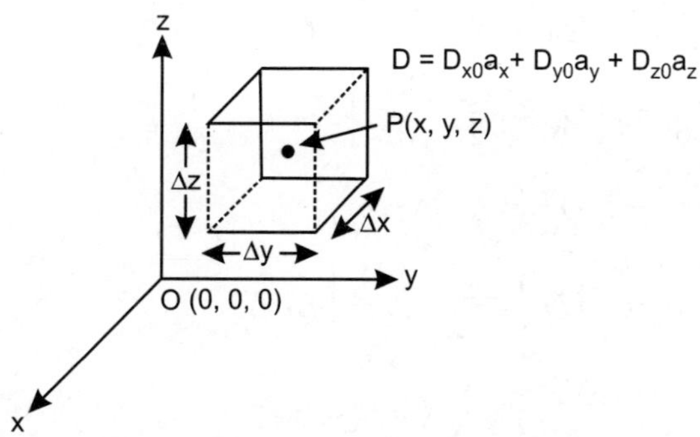

Fig. 3.24. Location of a point in (z, y, z)

Let us consider point $P(x, y, z)$ located by cartesian coordinate system. The value D at the point P can be expressed

$$D = D_{x_0} a_x + D_{y_0} a_y + D_{z_0} a_z$$

Let us choose close surface by Δx, Δy, Δz, and apply the Gauss Law

$$\oint \overline{D}\,\overline{d}s = Q$$

Now overall integral can be broken up into six integrals as follows:

$$\oint_S \overline{D}\,\overline{d}s = \int_{\text{front}}^{\overline{D}\,\overline{d}s} + \int_{\text{back}}^{\overline{D}\,\overline{d}s} + \int_{\text{left}}^{\overline{D}\,\overline{d}s} + \int_{\text{right}}^{\overline{D}\,\overline{d}s} + \int_{\text{top}}^{\overline{D}\,\overline{d}s} \int_{\text{bottom}}^{\overline{D}\,\overline{d}s}$$

Now selecting one surface

$$\int_{\text{front}} = D_{\text{front}} \cdot \Delta S_{\text{front}}$$

$$= D_{\text{front}} . \Delta y\, \Delta z\, dx$$

$$= D_{x\text{ front}}\, \Delta y\, \Delta z\, \text{dx}$$

Now D_{front} can be calculated

$$D_{x\text{ front}} = D_{x_0} + \frac{\Delta x}{2} \times \text{rate of change of } D_x \text{ with x}$$

$$= D_{x_0} + \frac{\Delta x}{2} \times \frac{\partial D_x}{\partial x}$$

So
$$\int_{\text{front}} = D_{x\text{ front}} . \Delta y\, \Delta z . a_x = \left[D_{x_0} + \frac{\Delta x}{2}\frac{\partial D_x}{\partial x}\right] \Delta y\, \Delta z\, a_x$$

$$\int_{\text{front}} = D_{x_0}\, \Delta y\, \Delta z\, a_x + \frac{\Delta x}{2} . \Delta y\, \Delta z\, \frac{\partial D_x}{\partial x} a_x$$

$$\int_{\text{back}} = D_{\text{back}}, \Delta S_{\text{back}}$$

$$= D_{\text{back}} . \Delta y . \Delta z\, a_x$$

$$= -D_{\text{back}}\, \Delta y\, \Delta z\, a_x$$

$$D_{\text{back}} = \left[D_{x_0} - \frac{\Delta x}{2} . \frac{\partial D_x}{\partial x}\right]$$

$$\int_{\text{back}} = -D_{\text{back}} . \Delta y . \Delta z . a_x$$

$$= -\left[D_{x_0} - \frac{\Delta x}{2} . \frac{\partial D_x}{\partial x}\right] \Delta y\, \Delta z\, a_x$$

Now
$$\int_{\text{front}} + \int_{\text{back}} = \frac{\partial D_x}{\partial x} . \Delta x . \Delta y . \Delta z$$

Similarly
$$\int_{\text{left}} + \int_{\text{right}} = \frac{\partial D_y}{\partial y}\, \Delta x\, \Delta y\, \Delta z$$

Similarly
$$\int_{\text{top}} + \int_{\text{bottom}} = \frac{\partial D_z}{\partial z}\, \Delta x\, \Delta y\, \Delta z$$

So
$$\int_{\text{front}} + \int_{\text{back}} + \int_{\text{left}} + \int_{\text{right}} + \int_{\text{top}} + \int_{\text{bottom}}$$

$$= \frac{\partial D_x}{\partial x}\Delta x\,\Delta y\,\Delta z + \frac{\partial D_y}{\partial y}\Delta x\,\Delta y\,\Delta z + \frac{\partial D_z}{\partial z}\Delta x\,\Delta y\,\Delta z$$

$$= \left(\frac{\partial D_x}{\partial x} + \frac{\partial D_y}{\partial y} + \frac{\partial D_z}{\partial z}\right)\Delta x\,\Delta y\,\Delta z = \oint D\,ds = Q$$

$$= \left(\frac{\partial D_x}{\partial x} + \frac{\partial D_z}{\partial y} + \frac{\partial D_z}{\partial z}\right)\Delta v = \oint D\,ds = Q$$

So charge enclosed by the volume Δv is

$$= \left(\frac{\partial D_x}{\partial x} + \frac{\partial D_y}{\partial y} + \frac{\partial D_z}{\partial z}\right) \times \text{volume } \Delta v$$

$$\text{or } \left(\frac{\partial D_x}{\partial x} + \frac{\partial D_y}{\partial y} + \frac{\partial D_z}{\partial z}\right) = \oint \frac{D\,ds}{\Delta V} = \frac{Q}{\Delta v} \qquad \text{as } \Delta v \to 0$$

$$\left(\frac{\partial D_x}{\partial x} + \frac{\partial D_y}{\partial y} + \frac{\partial D_y}{\partial z}\right) = \lim_{\Delta v \to 0} \oint \frac{Dds}{\Delta v}$$

$$= \lim_{\Delta u \to v}\left(\frac{Q}{\Delta v}\right) = \rho_v$$

$$\left(\frac{\partial D_x}{\partial x} + \frac{\partial D_y}{\partial y} + \frac{\partial D_z}{\partial z}\right) = \rho_v \qquad \text{...(3.33)}$$

$$\nabla.D = \rho_v$$

$$\text{div } D = \rho_v \qquad \text{...(3.34)}$$

called Maxwell's first equation.

Polarization in Dielectric

The common property of all dielectric whether they are in solid, liquid or air in nature is to store electric energy. This storage of energy takes place due to shift in the relative positions of internal, bound positive charges against their normal molecular and atomic forces. And this charge displacement mechanism differs in various dielectric materials.

Some dielectrics are called polar molecular, has permanent displacement of charges (positive and negative). A strong electric field produce additional displacement. A non polar dielectrics molecules has zero displacement in the absence of electric fields and produces displacement after the applications of the external electric fields.

The dipole moment of either type material is given by

$$\overline{P} = Q\overline{d}$$

where Q = positive charge

$\overline{d}$ = vector from –ve to + ve charge

If there are n dipoles in per unit volume, then $\overline{P} = n.\,Q.\,\overline{d}$ ...(3.35)

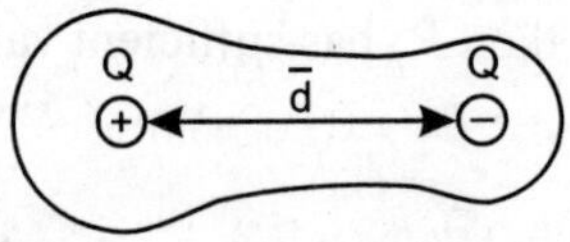

Fig. 3.25. Geometry of a dipole moment

Dielectric and Dielectric Constant

Dielectric material like insulator has no free electron to move even after the application of electric field. This characteristic of a dielectric material is called its permitivity or dielectric constant and donoted by ε. Dielectric constant of any material is more than the dielectric constant of the vacuum or air ($\varepsilon_r = 1$).

Dielectric constant of any material can be represented in terms of relative dielectric constant and dielectric constant of the free space, *i.e.*,

$$\varepsilon = \varepsilon_r \varepsilon_0 \text{ (F/m)} \quad \text{...(3.36)}$$

or

$$\varepsilon_r = \frac{\varepsilon}{\varepsilon_0}$$

where ε_r = relative dielectric (relative permitivity)

ε = total dielectric or absolute dielectric or permitivity

ε_0 = permitivity of the free space

$$= \frac{1}{36\pi \times 10^{-9}} \text{ F/m} = 8.854 \times 10^{-12} \text{ F/m}$$

For example ε_r = 5 for mica

= 1 for air and vacuum.

Here dielectric are more insulating.

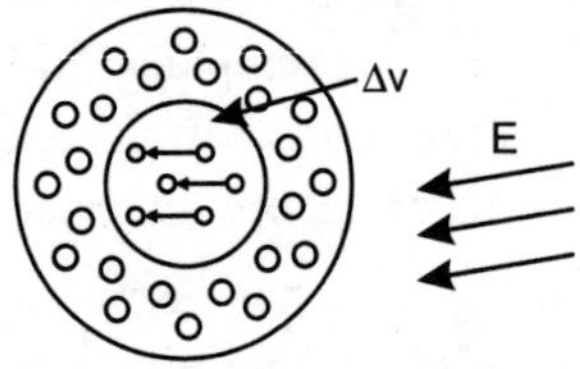

Fig. 3.26. Alignment of molecules along E.

[Note: Electric field align the molecules to some extent in the same direction, then in Δv volume, the dipoles = $n\Delta v$].

The total dipole moment

$$P_T = \sum_{n=1}^{n\Delta v} p_i = P_i$$

$$= P_1 + P_2 + P_3 + P_4 + \ldots P_i + P_{n\Delta v}$$

If alignment is in same direction then P_T has sufficient value; however if orientation is random then $P_T = 0$.

Polarization: Dipole moment per unit volume when volume shrinks to zero. *i.e.*, P = dipole moment/unit volume when volume, $\Delta v \to 0$.

$$= \lim_{\Delta v \to 0} \frac{P_T}{\Delta v}$$

$$= \lim_{\Delta v \to 0} \frac{1}{\Delta v} \sum_{i=1}^{n} P_i$$

Unit of P is columbs/sq. meter.

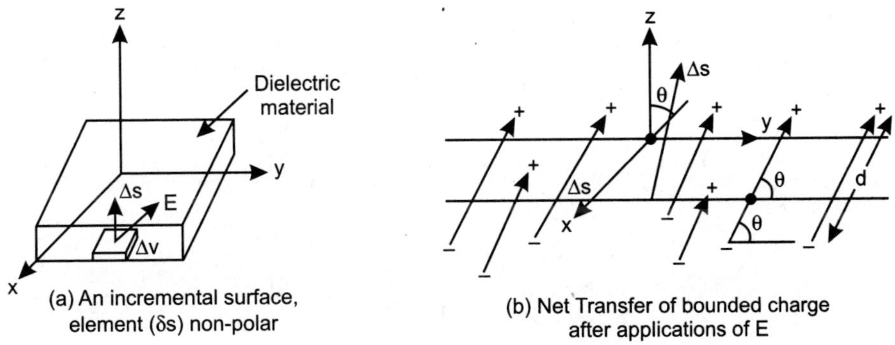

Fig. 3.27. Influence of E on dielectric material

Before application of electric field dipole moment P = 0. Dipole moment after the application of electric field can be calculated as if there is n molecules/ m^3.

The net total charge which cross the surface

$$nQd \cos \theta \, \Delta S$$

$$\Delta Q_b = nQd \cos \theta \, \Delta S \qquad b \text{ indicates bound charges.}$$

In terms of polarization $\int \Delta.Q_b = \int \rho \Delta S \qquad \Delta \to 0, \Delta = d$

$$Q_b = \int \Delta Q_b = \int d\, Q_b = \int \rho\, ds$$

Net increase charge in bound charge inside the dielectric is

$$Q_b = -\int \Delta Q_b = -\int dQ = -\oint_S \rho\, dS$$

$$\Delta \to 0, \Delta \to d$$

But from Gauss law

$$Q_T = \psi_T = \oint_S \overline{D}\ \mathrm{dS}$$

$$= \oint_S \varepsilon E\, dS \qquad D = \text{electric flux density}$$

As $D = \varepsilon E$ $\qquad = \varepsilon_r \varepsilon_0 E$

$$= \oint \varepsilon_0 E dS$$

$$Q_T = Q + Q_b \quad \text{...(3.37)}$$

where, Q = total free charge

Q_b = bounded charge

From eqn. (3.37)

$$Q = Q_T - Q_b$$

$$= \oint_S \varepsilon_0 E dS + \oint_S q\, dS$$

$$Q = \oint_S (\varepsilon_0 E + q) dS$$

Now D defined as

$$D = q + \varepsilon_0 E$$

$$Q = \oint_S D\, ds$$

Now

$$Q_b = \int_v q_b$$

$$Q = \int_v q\, dv$$

$$Q_T = \int_v q_T\, dv$$

Since from divergene theorem;

$$-\oint_S \rho\, ds = \oint_V (\nabla . V) dV \Rightarrow Q_b = \int_b \rho_b \text{ dV}$$

$$\nabla . \rho = -\rho_b$$

and

$$\nabla . D = \rho_v$$

In particular for feroelectric materials

$$P = X_e\, \varepsilon_0\, E$$

X_e = electric susceptibility

$$D = \varepsilon_0 E + \text{P}$$

$$= \varepsilon_0 E + X_e \varepsilon_0 E$$

$$= (1 + X_e)\, \varepsilon_0 E$$

$$D = (\varepsilon_r\, \varepsilon_0 E) \quad \text{...(3.38)}$$

where $\varepsilon_r = 1 + X_e$, is known as relative permitivity or dielectric constant

Hence

$$\text{D} = \varepsilon_0 \varepsilon_r\, \text{E} = \varepsilon \text{E}.$$

where $\varepsilon = \varepsilon_0\varepsilon_r$ E = εE, For isotropic uniform anstropic material $\varepsilon = \varepsilon_0\varepsilon_r$

For anstropic material, D and E are represented in different components;

$$D_x = \varepsilon_{xx} E_x + \varepsilon_{xy} E_y + \varepsilon_{xz} E_z$$
$$D_y = \varepsilon_{yx} E_x + \varepsilon_{yy} E_y + \varepsilon_{yz} E_z$$
$$D_z = \varepsilon_{zz} E_z + \varepsilon_{zy} E_y + \varepsilon_{zz} E_z \quad ...(3.39)$$

Relaxtion Time

Since Ohm's law in point form is,

$$J = \sigma E \quad ...(3.40)$$
$$J \propto E$$

Thus, according to ohm's law the current density (J) is proportional to the electric field strength (E).

Where, σ is called conductivity of the metal/conductor. Reciprocal of conductivity is called resistivity ρ.

$$\sigma = \frac{1}{\rho} \quad ...(3.41)$$
$$\rho = \frac{1}{\sigma} \quad ...(3.42)$$

One important feature of ohm's law that J is constant till electrical field strength is constant.

Let there are n free electrons/cubic-m. In the absent of electric field they moves in zig-zag fashion net movement is zero so no, resultant current flows. Now if an external electrical field is applied (E) along the x direction. Then force on electrons,

$$F = -eE \quad ...(3.43)$$

But from Newton's Law

$$\overline{F} = ma \quad ...(3.44)$$

where m mass of one electron therefore both must be equal,

$$ma = -eE$$
$$m\frac{d^2x}{dt^2} = -eE$$
$$\frac{d^2x}{dt^2} = \left(-\frac{e}{m}\right)E$$
$$\Rightarrow \quad \frac{d}{dt}\left(\frac{dx}{dt}\right) = \left(-\frac{e}{m}\right)E$$
$$\Rightarrow \quad \frac{d}{dt}(v_x) = \left(-\frac{e}{m}\right)E \quad ...(3.45)$$

v_x called drift velocity of the electron.

Now if at $t = 0$, electric field is zero, then drift velocity will reduce to zero exponentialy as t increases.

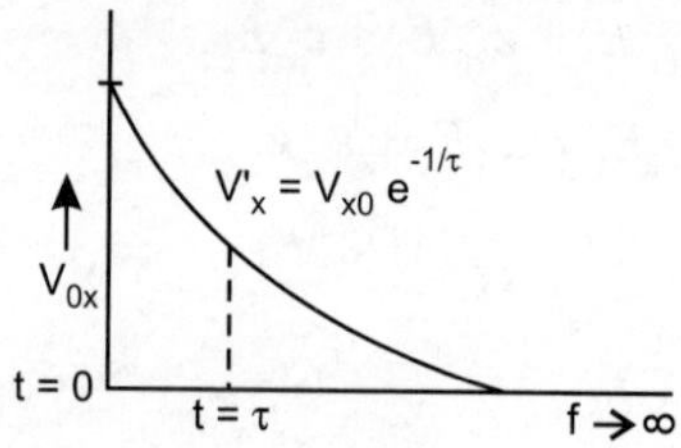

Fig. 3.28. Variation of drift velocity with time τ

where 'τ' is called relaxation time of the electron, and if $t = \tau$

$$V_x = V_{x_0} e^{-1} = \frac{V_{x_0}}{e}$$

$$V_x = \frac{V_{x_0}}{e} = V_{x_0}\left(\frac{1}{e}\right) \qquad \text{...(3.46)}$$

That is the relaxation time of the electron is the time at which the drift velocity, becomes $\left(\frac{1}{e}\right)$ times of the initial drift velocity.

Again from eqn. (3.45)

$$\frac{dv_x}{dt} = \left(-\frac{e}{m}\right)E$$

$$\int dv_x = \int_0^{\tau}\left(-\frac{e}{m}\right)Edt$$

$$v_x = -\frac{e}{m} E \int_0^{\tau} dt = -\frac{e}{m} E[t]_0^{\tau}$$

$$= -\frac{e}{m} E(\tau - 0)$$

$$v_x = -\frac{e}{m} E.\tau$$

$$v_x = \left(-\frac{e\tau}{m}\right)E \qquad \text{...(3.47)}$$

So, $\qquad v_x \propto E$

$$v_x = \mu_e .E \qquad \text{...(3.48)}$$

where μ_e is the permeability of the electron and 300 m^2/v.s for electrons and 500 m^2/v.s for holes. From eqn. (3.47) & (3.48) absolute value of μ_e will be:

$$\mu_e = \frac{e\tau}{m}$$

where μ_e is also called mobility of the electron.

But we know that $\qquad J = nqv_x$

$$= n.ev_x = -nev_x \text{ replacing } v_x \text{ from (3.47)}$$

$$= -ne \cdot \frac{e\tau}{m} E$$

$$= \left(\frac{ne^2\tau}{m}\right) E$$

As $$J = \sigma E$$

$$\sigma = \frac{ne^2\tau}{m}$$

or $$\frac{1}{\rho} = \frac{ne^2\tau}{m}$$

$$\rho = \frac{m}{ne^2\tau} \qquad ...(3.49)$$

So $$\rho = \frac{1}{\tau} \Rightarrow \sigma \propto \tau$$

That is, if resistivity is high then electrons has lower value of relaxation time.

Poisson's Equations and Laplace Equations

Poission equation is another important relationship between charge density and permitivity for a homogeneous medium. From Gauss's law;

Divergence of electric flux is equal to volume charge density

i.e., $$\text{div } D = \rho_u$$

$$\nabla.D = \rho_u \qquad ...(3.50)$$

where ∇ = Laplace operator and it is a vector quantity in 3-D and defined as:

$$\nabla = \frac{\partial}{\partial x} a_x + \frac{\partial}{\partial y} a_y + \frac{\partial}{\partial z} a_x \qquad ...(3.51)$$

in which a_x, a_y and a_z are unit vectors along the *x, y, z* axes.

and we also know that flux density and electric field are related as

$$D = \varepsilon E$$

and

or (–ve) gradient of voltage = electric field

$$-\text{grad V} = \text{E}$$

$$-\nabla.\text{V} = \text{E}$$

So $$\nabla.\text{D} = \nabla.\ \varepsilon\text{E}$$

$$= \nabla\varepsilon.\nabla V$$

$$-\varepsilon\nabla^2 V = \rho_v$$

$$\nabla^2 V = -\frac{\rho_v}{\varepsilon} \quad \text{...(3.52)}$$

The above equation is called Poisson's equation

∇^2 is called double del.

Since, $$\nabla = \frac{\partial}{\partial x}a_x + \frac{\partial}{\partial y}a_y + \frac{\partial}{\partial z}a_z$$

So, $$\nabla^2 = \nabla.\nabla$$

$$= \left(\frac{\partial}{\partial x}a_x + \frac{\partial}{\partial y}a_y + \frac{\partial}{\partial z}a_z\right).\left(\frac{\partial}{\partial x}a_x + \frac{\partial}{\partial y}a_y + \frac{\partial}{\partial z}a_z\right)$$

$$= \left(\frac{\partial^2}{\partial x^2} + \frac{\partial^2}{\partial y^2} + \frac{\partial^2}{\partial z^2}\right) \qquad [\text{Since } \overline{A}.\overline{B} = A_xB_x + A_yB_y + A_zB_z]$$

So equation (3.52) can be written as

$$\nabla^2 V = -\frac{\rho_v}{\varepsilon}$$

$$\left(\frac{\partial^2}{\partial x^2} + \frac{\partial^2}{\partial y^2} + \frac{\partial^2}{\partial z^2}\right) V = -\frac{\rho_v}{\varepsilon}$$

$$\left(\frac{\partial^2 V}{\partial x^2} + \frac{\partial^2 V}{\partial y^2} + \frac{\partial^2 V}{\partial z^2}\right) = -\frac{\rho_v}{\varepsilon} \quad \text{...(3.53a)}$$

called general form of the Poissions equation.

Note that volume charge density $\rho_v = 0$ then

$$\nabla^2.V = 0$$

or $$\frac{\partial^2 V}{\partial x^2} + \frac{\partial^2 V}{\partial y^2} + \frac{\partial^2 V}{\partial z^2} = 0 \quad \text{...(3.53b)}$$

The equation (3.53*b*) is called Laplace equation.

Here ∇^2 called Laplacian of V.

The derivatives $\frac{\partial}{\partial x}, \frac{\partial}{\partial y}, \frac{\partial}{\partial z}$ are called partial derivatives. Laplace and Poisson equations can also be written in cartesian, cylindrical and spherical coordinates, as follows:

$$\nabla^2 V = \frac{\partial^2 V}{\partial x^2} + \frac{\partial^2 V}{\partial y^2} + \frac{\partial^2 V}{\partial y^2} = -\frac{\rho_v}{\varepsilon} \quad \text{...(3.54)}$$

$$= \frac{1}{\rho}\frac{\partial}{\partial \rho}\left(\rho\frac{\partial V}{\partial \rho}\right) + \frac{1}{\rho^2}\frac{\partial^2 V}{\partial \phi^2} + \frac{\partial^2 V}{\partial z^2} = -\frac{\rho_v}{\varepsilon} \quad \text{...(3.55)}$$

$$\nabla^2 V = \frac{1}{r^2}\frac{\partial}{\partial r}\left(r^2\frac{\partial V}{\partial r}\right) + \frac{1}{\partial^2 \sin\theta}\cdot\frac{\partial}{\partial \theta}\left(\sin\frac{\partial V}{\partial \theta}\right) + \frac{1}{r^2 \sin\theta}\frac{\partial^2 V}{\partial \phi} \quad \text{...(3.56)}$$

Conductors, Dielectrics and Capacitors

After defining current and current density and developing the continuity equation, we shall now consider the conductor and present ohm's law, both in its micro-scopic and macro scopic forms. After brief consideration of a general semiconductor, we shall investigate the polarization of dielectric material or the dielectric constant and important engineering parameters. Having both dielectric and conductor, we may then put them together to form capacitors.

Current and Current Density

Electric charges in motion constitute the current, or rate of change of charge is known as current. The unit of current is ampere and it can be given as

$$I = \frac{dQ}{dt} \qquad \text{...(3.57)}$$

In field theory we are usually interesed in events occuring at a point rather than within some large region and we shall find the concept of the current density, measured in amperes per square meter (A/m^2). Current density (J) is a vector quantity.

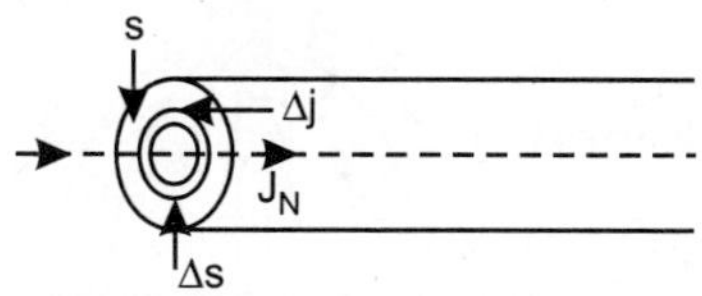

Fig. 3.29 Direction of In on a conductor

The incremental current ΔI crossing an incremental surface ΔS normal to the current density is

$$I = \int_S J.dS$$

$$\frac{\Delta I}{\Delta S} = J_N$$

$$J_N \Delta S = \Delta I$$

Total current

$$I = \int \Delta I = \int J_N \Delta S \qquad \text{...(3.58)}$$

Current density may be related to the velocity of volume charge density, at a point (drift velocity of the electrons)

$$\rho_v = \frac{\Delta Q}{\Delta v}$$

$$= \frac{\Delta Q}{\Delta s\, \Delta x}$$

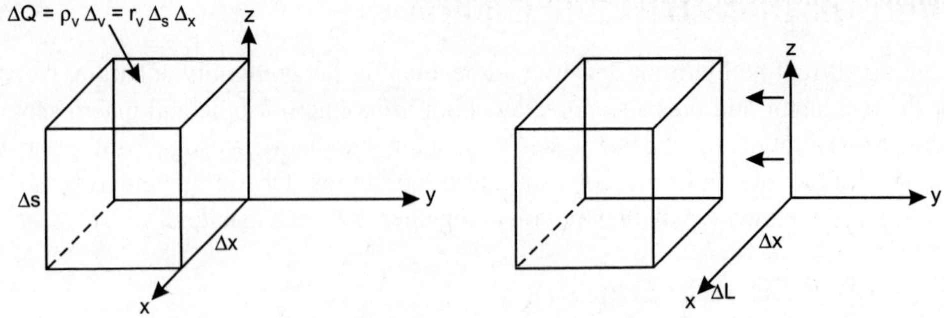

Fig. 3.30. Incremental dimensions

$$\Delta Q = \rho_v \,\Delta v = \rho_v \Delta S\, \Delta x$$

$$\rho_L = \frac{dQ}{dl}$$

and $$\rho_S = \frac{dQ}{ds}$$

$$\rho_s = \frac{dQ}{dv} \quad \text{for volume charge density.}$$

$$\Delta I = \frac{\Delta Q}{\Delta t} = \frac{\rho_v \Delta v}{\Delta t}$$

$$= \rho_v \frac{\Delta s\, \Delta x}{\Delta t}$$

$$= \rho_v \Delta_s \left(\frac{\Delta x}{\Delta t}\right)$$

$$\Delta I = \rho_v \Delta s v_x$$

$$V_x = \text{drift velocity in } x \text{ direction}$$

Now current density,

$$J = \frac{\Delta I}{\Delta s} = \frac{\rho_v \Delta s v_x}{\Delta s}$$

$$J = \rho_v v_x$$

$$J = J_x a_x + J_y a_y + J_z a_z$$

$$J_x = \rho_x v_x$$

$$J_y = \rho_v \, v_y$$

$$J_z = \rho_v \, v_z$$

hence $$J = \rho_v v \qquad \ldots(3.59)$$

when charge is in motion they constitute the current which is known as convection current & $J = \rho v_v$ called convection current density. The convection current density linearly related with change density as well as with velocity.

Example 7: The vector current density is given as

$$J = \frac{4}{r^2} \cos \theta a_r + 20e^{-2r} \sin \theta a_\theta - r \sin \theta \cos \phi a_\phi$$

(*a*) Find J at $r = 3, \theta = 0, \phi = \pi$,

(*b*) Find the total current passing through the spherical coordinate $r = 3, 0° \theta = 0° \leq \phi < 2\pi$ in the a_r direction.

Solution: $$J = \frac{4}{r^2} \cos \theta a_r + 20e^{-2r} \sin \theta a_\theta - r \sin \theta \cos \phi a_\phi$$

(*a*) putting $r = 3, \theta = 0, \phi = \pi$ in above eqn.

$$J = \frac{4}{3^2} \cos 0 a_r + 20e^{-2\times 3} \sin 0 a_\theta - 3 \sin 0 \cos \pi a_\phi$$

$$= \frac{4}{9} a_r + 0 + 0$$

$$= \frac{4}{9} a_r$$

$$J = 0.44 a_r$$

(*b*) $\Delta S = r\, dr.d\phi$

$$S = \int \Delta S = \int_S r\, dr\, d\phi$$

$$= \int_0^3 r\, dr \int_0^{2\pi} d\phi$$

$$= \left[\frac{r^2}{2}\right]_0^3 2\pi$$

$$= \frac{9}{2} \times 2\pi = \frac{18\pi}{2} = 9\pi$$

$$J = \frac{4}{r^2} \cos \theta a_r$$

$$= \frac{\frac{4}{r^2} \cos \theta}{9\pi} = \frac{4}{9} \cos \theta .\, a_v = \frac{4}{9}$$

$$I = J \times S = \frac{4}{\not 9} \times \not 9 \times = 3.14 \times 4$$

$$= 12.56 \text{ A}$$

Continuity of Current and Continuity Equations : The principle of conservation of charge states simply that charges can be neither created nor destroyed, although equal amounts of positive and negative charges may be simultaneously created.

Let a given region bounded by a closed surface, then current through the closed surface is,

$$I = \oint J ds$$

and this outward flow of positive charge must be balanced by decrease of positive charge within the closed surface.

The basic property of electrical charges implies that a charge can neither be created nor be distroyed. If a charge disappears from one point, it must be reappear at another point. This is known as principle of conservation of charge.

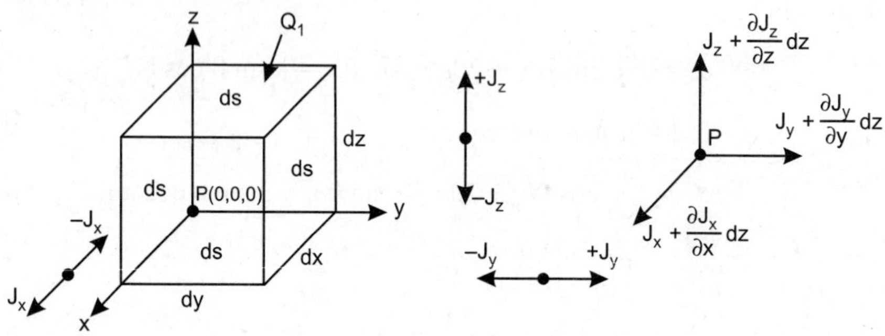

Fig. 3.31. Representation of conversion of charge

$$dv = dx\,dy\,dz$$

Let us consider a small volume element dv as shown in Fig. 3.31 is placed inside a conducting medium. The current density is a vector having same direction as current. In general current density has three rectangular components which vary with change of position.

Since integral of the normal components of the current density over a closed surface is zero, *i.e.*,

$$\oint JdS = 0$$

So

$$\nabla.J = 0$$

$$\oint_S Jds = \oint_V (\nabla.J)dv$$

and current through the closed surface, $\oint_V (\nabla.J)dV = \frac{dI}{ds} = \text{J}$

$$I = \oint_S J.\,dS \qquad ...(3.60)$$

the outward flow of positive charge must be balanced by a decrease positive charge (or increase of negative charge) within the closed surface. If charge in closed surface is Q_i then rate of decrease is,

$$\frac{-dQ_i}{dt} \qquad ...(3.61)$$

and from principle of charge conservation.

and from eqn. (3.60) & (3.61)

$$I = \oint_S J.dS = -\frac{dQ}{dt} \qquad ...(3.62)$$

Above equation is continuity equation. Integral form and differential or point form of this equation can be obtained as

$$\nabla.A = \frac{\partial A_x}{\partial x} + \frac{\partial A_x}{\partial y} + \frac{\partial A_z}{\partial z} = \lim_{dv \to 0} \int_S \frac{AdS}{dv}$$

$$\nabla.A = \frac{\int Ads}{dv}$$

Replace A by J

$$\nabla.J = \frac{\int_S Jds}{dv}$$

$$\int_v (\nabla.J)dv = \int_S JdS$$

Since $$\nabla.J = \frac{\partial \rho_u}{\partial t}$$

then $$\oint_S J.dS = \int_V (\nabla.J)dv$$

and $$\nabla.J = -\frac{\partial \rho_v}{\partial t}$$

then $$\oint_S JdS = -\frac{d}{dt}\int \partial\rho_v dv$$

although it is true for any volume but it is more true for the incremental volume, $\Delta v \to 0$

$$(\nabla.J)\,\Delta v = -\frac{\partial \rho_v}{\partial t}\,\Delta v$$

$$\nabla.J = -\frac{\partial \rho_u}{\partial t} \text{ continuity equation} \qquad ...(3.63)$$

Example 8. Calculate the capacitance per km of the length of air filled coaxial line with an inner conductor of 3 mm diameters and outer conductor of inside diameters of 1 cm.

Solution. We know the formula for co-axil conductor; is

$$C = \frac{2\pi\varepsilon l}{\log_e\left(\frac{b}{a}\right)}$$

or $$\frac{C}{l} = \frac{2\pi\varepsilon}{\log_e\left(\frac{b}{a}\right)}$$

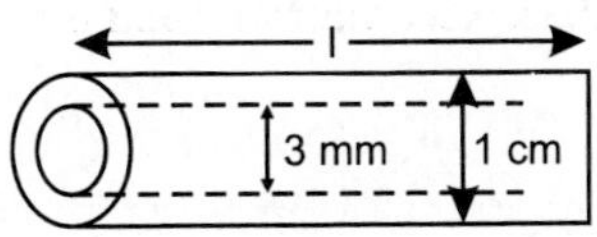

Fig. 3.32. Geometry for example 8

Given that

$$a = \frac{3\,mm}{2} = 1.5 \text{ mm} = 0.15 \text{ cm}$$

$$b = 1 \text{ cm} = \frac{1}{2} \text{ cm} = 0.5 \text{ cm}$$

$$\varepsilon = 8.54 \times 10^{-12} \text{ F/m for air}$$

$$l = 1000 \text{ m}$$

So $$C = \frac{2\pi \times 8.854 \times 10^{-12}}{\log_e\left(\frac{0.5}{0.15}\right)} = \frac{2\pi \times 8.854 \times 10^{-12}}{2.303 \log_{10}\left(\frac{0.5}{0.15}\right)}$$

$$C = 46.2 \text{ PF/km}$$

Note $$\text{loge } x = \frac{\log_{10} x}{\log 10^e} = \frac{\log_{10} x}{\log_{10}(2.71818)}$$

Example 9. A thunder cloud above the ground set up a vertical electric field of 40 V/m. A rain drop carrying a charge of 0.1 μF lies in this field. What is the electrostatic force excerted on the rain drop?

Solution. We know that force on a charge can be given as;

$$F = -\text{eE}$$

Given, $$E = 40 \text{ V/m}$$

$$e = q = 0.1\mu C = 0.1 \times 10^{-6}$$

$$= 1 \times 10^{-7} \text{ C}$$

So $$F = -1 \times 10^{-7} \times 40$$

$$= -4 \times 10^{-6} \text{ Newton}$$

$$F = -4 \times 10^{-6} \text{ N}$$

Example 10. Two point charges of 1 C each and of same magnitude are separated 1 mm apart. Calculate the magnitude of the repulsive force.

Solution. Given $Q = 1\ C$, $Q_2 = 1\text{C}$

and $$r = 1 \text{ mm} = 1 \times 10^{-3} \text{ m}$$

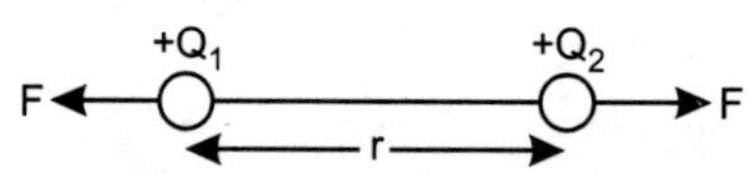

Fig. 3.33. Geometry for example 10

From columb's Law

$$F = \frac{1}{4\pi\varepsilon_0} \frac{Q_1 Q_2}{r^2}$$

$$= \frac{1}{4\pi \times \frac{1}{36\pi} \times 10^{-9}} \times \frac{1 \times 1}{(1 \times 10^{-3})^2}$$

$$= \frac{36\pi}{4\pi \times 10^{-9}} \frac{1}{1 \times 10^{-6}} = \frac{9}{10^{-15}}$$

$$= 9 \times 10^{15} \text{ N}$$

$$F = 9 \times 10^{15} \text{ N}$$

Numerical 11. Two α particles are separated by a distance of 1 mm and each are having a charge of 3.2×10^{-9} C. Determine the repulsive force.

Solution. From Newton equation.

$$F = \frac{Q_1 Q_2}{4\pi\varepsilon r^2}$$

$$= \frac{3.2 \times 10^{-9} \times 3.2 \times 10^{-9}}{4\pi \times \frac{1}{36\pi} \times 10^{-9} \times (10^{-3})^2}$$

$$= \frac{(3.2 \times 10^{-9})^2 \times 9}{10^{-9} \times 10^{-6}}$$

$$= \frac{9.216 \times 10^{-17}}{10^{-15}}$$

$$= 9.216 \times 10^{-2}$$

$$F = 0.09216 \text{ N}$$

Example 12. Calculate the electrostatic field intensity at a distance of 0.20 meters from a charge of 2 μC in vaccum.

Solution. We know that,

$$E = \frac{1}{4\pi\varepsilon r^2} a_r$$

$$= \frac{1}{4\pi\varepsilon_0} \frac{Q}{r^2} a_r \text{ for free space or air}$$

$$= \frac{1}{4\pi \times \frac{1}{36\pi} \times 10^{-9}} \times \frac{2 \times 10^{-6}}{(0.20)^2} a_r$$

$$= \frac{9 \times 10^9 \times 2 \times 10^{-6}}{4 \times 10^{-2}}$$

$$\text{E} = 4.5 \times 10^5 \text{ C/m}$$

$$\overline{E} = 4.5 \times 10^5 a_r \text{ C/m}$$

Example 13. What is meant be permitivity of free space? compute the ratio between the electrostatic repulsion and gravitational attraction between two electrons. Given Q = 1.6×10^{-31} C

$$G = 6.67 \times 10^{-11} \text{ Nm}^2/\text{kg and m} = 9.1 \times 10^{-31} \text{ kg}$$

Solution. From Columb's law

$$F_e = \frac{Q_1 Q_2}{4\pi\varepsilon_0 r^2}$$

$$= \frac{1.6 \times 10^{-31} \times 1.6 \times 10^{-31}}{4\pi \times \frac{1}{36\pi} \times 10^{-9} \times r^2}$$

Now from Newton's gravitational force formula;

$$F_a = G\frac{m_1 m_2}{r^2} = \frac{6.67 \times 10^{-11} \times 9.1 \times 91 \times 10^{-31} \times 10^{-31}}{r^2}$$

$$\text{Now, } \frac{\text{electrostatic repulsion } F_e}{\text{gravitational attraction } F_G} = \frac{9 \times 2.56 \times 10^{44}}{6.67 \times (9.1)^2}$$

$$= 4.171 \times 1042$$

As we know that insulating property of any material can be represented in terms of its permitivity which also called dielectric constant. The permitivity of free space is the ability to insulate the medium prtmitivity of any medium and can be given as

$$8.854 \times 10^{-12} \text{ F/m}$$

Example 14. Calculate the force of interaction between two charges of values 4×10^{-8} C and 6×10^{-5} C and space 10 cm apart in free space. What is the force of interaction if the medium is Kerosone oil?

Solution. By Columb's law

$$F = \frac{1}{4\pi\varepsilon}\frac{Q_1 Q_2}{r^2}$$

$$= \frac{1}{4\pi\,\varepsilon_0} \cdot \frac{Q_1 Q_2}{r^2} \qquad \text{(in free space)}$$

$$= \frac{1}{4\pi \times \frac{1}{36\pi} \times 10^{-9}} \times \frac{4 \times 10^{-8} \times 6 \times 10^{-5}}{(10 \times 10^{-2})^2}$$

$$= 9 \times 10^9 \times \frac{24 \times 10^{-11}}{10^{-2}}$$

$$= 9 \times 24 \times 10^{-2}$$

$$\text{F} = 2.16 \text{ N}$$

Now if charges are in Kerosene oil,

$$F_1 = \frac{1}{4\pi\varepsilon}\frac{Q_1 Q_2}{r^2} = \frac{1}{4\pi\varepsilon_r \varepsilon_0}\frac{Q_1 Q_2}{r^2}$$

$$= \frac{1}{4\pi \times \frac{1}{36\pi} \times 10^{-9} \times 2} \times \frac{4 \times 10^{-8} \times 6 \times 10^{-5}}{(0.1)^2}$$

$$= \frac{9\times10^9}{2}\times\frac{24\times10^{-23}}{1\times10^{-2}}$$

$$= 108 \times 10^{-2}$$

$$F = 1.08\ N$$

Example 15. Two charged particles each of mass m and charge q are suspended by strings each of length '*l*' from the same points. Prove that the indication of each string to the vertical axis is given by

$$q^2 \cos\theta = 16\pi\varepsilon_0\, mgl^2 \,.\, \sin^3\theta$$

and $$\frac{\tan^3\theta}{1+\tan^2\theta} = \frac{q^2}{16\pi\varepsilon_0 mgl^2}$$

Solution.

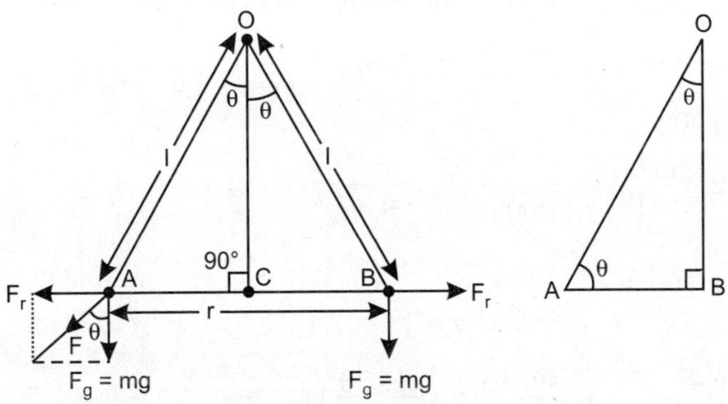

Fig. 3.34. Geometry for example 15

Force of repulsion between the similar charges.

$$F_r = \frac{1}{4\pi\varepsilon}\frac{q_1 q_2}{r^2}$$

r = distance between two point charges

Since each particles has mass m so gravitation force of each will be;

$$F_g = mg$$

Now from trangle OAC

$$\frac{AC}{CO} = \sin\theta$$

$$\frac{\frac{r}{2}}{l} = \sin\theta$$

$$r = 2l\sin\theta \qquad ...(3.64a)$$

So $$F_r = \frac{q^2}{4\pi\varepsilon_0(2l\sin\theta)^2} = \frac{q^2}{4\pi\varepsilon_0\, 4l^2\sin^2\theta}$$

$$F_r = \frac{q^2}{16\pi\varepsilon_0 l^2 \sin^2\theta} \qquad \text{...(3.64b)}$$

Resultant force

$$F = \sqrt{F_r^2 + F_g^2}$$

and
$$\tan\theta = \frac{F_r}{F_g}$$

$$= \frac{\dfrac{q^2}{16\pi_0 l^2 \sin^2\theta}}{mg}$$

$$\frac{\sin\theta}{\cos\theta} = \frac{q^2}{16\pi\varepsilon_0 l^2 \sin^2\theta\, mg}$$

$$q^2 \cos\theta = 16\pi\varepsilon_0 l^2 \sin^3\theta.\text{mg} \qquad \textbf{Proved}$$

Now for second part

$$\frac{\sin\theta}{\cos\theta} = \frac{q^2}{16\pi\varepsilon.\, l^2 \sin^2\theta.mg}$$

$$\frac{\sin^3\theta}{\cos\theta} = \frac{q^2}{16\pi\varepsilon l^2\, mg}$$

$$\frac{q^2}{16\pi\varepsilon l^2\, mg} = \frac{\sin 3\theta}{\cos\theta} \times \left(\frac{\cos 3\theta}{\cos 3\theta}\right)$$

$$= \left(\frac{\sin 3\theta}{\cos 3\theta}\right).\left(\frac{\cos 3\theta}{\cos\theta}\right)$$

$$= \tan^3\theta.(\cos 2\theta)$$

$$= \frac{\tan 3\theta}{\sec^2\theta}$$

$$= \frac{\tan 3\theta}{1+\tan^2\theta}$$

Since
$$\sec^2\theta = \tan^2\theta + 1$$

So
$$\frac{q^2}{16\pi\varepsilon mgl^2} = \frac{\tan^3\theta}{1+\tan^2\theta} \qquad \text{Hence proved.}$$

Example 16. A positive charge of magnitude 4 μC is situated in air at the origin of a rectangular coordinate system. Calculate the electric field strength at the point at the *z*-axis away 8 meters from the origin on the *x*-axis.

Solution.
$$E = \frac{1}{4\pi\varepsilon}\cdot\frac{Q.Q}{r^2}a_z = \left(\frac{9\times10^9\times4\times10^{-6}}{64}\right)a_z$$

$$= \left(\frac{9}{16} \times 10^3\right) a_z$$

$$= 5.625 \times 102 \text{ V/m } a_z$$

$$E = 562.5 \text{ V/maz}$$

Example 17 (*a*). A negative charge of magnitude 2 μC is situated in air origin and two + Ve point charges of 1.0 μC each are at points y = ± 2 m. Calculate the electric field strength at a point 4 from the origin on the *x*-axis.

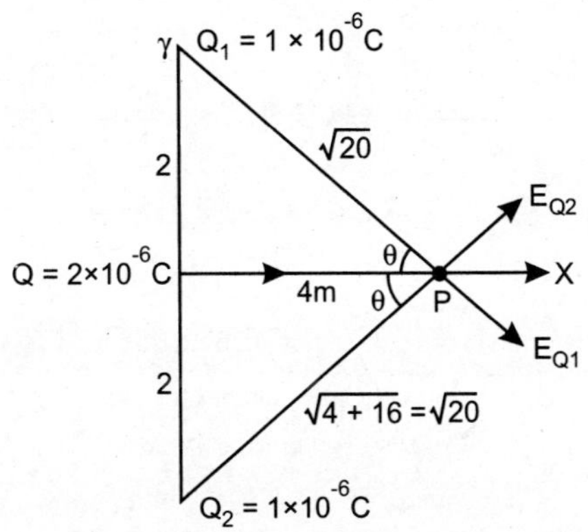

Fig. 3.35. Geometry of example 17 (*a*)

Solution.

$$E_Q = \frac{-Q}{4\pi\varepsilon r^2} a_x$$

$$= \frac{-2 \times 10^{-6} . a_x}{4\pi \times \frac{1}{36\pi} \times 10^{-9} (4)^2}$$

$$= \frac{-2 \times 9 \times 10^9}{61} a_x$$

$$= \frac{-18}{16} a_x$$

$$E_Q = -1.125 a_x$$

$$E_{Q1} = \frac{Q_1}{4\pi\varepsilon_0 r^2} = \frac{1 \times 10^{-6}}{4\pi \times \frac{1}{36\pi} \times 10^{-9}} - \frac{1}{(20)}$$

$$= \frac{9 \times 10^3}{20} = 4.5 \times 10^{-2}$$

In *x* direction

$$\overline{E_{Q1}} = E_{Q1} \cos \theta . a_x$$

$$= 4.5 \times 10^2 \, a_x \cos \theta$$

Similarly

$$\overline{E_{Q2}} = 4.5 \times 10^2 \, a_x \cos \theta$$

Now resultant

$$E = \overline{E_Q} + \overline{E_{Q1}} + \overline{E_{Q2}}$$

$$= -1.125 \times 10^3\, a_x + \left(4.5 \times 10^2 \frac{4}{\sqrt{20}} + 4.5 \times 10^2 \times \frac{4}{\sqrt{20}}\right) a_x$$

$$E = -320 a_x \text{ V/m}$$

Example 17(*b*). The concentrated charged of 0.25 μC are placed at the vertix of an equilateral triangle whose side is 100 mm. Determine the magnitude and direction of the resultant force on one charge due to the two charges.

Solution. $$F_1 = \frac{1}{4\pi\varepsilon}\frac{Q_1 Q_2}{r^2} = 9 \times 10^9 \times \frac{0.25 \times 0.25 \times 10^{-12}}{0.1 \times 1}$$

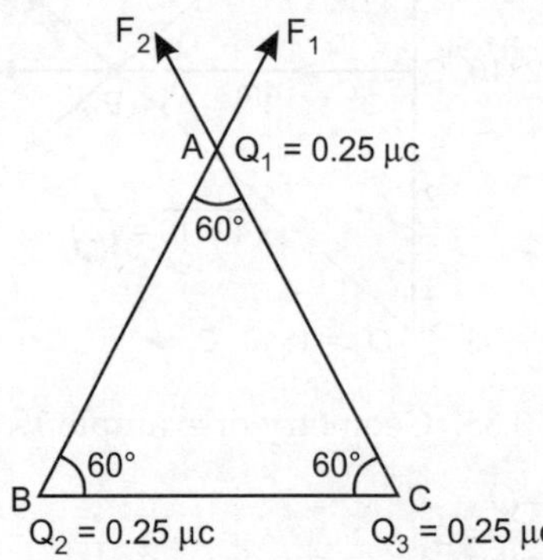

Fig. 3.36. Geometry of example 17 (*b*)

$$= 9 \times 625 \times (10^{-16} \times 10^{+2} \times 10^9)$$

$$F_1 = 5525 \times 10^{-5} \text{ N}$$

Similarly $$F_2 = 56.25 \times 10^{-34}$$

Now Resultant

$$F = \sqrt{F_1^2 + F_2^2 + 2F_1F_2 \cos\theta}$$

$$= \sqrt{(56.25 \times 10^{-3})^2 + (56.25 \times 10^{-3})^2 + 2 \times 56.25 \cos\theta}$$

$$= 56.25 \times 10^{-3} \sqrt{1 + 1 + 2 \times \frac{1}{2}} = 56.25 \times 10^{-3} \sqrt{3}$$

$$F = 97.425 \times 10^{-3} \text{ N}$$

Example 18. The three point charges $q_1 = + 10^6\ C$, $q_2 = -1 \times 10^{-6}\ C$ and $q_3 = 0.5 \times 10^{-6}\ C$ are placed in the air at the corners of an equilateral triangle of 50 cm side. Determine the magnitude and direction of the force on q_3.

Solution.

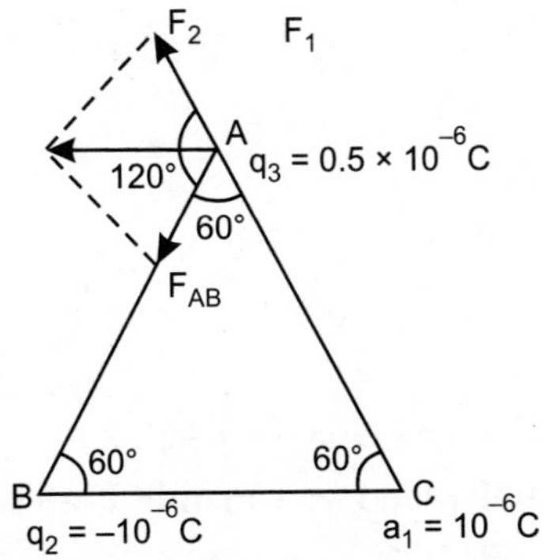

Fig. 3.37. Geometry of example 18

Force due to q_2 and q_3

$$F_{AC} = \frac{Q_1 Q_2}{4\pi\varepsilon_0 r^2} = \frac{(-1\times10^{-6}\times0.5\times10^{-6})}{4\pi\times\frac{1}{36\pi}\times10^{-9}\times(0.50\times0.5)}$$

$$= \frac{-9\times10^{9}\times0.5\times10^{-12}}{0.25}$$

$$= \frac{-9\times10^{-4}\times5\times10^{2}}{25}$$

$$= -1.8\times10^{-2}$$

$$F_{AC} = -0.018\text{N from A to B.}$$

Now force on q_3 due to q_1

$$F_{CA} = \frac{q_1 q_3}{4\pi\varepsilon_0 r^2}$$

$$= \frac{10^{-6}\times0.5\times10^{-6}}{4\pi\times\frac{1}{36\pi}\times10^{-9}\times(0.25)}$$

$$= \frac{9\times10^{9}\times0.5\times10^{-12}}{0.25}$$

$$= \frac{9\times10^{-4}\times5}{25}\times10^{2}$$

$$= \frac{9}{5}\times10^{-2}$$

$$= 1.8\times10^{-2}$$

$$= 0.018\text{ N from C to A}$$

$$\text{Resultant F} = \sqrt{F_1^2+F_2^2+2F_1F_2}\cos\theta$$

$$= 0.0018\text{ N parallel to BC}$$

General Procedures for Solving Poissions and Laplace Equations

Capacitance and Resistance

We known Laplace and poisson's equations are second order partial differential equation. There are number of methods to solve these expressions but direct integration method is simplest methods for solving of these type of expressions. Direct methods are applicable for the problems which are only one dimensional.

1. Fields and Capacitance between two Parallel Plaies:

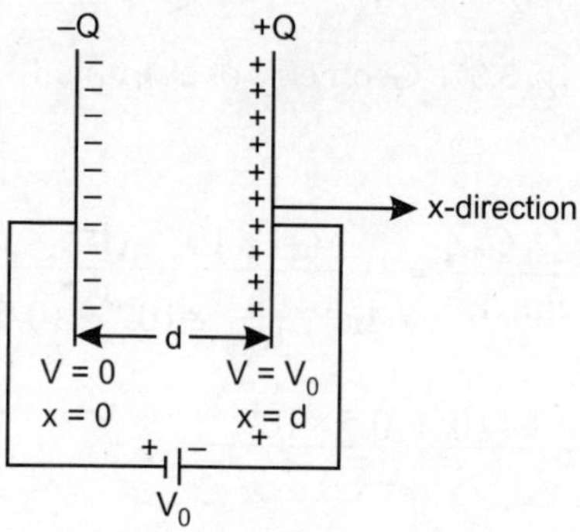

Fig. 3.38. Charge distribution between two plates

From the Laplace equation

$$\frac{\partial^2 V}{\partial x^2} + \frac{\partial^2 V}{\partial y^2} + \frac{\partial^2 V}{\partial z^2} = 0 \quad ...(3.65)$$

Since variation of voltage is only in x direction, so Laplace equation reduce to

$$\frac{\partial^2 V}{\partial x^2} = 0 \quad ...(3.66)$$

Partial derivative can be replace by ordinary derivative of V as it is not the function of y and z

$$\frac{d^2 V}{dx^2} = 0$$

integrating both side

$$\int \frac{d^2 V}{dx^2} = \int (0)$$

$$\int \frac{d}{dx}\left(\frac{dV}{dx}\right) = \int 0dx$$

$$\frac{dV}{dx} = \text{A, constant}$$

$$\int dV = \int Adx$$

$$V = Ax + K$$

K is integration constant.

Since A and K are constant they can be calculated as.

When $x = 0, V = 0$

$$0 = A(0) + K$$

$$0 = 0 + K$$

$$K = 0$$

Now when $x = d, V = V_0$

$$V_0 = Ad$$

$$A = \frac{V_0}{d}$$

So $$V = \frac{V_0}{d} x \text{ Volts} \qquad ...(3.67)$$

Now if we select the boundary condition in more general

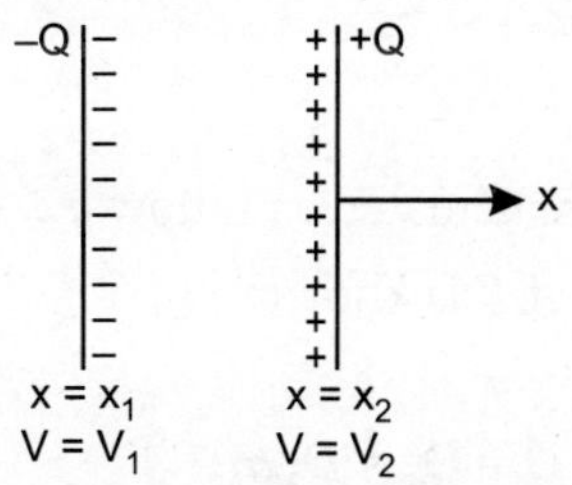

Fig. 3.39. Charge distribution under boundary condition

$$\frac{dV}{dx} = A$$

$$\int dV = \int A\, dx$$

$$V = Ax + K$$

When $x = x_1\ V = V_1$

$$V_1 = Ax_1 + K \qquad ...(3.68)$$

and when $x = x_2\ V = V_2$

$$V_2 = Ax_2 + K \qquad ...(3.69)$$

Now subtraction equation (3.68) and (3.69)

$$V_2 - V_1 = A(x_2 - x_1)$$

$$A = \frac{V_2 - V_1}{(x_2 - x_1)}$$

So $$V = \frac{V_2 - V_1}{(x_2 - x_1)} x + K \qquad ...(3.70)$$

Now multiply eqn. (3.68) by x_2 and (3.69)x_z

$$V_1x_2 = Ax_2x_1 + x_2K \quad ...(3.71)$$

$$V_2x_1 = Ax_1x_2 + x_1K \quad ...(3.72)$$

Now subtracting (3.71) from (3.72)

$$V_2x_1 - V_2x_2 = (x_1 - x_2)K$$

$$K = \frac{V_2x_1 - V_1x_2}{(x_1 - x_2)}$$

Now replacing the value of A and K in eq. (3.68);

$$V = \left(\frac{V_2 - V_1}{x_2 - x_1}\right)x + \frac{V_2x_1 - V_2x_2}{(x_1 - x_2)} \quad \text{(general solution)}$$

Since capacitance is equal to

$$C = \frac{Q}{V}$$

or

$$Q = CV$$

Now charge on the plates can be calculated as follows:

(*i*) Calculate E from $E = -\nabla V$ if V is known.

(*ii*) Calculate D from $D = \varepsilon E$

(*iii*) Calculate D on either plate $D = D_S = D_N a_N$

(*iv*) Recognize that $p_S = D_N$.

(*v*) Now find Q by a surface integration over the plate

$$Q = \int_S \rho_S ds$$

$$Q = \oint Dds$$

$$V = V_0 \frac{x}{d}$$

So, from divergence of V, $E = -\nabla V$

$$E = -\left(\frac{\partial}{\partial x} ax + \frac{\partial}{\partial y} + \frac{\partial}{\partial z} a_z\right)V$$

$$= -\left(\frac{\partial V}{\partial x} a_x + \frac{\partial V}{\partial y} a_y + \frac{\partial V}{\partial z} a_z\right)$$

Since

$$\frac{\partial}{\partial y} = \frac{\partial}{\partial z} = 0 \quad \text{(since variation of field is only in } x \text{ direction)}$$

$$E = -\frac{\partial V}{\partial x} a_x.$$

$$= -\left(V_0 \frac{x}{d}\right)a_x$$

$$= -\frac{V_0}{d}\left(\frac{dx}{dx}\right)a_x$$

$$E = -\frac{V_0}{d}a_x$$

Now $$D = \varepsilon E$$

$$D = -\frac{\varepsilon V_0}{d}a_x$$

$$D_S = D_x = D_x a_n$$

$$= D_N a_x$$

$$= -\frac{\varepsilon V_0}{d}d_x\, a_x$$

$$= -\frac{\varepsilon V_0}{d}$$

$$D_n = -\left(\frac{\varepsilon V_0}{d}\right) = \rho_S$$

$$Q = \oint \rho_S ds = \frac{\varepsilon V_0}{d}.\oint_S ds$$

$$= \frac{\varepsilon V_0}{d}.A$$

$$Q = \frac{\varepsilon V_0 A}{d}$$

$$C = \frac{\varepsilon A}{d} \quad ...(3.73)$$

Hence, $\frac{Q}{V_0} = \frac{\varepsilon A}{d}$ is capacitance between plates.

Laplace Equation in Cylindrical Coordinate

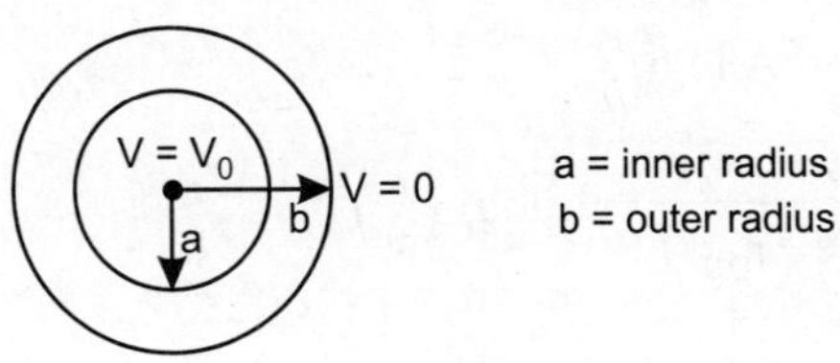

Fig. 3.40. Font view of a cylindrical conductor

Now Laplace equation in Cylindrical coordinate system is,

$$\nabla^2 V = \frac{1}{\rho}\frac{\partial}{\partial \rho}\left(\rho\frac{\partial V}{\partial \rho}\right) + \frac{1}{\rho^2}\frac{\partial^2 V}{\partial \phi^2} + \frac{\partial^2 V}{\partial z^2}$$

Since there is single coordinate system

$$\nabla^2 V = \frac{1}{\rho}\frac{\partial}{\partial \rho}\left(\rho \frac{\partial V}{\partial \rho}\right)$$

$$= \frac{1}{r}\frac{\partial}{\partial r}\left(r\frac{\partial V}{\partial r}\right) = 0$$

$$\int \frac{\partial}{\partial r}\left(r\frac{\partial V}{\partial r}\right) = \int 0$$

$$r\frac{\partial V}{\partial r} = A$$

$$\frac{\partial V}{\partial r} = \frac{A}{r}$$

$$\int \partial V = \int \frac{A\partial r}{r}$$

$$V = \text{A} \log_e r + B$$

at $r = a, V = 0$ $\quad b > a$

$$r = b, V = V_0$$

$$V = \log_e a + B$$

$$0 = \text{A} \log_e b + B$$

$$B = -\text{A loge } b$$

$$V = \text{A loge } a - A.\log_e b$$

$$V = \text{A} \log_e \left(\frac{a}{b}\right)$$

or

$$A = \frac{V_0}{\log_e\left(\frac{a}{b}\right)}$$

$$V = \text{A.} \log_e r + B$$

$$= A \log_e r - \text{A} \log_e b$$

$$= \text{A} (\log_e r - \log_e b)$$

$$= \text{A} \log\left(\frac{r}{b}\right)$$

$$= \frac{V}{\log_e\left(\frac{a}{b}\right)}.\log_e\left(\frac{r}{b}\right)$$

$$V = \frac{V_0 \log_e\left(\frac{r}{b}\right)}{\log_e\left(\frac{a}{b}\right)} \qquad ...(3.74)$$

Now for calculations of charge the following steps can be follows;

$$E = -\nabla.V$$

$$= -\left[\frac{\partial V}{\partial r}a_r + \frac{\partial V}{\partial \phi}a_\phi + \frac{\partial V}{\partial z}a_z\right]$$

$$= -\frac{\partial V}{\partial r}a_r$$

$$= -\frac{\partial}{\partial r}\left[\frac{V_0 \log_e \frac{r}{b}}{\log\left(\frac{a}{b}\right)}\right]a_r$$

$$= -\frac{V_0}{\log\left(\frac{a}{b}\right)}\frac{\partial}{\partial r}.\log_e\left(\frac{r}{b}\right)$$

$$= -\frac{V_0}{\log\left(\frac{a}{b}\right)}.\frac{1}{\left(\frac{r}{b}\right)}\frac{\partial}{\partial r}\left(\frac{r}{b}\right)$$

$$= \left[\frac{V_0}{\log_e\left(\frac{a}{b}\right)}\frac{b}{br}\right]$$

$$E = -\left[\frac{V_0}{r\log_e\left(\frac{a}{b}\right)}\right]a_r$$

Now $$D_n = \varepsilon E$$

$$r = a$$

$$= \frac{\varepsilon V_0}{a\log_e\left(\frac{a}{b}\right)}a_r a_r = \rho_S$$

$$Q = \int \rho_S ds = \int \frac{\varepsilon V_0}{a\log_e\left(\frac{a}{b}\right)}dS$$

$$= \frac{\varepsilon V_0}{a\log\left(\frac{b}{a}\right)}.\int ds$$

$$= \frac{\varepsilon V_0}{a\log\left(\frac{a}{b}\right)}2\pi al$$

$$Q = \frac{2\pi\varepsilon l V_0}{a.\log_e\left(\frac{b}{a}\right)} \quad ...(3.75)$$

So $$C = \frac{Q}{V_0} = \frac{2\pi\varepsilon l}{a\log\left(\frac{b}{a}\right)}$$

$$\frac{C}{l} = \frac{2\pi\varepsilon}{a \log\left(\frac{b}{a}\right)} \text{F/m}$$

It is the capacitor/unit lengths.

SOLVED EXAMPLES

Example 19. Point charge 1 mC and -2 mC located at (3, 2, –1) and (–1, –1, 4), respectively. Calculate the electric force on a 40 nC located at (0, 3, 1) and the electric field intensity at that point.

Solution:
$$F = \sum_{k=1,2} \frac{QQ_k}{4\pi\varepsilon_0 R^2} a_r = \sum_{k=1,2} \frac{QQ_k}{4\pi\ \varepsilon_0 |r - r_k|^3}$$

$$= \frac{Q}{4\pi\varepsilon_0}\left[\frac{10^{-3}(-3a_x + a_y + 2a_z)}{|(0,3,1)-(3,2,1)|^2}\right] - 2\times10^{-3}\left[\frac{a_x + 4a_y - 3a_z}{|(0,3,1)-(-1-1-4)|^2}\right]$$

$$= \frac{10^{-3}\times10\times10^{-9}}{\frac{4\neq\times10^{-9}}{36\neq}}\left[\frac{(-3a_x + a_y + 2a_z)}{(9+1+4)^{3/2}} - \frac{2(a_x + 4a_y - 3a_z)}{(1+16+9)^{3/2}}\right]$$

$$= 9\times 10\text{–}2\left[\frac{-3a_x + a_y + 2a_z}{14\sqrt{14}} + \frac{(-2a_x - 8a_y + 6a_z)}{26\sqrt{26}}\right]$$

$$F = (-6.507\ a_x - 3.817a_y + 7.506a_z)\ \text{mN}$$

and also electric fields

$$E = \frac{F}{2} = \frac{(-6.507a_x - 3.817a_y + 7.566a_z)}{10\times10^{-9}} \times 10^{-3}$$

$$= \frac{-6.507a_x - 3.817a_y + 7.506a_z}{10\times10^{-6}}$$

$$E = (-650.7a_x - 381.7a_y + 750.6a_z)\ \text{kV/m}$$

Example 20. A circular ring of a radius 'a' carries a uniform charge ρ_2 C/m and placed on the *x-y* plane with axis the same as the *z* axis as given in below Fig. 3.41.

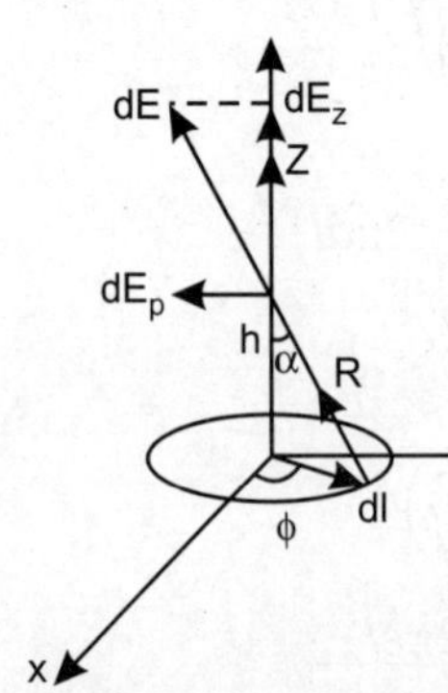

Fig. 3.41. Geometry of example 20

(*a*) Calculate what value of *h* gives the maximum value of *E*.

(*b*) Show that electric field at *h* height can be given as

$$E = \frac{\rho_2\, a.h}{2\varepsilon_0 [h^2 + a^2]^{3/2}} \text{az}$$

(*c*) If total charge at the ring is *Q*, find *E* as a → 0.

Solution: As given in Fig. 3.42,

$$dl = a.\, d\phi$$

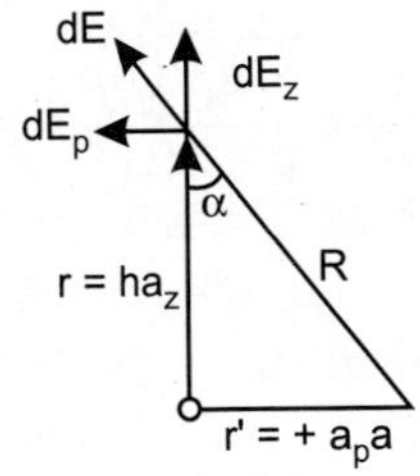

Fig. 3.42. Components of 'E'

Let unit vector in ρ, ϕ, z direction is

$$a_\rho, a_\phi \text{ and } a_z$$

$$|\overline{R}| = r - r$$

$$= ha_z - a_p a$$

$$R = |\overline{R}| = (h^2 + a^2)^{1/2}$$

$$a_R = \frac{\overline{R}}{|R|} = \frac{ha_z - aa_p}{(a^2 + h^2)^{3/2}}$$

Hence $$E = \int \frac{\rho_L\, dl}{4\pi\, \varepsilon_0\, R^2} . a_R$$

$$= \rho_L \frac{1}{4\pi\varepsilon_0} \int dl \frac{\overline{R}}{|R|^3}$$

$$= \frac{\rho_L}{4\pi\varepsilon_0} \int_0^{2\pi} \frac{ha_z - aa_p}{(a^2 + h^2)^{3/2}} .a\, d\phi$$

= (due to symmetry a_ρ component is zero)

$$= \frac{\rho_L .ah\, a_z}{4\pi\, \varepsilon_0 (h^2 + a^2)^{3/2}} \int_0^{2\pi} d\phi$$

$$= \frac{\rho_L\, ah\, a_z}{4\pi\, \varepsilon_0 (h^2 + a^2)^{3/2}} \times [\phi]_0^{2\pi}$$

$$= \frac{\rho L .ah\, a_z}{4\pi\, \varepsilon_0\, (h^2 + a^2)^{3/2}} \cdot [2\pi - 0]$$

$$= \frac{\rho L\, a.h.a_z}{2\varepsilon_0\, (h^2 + a^2)^{3/2}}$$

$$E = \frac{\rho_L .ah\, a_z}{2\varepsilon_0\, (h^2 + a^2)^{3/2}} \qquad ...(3.76)$$

Now differentiating E w.r.t. h

$$\frac{dE}{dh} = \frac{\rho_L a}{2\varepsilon_0}\left[\frac{(h^3 + a^2)^{3/2}\,(1) - \frac{3}{2} h.\, 2h\,(h^2 + a^2)^{1/2}}{(h^2 + a^2)^3}\right] \qquad ...(3.77)$$

From max E, $\frac{dE}{dh}$ must be zero

i.e., $$\frac{dE}{dh} = 0$$

So $$0 = \frac{\rho_L .a}{2\varepsilon_0}\left[\frac{(h^2 + a^2)^{3/2} - \frac{3}{2} h.\, 2h\,(h^2 + a^2)^{1/2}}{(h^2 + a^2)^3}\right]$$

$$(h^2 + a^2)^{3/2} - \frac{3}{2} h.\, 2h\,(h^2 + a^2)^{1/2} = 0$$

$$(h^2 + a^2)^{3/2} - 3h^2\,(h^2 + a^2)^{1/2} = 0$$

$$(h^2 + a^2)^{3/2} = 3h^2\,(h^2 + a^2)^{1/2}$$

$$(h^2 + a^2)^3 - 3h^2\,(h^2 + a^2) = 0$$

$$(h^2 + a^2)\,[(h^2 + a^2)^2 - 3h^2] = 0$$

$$h^2 + a^2 = 0$$

$$h = \pm a$$

and $$(h^2 + a^2)^2 = 3h^2$$

$$h^2 + a^2 = \sqrt{3}\, h$$

$$h = \pm \frac{9}{\sqrt{2}} \text{ for maximum value}$$

Since it is line charge density;

$$\rho_L = \frac{Q}{2\pi a}$$

$$E = \frac{Qh}{4\pi\, \varepsilon_0\, (h^2 + a^2)^{3/2}}\, a_z$$

as $a \to 0$

$$E = \frac{Q}{4\pi\, \varepsilon_0\, h^2} .a_z$$

Example 21. A charge distribution with spherical symmetry has density.

$$\rho_v = \begin{cases} \dfrac{\rho_0 r}{R} & ; 0 \le r \le R \\ 0 & ; r > R \end{cases}$$

Determine E every where as well as D.

Solution: From the Gauss's law

$$\int D\, ds = Q_{enc} = \int \rho_u \,.\, dv$$

$$\varepsilon_0 E_r .\, 4\pi r^2 = Q_{enc} = \int_0^r \int_0^\pi \int_0^{2\pi} \rho_u\, r^2 \sin\theta .\, d\theta\, d\phi\, dr$$

$$= \frac{\rho_0 \pi r^4}{R}$$

So, $$E_r = \frac{\rho_0 \pi r^2}{4\varepsilon_0 R} a_r$$

and $$D = \varepsilon_0 \mathrm{E}$$

$$= \frac{\rho_0 \pi r^2}{4R} a_r$$

For $r > \mathrm{R}$

$$\varepsilon E_r 4\pi r^2 = Q_{enc} = \int_0^r \int_0^\pi \int_0^{2\pi} \rho_u\, r^2 .\sin\theta\, d\theta\, d\phi\, dr$$

$$= \int_0^R \int_0^\pi \int_0^{2\pi} \frac{\rho_0 r}{R} 4\pi r^2 .dr$$

$$= \pi\rho_0 R^3$$

So, $$E_r = \frac{\rho_0 R^3}{4\varepsilon_0 r^2} a_r$$

and $$D = \varepsilon E = \frac{\rho_0 R^3}{4r^2} a_r$$

EXERCISES

1. A current density in a region is expressed as $J = -10^4 \nabla.\mathrm{V}$ A/m^2, where $= 10e^{-x} \sin yV$. (*a*) Find the current in the a_x direction crossing the surface $x = 1$ and bounded by $y = 0, y = 2$, $z = 0$ and $z = 3$. (*b*) Find the total current leaving the cube $0 < x, y, z < 1$ by integrating J. *ds* over the surface of the cube
2. A certain current density is given as $J = \left[\dfrac{100.\cos\theta}{r^2 + 1}\right] a_y$ A/m^2.

   How much current could flow through the spherical cup $r = 4, 0 < \theta < \dfrac{2\pi}{3} 0 < \phi < 2\pi$.
3. Let $V = 200\, x^{4/3}$ V for $x > 0$ with $\varepsilon = \varepsilon_0$.

(*a*) Find *E*, *D* and ρ_v as a function of *x* (*b*) If the velocity of charge density is $v_x = 3 \times 10^6 x^{2/3}$ m/sec. Find J_x at $x = 0$ and $x = 1$ m.

4. Given the potential field $V = \dfrac{200 \sin\theta . \sin\phi}{r^2} . v.$

(*a*) Find the equation of the conductor surface on which $V = 100$ volt (*b*) Find *E* at point *P* (*r*, 60° 60°) on the conductor surface (*c*) Find ρ_s at point *P* assume $\varepsilon = 4\pi \times 10^{-7}$ F/m adjacent to the surface.

5. The potential field, $V = 2x^2 + 4y - 2z^2$ volt, exist in the free space surrounding a perfectly conducting surface point *P* (4, 3, 2) lies on the surface.

(*a*) Give the equation of the surface (*b*) Find a unit outward vector normal to the surface at *P*, assuming the origin is inside the surface (*c*) Find ρ_s on the surface at *P* (*d*) Find $|E_p|$ and E_p.

6. Given the field $E = 8x. y (z + 1)a_x + 4x^2 (z + 1)a_y + 4x^2 ya_z$ V/m in free space and the information that point *P* (2, 3,4) lies on a conducting surface, find (*a*) The magnitude of ρ_s at *P* (*b*) ρ_v adjacent to the conductor surface *P* (*c*) V_{PQ} if *Q* is at the origin and *E* is conservative.

7. Given the potential field $V = 100e^{-50x} \sin 50y$ volt in free space.

(*a*) Show that $\nabla . D = 0$ (*b*) Show that $y = 0$ is an equipotential surface.

8. A circular ring $P = 4$ *m*, $z = 2$, carries a uniform charge density of 10 nc/m and the plane $z = 0$ is perfect conducting. Assuming the region $z > 0$ is free space. Find *V* at *P* (0, 0, *z*), $z > 0$.

9. A certain homogeneous isotropic, dielectric material has a relative permitivity of 1.8, if $V = 4000y$ volt. Find (*a*) *E*, *D* and *P* (*b*) ρ_u ρ_b and ρ_T.

10. The polarization within the region having $\varepsilon_R = 2.7$ has the uniform value $P = -0.2\ a_x + 0.7\ a_y + 0.4\ a_z$, $\mu C/m^2$. Find (*a*) $\overline{E}$ and $\overline{D}$ (*b*) Magnitude of the voltage gradient.

11. A vector directed from region 1 to region 2 at the planar boundary between perfect dielectric is given as $a_{NIZ} = -\frac{2}{7} a_x + \frac{3}{7} a_y + \frac{5}{7} a_z$. Assume $\varepsilon_{R_1} = 3$, $\varepsilon_{R_2} = 2$ and $E = 90a_x + 70a_y + 50a_z$. V/m. Find E_2.

12. Two infinite slab of perfect dielectric are parallel and surrounded by free space, as shown in Fig. 3.43. In region 3, $E_3 = 3a_z$ V/m and $P_3 = 15a_z$ PC/m² if $P_2 = 20\ a_z$ PC/m². Find (*a*) ε_{R_2} (*b*) ε_{R_2} (*c*) E_1 (*d*) Total voltage across the region 2 and 3.

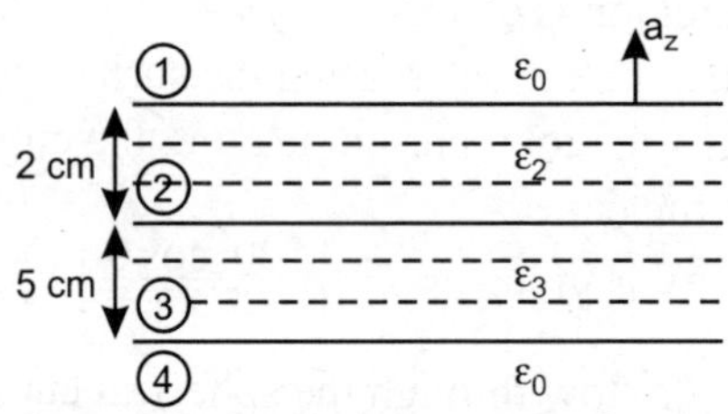

Fig. 3.43. Geometry of question 12

13. In given Fig. 3.44 let $D_1 = 2a_x + 5a_y - 3a_z$ nC/m^2 and find (*a*) D_2 (*b*) D_{N_2} (*c*) P_{t_2} (*d*) The energy density in each region.

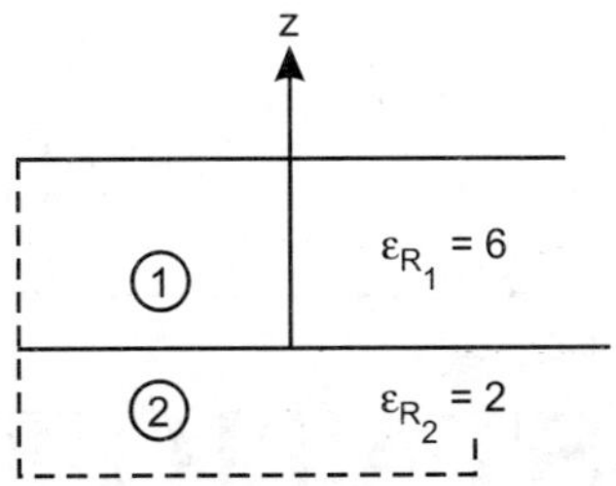

Fig. 3.44. Geometry of question 13

- "Intellectual growth should commence at birth and cease only at death"

Albert Einstein

Chapter **4**

MAGNETO STATIC FIELDS

4.1 INTRODUCTION

When charges are in motion produce current or anal if this current is dc (unidirectional) it produces the magnetic field called magneto static field or Magnetic field produces by steady state current is called magneto static fields.

Biot-Savart law : Biot-Savart law is used to explain magnetic field generated by a current carrying conductor.

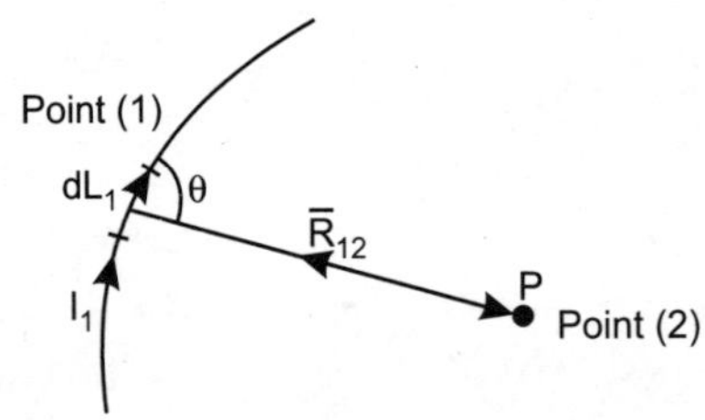

Fig. 4.1. A current carrying conductor

Here $\overline{dL_1}$ is called differential vector length (Fig. 4.1)

According to Biot-savart law, at any point P the magnitude of the magnetic field intensity (HA/m) produced by a differential element $\overline{dL}$, is proportional to product of current, magnitude of differential length and sine of the angle laying between the current element and a line connecting the conductor to that point and inversely proportional to the distance between current element and that point, *i.e.,*

$$dB \propto dl \,.\, I \qquad ..(4.1)$$

$$dB \propto \sin\theta \qquad ...(4.2)$$

$$dB \propto \frac{1}{r^2} \qquad ...(4.3)$$

Therefore, by combing $dB \propto \frac{I\,dl \sin\theta}{r^2}$

$$dB = \frac{\mu}{4\pi} \cdot \frac{I\,dl \sin\theta}{r^2} \qquad ...(4.4)$$

where $\frac{\mu}{4\pi}$ called proportionality constant.

In which μ is called permeability of medium and given by $\mu = \mu_r\,\mu_0$

In which permeability of free space or vacuum

$$\mu_0 = 4\pi \times 10^{-7}\ \text{H/m}$$

$$\mu_r = \text{relative permeability} = 1 \text{ for free space or vacuum.}$$

From equation (4.4) we can calculate total H as

$$\int dB = B = \int \frac{\mu\,dl}{4\pi}\frac{\sin\theta}{r^2} = \frac{\mu I}{4\pi}\int \frac{dl \sin\theta}{r^2} \qquad ...(4.5)$$

From eqn. (4.5) we now can calculate the magnetic flux density B (just like electric flux density)

$B = \frac{\Psi}{A}$ flux passing through any cross section area perpendicular to the area

$$= \mu H$$

$$\Rightarrow \qquad H = \frac{B}{\mu} = \frac{1}{4\pi}\int \frac{dl \sin\theta}{r^2}$$

Now in vector form $dH = \frac{I\overline{dL} \times a_R}{4\pi R^2} = \frac{I\overline{dL} \times \overline{R}_{12}}{4\pi R^3}$

The unit of H is (A / m)

or in more general form, if we consider all element at conductor with subscript and the point by 2.

Then

$$dH_2 = \frac{I_1\,\overline{dL}_1 \times a_{12}}{4\pi R_{12}^2} \qquad ...(4.6)$$

The above eqn. sometimes called Amper′ s Law for Current Element.

The direction of magnitude field identensity is normal to the plane containing the differential filament and the line drawn from the current filament to the point P.

This law can not be experimently prove since we can not isolate the current element.

Biot Savarat law can also be written is terms of current density instead of current as follows;

$$\frac{I}{dS} = J$$

$$I = \text{Jds}$$

$$I\,dL = J\,ds\,.\,dL$$

$$I\,dL = J\,dV \qquad \text{(since } dV = dl\,.\,ds\text{)}$$

Now replacing $I\,dL$ in equation (4.6), gives;

$$dH_2 = \int_V \frac{J\,dV \times a_{12}}{4\pi.R_{12}^2}$$

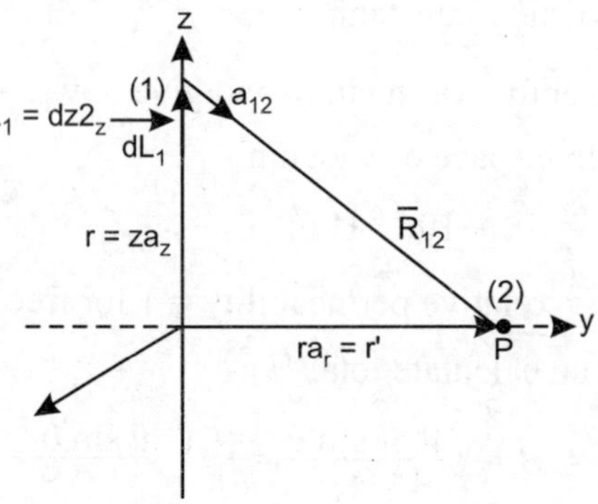

Fig. 4.2. Geometry to describe Bio-Savart law.

$$\overline{R}_{12} = r - r' = (ra_z - za_z),\ R_{12}^2 = (r^2 + z^2)$$

$$a_{12} = \frac{(ra_r - za_z)}{\sqrt{r^2 + z^2}}$$

$$dL = dza_z$$

$$dH_z = \frac{I.dza_z}{(r^2 + z^2)} \times \frac{(ra_r - za_z)}{\sqrt{r^2 + z^2}}$$

$$dH_z = \frac{I\,dza_z \times (ra_r - za_z)}{(r^2 + z^2)^{3/2}}$$

$$\int dH_z = H_z = \int_{-\infty}^{\infty} \frac{Idza_z}{(r^2 + z^2)^{3/2}} \times (r\,a_z - az_z)$$

Let $$z = r \tan\theta$$

$$dz = r \sec^2 d\theta$$

So, $$H_z = \frac{I}{2\pi r}\, a_\phi \qquad \text{...(4.7)}$$

Wave and its Applications

Electrostatic fields are generated with stationary charges and denoted by $E\,(x, y, z)$. However, if charge is in motion with constant velocity it generates magnetostatic field. They are denoted by $H\,(x, y, z)$ and $B\,(x, y, z,)$ and called static magnetic intensity (A /m) and magnetic flux density (W / m^2) respectively.

Time varying fields or waves are usually due to accelerated charges or time-varying currents such as sinewave pulse form of currents. Any pulsating current will produce radiation (time varying fields). They are denoted as $E(x, y, z, t)$. The main application of wave in wireless communication.

Faraday′ s Law : According to Faraday′ s Law a static magnetic field produces no current flow but a time varying field induced voltage called electromotive force or simply emf in a closed circuit which causes a flow of current.

"The induced emf in any closed circuit is equal to the time rate of change of the magnetic flux linkage by the circuit". Therefore,

$$V_{emf} = -\frac{d\lambda}{dt} = -N\frac{d\psi}{dt} \qquad ...(4.8)$$

Where N = No. of loops or turn in the circuit

$\lambda = N\psi$ total flux

ψ = magnetic flux through single loop or coil.

Negative sign shows that induced voltage acts in a such way to oppose the flux producing it. This is called lenz′s law.

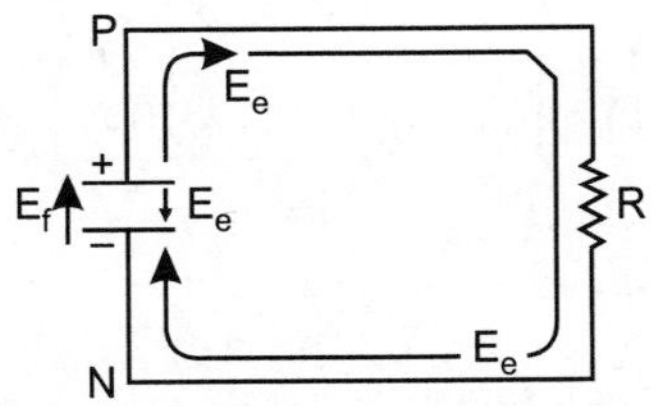

Fig. 4.3. Circuit of Faraday′s law

Electric fields due to static charge are only fields, in this case electric flux line start and end on the charges.

The total field at any point is

$$E = E_f + E_e$$

where E_f is zero outside the battery. E_f and E_e are opposite in the battery. and E_e opposite in battery to that of outside. Integrating throught the loops as;

$$\oint_L Edl = \oint_l (E_f + E_e)dl$$

$$= \oint E_f dl + \oint E_e dl$$

$$= \oint E_f dl + 0 \qquad \text{because } E_e \text{ is constant}$$

So, $$\oint_L Edl = \oint_N^P E_f dl = IR$$

$$V_{emf} = +\oint_N^P E_f dl = -\oint_N^P E_e dl = IR$$

within the battery.

Transformer and Motional emf

From Faraday's Law we can relate electric and magnetic fields. Because it is only the law which relates magnetic fields with electric fields. From Faraday's law

$$V_{emf} = -\frac{d\psi}{dt} \qquad \text{...(4.10) [}N = 1 \text{ for single turn coil)}$$

But magnetic flux density passing through a given surface *ds* is defined as;

$$B = \frac{d\psi}{dS} \Rightarrow BdS = d\psi$$

$$\int d\psi = \int BdS$$

$$\psi = \int_S BdS \qquad \text{...(4.11)}$$

Now from eqn. (4.10) and (4.11) replacing ψ in (4.10)

$$V_{emf} = -\frac{d\psi}{dt}$$

$$= -\frac{d}{dt}\int_S BdS$$

$$V_{emf} = -\int_S \frac{dB}{dt} \qquad \text{...(4.12)}$$

Eqn. (4.12) shows that time varying magnetic fields produce electric field or vice-versa.

The variation of flux with time can be given as:

(*i*) By having stationary loop in a time-varying fields.
(*ii*) By having a time varying loop area in a static fields.
(*iii*) By having a time-varying in a time-varying fields.

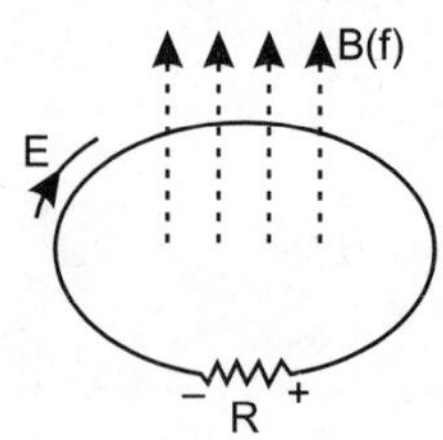

Fig. 4.4. Loop placed in B

Transformer EMF : If there is a stationary loop in a time varying magnetic field *B* then from equation (4.12):

$$V_{emf} = \oint_L Edl = -\oint_S \frac{\partial}{\partial t} Bds \qquad ...(4.13)$$

This emf is induced by the time varying current which produces a time varying field in a stationary loop is called as transformer emf, since it is due to transformer action.

From the Stoke's theorem

$$\oint_V Edl = \oint_S (\nabla \times E)\, dx \qquad ...(4.14)$$

From eqn. (4.13) and (4.14) since LHS is equal, so right hand side is will be also equal.

$$\int_S (\nabla \times E)\, dS = -\int_S \left(\frac{dB}{dt}\right) dS \qquad ...(4.15)$$

If both integrals are equal then integrands are also equal so,

$$\nabla \times E = -\frac{\partial B}{\partial t} \qquad ...(4.16)$$

It is called Maxwell's equation for time-varying fields

If loop is moving and *B* is fixed (Motional EMF) : When a conducting loop is moving in fixed magnetic field i.e.,

$$\frac{\partial B}{\partial t} = 0 \qquad ...(4.17)$$

this induced emf in loop is called, motional EMF. If the velocity of the moving charge is v then force on charge Q.

$$F_m = Q\,(\vec{\nabla} \times \vec{B}) \qquad ...(4.18)$$

Now motinal electric field can be defined as

$$E_m = \frac{F_m}{Q}$$

Since $$F_m = QE_m$$

From eqn. (4.17) and (4.18) we can give as

$$QE_m = Q\,(v \times B)$$

or $$E_m = v \times B$$

If conducting loop has large number of free electrons, moving with uniform velocity v the induced emf can be given as

$$V_{emf} = -\oint E_m dl = -\oint (v \times B)\, dl$$

This emf is called motional or flux cutting emf because it is due to motional action of coil or loop only. The example of motional emfs are: Electrical machine motor, generater etc.

Now from Stroke's theorem.

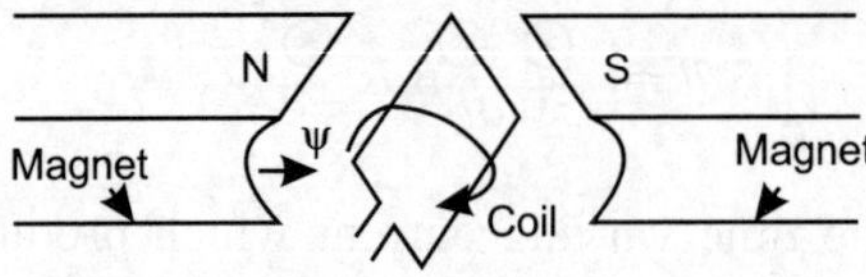

Fig. 4.5. A coil in the magnets region.

$$\oint A\,dl = \oint_S (\nabla \times A)\,ds$$

Let $A = E_m$

$$\oint E_m dl = \oint_S (\nabla \times E_m)\,ds \quad ...(4.22)$$

$$= \oint_S \nabla \times (v \times B)\,ds$$

From eqn. (4.21) and (4.22)

$$-\oint (v \times B)\,dl = \oint_S (\nabla \times v \times B)\,ds \quad ...(4.23)$$

$$-\oint_e (v \times B)\,dl = \oint_S (\nabla \times v \times B)\,ds \quad ...(4.24)$$

So from (4.23) and (4.24), $\nabla \times E_m = \nabla \times v \times B$

This type of emf is called motional emf or flux cutting emf. Another example of motional emf is where rod is moving between a pair of rail. Here B and v are perpendicular.

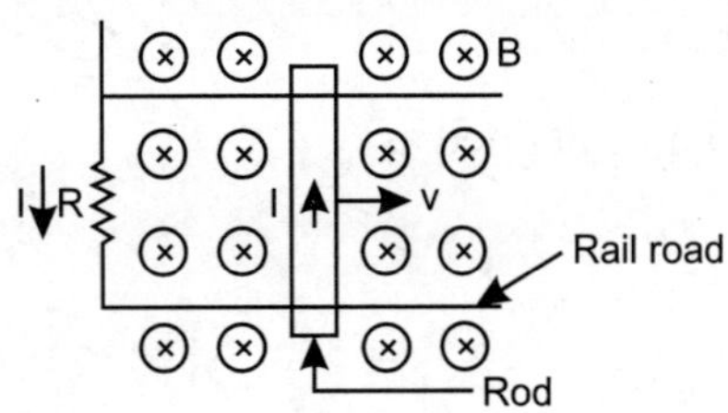

Fig. 4.6. Moving rod between a pair of rail.

$$F_m = I\,(l \times B)$$

or in magnitude form, $F_m = IlB$ ...(4.25)

Example 1 : A conducting bar slide freely over two conducting rail as in shown Fig. 4.7. Calculate induced voltage in the bar.

(*a*) If the bar is stationary at $y = 8$ cm $B = 4 \cos 10^6\, t a_z$, Wb / m²

(*b*) If the bar is slides at a velocity $V = 20\, a_y$ m/sec, $B = 4\, a_z$ Wb / m²

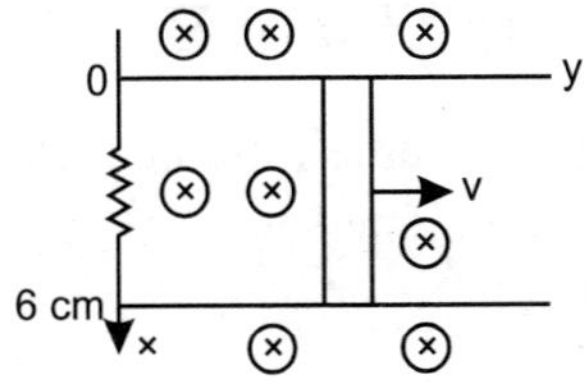

Fig. 4.7. Geometry of example 1.

Solution. (*a*) When loop is stationary B is varying

$$V_{emf} = -\int \frac{\partial B dS}{dt} = -\int_0^x \int_0^y \frac{dB}{dt} a_x \cdot a_y$$

$$= -\int_0^{0.06} \int_0^{0.08} \frac{d}{dt} 4 \cos(106\, t)\, a_z \cdot a_x \cdot a_y$$

$$= -\int_0^{0.06} \int_0^{0.08} 4 - \sin 10^6\, t a_z\, (10^6)\, a_x \cdot a_y$$

$$= 4 \times 10^6 \cos 10^6\, t \times 10^3 \int_0^{0.06} \int_0^{0.08} a_x \cdot a_y \cdot a_z$$

$$= 4 \times 10^6 \cos 10^6\, t.\, [x]_0^{0.06}\, [y]_0^{0.08}\, a_z \times 10^{-3}$$

$$= 4 \times 10^6 \cos 10^6\, t\, [0.06 - 0]\, [0.08 - 0]\, a_z \times 10 - 3$$

$$= 4 \times 10^6\, t\, (0.06 \times 0.08)\, a_z \cos 10^6\, t \times 10^{-3}$$

$$= 4 \times 10^6 \times 48 \times 10^{-4}\, a_z \cos 10^6\, t \times 10^{-3}$$

$$= 19.2 \times \cos 10^6\, t a_z$$

$$V_{emf} = 19.2 \cos 10^6\, t a_z \text{ volt}$$

(*b*) In case of motional emf

$$V_{emf} = \int (v \times B)\, \overline{dl}$$

$$= \int_0^{0.06} (20 a_y \times 4 a_z) \times 10^{-3}\, dl$$

$$= \int_0^{0.06} 80 a_x \cdot dl\, a_x \times 10^{-9}$$

$$= 80\, [d_x]_0^{0.06}$$

$$= 80 \times 0.06 \times 10^{-3}$$

$$V_{emf} = 4.8 \text{ m volt}$$

Moving loop in Time Varying Field B

If there both conditions are happening in same time *i.e.,* Transformer emf and motional emf are present. Both induced voltages can be combined as:

$$V_{emf} = \int_S \frac{dB}{dt} ds + \int_L (v \times B)\, \overline{dl}$$

or in differential form

$$V_{emf} = \nabla \times E = -\frac{\partial B}{\partial t} + \nabla \times (v \times B)$$

It is a general case of electromagnetic emf induction.

Example 2 : The loop given in Fig. 4.8 is inside a uniform magnetic fields B = 50 a_x / m^2. If side dc of the loop cuts the flux lines at the frequency of 50 Hz and loop lies in the *yz* plane at $t = 0$. Find

(*a*) Induced emf at $t = 1$ m sec

(*b*) The induced current at $t = 3$ m sec

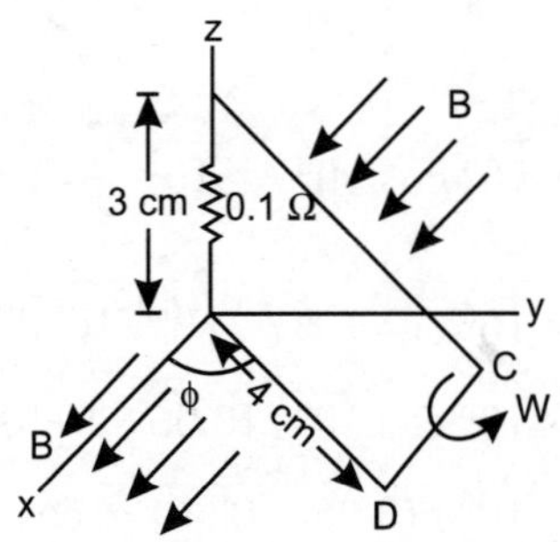

Fig. 4.8. Geometry of example 2.

Solution: As the value of *B* is constant and loop is moving

So, from second case induced emf is

$$V_{emf} = -\oint_l (V \times B) . \overline{dl}$$

$$= \oint (\overline{v} \times \overline{B}) . \overline{dl}$$

Here

$$\overline{v} = r\omega \ (r = \text{radius})$$

$$\overline{v} = rwa_\phi \text{ in vector form}$$

$$\overline{B} = 0.50 a_x \times 10^{-3} \text{ Wb / m2}$$

Now from the relationship between coordinate components i.e.

$$\begin{bmatrix} A_\rho \\ A_\phi \\ A_z \end{bmatrix} = \begin{bmatrix} \cos\phi & \sin\phi & 0 \\ \sin\phi & \cos\phi & 0 \\ 0 & 0 & 1 \end{bmatrix} \begin{bmatrix} A_x \\ A_y \\ A_z \end{bmatrix}$$

$$\begin{bmatrix} B_\rho \\ B_\phi \\ B_z \end{bmatrix} = \begin{bmatrix} \cos\phi & \sin\phi & 0 \\ \sin\phi & \cos\phi & 0 \\ 0 & 0 & 1 \end{bmatrix} \begin{bmatrix} 50\times10^{-3} \\ 0 \\ 0 \end{bmatrix}$$

$$\begin{bmatrix} B_\rho \\ B_\phi \\ B_z \end{bmatrix} = \begin{bmatrix} 50\times10^{-3}\cos\phi \\ -50\times10^{-3}\sin\phi \\ 0 \end{bmatrix}$$

$$B_\rho = 50 \times 10^{-3} \cos \phi$$

$$B_\phi = -50 \times 10^{-3} \sin \phi \text{ and } B_z = 0$$

So $$\overline{B} = B_\rho a_\rho + B_\phi a_\phi + B_z a_z$$

So $$v \times B = \begin{bmatrix} a_\rho & a_\phi & a_z \\ 0 & 0 & 0 \\ 50\times10^{-3}\cos\phi & -50\times10^{-3}\sin\phi & 0 \end{bmatrix}$$

$$a_\rho (0-0) - a_\phi (0-0) + a_z (\rho\omega - 50 \times 10^{-3} \sin \phi)$$

$$= -50 \times 10^{-3} (\rho\omega) \cos\phi a_z$$

$$= -50 \times 10^{-3} (\rho\omega) \cos\phi a_z$$

So, $(v \times B) \times \overline{dl}$

$$= -50 \times 10^{-3} \rho\omega \cos\phi_0 \oint_{z=0}^{0.03} (d_z \cdot a_z)\, a_z$$

$$= -50 \times 10^{-3} \times 4 \times 10^{-2} \times (2\pi f) \left[z\right]_0^{0.03}$$

$$= -6\pi \cos\phi\, mV$$

or $$V_{emf} = -6\pi \cos\phi\, mV$$

Now ϕ can be calculated as

$$\frac{d\phi}{dt} = \omega$$

$$\int d\phi = \int \omega dt$$

$$\phi = \omega t + K$$

When $$t = 0, k = \frac{\pi}{2}$$ (Since loop is in *yz* plane)

$$\frac{\pi}{2} + 0 + K \Rightarrow K = \pi/2$$

So $$\phi = \omega t + \frac{\pi}{2}$$

So $$V_{emf} = 6\pi \cos \phi$$

at $t = 0$,

$$-6\pi.\cos\left(\omega t + \frac{\pi}{2}\right) = -6\pi.\cos\left(\omega \times t + \frac{\pi}{2}\right)$$

$$= -6\pi - f_n\,\omega t$$

$$= 6_\pi f_n\,(2\pi f t)$$

$$= 6\pi f_n\,(2\pi \times 50t)$$

$$= 6\pi \times f_n\,(100 f_n) \times 60 f_n\; 100\pi \times 10^{-3}$$

$$= 60 f_n + 0.5$$

$$= 60 \times 5.825 \text{ mV}$$

Similarly at $t = 6 \times 10^{-8}$ current (I) :

$$I = \frac{V_{emf}}{R} = \frac{6\pi\; f_n\,(100\,\pi t)}{0.1} = 60\pi f_n\; 100\;\pi t$$

$$= 60 f_n\,(100 \times 3 \times 10^{-3})$$

$$I = 0.1525 \text{ A}$$

Displacement Current

The displacement current is defined as time rate of change of electric field intensity of electric flux density. From the Faraday Law we have achieved the Maxwell's second equation as

$$\nabla \times E = -\frac{\partial B}{\partial t} \qquad ...(4.26)$$

Means time-changing magnetic field generates electric fields. The electric field has special property. The line integral around any closed path of electric field does not zero. But for magnetic field it is valid as from Ampere's circuital law.

$$\nabla \times H = J \qquad ...(4.27)$$

Taking divergence of above eqn. (4.27)

$$\text{div}\,(\nabla \times H) = \text{div}\,J$$

$$\nabla\,.\,(\nabla \times H) = \nabla\,.\,J \qquad ...(4.28)$$

But the divergence of a curl is zero. So

$$\nabla\,.\,(\nabla \times H) = 0 \qquad \text{Means } \Delta.J = 0 \;...(4.29)$$

But from continuity

$$\nabla\,.\,J = -\frac{\partial \rho_v}{dt} \qquad ...(4.30)$$

It means $-\frac{\partial \rho_v}{dt} = 0$

Which is not possible and unrealistic. So eqn. (4.26) can be modify as

$$\nabla \times H = J + J_d \quad ...(4.31)$$

Now taking divergence of eqn. (4.31)

$$\nabla . (\nabla \times H) = 0 = \nabla . J + \nabla . J_d$$

or $$\nabla J + \nabla . J_d = 0$$

$$\nabla . J = -\nabla . J_d \qquad \nabla . J = -\frac{\partial P_v}{dt}$$

$$\nabla . J_d = \nabla . J = \frac{\partial \rho_v}{dt} \quad ...(4.32)$$

From Maxwell's third eqn.

$$\nabla . D = \rho_v \quad ...(4.33)$$

Putting it in eqn. (4.32)

$$\nabla . J_d = \frac{\partial}{\partial t} \rho_v = \frac{\partial}{\partial t} (\nabla . D)$$

$$\nabla . J_d = \nabla . \frac{\partial D}{\partial t}$$

$$J_d = \frac{\partial D}{\partial t} \quad ...(4.34)$$

$$J_d = \frac{\partial \varepsilon E}{\partial t}$$

So eqn. (4.27) can be modified as $J_d = \frac{\varepsilon \partial E}{\partial t}$

$$\nabla \times H = J + J_d = J + \frac{\partial D}{\partial t}$$

$$\nabla \times H = J + \frac{\partial D}{\partial t} \quad ...(4.35)$$

where $J_d = \frac{\partial D}{\partial t}$ called displacement current density and J is called conduction current density ($J = \sigma E$). The insertion of J_d is major contribution of the Maxwell's equation. Without the term J_d the EMW propagation would be impossible. At low frequency J_d is neglected. However at high frequency it would be comparable.

Based on the displacement current density, the displacement current can be calculated as

$$I_d = \int_S I_d dS = \int_S \frac{\partial D}{\partial t} . dS \quad ...(4.36)$$

Now Maxwell's first equation can be given as

$$\nabla \times H = J + \frac{\partial D}{\partial t}$$

$$\int_S (\nabla \times H)\, ds = \int_S \left(J + \frac{\partial D}{\partial t}\right)$$

$$\int_S (\nabla \times H)\, ds = \int_S J\, ds + \int \left(\frac{\partial D}{\partial t}\right) ds \qquad ...(4.37)$$

From Stokes theorem $\oint_S H \,.\, dl = \int_S (\nabla \,.\, H)\, ds$ ...(4.38)

From (4.37) and (4.38) $\oint_H H \,.\, dl = \mathrm{I} + \mathrm{Id} = \mathrm{I} + \int_S \frac{\partial D}{\partial t} ds$ ...(4.39)

Displacement current is associated with time varging electric field and exists in all imperfect conductor carrying time varying conduction current.

The physical meaning of this displacement current is that a changing electric field makes a changing magnetic field. To find the direction of the magnetic field produced by this effect we use our usual right-hand rule for currents : point your thumb in the direction of the displacement current (the direction of (dE / dt), and your fingers will curl around in the direction of the magnetic field. (Note that if the electric field is decreasing in time, then the negative time derivative gives a displacement vector that points opposite to E).

Example 3. A parallel plate capacitor with plate are 5 cm^2 and plate separation of 3 mm has a voltage 50 sin 10^3 tV applied to its plate. Calculate displacement current assuming $\varepsilon = 2\varepsilon_0$.

Solution. Since

$$J_d = \frac{\partial D}{\partial t}$$

$$D = \varepsilon E$$

$$D = \varepsilon \frac{V}{d} = 2\frac{\varepsilon_0 V}{d}$$

$$= \frac{2\varepsilon_0}{d}(50 \times \sin 10^3\, t)$$

$$D = \frac{100\, \varepsilon_0}{d} \sin 10^3\, t$$

Now

$$J_d = \frac{\partial D}{\partial t} = \frac{\partial}{\partial t}\left(\frac{100\, \varepsilon_0 \sin 10^3\, t}{d}\right)$$

$$= \frac{100\, \varepsilon_0}{d} \,.\, \cos 10^3\, t \,.\, 10^3$$

$$J_d = \frac{100\, \varepsilon_0}{d} \,.\, \cos 10^3\, t$$

So

$$I_d = \frac{J_d}{S} = \frac{10^5\, \varepsilon_0}{dS} \,.\, \cos 10^3\, t$$

$$= \frac{50 \times 10^{-4} \times 10^5\, \varepsilon_0}{3 \times 10^{-3}} \cos 10^{3t} = \frac{50}{3} \times 1.3 \times \frac{1}{36\pi} \times 10^{-9} \cos \omega t$$

$$= \frac{50 \times 10^{-5}}{108\pi} \cos 10^{3t}$$

$$= 0.0147 \times 10^{-5} \cos 10^{3t}$$

$$I_d = 147.\cos 10^3 \, (nA)$$

Example 4. In a free space $E = 20\ (\omega t - 50x)\ a_y\ V/m$ calculate (*a*) J_d (*b*) H (c) ω.

Solution. (*a*) $$J_d = \frac{\partial D}{\partial t} = \frac{\partial \varepsilon_0 E}{\partial t} = \frac{\varepsilon_0\, \partial 20 \cos(\omega t - 50x) a_y}{\partial t}$$

$$= 20\varepsilon_0 - \sin(\omega t - 50x)\, w$$

$$J_d = -20\varepsilon_0 \omega \sin(\omega t - 50x) a_y \text{ A / m}$$

(*b*) $$\nabla \times H = J + J_d$$

$$\nabla \times H = J_d$$

$$\begin{bmatrix} a_x & a_y & a_z \\ \frac{\partial}{\partial x} & \frac{\partial}{\partial y} & \frac{\partial}{\partial z} \\ H_x & H_y & H_z \end{bmatrix} = [0 - 20\varepsilon_0 \omega \sin(\omega t - 50x) a_y]$$

$$a_x\left(\frac{\partial H_z}{\partial y} - \frac{\partial H_y}{\partial z}\right) - a_y\left(\frac{\partial H_z}{\partial x} - \frac{\partial H_x}{\partial z}\right) + a_z\left(\frac{\partial H_z}{\partial x} \frac{\partial H_x}{\partial y}\right)$$

$$\frac{\partial H_z}{\partial y} - \frac{\partial H_y}{\partial z} = 0 \Rightarrow \frac{\partial H_z}{\partial y} = \frac{\partial H_y}{\partial_z}$$

$$\frac{\partial H_y}{\partial x} - \frac{\partial H_x}{\partial y} \Rightarrow \frac{\partial H_y}{\partial x} = \frac{\partial H_x}{\partial y}$$

$$-\left(\frac{\partial H_z}{\partial x} - \frac{\partial H_x}{\partial z}\right) = -20\varepsilon_0\ \omega.\sin(\omega t - 50x)$$

$$\frac{\partial H_z}{\partial x} + \frac{\partial H_x}{\partial z} = -20\varepsilon_0 \omega \sin(\omega t - 50x)$$

$$\frac{\partial H_z}{\partial x} + \frac{\partial}{\partial z} \frac{H_y}{\partial x} \partial_y = -20\ \varepsilon_0 \omega \sin(\omega t - 50x)$$

$$\frac{\partial H_z}{\partial x} + \frac{\partial y}{\partial x} . \frac{\partial H_z}{\partial y} = -20\ \varepsilon_0 \cos\omega \sin(\omega t - 50x)$$

$$2\frac{\partial H_z}{\partial x} = -20\ \varepsilon_0 \cos\omega\ (\omega t - 50x)$$

$$\frac{\partial H_z}{\partial x} = -10\ \varepsilon_0 \cos\omega\ (\omega t - 50x)$$

$$\int \partial H_z = -\int 10\ \varepsilon_0 \cos\omega\ (\omega t - 50x)$$

$$H_z = -10\ \varepsilon_0 \omega \frac{\cos(\omega t - 50x)}{-50}$$

$$H_z = 0.2\ \varepsilon_0\ \omega\ (\omega t - 50x)$$

So $$H = H_z a_z = 0.2\,\varepsilon_0 \omega \cos(\omega t - 50x)$$

(c) $J_d = \varepsilon \dfrac{\partial D}{dt}$

$$J_d = j\omega\varepsilon E = -20\,\varepsilon\omega \sin(\omega t - 50x)$$

Maxwell's Equations in Final Form

James Clerk Maxwell's is the founder of electromagnetic theory. It is Maxwell's equation due to which electromagnetic waves were discovered. For the static Fields Maxwell's equation can be given as following,

For Static Field

Differential form Integral form :

$\nabla . D = \rho_v \rightarrow$ Gauss law $\rightarrow \oint_S \overline{D} . \overline{dS} = \int \rho_v dV$

$\nabla . B = 0 \rightarrow$ Nonexistance of Magnetic dipole $\rightarrow \oint_S BdS = 0$

$\nabla \times E = 0 \rightarrow$ Conservation of Electrons fields $\rightarrow \oint_C Edl = 0$

$\nabla \times H = J \rightarrow$ Amper's law $\oint_C Hdl = \oint_S JdS$

The general form of Maxwell equation is time varying fields and it can be gives as;

$$\nabla . D = \rho_v \quad ...(4.40a)$$

$$\nabla . B = 0 \quad ...(4.40b)$$

$$\nabla \times E = -\frac{\partial B}{\partial t} \quad ...(4.40c)$$

$$\nabla \times H = J + \frac{\partial D}{\partial t} \quad ...(4.40d)$$

Now in integral form

$$\oint_S Dds = \int_V \rho dv \quad ...(4.41a)$$

$$\oint_S BdS = 0 \quad ...(4.41b)$$

$$\oint_C Edl = -\frac{\partial}{\partial t}\int_S BdS \quad ...(4.41c)$$

$$\oint_L Hdl = \int_S \left(J + \frac{\partial D}{dt}\right) ds \quad ...(4.41d)$$

From eqn. $\nabla \cdot D = \rho_v$ or $\int_S D ds = \int_V \rho_v dv$ it is cleared that the charge density is the source of electric flux lines, and every coulomb's charge must have one coulomb's of electric flux diverging from it. From eqn. $\nabla \cdot B = 0$ magnetic flux is always found in closed loops and never diverges from a point charge or point source.

The Auxilary equations relating E and H with D, B can be gives as

$$D = \varepsilon E \qquad ....(4.42a)$$

$$B = \mu H \qquad ...(4.42b)$$

For the linear, isotropic and homogeneous mediums the above equations can be given as,

$$D = \varepsilon_0 E + P$$

$$B = \mu (H + M) \qquad ...(4.45a)$$

In which for linear materials $P = \alpha_m \varepsilon_0 E$ ...(4.45b)

and $M = \alpha_m H$ ...(4.45c)

where M called magnetization and it is defined as magnetic dipole moment / unit volume.

It happend due to bounded charges. (orbital electrons) and $m = I_b \cdot dS$,

$$I_b = \text{bounded current}$$

$$I_b = \oint M \cdot \overline{d}L$$

α_e called magnetic susceptibility,

So $B = \mu_0 (H + M)$

$$= \mu_0 (H + \alpha_m H)$$

$$= \mu_0 (1 + \alpha_m) H$$

$$= \mu_0 \mu_r H$$

Where $\mu_r = 1 + \alpha_m$ ...(4.46*a*)

Similarly $D = \varepsilon_0 E + P$

$$= \varepsilon_0 E + \alpha_e \varepsilon_0 E$$

$$= \varepsilon_0 E (1 + \alpha_e)$$

$$= \varepsilon_0 E \varepsilon_r$$

$$\varepsilon_r = 1 + \alpha_e \qquad ...(4.46b)$$

α_e called electrical susceptibility

Now complete force eqn. can be gives as:

$$f = \rho_v (EA \times B) \qquad ...(4.47)$$

Electromagnetism (Magnetostatic)

When charges are in motion they generates the magnetic fields. If current is direct then generated magnetic fields is called magneticstatic fields. For example a wire carrying the current I has magnetic field surrounding it.

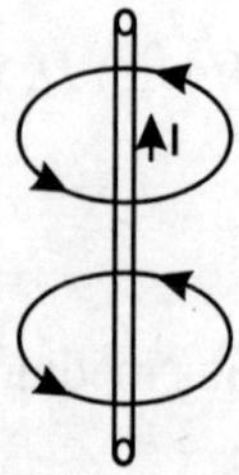

Fig. 4.9. Magnetic field around wire

Electric and magnetic fields are fundamentally fields of forces which originated from electric charges. The charges moving with constant velocity produce the magnetostatic field but if these charges accelerated produce both time varying electric and magnetic fields called electromagnetic fields.

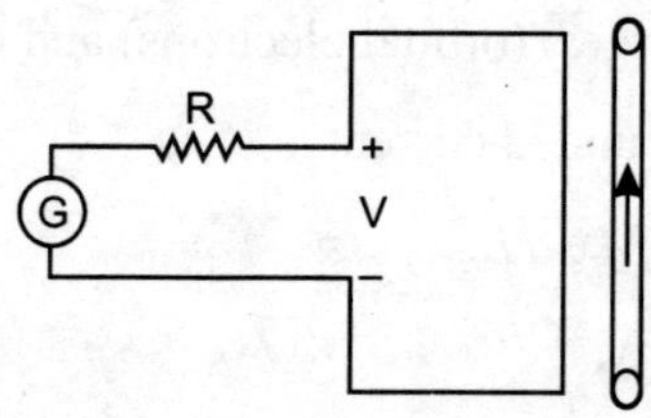

Fig. 4.10. Example for producing EM fields

We get

$$I = \frac{V}{R}$$

$$I = \frac{dQ}{dt}$$

$$dQ = I\,dt$$

$$Q = \int dQ = \int I\,dt$$

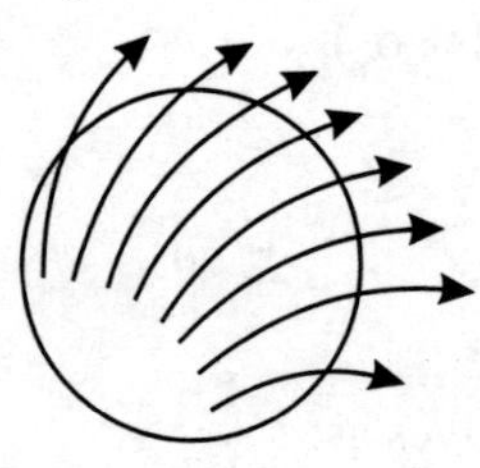

Fig. 4.11. Magnetic flux thought a loop

The magnetic flux through a closed loop can be defined as
'Time integral of voltage induced in the loop throughout'.
The interval during the magnetic field is being establish, hence

$$\phi = \pm \int_0^t V \, . \, dt \qquad ...(4.48)$$

Interval $(0 - t)$ is the time required to establish the fields or decrease to zero.

So, differentiating (4.48) with respect to time, gives

$$\frac{d\phi}{dt} = \pm V \qquad ...(4.49)$$

$$\Rightarrow \qquad \frac{d\phi}{dt} = -V$$

called Faraday's law of Induction.
Negative sign means when flux increases in positive direction, the induced emf reduces.
Unit of flux weber or volt.

Or

Time rate of decrease of the total magnetic flux linking in the circuit is equal to the total emf induced in the circuit.

If loop has N turns

$$V = -N \, . \, \frac{d\phi}{dt} \qquad ...(4.50)$$

Now the important parameter E and H can be calculated as

$$V_e = \oint E dl \qquad ...(4.51a)$$

Here $\qquad E_e \neq -\nabla V_e$ is not applicable $\qquad ...(4.51b)$

Now total flux is equal to the flux density B over surface bounded by the circuit

$$\phi = \int_S B ds \qquad ...(4.51c)$$

So $\qquad V = \oint E dl = -\frac{d}{dt} \iint_S B ds \qquad ...(4.51d)$

relates the electric and magnetic fields.
Eq. (4.51*d*) states that changing magnetic field will induce electric field.

Application of Ampere's Law

The ampere's law has the following applications

(*i*) Ampere's Law in integral form, i.e.,

$$\oint H dl = I_{\text{enclosed}}$$

or $$\oint \frac{B}{\mu_0} dl = I_{enc}$$

$$\oint Bdl = \mu_0 I_{enclosed}$$

which can be used to calculate the current in a given conductor.

(*ii*) Closed path solution of the Ampere's Law give us ability to construct the Amperian path.

(*iii*) In general the solution of integral equations must be obtained using numerical technique.

(*iv*) Calculation of Magnetic field intensity of an infinity long carrying conductor.

$$\oint Hdl = I$$

$$\oint H_\phi \,.\, rd\phi = \int_0^{2\pi} H_\phi \,.\, rd\phi$$

$$H_\phi \,.\, r \,.\, [\phi]_0^{2\pi} = I$$

$$H_\phi r\,(2\pi - 0) = I$$

$$H_\phi = \frac{1}{4\pi r}$$

or $$\frac{B}{\mu} = \frac{I}{4\pi r}$$

$$B = \frac{\mu}{4\pi}\frac{I}{r} \qquad ...(4.52)$$

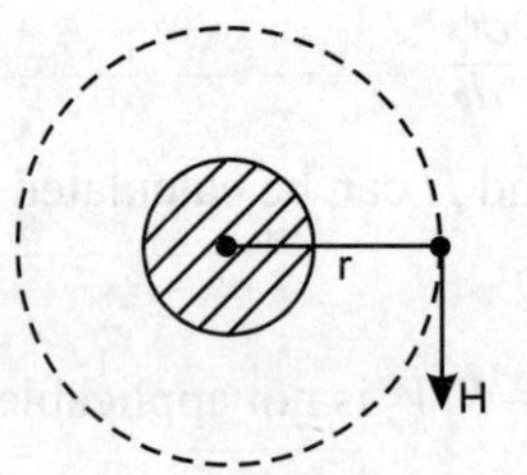

Fig. 4.12 Conductor placed in *z* direction (case 1)

Magnetic flux density inside on impurity by conductor containing an uniform current density can be calculated as follows :

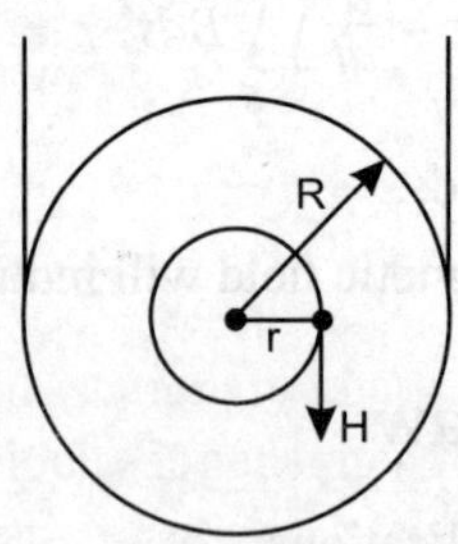

Fig. 4.13. Conductor placed in H field (case 2).

$$I \propto A$$

$$I_r \propto A_r$$

$$I \propto A_R$$

$$\frac{I_r}{I} = \frac{K\,\pi r^2}{K\,\pi R^2}$$

$$I_r = \frac{r^2}{R^2} I \qquad \text{...(4.53a)}$$

The magnetic field at radius r

$$H = \frac{I_r}{2\pi r} a_\phi$$

$$= \frac{r^2}{R^2}\frac{I}{2\pi r} a_\phi$$

$$H = \frac{Ir}{2\pi R^2} a_\phi \qquad \text{...(4.53b)}$$

So,

$$B_\phi = \frac{\mu H}{A} = \frac{\mu I_r}{2\pi R^2} a_r \qquad \text{...(4.54)}$$

Uniqueness Theorem of Electrostatic Fields

Any solution to Laplace's equation or Poisson's equation which also satisfy the boundary conditions must be the only solution that exist. It is unique, *i.e.,* only possible solution.

Proof : Let there are two values of V_1 and V_2 which satisfy the Laplace's equation

$$\nabla^2 V_1 = 0 \qquad \text{...(4.55a)}$$

$$\nabla^2 V_2 = 0 \qquad \text{...(4.55b)}$$

everywhere bounded by the surface. Now subtracting equation (4.55*a*) from (4.55*b*);

$$\nabla^2 V_2 - \nabla^2 V_1 = 0$$

$$\nabla^2 (V_2 - V_1) = 0 \qquad \text{...(4.56)}$$

Each solution must satisfy the boundary condition and if we represent the given potential value on the boundary by V_b then the value of V_1 on the boundary V_{1b} and value V_2 on boundary is V_{2b} and both must be identical, *i.e.,*

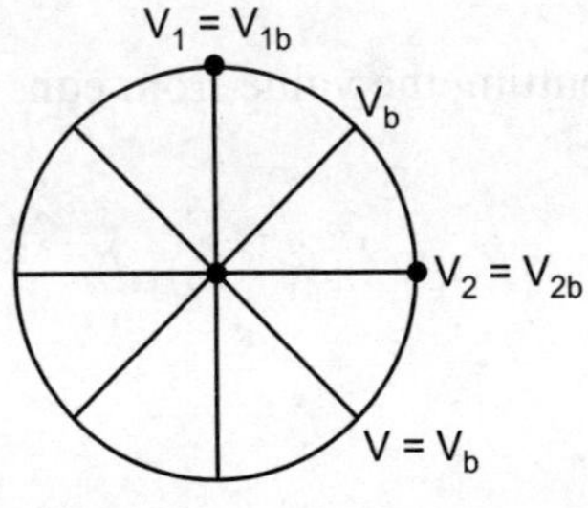

Fig. 4.14. Value of V_1 & V_2 at boundary

$$V_{1b} = V_{2b} = V_b$$

or $$V_{1b} - V_{2b} = 0 \qquad ...(4.57)$$

Now from the vector identify

$$\nabla . (VD) = V(\nabla . D) + D(\nabla . V) \qquad ...(4.58)$$

where V is scalar and D is called vector.

Let $$V = V_1 - V = \text{scalar}$$

$$D = \nabla (V_1 - V_2) = \text{vector}$$

Then from (4.58) :

$$\nabla . [(V_1 - V_2) \nabla . (V_1 - V_2)] = (V_1 - V_2) [\nabla . \nabla (V_1 - V_2] + \nabla (V_1 - V_2) [\nabla . (V_1 - V_2)] \qquad ...(4.59)$$

$$\nabla [(V_1 - V_2) \{\nabla (V_1 - V_2)\}] = (V_1 - V_2) [\nabla^2 (V_1 - V_2)] + [\nabla (V_1 - V_2)]^2$$

Now integrating the above eqn. (4.59), over the volume bounded by the closed surface

$$\int_V \nabla [(V_1 - V_2) \{\nabla (V_1 - V_2)\}] dv = \int_V [(V_1 - V_2) [\nabla^2 (V_1 - V_2)] + (\nabla (V_1 - V_2)]^2 dV$$

$$= \int_V (V_1 - V_2) [\nabla^2 (V_1 - V_2)] dV + \int_V \nabla (V_1 - V_2)] dV \qquad ...(4.60)$$

Now applying the divergence theorem to convert volume integral to surface integral.

$$\int (\nabla . D) dV = \oint_S D ds \qquad ...(4.61)$$

Now left handside of (4.61) equation:

$$\int_V \nabla . [(V_1 - V_2) \{\nabla (V_1 - V_2\}] dV = \int_S [(V_1 - V_2) \{\nabla (V_1 - V_2\}] ds \qquad ...(4.62)$$

But from boundary condition

$V_{1b} = V_{2b}$ putting in (4.62) equation:

$$\oint_S (V_{1b} - V_{2b}) \{\nabla . (V_{1b} - V_{2b})\} dS = \oint_S 0 \{\nabla (V_{1b} - V_{2b})\} ds = 0$$

Now from eqn. (4.60) 1st term i.e.

$\int_V (V_1 - V_2) [\nabla . \nabla (V_1 - V_2)] dV$, putting the value from eqn. (4.56)

$$\int_V (V_1 - V_2) [\nabla^2 (V_1 - V_2)] dV$$

$$\int_V (V_1 - V_2)(0) dV = 0$$

$$\int_V [\nabla (V_1 - V_2)]^2 dV = 0 \qquad ...(4.63)$$

Now there are two reasons why an integral may be zero. Either integrand $[\nabla (V_1 - V_2)]^2$ is zero or integral is positive somewhere and negative at other region and net contribution is zero. And because $[\nabla (V - V_2)]^2$ never be negative so only first condition is applicable i.e.

$$[\nabla (V_1 - V_2)]^2 = 0$$

$$[\nabla (V_1 - V_2)] = 0$$

or $$V_1 - V_2 = \text{constant in volt} \qquad ...(4.64)$$

Because zero gradient means $V_1 - V_2$ cannot charge w.r.t. any direction of V.

Now from eqn. (4.64) if

If $$V_1 - V_2 = 0$$

Then $V_1 = V_2$ which is the uniqueness theorem.

The same proof is also applicable for the Poisson's equation as;

$$\nabla^2 V = \frac{\rho_v}{\varepsilon}$$

$$\nabla^2 V_1 = \frac{\rho}{\varepsilon}$$

$$\nabla^2 V_2 = \frac{\rho}{\varepsilon}$$

$$\nabla^2 V_1 - \nabla^2 V_2 = 0$$

$$\nabla^2 (V_1 - V_2) = 0$$

Now it is same as eqn. (4.56) Rest derivation is same as previous.

Example 5. Find magnetic Field due to finite length conductor: placed z-axis, as shown ing Fig. 4.15.

Solution: Let $$z = r \cot \theta$$

$$dz = -r \operatorname{cosec}^2\theta d\theta$$

if $$z = -\infty$$

$$\cot \theta = -\infty$$

$$\theta = \cot^{-1}(-\infty) = \pi$$

$$\theta = \cot^{-1}(\infty) = \pi$$

$$\theta = \cot^{-1}(\infty) = 0$$

We know that

$$H_Z = \int \frac{I.r.dz}{4\pi]r^2 + z^2]^{3/2}} . a_\phi$$

$$H_Z = \frac{I}{4\pi} \int_1^2 \frac{-r \operatorname{cosec}^2 \theta d\theta}{(r^2 + r^2 \cot^2 \theta)^{3/2}} a_\phi = \frac{I}{4\pi} \int_1^2 -\frac{r}{r^3} \frac{\operatorname{cosec}^2 \theta d\theta}{\operatorname{cosec}^3 \theta} d_\phi \qquad \begin{matrix} 1 - \alpha_1 \\ 2 - \alpha_2 \end{matrix}$$

$$= -\frac{I}{4\pi r} \int_1^2 \frac{1}{\operatorname{cosec}} d\theta . a_\phi$$

$$= -\frac{I}{4\pi r} \int_1^2 (\sin \theta) d\theta . a_\phi \qquad ...(4.65a)$$

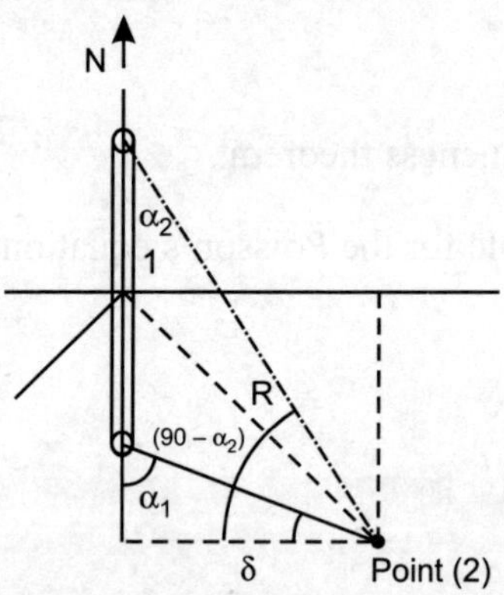

Fig. 4.15 Geometry for example 5.

$$H_Z = \frac{I}{4\pi r} (\cos \alpha_2 - \cos \alpha_1) a_\phi \qquad ...(4.65b)$$

There an two cases:

(*a*) If $\alpha_1 = 90°$, $\alpha_2 = 0°$

$$H = \frac{I}{4\pi v} . a_\phi$$

(*b*) $\alpha_1 = 180°$

$$\alpha_2 = 0$$

$$H = \frac{I}{2\pi r} a_\phi$$

Example 6. Find the total magnetic field at point '*P*' by the conductors an arranged in Fig. 4.16.

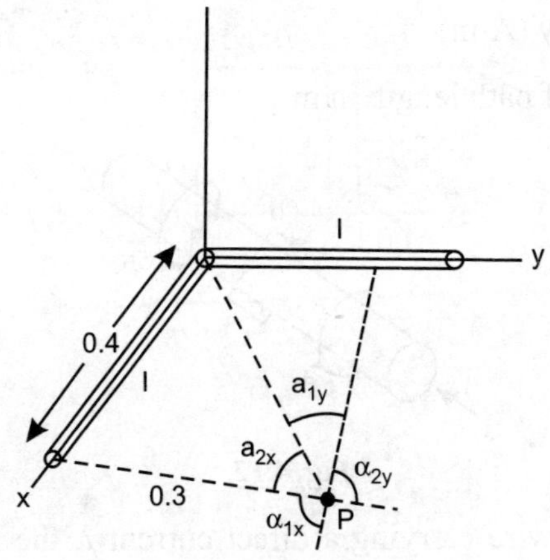

Fig. 4.16

$$\tan\theta = \left(\frac{0.4}{0.3}\right)$$

$$\theta = \alpha_{2x} = 53.1°$$

$$\alpha_{1x} = -90°$$

we know that
$$H = \frac{I}{4\pi r}(\cos\alpha_2 - \cos\alpha_1)$$

$$= \frac{I}{4\pi r}[\cos(90° - a_{2x}) - \cos(90 - a_{1x})$$

$$= \frac{I}{4\pi r}[\sin a_{2x} - \sin a_{1x}]$$

$$\text{So, } \frac{8}{4\pi.0.3}(\sin 53.1 + 1) = \frac{12}{\pi}a_\phi$$

$$\alpha_{1y} = -\tan\left(\frac{0.3}{0.4}\right) = -36.9$$

$$\alpha_{2y} = 90°$$

So,
$$H_{2y} = \frac{8}{4\pi \times 0.4}(\sin 90 + \sin 36.9) = \left(-\frac{8}{\pi}\right)a_\phi$$

$$H = H_{2x} + H_{2y} = \frac{-12}{\pi}a_\phi - \frac{8}{\pi}a_y = -\frac{-20}{\pi}a_\phi$$

Ampere's Circutal Law

It states that

"Line integral of magnetic field intensity (H) about any path closed path is exactly equal to the direct current enclosed by that path mathematically i.e.,

$$I = \oint H\,dl \qquad ...(4.66)$$

where $I = dc$ current enclosed in A.

H = Magnetic field intensity (A/m)

dl = infinitismall element of path length in m

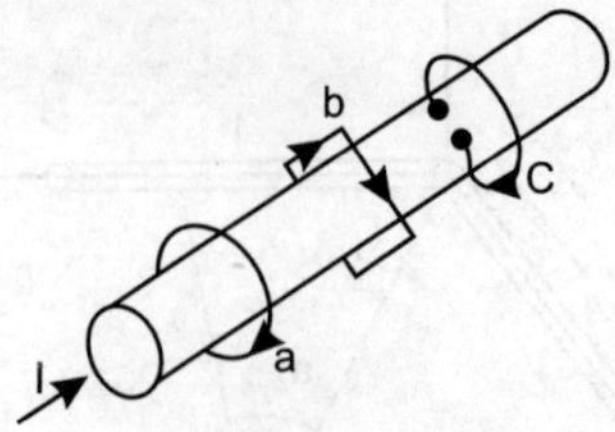

Fig. 4.17

In which fig 4.17 circular wire carrying a direct current I, the line integral at H about path lettered a, b and c, a & b gives the exactly I current but both c gives less than I and a portion of the total current which is enclosed by path c path a and b gives the same answer.

The Ampere's circutal Law is analogous to Gauss Law Gauss Law used to find total charge where as amperes usual to find total current.

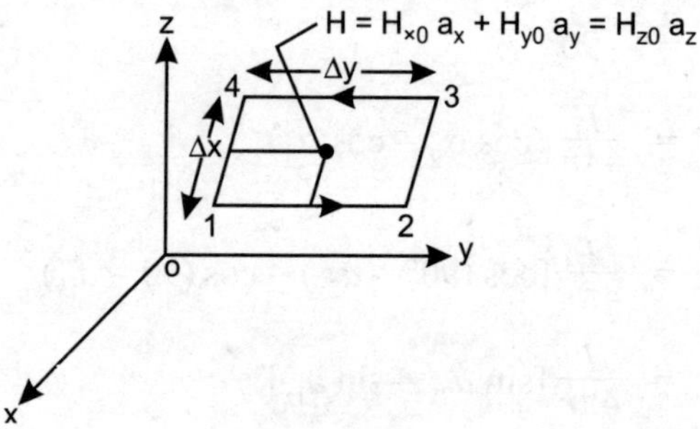

Fig. 4.18

The closed line integral of H about closed path is then approximately equal to sum of the four values of $H\nabla L$ one each side. We will choose the direction 1 – 2 – 3 – 4 which will corresponds to direction of current in a_z direction. So,

$$(H \,.\, \Delta L)_{1-2} = H_{y\,1-2}\Delta y$$

Now $H_{y\,1-2}$ can be given in terms of reference value

$$H_{y12} = H_{y0} + \frac{\partial H_y}{\partial x} \cdot \left(\frac{1}{2}\,\Delta x\right)$$

So $$(\overline{H} \,.\, \overline{\Delta L})_{1-2} = \left[H_{y0} + \frac{\partial H_y}{\partial x}\,\frac{1}{2}\,\Delta x\right]\Delta y \qquad \text{...(4.67}a\text{)}$$

Along the next section of path

$$(H\,\Delta L)_{2-3} = H_{x\,2-3} - \Delta x$$

Now $$H_{x\,1-2} = \left(Hx_0 + \frac{\partial H_x}{\partial y}, \frac{\Delta y}{2}\right)$$

So $\quad (\overline{H}\ \Delta\overline{L})_{2-3} = -\left(Hx_0 + \frac{\partial H_x}{\partial y}\frac{1}{2}\Delta y\right)\Delta x$ ...(4.67*b*)

Similarly $\quad (\widehat{\overline{H}\ \Delta\overline{L}})_{2-3} = H_{y\,1-2}\Delta y$

$$H_{y\,1-2}\Delta y = -\left[H_{y0} + \frac{\partial H_y}{\partial x}\cdot\frac{1}{2}\Delta x\right]\Delta y \quad ...(4.67c)$$

Similarly,

$$(\overline{H}\cdot\Delta\overline{L})_{4-1} = H_{x12}\Delta x$$

$$H_{x12} = \left(H_{x0} + \frac{\partial H_x}{\Delta y}\cdot\frac{1}{2}\Delta y\right)$$

So $\quad (\overline{H}\ \Delta\overline{L})_{4-1} = \left(H_{x0} + \frac{\partial Hx}{\partial y}\Delta\frac{1}{2}\Delta y\right)\Delta x$ ...(4.67*d*)

Hence, for closed loop

$$(H\,\Delta L)_{12} + (H\,\Delta L)_{2-3} + (H\,\Delta L)_{3-4} + (H\,\Delta L)_{4-1}$$

$$H_{y0}\Delta y + \frac{1}{2}\frac{\partial H_y}{\partial x}\Delta x\cdot\Delta y - H_{x0}\Delta x - \frac{1}{2}\frac{\partial H_x}{\partial y}\Delta y\cdot\Delta x - H_{y0}\Delta y - \frac{1}{2}\frac{\partial H_y}{\partial x}\Delta x\,\Delta y$$

$$+ H_{x0}\Delta x + \frac{1}{2}\frac{\partial H_y}{\partial y}\Delta x\cdot\Delta y$$

$$\frac{\partial H_y}{\partial x} - \partial H = \left(\frac{\partial H_y}{\partial x} - \frac{\partial H_x}{\partial y}\right)\Delta x\,\Delta y$$

$$\oint H\cdot dl = \left(\frac{\partial H_y}{\partial x} - \frac{\partial H_x}{\partial y}\right)\Delta x\,\Delta y = J_z\,\Delta x\,\Delta y$$

Since $\quad I = \int Jds$

$$J_z = \frac{\left(\frac{\partial H_y}{\partial x} - \frac{\partial H_x}{\partial y}\right)}{\Delta x\,\Delta y} = \oint\frac{H\,dL}{\Delta x\,\Delta y}$$

or

$$J_z^a = \left(\frac{\partial H_y}{\partial x} - \frac{\partial H_x}{\partial y}\right) = \lim_{\substack{\Delta y\to 0\\ \Delta z\to 0}}\oint\frac{H\cdot dl}{\Delta x\,\Delta y} \quad ...(4.68a)$$

If we choose the path in $y - z$ plane then

$$J_x = \left(\frac{\partial H_z}{\partial y} - \frac{\partial H_y}{\partial z}\right) = \lim_{\substack{\Delta y\to 0\\ \Delta z\to 0}}\oint\frac{H\cdot dl}{\Delta y\,\Delta z} \quad ...(4.68b)$$

and in $z\,x$ plane

$$J_y = \left(\frac{\partial H_x}{\partial z} - \frac{\partial H_x}{\partial x}\right) = \lim_{\substack{\Delta z\to 0\\ \Delta x\to 0}}\oint\frac{H\cdot dl}{\Delta z\,\Delta x} \quad ...(4.68c)$$

As
$$J = J_x a_x + J_y a_y + J_z a_z$$
$$= \left(\frac{\partial H_z}{\partial y} - \frac{\partial H_y}{\partial z}\right) a_x + \left(\frac{\partial H_x}{\partial z} - \frac{\partial H_z}{\partial x}\right) a_y + \left(\frac{\partial H_y}{\partial x} - \frac{\partial H_x}{\partial y}\right) a_z$$
$$= \begin{vmatrix} a_x & a_y & a_z \\ \frac{\partial}{\partial x} & \frac{\partial}{\partial y} & \frac{\partial}{\partial z} \\ H_x & H_y & H_z \end{vmatrix} = J \quad \text{...(4.69a)}$$
$$\nabla \times H = J \quad \text{...(4.69b)}$$
$$\text{curl of } H = \text{curl } H = J \quad \text{...(4.69d)}$$

Above eqn. called differential form of Ampere's circutal law.

In move general form
$$\nabla \times H = J + \frac{\partial D}{\partial t} \quad \text{...(4.70)}$$
where $\frac{\partial D}{\partial t}$ is displacement which is present only in time varying field, but in steady state, $\frac{\partial D}{\partial t} = 0$

Example 7. A parallel-plate capacitor, with plate area of 4 cm^2 and separation of 2.5 mm has a voltage 50 sin 10^3 *t V* applied to its plates. Find *Id*, assuming $\varepsilon = 1.5\varepsilon_0$.

Solution:
$$Id = Id.S = \frac{\varepsilon A}{d}\frac{dV}{dt} = 1.5\frac{\varepsilon_0 A}{d}\, d\,(50 \sin 10^3 t)$$
$$= 1.5 \times 8.854 \times 10^{-12} \times \frac{4 \times 10^{-4}}{2.5 \times 10^{-3}} \times 10^3 \times 50 \cos 10^3 .t$$
$$= 106.25 \cos 10^3 t \quad (nA)$$

EXERCISES

1. A thin ring of radius 5 cm is placed on plane $z = 1$ cm so that its centre is at (0, 0, 1 cm). If the ring carries 50 mA and along a_ϕ, find H at,
   (*a*) (0, 0, – 1 cm) (*b*) (0, 0, 10 cm)
   [**Hint :** (a) 400 a_z mA/m (b) 57.3 a_z mA/m]
2. Planes $z = 0$ and $z = 4$ carry current, $I = -10\, a_z$ A/m and I = 10 a_z A/m, respectively. Determine H at,
   (*a*) (1, 1, 1) (*b*) 0, – 3, 10
   [**Hint :** (a) 10 a_y A/m (b) 0 A/m]
3. Given that $H_1 = -2a_x + 6a_y + 4a_z$ A/m in region $y - x - 2 \le 0$, where $\mu_1 = 5\mu_0$ calculate,
   (*a*) M_1 and B_1
   (*b*) H_2 and B_2 in region $y - 2 - 2 > 0$, where $\mu_2 = 2\mu_0$.
   [**Hint :** (a) $M_1 = -8a_x + 24a_y + 16a_z$ A/m
   (b) $H_2 = -8a_x + 12a_y + 4a_y$ A/m, $B_2 = -20.11a_y + 30.16a_y + 60.05a_z$ μWb/m^2

• "Failure will never overtake me if my definition to succeed is strong enough"

APJ Abdul Kalam

Chapter 5

MAXWELL'S EQUATIONS AND UNIFORM PLANE WAVE

Maxwell's Equation

James Clerk Maxwell's describes the relationship between electric and magnetic fields, through the equation which can explain the radiation from the antenna. These equation are,

$$\nabla \times H = J + \frac{\partial D}{\partial t} \qquad ...(5.1)$$

$$\nabla \times E = -\frac{\partial B}{\partial t} \qquad ...(5.2)$$

$$\nabla . D = \rho_v \qquad ...(5.3)$$

$$\nabla . B = 0 \qquad ...(5.4)$$

where H = magnetic field intensity, A/m

J = conduction current density = σE, A / m^2

D = displacement current density A /m^2, $D = \varepsilon E$

E = electric field strength, V / m

B = magnetic field density, wb / m^2 = uH

ρ_v = volume charge density, C / m^3

above equations are called differential form of Maxwell' s equation. Now if

$$B = B_o e^{j\omega t} \qquad ...(5.5a)$$

$$E = E_o e^{j\omega t} \qquad ...(5.5b)$$

Then $$\frac{\partial B}{dt} = j\omega B_0 e^{j\omega t} = j\omega(B_0 e^{j\omega t}) = j\omega B \qquad ...(5.6a)$$

$$= j\omega B \Rightarrow \frac{2B}{2t} = j\omega B$$

Similarly for E. $$\frac{\partial E}{dt} = j\omega E \qquad ...(5.6b)$$

Nov in phaser form or frequency domain, the Maxwell's equation is

1st Maxwell's equation

$$\nabla \times H = \text{curl H} = J + \frac{\partial D}{\partial t}$$

$$= \sigma E + \varepsilon \frac{\partial E}{\partial t}$$

$$= \sigma E + j\omega\varepsilon E \qquad ...(5.7)$$

$$\nabla \times H = (\sigma + j\omega\varepsilon)\, E$$

2nd Maxwell's equation

and $$\text{curl } E = \nabla \times E = -\frac{\partial B}{\partial t}$$

$$= -j\omega B$$

or $$= -j\omega\mu\, H$$

$$\nabla \times E = -j\omega\mu\, H \qquad ...(5.8)$$

where μ is called permeability of mediums, if it is free space then it becomes

$$\mu_0 = 4\pi \times 10^{-7} \text{ H/m}$$

and in other medium $\mu = \mu_r\, \mu_o$

where μ_r called relative of the medium = $\varepsilon_r\, \varepsilon_o$, if medium is free space then

$$\varepsilon = \varepsilon_o, \varepsilon_r = 1$$

$$\varepsilon_r = \text{is relative permitivity.}$$

3rd Maxwell's equation

$$\nabla . D = \rho_v \qquad ...(5.9)$$

and

4th Maxwell's equation

$$\nabla . B = 0. \qquad ...(5.10)$$

Maxwell's 1st equation

Ampere's Law forms the basis for Maxwell's equation, and it states that, line integral of magnetic field intensity H around the closed loop contour is equal to current to the enclosed. The current referred is total current and equal to the sum of conduction and displacement currents. Mathematically

$$\oint H.dl = \int_S (J_c + J_d)\, ds$$

$$= \oint s\left(\sigma E + \frac{\partial D}{\partial t}\right) ds$$

$$\oint H.dl = \int_S \left(J + \frac{\partial D}{\partial t}\right) ds \qquad ...(5.11)$$

By applying Stroke's Theorem

$$\int_S (\nabla \times H) ds = \oint H\, dl$$

$$\nabla \times H = J + \frac{\partial D}{\partial t} \qquad ...(5.12)$$

$$\oint_S (\nabla \times H) ds = \int_S \left(J + \frac{\partial D}{\partial t}\right) ds \qquad ...(5.13)$$

Now again Stroke Theorem

$$\oint_S H\, ds = \int_S (\nabla \times H) ds$$

$$\oint_l H\, ds = \int_S \left(J + \frac{\partial D}{\partial t}\right) ds \qquad ...(5.14)$$

Equation (5.14) is integral form of 1st Maxwell's equation.

In other word the magnetic field intensity around any closed path is equal to the sum of conduction current and displacement current though any surface closed by path.

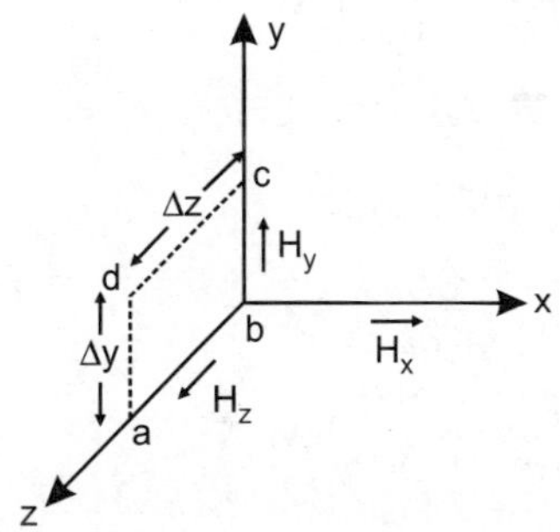

Fig. 5.1. Components of H-field for a closed loop.

Equation (5.14) can be proved from Ampere's Law, according to Ampere's law the line integral of H (magnetic) around any closed path is equal the current enclosed by the path, *i.e.*,

$$\oint H\, dl = 1$$

From Fig. 5.1:

$$\oint H\, dl = \int_{ab} H\, dl + \int_{bc} H\, dl + \int_{cd} H\, dl + \int_{da} H\, dl$$

$$-H_z \Delta z + H_y \Delta y + \left(H_z + \frac{\partial H_y}{\Delta y}\Delta y\right)\Delta z - \left(H_y + \frac{\partial H_y}{\partial z}\Delta z\right)\Delta y$$

$$\text{Since, } \left(H_{cd} = H_z + \frac{\partial H_z}{\partial y}\Delta y\right)$$

and

$$H_{da} = -H_y - \frac{\partial H_y}{\partial z}\Delta z = -\left(H_y + \frac{\Delta H_y}{\partial z}\right), \oint_{abcd} (H_z \Delta z + H_y \Delta y +$$

$$H_z \Delta z + \frac{\partial H_z}{\partial y}\Delta y \Delta z - Hy\, \Delta y - \frac{\partial H_y}{\Delta z}\Delta z \Delta y)$$

$$= \left(\frac{\partial H_z}{\partial y}\Delta y \Delta z - \frac{\partial H_y}{\partial z}\Delta z \partial y\right)$$

$$\oint_{abcd} Hdl = \left(\frac{\partial H_z}{\partial y} - \frac{\Delta H_y}{\Delta z}\right)\Delta y \Delta z \qquad ...(5.15)$$

Loop perpendicular to y and z x,

$$\oint_{abcd} Hdl = \left[\frac{\partial H_x}{\partial z} - \frac{\Delta H_z}{\partial x}\right] \Delta x \Delta y \qquad ...(5.16)$$

Similarly loop perpendicular to z, and xy

$$\oint_{abcd} Hdl = \left[\frac{\partial H_y}{\partial x} - \frac{\partial H_x}{\partial y}\right] \Delta x \Delta y \qquad ...(5.17)$$

There fore from equations (5.15), (5.16) and (5.17), we get:

$$\begin{aligned}\oint Hdl &= \left[\oint_{abcd} Hdl\right]_x a_x + \left[\oint_{abcd} Hdl\right]_y a_y + \left[\oint_{abcd} Hdl\right]_z a_z \\ &= \left[\frac{\partial H_z}{\partial y} - \frac{\partial H_y}{\partial z}\right] \Delta y\, \Delta z\, a_x + \left[\frac{H_x}{\partial z} - \frac{\partial H_y}{\partial x}\right] \\ &= \left[\frac{\partial H_y}{\partial x} - \frac{\partial H_x}{\partial y}\right] \Delta x\, \Delta y\, a_x \qquad ...(5.18)\end{aligned}$$

and also

$$\begin{aligned}\oint_{abcd} Hdl &= \Delta I = J\, \Delta y \Delta z \\ &(J_c + J_d)\, \Delta y \Delta z \\ &= \left(J_{cx} + \frac{\partial D_x}{\partial t}\right) \Delta y\, \Delta z \\ &= \left(\frac{\partial H_z}{\partial y} - \frac{\partial H_y}{\Delta z}\right) \Delta y\, \Delta z \qquad ...(5.19)\end{aligned}$$

Form (5.18) and (5.19):

$$\begin{aligned}\left(J_{ex} + \frac{\partial D_x}{\partial t}\right) &= \left(\frac{\partial H_z}{\Delta y} - \frac{\partial y_y}{\partial z}\right) \\ \left(J_{cy} + \frac{\partial D_y}{\partial t}\right) &= \left(\frac{\partial H_x}{\partial z} - \frac{\partial H_y}{\partial x}\right) \\ \left(J_{cy} + \frac{\partial D_z}{\partial t}\right) &= \left(\frac{\Delta H_y}{\partial x} - \frac{\partial H_x}{\partial y}\right)\end{aligned}$$

On simplification:

$$\begin{aligned}J + \frac{\partial D}{\partial t} &= \begin{vmatrix} a_x & a_y & a_z \\ \frac{\partial}{\partial x} & \frac{\partial}{\partial y} & \frac{\partial}{\partial z} \\ H_x & H_y & H_z \end{vmatrix} \\ &= \Delta \times \mathrm{H} \\ \Delta \times H &= J + \frac{\partial D}{\partial t} \qquad ...(5.20)\end{aligned}$$

Equation (5.20) is the Maxwell's 1st equation.

Maxwell's 2nd Equation

Maxwell's second equation is an expression of Faraday's Law and give quantitative relation between the emf induced in a closed circuit and the changing magnetic field that produces it. According to this law; the emf induced in a closed circuit is equal to the rate of change of magnetic flux linking the circuit Mathematically,

$$e = -\frac{d\phi}{dt}$$

where, e = induced electromotive force (emf)

ϕ = magnetic flux

t = time

Negative sign here indicates that direction of induced emf opposes the cause that produces it and if conductor has N turns;

$$I = -N.\frac{d\psi}{dt}$$

Since $\phi = BA$

or $d\psi = B.\ ds$

$$\phi = \int d\psi = \int B.ds$$

replacing it in above equation

$$e = \frac{\partial}{\partial t}\int B.ds$$

and also $e = \oint E\ dl$

so, $e = \oint E\ dl = \frac{\partial}{\partial t}\int B.\ ds$

or $$\oint Edl = -\frac{\partial}{\partial t}\oint_s Bds \qquad ...(5.21a)$$

it is integral form of Maxwell's equation.

Now , from Stroke theorem

$$\oint Edl = \int (\nabla \times E)\ ds$$

or $$(\nabla \times E) = -\frac{\partial}{\partial t}B$$

Maxwell's equation of differential form.

$$\nabla \times E = -\frac{\partial B}{\partial t}$$

$$\oint Edl = -\int \frac{\partial B}{\partial t}.ds \qquad ...(5.21b)$$

i.e., Electromotive force around a closed path is equal the time derivative of magnetic displacement through any surface bonded by the path or electric potential around a closed path is equal to the magnetic current through the path.

$$\oint Edl = -\frac{\partial \phi_y}{\partial t}$$

$$- \frac{\partial}{\partial t} B_y \Delta y \,.\, \Delta z \qquad \text{(As } \phi_y = B_y \Delta y.\Delta z\text{)}$$

$$\oint_{abcd} Edl = -\frac{\Delta B_y}{\Delta t} \Delta y \Delta z \qquad ...(5.22a)$$

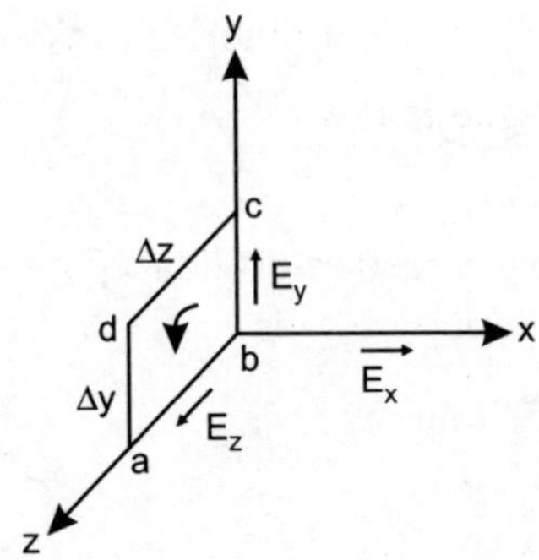

Fig. 5.2. Components of E for a closed loop

Similarly if loop is taken perpendicular to *y* and *z* axis then;

$$\left[\oint_{abcd} Edl\right]_y = -\frac{\partial B_z}{\partial t} \Delta x \,.\Delta z \qquad ...(5.22b)$$

and $$\left[\oint_{abcd} Edl\right]_z = -\frac{\partial B_x}{\partial t} \Delta x \,.\, \Delta y \qquad ...(5.22c)$$

so, $$\oint Edl = \int_{ab} E\,dl + \int_{bc} E\,dl + \int_{cd} Edl + \int_{da} Edl \qquad ...(5.23)$$

$$\int_{ab} Edl = -E_z \Delta z$$

$$\int_{bc} Edl = E_y \Delta y$$

$$\int_{cd} E = \left(+E_z + \frac{\partial E_z \Delta y}{\partial y}\right) \Delta z$$

$$\int_{da} Edl = -\left[E_y + \frac{\partial E_y}{\partial z} \Delta z\right] \Delta y$$

$$\int_{abcd} Edl = \left(\frac{\partial E_z}{\partial y} - \frac{\partial E}{\partial z}\right) \Delta y \Delta z \qquad ...(5.24a)$$

If loop is taken around the *y* ;

$$\int_{abcd} Edl = \left(\frac{\partial E_x}{\partial z} - \frac{\partial E_z}{\partial x}\right) \Delta z \Delta x \qquad ...(5.24b)$$

Similarly $$\int_{abcd} Edl = \left(\frac{\partial E_y}{\partial x} - \frac{\partial E_x}{\partial y}\right) \Delta x \Delta y \qquad ...(5.24c)$$

Now after vector addition of equations (5.22), (5.23) and (5.24), we get

$$\left(-\frac{\partial B_x}{\partial t} \Delta y \Delta z\right) a_x + \left(-\frac{\partial B_y}{\partial t} \Delta z \Delta x\right) a_y + \left(\frac{\partial B_z}{\partial t} \Delta x \Delta y\right) a_z$$

$$= \left[\left(\frac{\partial E_z}{\partial y} - \frac{\partial E_y}{\partial z}\right) \Delta y \Delta z\right] a_x - \left[\left(\frac{\partial E_x}{\partial z} - \frac{\partial E_z}{\partial x}\right) \Delta x \Delta y\right] a_y + \left[\left(\frac{\partial E_y}{\partial x} - \frac{\partial E_x}{\partial y}\right) \Delta x \Delta y\right] a_z$$

$$\Rightarrow\left(\frac{\partial B_x}{\partial t}ax+-\frac{\partial B_y}{\partial t}a_y\frac{\partial B_z a_z}{\partial t}\right)=\left(\frac{\partial E_z}{\partial y}-\frac{\partial E_z}{\partial z}\right)a_x-\left(\frac{\partial E_x}{\partial z}-\frac{\partial E_z}{\partial x}\right)ay+\left(\frac{\partial E_y}{\partial x}-\frac{\partial E_x}{\partial y}\right)a_z$$

On simplification:

$$\frac{\partial}{\partial t}(B_x a_x + B_y a_y + B_z a_z) = \begin{vmatrix} a_x & a_y & a_z \\ \frac{\partial}{\partial x} & \frac{\partial}{\partial y} & \frac{\partial}{\partial z} \\ E_x & E_y & E_z \end{vmatrix}$$

$$-\frac{\partial B}{\partial t} = \nabla \times \mathrm{E}$$

$$\nabla \times E = -\frac{\partial B}{\partial t} \text{ it is Maxwell's 2}^{\text{nd}} \text{ equation} \quad ...(5.25)$$

Third Equation in Differential Form

Maxwell's third equation is mathematical interpretation of Gauss law which states that "The total electric flux through any closed surface surrounding charges is equal to the total charge enclosed".

i.e., $$\int D.ds = \int \rho_v \, dv \quad ...(5.26a)$$

Now from divergence theorem

$$\int_S D.ds = \int_V (\nabla.D)\,dv \quad ...(5.26b)$$

From eqn. (5.26*a*) and (5.26*b*)

$$\int_V (\nabla.D)\,dv = \int \rho_v \, dv$$

$$\nabla.D = \rho_v$$

Maxwell's fourth equation

Maxwell's fourth equation signifies that there cannot be isolated magnetic poles on which magnetic lines can be terminated and that magnetic lines are continuous.

Mathematically,

$$\nabla.B = 0$$

If we apply Divergence Theorem

$$\int_S B ds = \int_V (\nabla.B)\,dv$$

So, $$\int_S B\,ds = 0$$

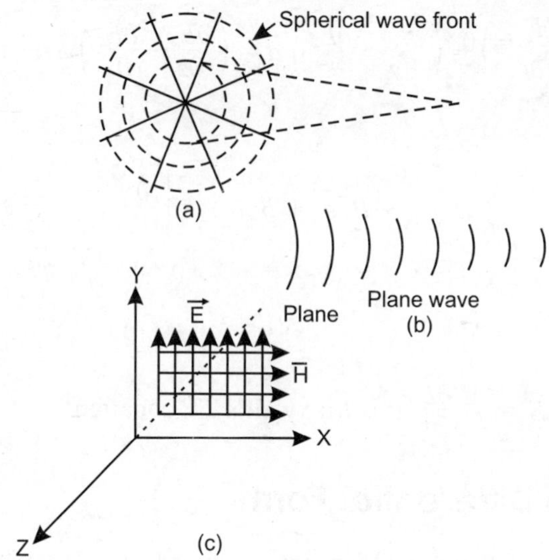

Fig. 5.3 Different wavefronts

Plane Wave and Uniform Plane wave in Free or Empty Space : An electromagnetic wave originates from a point in free space, spread out uniformly in all the directions and it forms a spherical wave front. An observation at a great distance from source is able to observe only a small portion of wave in his immediate vicinity and it appears to him as plane wave just as ocean appears flat to a person who can see only a few miles around him. Free space specified by $\sigma = 0$ and $\rho = 0$.

Plane wave has the following properties :

(*i*) At every point in space the electric field vector and magnetic field vector are perpendicular to each other and also to the direction of propagation, (Fig. 5.4).

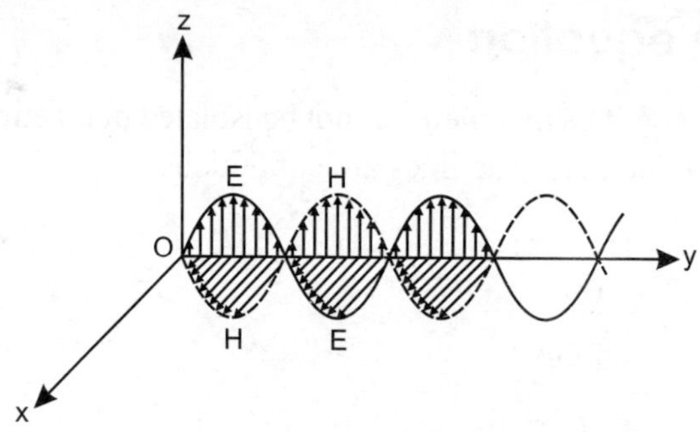

Fig. 5.4. Vector representation of an EM wave

(*ii*) A uniform plane wave is one in which E and H lies in a plane and have same values for every wave in that plane at any fixed instant. If wave propagates in z direction then,

$$\frac{\partial E}{\partial x} = 0$$

$$\frac{\partial H}{\partial y} = 0$$

$$\frac{\partial E}{\partial y} = 0$$

$$\frac{\partial H}{\partial x} = 0$$

(*iii*) Velocity of wave in free space

$$\upsilon = c = \frac{1}{\sqrt{\mu_0 \varepsilon_0}}$$

$$= \frac{1}{\sqrt{4\pi \times 10^{-7} \times \frac{1}{36\pi} \times 10^{9}}} = 3 \times 10^{10} \text{ .cm / sec}$$

(*iv*) E and H oscillates in phase and ratio of their amplitudes are constant,

$$\frac{E_x}{H_y} = -\frac{E_y}{H_x} = \sqrt{\frac{\mu}{\varepsilon}} = \eta = 120\,\pi\,\Omega = 376\,\Omega$$

called free space impedance.

(*v*) Whatever may be the frequency, the electromagnetic wave travesl in space with velocity of light. (3×10^8 m/s).

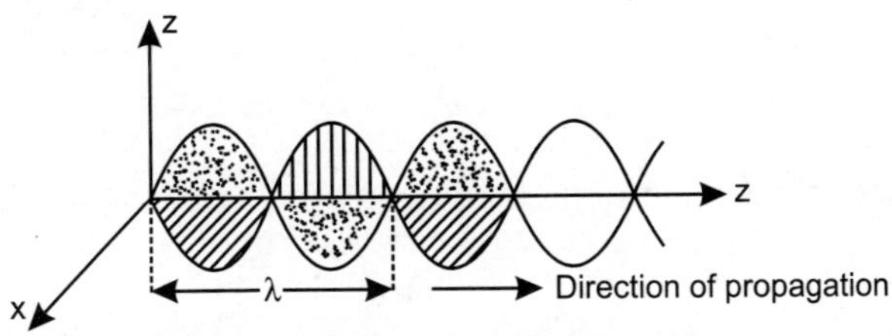

Fig. 5.5. Eletromagnetic wave

(*vi*) A uniform plane wave is also called as *TEM* (transverse electromagnetic), since both fields: Electric and magnetic fields are transverse (perpendicular).

(*vii*) Uniform plane wave have constant magnetitude and phase in the medium.

Electromagnetic Wave Equations (For a conducting Medium)

From Maxwell's four equations,

$$\nabla \times H = J + \frac{\partial D}{\partial t} = \sigma E + \frac{\varepsilon \partial E}{\partial t} \qquad \text{...(5.27a)}$$

$$\nabla \times E = -\frac{\partial B}{\partial t} = -\mu \frac{\partial H}{\partial t} \qquad \text{...(5.27b)}$$

$$\nabla . B = 0 \qquad \text{...(5.27c)}$$

$$\nabla . D = \rho_v \qquad \text{...(5.27d)}$$

Taking curl of equation (5.27*b*);

$$\text{Curl of } \nabla \times E = \text{curl of} \left(-\mu \frac{\partial H}{\partial t}\right)$$

$$\underset{\nabla\times\nabla\times E}{A\ \ B\ \ C} = \nabla \times -\mu \frac{\partial H}{\partial t}$$

$$= -\mu \frac{\partial}{\partial t} (\nabla \times H)$$

Since, $A \times B \times C = (A . B) B - (A . B) C$

So, $(\nabla . E) \nabla - (\nabla . \nabla) E = -\mu \frac{\partial}{\partial t} (\nabla \times H)$

$$(\nabla . E) \nabla - \nabla^2 E = -\mu \frac{\partial}{\partial t} (\nabla . H)$$

$$= \left(\frac{\rho_v}{\varepsilon}\right) \nabla - \nabla^2 \text{E} = \mu \frac{\partial}{\partial t} \left[\sigma E + \frac{\varepsilon \partial E}{\partial t}\right]$$

$$\nabla . D = \rho_v$$

$$\nabla . \varepsilon E = \rho_v$$

$$\nabla . E = -\frac{\rho_v}{\varepsilon} \qquad ...(5.28)$$

Since there is no net charge within conductor always, but it may be on the surface so charge density $\rho_v = 0$, so

$$\nabla^2 E = \mu\sigma \frac{\partial E}{\partial t} + \mu\varepsilon \frac{\partial^2 E}{\partial t^2} \qquad ...(5.29)$$

Equation (5.29) called wave equation in conducting medium. In phasor form let $E = E_o e^{j\omega t}$

so, $\frac{\partial E}{\partial t} = j\omega E$ and $\frac{\partial E}{\partial t^2} = -\omega^2 E$

$$\nabla^2 E = \mu\sigma j\omega E - \mu\varepsilon\omega^2 E$$

Where for different medium σ, μ , ε are different, where as for :

(*i*) Homogeneous medium ε, μ, σ are constant throughout out the medium (free space).
(*ii*) Non-humogeneous medium. σ, μ, σ are not constant.
(*iii*) Isotropic medium ε is scalar constant (free space).
(*iv*) Aniotropic ε is not a scalar constant.

The different medium can be specified as, follows;

$\frac{\sigma}{\omega\varepsilon} > 1$ conducting medium

$\frac{\sigma}{\omega\varepsilon} = 1$ quasi conducting medium

$\frac{\sigma}{\omega\varepsilon} < 1$ insulating medium

Mutually perpendicular relationship between *E* and *H* in uniform plane wave.

Now from Maxwell's second equation.

$\nabla \times H \left(J + \frac{\partial D}{\partial t}\right)$, If the medium is free space, then $\sigma = 0$ and $J = \sigma E = 0$

then, $$\nabla \times H = \frac{\partial D}{\partial t} \qquad ...(5.30)$$

in matrix form, $$\begin{vmatrix} a_x & a_y & a_z \\ \frac{\partial}{\partial x} & \frac{\partial}{\partial y} & \frac{\partial}{\partial z} \\ H_x & H_y & H_z \end{vmatrix} = +\left(\frac{\partial D_x}{\partial t}\right) a_x + \left(\frac{\partial D_y}{\partial t}\right) a_y + \left(\frac{\partial D_z}{\partial t}\right) a_z$$

Since, $\frac{\partial}{\partial x} = 0 \; \frac{\partial}{\partial y} = 0$ ...(5.31)

E_z and $H_z = 0$

then, $\frac{\partial}{\partial z}(a_x H_y - a_y H_x) = \frac{\partial D_x}{\partial t} a_x + \left(\frac{\partial D_y}{\partial t}\right) a_y \; \frac{\partial H_y}{\partial z} = \frac{\partial D_x}{\partial t} = \frac{\partial \varepsilon E_x}{\partial t} = \varepsilon \frac{\partial E_x}{\partial t}$

$$\frac{\partial H_y}{\partial z} = \varepsilon \frac{\partial E_x}{\partial t} \quad ...(5.32a)$$

$$\frac{\partial H_x}{\partial z} = \varepsilon \frac{\partial E_y}{\partial t} \quad ...(5.32b)$$

Let $E_y = f_1(z - ct) + f_2(z + ct)$

where $f_1(z - ct) =$ incident wave

$f_2(z - ct) =$ reflected wave

$$E_y = f_1(z - ct) = F \text{ (say)}$$

$$\frac{\partial E_y}{\partial t} = f_1(z - ct)(-c)$$

$$\frac{\partial E_y}{\partial t} = -cF'$$

$$\frac{\partial H_x}{\partial z} = \varepsilon . -cF'$$

$$\frac{\partial H_x}{\partial z} = c\,\varepsilon F'$$

$$\frac{\partial H_z}{\partial z} = c\,\varepsilon F'$$

Integrating both side

$$\int \frac{\partial H_z}{\partial z} \partial z = \int c\,\varepsilon F'\,dx$$

$$H_x = c\,\varepsilon F + K$$

Wave K called integration constant.

Since static field has no meaning in time veriying field, hence $K = 0$

$$H_x = c\,\varepsilon\,F$$

$$H_x = c\,\varepsilon\,Fy$$

$$\frac{E_y}{H_x} = \frac{1}{c\varepsilon} = \frac{\sqrt{\mu\varepsilon}}{\varepsilon} \qquad c = \frac{1}{\sqrt{\mu\varepsilon}}$$

$$\frac{E_y}{H_x} = \frac{\sqrt{\mu}}{\varepsilon} \quad ...(5.33a)$$

Similarly if $E_x = f(z - cf) = F$

$$\frac{\partial E_x}{\partial t} = -cF'$$

$$\frac{\partial H_y}{\partial z} = -\varepsilon c F$$

$$H_y = -\varepsilon c F + K$$

constant, taking $K = 0$ for time varying field.

$$H_y = -\varepsilon c F$$

$$H_y = -\varepsilon\, C \times E_x$$

$$\frac{E_x}{H_y} = -\varepsilon C = -\frac{\sqrt{\mu}}{\varepsilon} = \frac{E_x}{H_y} = -\frac{\sqrt{\mu}}{\varepsilon} \quad \text{...(5.33}b\text{)}$$

$$E = |E| = \sqrt{E_x^2 + E_y^2 + E_z^2}$$

$$= \sqrt{E_x^2 + E_y^2 + 0}$$

$$E = \sqrt{E_x^2 + E_y^2}$$

Similarly,

$$H = \sqrt{H_x^2 + H_y^2}$$

$$E = \sqrt{\frac{\mu}{\varepsilon} H_x^2 + \frac{\mu}{\varepsilon} H_y^2}$$

$$= \sqrt{\frac{\mu}{\varepsilon}}\,.\,H$$

$$\frac{E}{H} = \sqrt{\frac{\mu}{\varepsilon}} = \acute{\eta} = \frac{\text{total electric field}}{\text{total magnetic field}} \quad \text{...(5.34)}$$

$$\frac{V/m}{A/m} = \sqrt{\frac{\mu}{\varepsilon}}$$

$$\frac{V}{A} = \frac{V}{I} = \acute{\eta} = \sqrt{\frac{\mu}{\varepsilon}} = \Omega$$

it is called intrinsic impedance of the medium (non - conducting medium)

Case. 1. If both E and H are perpendicular

$$E\,.\,H = |E|\,|H|\,cos\,\theta$$

$$(a_x E_x + a_y E_y)\,.\,(H_y a_x + H_y a_y) = \sqrt{H_x^2 + H_y^2}\,\sqrt{E_x^2 + E_y^2} \times \cos\theta$$

$$E_x H_x + E_y H_y = \sqrt{H_x^2 + H_y^2} \times \sqrt{E_x^2 + E_y^2}$$

$$\cos\theta = 0$$

$$\theta = 90^\circ$$

Vector product gives the direction of propagation

$$E \times H = \begin{vmatrix} a_x & a_y & a_z \\ E_x & E_y & 0 \\ H_x & H_y & 0 \end{vmatrix}$$

$$= a_x(0) + a_y(0) + a_z(E_x H_y - H_x E_y)$$

$$= a_z(-E_x \eta H_x H_y - \eta H_x H_y)$$

$$= -\eta a_z H^2 = (-\eta H^2)\, a_z$$

in which a_z show the direciton.

(*a*) Propagation of electromagnetic wave in conducting medium or lossy medium : Wave eqn. for conducting medium are

$$\nabla^2 E = \mu\sigma \cdot \frac{\partial E}{\partial t} + \mu\varepsilon \cdot \frac{\partial^2 E}{\partial t} \quad ...(5.35a)$$

$$\nabla^2 H = \mu\sigma \cdot \frac{\partial H}{\partial t} + \mu\varepsilon \cdot \frac{\partial^2 H}{\partial t^2} \quad ...(5.35b)$$

Since E and H are varying sinusoidaly, so

$$E = E_0 e^{j\omega t}$$

$$H = H_0 e^{j\omega t}$$

So
$$\frac{\partial E}{\partial t} = j\omega E$$

$$\frac{\partial^2 E}{\partial t^2} = j^2 \omega^2 E$$

$$E(z, t) = E(z) = [\cos\theta + j\sin\theta]$$

$$= E(z) = e^{j\theta}\ \frac{d\theta}{dt} = \omega$$

$$E = [\,E(z)\, e^{j\omega t}$$

$$\frac{Q}{t} = \omega\, E_0\,(z) e^{j\omega t}$$

Similarly for magnetic field

$$H = H_0\, e^{j\omega t}$$

$$\frac{\partial^2 H}{\partial t} = j\omega H$$

$$\frac{\partial^2 H}{\partial t^2} = +j^2 \omega^2 H$$

Replacing these values in above equation, gives:

$$\nabla^2 E = \mu\sigma j\omega + \mu\varepsilon j^2 \omega^2 E$$

$$= \mu j\omega\,(\sigma + j\omega\varepsilon)\, E$$

$$= j\omega\mu\,(\sigma + j\omega\varepsilon)\, E$$

$$\nabla^2 E = \gamma^2 E \quad ...(5.36a)$$

Similarly
$$\nabla^2 H = \gamma^2 H \quad ...(5.36b)$$

Equations (5.36) called wave eqn in phase form.

where,
$$\gamma = \text{propagation constant}$$

$$\gamma = \alpha + j\beta$$

$$\alpha = \text{attenulatin constant } (dB/m)$$

$$\beta = \text{Phase constant (rad / m)}$$

Propagation Constant (attenuation and phase constant)

Since,
$$\gamma^2 = j\omega\mu .(\sigma + j\omega\varepsilon)$$
$$\alpha + j\beta = \gamma = \sqrt{j\omega H(\sigma + j\omega\varepsilon)}$$
$$= \sqrt{j\omega\mu\left(1 + \frac{j\omega\varepsilon}{\sigma}\right)}$$

Since medium is conductive
$$\frac{\omega\varepsilon}{\sigma} << 1$$
$$\frac{\sigma}{\omega\varepsilon} >> 1$$
$$= j\omega\sqrt{\mu\varepsilon}\sqrt{1 + \frac{\sigma}{j\omega\varepsilon}}$$
$$(\alpha + \mathrm{j}\beta) = j\omega\sqrt{\mu\varepsilon}\left(1 - j\frac{\sigma}{j\omega}\right)^{1/2}$$
$$(\alpha + j\beta)^2 = j^2\omega^2\mu\varepsilon\left(1 + \frac{\sigma}{j\omega\varepsilon}\right)$$
$$\alpha^2 - \beta^2 + 2j\alpha\beta - \omega^2\mu\varepsilon\omega + j\omega\mu\sigma$$
$$\alpha^2 - \beta^2 = -\omega^2\mu\varepsilon$$
$$2\alpha\beta = \omega\mu\sigma$$
$$\alpha\beta = \frac{\omega\mu\sigma}{2}$$
$$\alpha = \frac{\omega\mu\sigma}{2\beta} \qquad ...(5.37a)$$
or
$$\beta = \frac{\omega\mu\sigma}{2\sigma} \qquad ...(5.37b)$$
$$\alpha^2 - \beta^2 = -\omega^2\mu\varepsilon$$
$$\alpha^2 - \left(\frac{\mu\omega\sigma}{2\alpha}\right)^2 = -\omega^2\mu\varepsilon$$
$$\alpha^2 - \frac{\omega^2\mu^2\sigma^2}{4\alpha^2} = -\omega^2\mu\varepsilon$$
$$4\alpha^2 - \omega^2\mu^2\sigma^2 = (\omega^2\mu\varepsilon)\,4\alpha^2$$
$$4\alpha^4 + (4\omega^2\mu\varepsilon).\,\alpha^2 = \omega^2\mu^2\sigma^2 = 0$$
$$\alpha^4 + \omega^2\mu\varepsilon\,.\alpha^2 = \frac{\omega^2\mu^2\sigma^2}{4}$$
$$\alpha 4 + \frac{2\omega^2\mu\varepsilon}{2}.\alpha^2 + \left(\frac{\omega^2\mu\varepsilon}{2}\right)^2 - \left(\frac{\omega^2\mu\varepsilon}{2}\right)^2 = \frac{\omega^2\mu^2\sigma^2}{4}$$
$$\left[\alpha + \frac{\omega^2\mu\varepsilon}{2}\right]^2 = \left(\frac{\omega^2\mu\varepsilon}{2}\right)^2 + \left(\frac{\omega\mu\alpha}{2}\right)^2$$

$$\Rightarrow \quad \left(\frac{\omega^2\mu\varepsilon}{2}\right)^2\left[1+\left(\frac{\omega\mu\varepsilon}{2}\right)^2\left(\frac{2}{\omega^2\mu\varepsilon}\right)^2\right]$$

$$= \left(\frac{\omega^2\mu\varepsilon}{2}\right)^2\left[1+\frac{\sigma^2}{\omega^2\varepsilon^2}\right]$$

Taking root of both side

$$\left[\alpha^2+\frac{\omega^2\mu\varepsilon}{2}\right] = \frac{\omega^2\mu\varepsilon}{2}\sqrt{\left(1+\frac{\sigma^2}{\omega^2\varepsilon^2}\right)}$$

$$\alpha^2 = \frac{\omega^2\mu\varepsilon}{2}\sqrt{\left(1+\frac{\sigma^2}{\omega^2\varepsilon^2}\right)}-\omega^2\mu\varepsilon$$

$$= \frac{\omega^2\mu\varepsilon}{2}\left[\sqrt{1=\frac{\sigma^2}{\omega^2\varepsilon^2}}-1\right]$$

$$\alpha = \omega\sqrt{\frac{\mu\varepsilon}{2}\sqrt{\left[1+\frac{\sigma^2}{\omega^2\varepsilon^2}\right]}-1} \quad \text{...(5.38a)}$$

Similarly for β

$$\beta = \omega\sqrt{\frac{\mu\varepsilon}{2}}\sqrt{\left(1+\frac{\sigma^2}{\omega^2\varepsilon^2}\right)}+1 \quad \text{...(5.38b)}$$

Phase Difine Velocity and Group Velocity

$$V_p = f \qquad \left[\beta = \frac{2\pi}{\lambda}\right]$$

$$V_p = \frac{f.2\pi}{\beta} = \frac{2\pi f}{\beta} = \frac{\omega}{\beta}$$

$$V_p = \frac{\omega}{\beta} = \frac{\omega}{\omega\sqrt{\frac{\mu\varepsilon}{2}\left[\sqrt{1+\frac{\sigma^2}{\omega^2\varepsilon^2}}\right]}}+1$$

$$V_p = \sqrt{\frac{\mu\varepsilon}{2}\left[\sqrt{1+\frac{\sigma^2}{\omega^2\varepsilon^2}}\right]}+1$$

Group velocity in which the velocity which propagates in the medium.

Mathematically, $\quad V_g = -\frac{\partial\omega}{\partial\beta}$

$$V_p V_g = \text{constant} = c^2 \qquad \text{(Velocity of light)}$$

If uniform wave propagate in z direction, then

$$\frac{\partial^2 E_x}{\partial z^2} = \gamma^2 E_x = \frac{\partial^2 E_y}{\partial z^2} = \gamma^2 E_y$$

$$E_y(z) = E_0\, e^{-rz}$$

$$E(z, t) = E_0\, e^{-rz}\, e^{j\omega t}$$

$$\bar{E}(z, t) = e^{-\alpha z}.\text{Real}\left[E_0 \cos(\omega t-\beta z)+\sin(\omega t-\beta z)\right]$$

Intrinsic impedance

$$\eta = \sqrt{\frac{Z}{\gamma}} = \sqrt{\frac{2\omega\mu}{(\sigma + j\omega\varepsilon)}}$$

$$= \sqrt{\frac{j\omega\varepsilon}{j\omega\varepsilon\left(1 + \frac{\sigma}{j\omega\varepsilon}\right)}}$$

$$= \sqrt{\frac{\mu}{\varepsilon}}\left(1 + \frac{\sigma}{j\omega\varepsilon}\right)^{-1/2}$$

$$\eta = \sqrt{\frac{\mu}{\varepsilon}}.\left(1 - \frac{\sigma}{2j\omega\varepsilon}\right) \qquad ...(5.39)$$

(*b*) Propagation of Plane electromagnetic wave in good conducting medium.

For good conductor $\left[\frac{\sigma}{j\omega\varepsilon} >>> 1\right]$

We know that wave equation on :

$$\nabla^2 E = j\omega\mu\,(\sigma + j\omega\varepsilon)\,E = \gamma^2 E$$

$$\nabla^2 H = j\omega\mu\,(\sigma + j\omega\varepsilon)H = \gamma^2 H$$

Now propagation constant for good conducting medium

$$\gamma = \sqrt{Z \cdot Y} = \sqrt{j\omega\varepsilon(\sigma + j\omega\varepsilon)}$$

$$= \sqrt{j\omega\varepsilon\; j\omega\varepsilon\left(1 + \frac{\sigma}{j\omega\varepsilon}\right)}$$

$$= \omega\sqrt{\mu\varepsilon}\left(\frac{\sigma}{j\omega\varepsilon}\right)^{1/2}$$

$$= \omega j\sqrt{\mu\varepsilon}\left(\frac{\sigma}{j\omega\varepsilon}\right)^{1/2}$$

$$= \sqrt{\frac{j^2\omega^2\mu\varepsilon\sigma}{j\omega\varepsilon}}$$

$$\gamma = \sqrt{\frac{j\omega\mu\varepsilon\sigma}{\varepsilon}} = \sqrt{j\omega\mu\sigma}$$

$$\gamma = \sqrt{j\omega\mu\sigma} \qquad ...(5.40)$$

Now Since $\sqrt{j} = \left(\frac{1+j}{\sqrt{2}}\right)$ $\qquad \because j^2 = \frac{1 - 1 + \frac{2j}{2}}{\sqrt{j^2}}$

$$\gamma = \sqrt{j\omega\mu\sigma} = \sqrt{\omega\mu\sigma}\,\sqrt{j}$$

$$= \sqrt{\omega\mu\sigma}\,.\left(\frac{1+j}{\sqrt{2}}\right)$$

$$y = \sqrt{\omega\mu\sigma}\left(\frac{1}{\sqrt{2}} + \frac{j}{\sqrt{2}}\right)$$

$$\alpha + j\beta = \sqrt{\omega\mu\sigma}\left(\frac{1}{\sqrt{2}} + \frac{i}{\sqrt{2}}\right)$$

$$\alpha = \frac{\sqrt{\omega\mu\sigma}}{2} = \frac{\sqrt{2\pi f\mu\sigma}}{2} = \sqrt{\pi f\mu\sigma}$$

$$\beta = \sqrt{\pi f\mu\sigma}$$

$$\alpha = \frac{\sqrt{\omega\mu\sigma}}{2}$$

so $$\alpha = \beta = \frac{\sqrt{\omega\mu\sigma}}{2} = \sqrt{\mu f\mu\sigma} \quad ...(5.41)$$

Intrinsic impedance

$$\eta = \sqrt{\left(\frac{j\omega\mu}{\sigma + j\omega\varepsilon}\right)^2}$$

$$= \sqrt{\frac{j\omega\mu}{\sigma\left(1 + \frac{j\omega\varepsilon}{\sigma}\right)}}$$

$$= \sqrt{\frac{j\omega\mu}{\sigma}} \cdot \left(1 + \frac{j\omega\varepsilon}{\sigma}\right)^{-1/2}$$

$$= \sqrt{\frac{j\omega\mu}{\sigma}} \cdot \left(1 + \frac{j\omega\varepsilon}{\sigma} + ...\right)$$

$$= \sqrt{\frac{j\omega\mu}{\sigma}} \cdot \left(1 - \frac{j\omega\varepsilon}{2\sigma}\right)$$

$$= \sqrt{\frac{j\omega\mu}{\sigma}} \cdot \left(1 - \frac{j\omega\varepsilon}{2\sigma}\right)$$

$$= \sqrt{\frac{\omega\mu}{\sigma}} \sqrt{i} \left(1 - \frac{j\omega\varepsilon}{2\sigma}\right)$$

$$= \sqrt{\frac{\omega\mu}{\sigma}} \left(\frac{1+i}{\sqrt{2}}\right)\left(1 - \frac{j\omega\varepsilon}{2\sigma}\right)$$

$$= \sqrt{\frac{\omega\mu}{\sigma}} \cdot \left[1 - \frac{j\omega\mu}{2\sigma} + j\frac{\omega\varepsilon}{2\sigma}\right]$$

$$\eta = \sqrt{\frac{\omega\mu}{2\sigma}}\left(1 - \frac{\omega\varepsilon}{2\sigma}\right) + j\left(1 - \frac{\omega\varepsilon}{2\sigma}\right)$$

$$\eta = R_0 + jX_0$$

$$R_o = \sqrt{\frac{\omega\mu}{2\sigma}}\left(1 + \frac{\omega\varepsilon}{2\sigma}\right) \approx \sqrt{\frac{\omega\varepsilon}{2\sigma}} \cdot \sqrt{\frac{\omega\mu}{2\sigma}}$$

$$R_o = \sqrt{\frac{\omega\mu}{2\sigma}}$$

$$\alpha_o = \sqrt{\frac{\omega\mu}{2\sigma}} \quad ...(5.42)$$

That is or a good conductor both $R_0 = \alpha_0$ are equal.

The real and imaginary part are equal means *E* and '*H*' are 45° in phase difference in their time phases which characteristic feature of the electromagnetic wave in metal.

Phase velocity $$\upsilon_p = \frac{\omega}{\beta} = \frac{\omega}{\sqrt{\frac{\omega\mu\sigma}{2}}}$$

$$\upsilon_p = \sqrt{\frac{2\omega}{\mu\sigma}} \qquad ...(5.43)$$

Group velocity $$V_p \,.\, V_g = c = \frac{1}{\sqrt{\mu_0\,\varepsilon_0}}$$

$$V_g = \frac{c}{V_p} = \frac{c}{\sqrt{\frac{2\omega}{\mu\sigma}}}$$

$$V_g = c\sqrt{\frac{\mu\sigma}{2\omega}} \qquad ...(5.44)$$

(*c*) **Propagation of Plane Electromagnetic Waves in Goo Dielectrics (dielectric having loss):** For good dielector

$$\frac{\sigma}{\omega\varepsilon} << 1$$

Now from Maxwell's 1st or 2nd equation

$$\nabla^2 E = j\omega\mu \,.\, (\sigma + j\omega\varepsilon)\, E$$

$$\nabla^2 H = j\omega\mu\, (\sigma + j\omega\varepsilon)\, H$$

Propagation Constant

$$\gamma = \sqrt{Z \cdot Y} = \sqrt{j\omega\mu(\sigma + j\omega\varepsilon)}$$

$$= \sqrt{j\omega\mu \cdot j\omega\varepsilon\left(1 - \frac{j\sigma}{\omega\sigma}\right)}$$

$$\gamma = \left[j\omega\sqrt{\mu\varepsilon}\sqrt{1 - \frac{j\sigma}{\omega\varepsilon}}\right] \qquad ...(5.45)$$

Now α and β

$$\alpha + j\beta = j\omega\,\sqrt{\mu\varepsilon}\,\left(1 - \frac{j\sigma}{\omega\varepsilon}\right)$$

$$= j\omega\,\sqrt{\mu\varepsilon}\left(1 - \frac{j\sigma}{2\omega}\right)^{1/2}$$

$$= j\omega\,\sqrt{\mu\varepsilon}\left(1 - \frac{j\sigma}{2\omega\varepsilon}\right)\left[(1+x)^n = 1 + nx + \frac{1}{2}nx\right]$$

$$= j\omega\,\sqrt{\mu\varepsilon} + \frac{\omega\sqrt{\mu\varepsilon}}{2\omega\varepsilon}\,.\sigma$$

$$\alpha + j\beta = \left(j\omega\sqrt{\mu\varepsilon} + \frac{\sigma}{2}\frac{\sqrt{\mu}}{\varepsilon}\right)$$

Have, $$\alpha = \frac{\alpha}{2}\sqrt{\frac{\mu}{\varepsilon}} \quad \text{and} \quad \beta = \omega\sqrt{\mu\varepsilon} \qquad ...(5.46)$$

Phase Velocity and Group Velocity

$$V_p = \frac{\omega}{\beta} = \frac{\omega}{\omega\sqrt{\mu\varepsilon}} = \frac{1}{\sqrt{\mu\varepsilon}}$$

$$V_p = \frac{1}{\sqrt{\mu\varepsilon}} \quad \text{...(5.47a)}$$

Now

$$V_g = \frac{c}{V_P} = c\sqrt{\mu\varepsilon}$$

$$V_g = c\sqrt{\mu\varepsilon} \quad \text{...(5.47b)}$$

Intrinsic Impedance

$$\eta = \sqrt{\frac{j\omega\mu}{\sigma + j\omega\varepsilon}} = \sqrt{\frac{j\omega\mu}{j\omega\varepsilon\left(1 - \frac{j\sigma}{\omega\varepsilon}\right)}}$$

$$\eta = \sqrt{\frac{\mu}{\varepsilon}}\left(1 + \frac{j\sigma}{2\omega\sigma}\right) = \left[\sqrt{\frac{\mu}{\varepsilon}} + j\frac{\sigma}{2\omega\varepsilon}\sqrt{\frac{\mu}{\varepsilon}}\right]$$

$$R_0 + jX_0 = \left(\sqrt{\frac{\mu}{\varepsilon}} + j\frac{\sigma}{2\omega\varepsilon}\sqrt{\frac{\mu}{\varepsilon}}\right)$$

$$R_0 = \sqrt{\frac{\mu}{\varepsilon}} \quad \text{...(5.47c)}$$

and

$$X_0 = \frac{\sigma}{2\omega\varepsilon}\sqrt{\frac{\mu}{\varepsilon}} \quad \text{...(5.47d)}$$

Depth of Penetration (Skin Effect)

It is defined as the depth in which the wave has been attenuated by an amount of $\left(\frac{1}{e}\right)$ times of its initial value and denoted by δ.

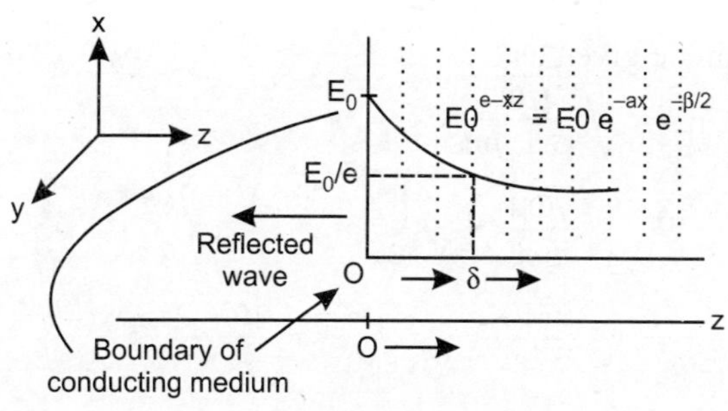

Fig. 5.6

Hence, at depth δ

$$E_1 = E_0\, e^{-\delta\alpha}$$

$$\frac{E_0}{e} = E_0\, e^{-\delta\alpha}$$

$$E_0\, e^{-1} = E_0\, e^{-\delta\alpha}$$

$$\alpha\delta = 1$$

$$\delta = \frac{1}{\alpha} = \frac{1}{\beta} \qquad \text{for conducting medium}$$

$$\delta = \frac{1}{\sqrt{\pi f \mu\sigma}} \qquad ...(5.48)$$

$$\delta \propto \frac{1}{\sqrt{f}}$$

i.e., as frequency increases skin depth decreases.

(D) Electromagnetic Wave Preparations in Lossless Dielectric

The lossless dielectric which is often called the good dielectric a perfect dielectric, is characteristerized by $\sigma = 0$. Hence wave equation can be given as

$$\Delta^2 E = \gamma^2 E$$

$$\Delta^2 H = \gamma^2 H$$

where,

$$\gamma^2 = j\omega\mu\,(\sigma + j\omega\varepsilon)$$

$$\alpha + j\beta = \sqrt{j\omega\mu\,(0 + j\omega\varepsilon)}$$

$$= \sqrt{j\omega\mu\,(j\omega\varepsilon)}$$

$$\sigma = 0 \qquad ...(5.49a)$$

$$\beta = \sqrt{-\omega^2\,\mu\varepsilon}$$

$$= \omega\sqrt{\mu\varepsilon}\,\sqrt{-1}$$

$$j\beta = j\omega\sqrt{\mu\varepsilon}$$

$$\beta = \omega\sqrt{\mu\varepsilon} \qquad ...(5.49b)$$

Intrinsic impedance can be given as

$$\eta = \sqrt{\frac{j\omega\mu}{\sigma + j\omega\varepsilon}}$$

$$= \sqrt{\frac{j\omega\mu}{j\omega\varepsilon}} = \sqrt{\frac{\mu}{\varepsilon}}$$

$$\eta = \sqrt{\frac{\mu_0\mu_r}{\varepsilon_0\,\varepsilon_e}} = \sqrt{\frac{\mu_0}{\varepsilon_0\mu_r}} = \frac{\mu_0}{\sqrt{\varepsilon_r}}$$

$$\eta = \frac{377\Omega}{\sqrt{\varepsilon_r}} \qquad ...(5.49c)$$

Phase velocity

$$V_P = \frac{\omega}{\beta} = \frac{\omega}{\omega\sqrt{\mu\omega}} = \frac{1}{\sqrt{\mu\varepsilon}}$$

$$= \frac{1}{\sqrt{\mu_0\mu_r\varepsilon_0\varepsilon_r}} = \frac{1}{\sqrt{\mu_0\varepsilon_0\varepsilon_r}}$$

$$= \frac{V_c}{\sqrt{\varepsilon_r}} = \frac{3\times10^8}{\sqrt{\varepsilon_r}}$$

$$V_p = \frac{3\times10^8}{\sqrt{\varepsilon_r}} \qquad \text{...(5.49}d\text{)}$$

Boundary Conditions

Since Maxwell's equations are in the form of differential rather than algebraic equations, boundary conditions must be applied to a given problem if a specific solution is required.

There are four basic rules for boundary conditions at the between two different materials. These conditions can be given as,

1. The lengential components of electric fields intensity are continuous across the boundary, *i.e.*, from Faraday's law

i.e.,
$$\int_e E.dl = 0$$

$$E_{t1}\,\Delta l = E_{t2}\,\Delta l = 0 \qquad \text{(see Fig. 5.7)}$$

$$\Delta l\,(E_{t1} - E_{t2}) = 0$$

$$\Delta l\,(E_{t1} - E_{t2}) = 0$$

$$\Delta l = 0$$

$$E_{t1} = E_{t2} \qquad \text{...(5.50)}$$

2. The normal components of electric flux density are discontinuous at the boundary by an amount equal to the surface-charge density on the boundary.

   It can be proceed by Gauses's law as,

$$\int_s D\,ds = D_{n1}\,\Delta S - D_{n2}\,\Delta S = \rho_s\,\Delta S \qquad \text{(see Fig. 5.7)}$$

$$\Delta S\,(D_{n1} - D_{n2}) = \rho_s\,\Delta S$$

or
$$D_{n1} - D_{n2} = \rho_s$$

or
$$D_{n1} = D_{n2} + \rho_s \qquad \text{...(5.51)}$$

3. The tangential components of the magnetic field intensity are discontinuous at the boundary by an amount equal to the surface current density on the boundary, *i.e.*, from Ampere's law

$$\int_e H.dl = H_{t1}\,\Delta l - H_{t2}\,\Delta l = J_S\,\Delta l \qquad \text{(see Fig. 5.7)}$$

$$= \Delta l\,(H_{t1} - H_{t2}) = J_S\,\Delta l$$

or
$$H_{t1} = J_s + H_{t2} \qquad \text{...(5.52)}$$

4. The normal components of the magnetic flux density are continuous across the boundary, *i.e.*, from Maxwell's equation,

$$\int B\,ds = B_{n1}\,\Delta S - B_{n2}\,\Delta S = 0,\ \Delta S\,(B_{n1} - B_{n2}) = 0 \qquad \text{...(see Fig. 5.7)}$$

So, $B_{n1} = B_{n2}$ ...(5.53)

Since $\Delta S = 0$

The above equation were proved from the given Fig. 5.7.

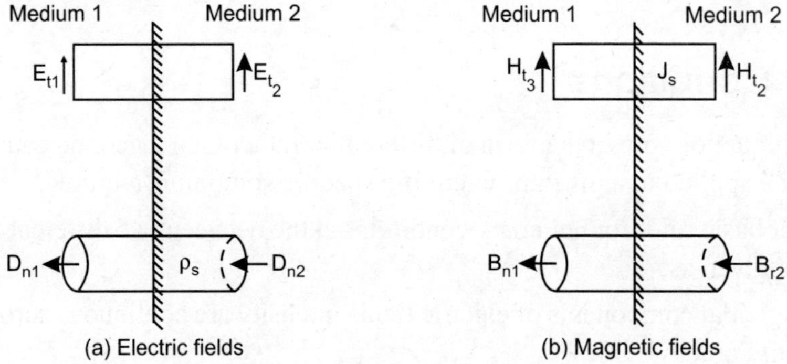

Fig. 5.7 Boundary conditions.

SOLVED EXAMPLES

Example 1. Find the depth of penetration of a mega cycle wave into copper which has $\sigma = 5.8 \times 10^7$ mhos /meter and permeability, permitivity that free space.

Solution : Given that $\sigma = 5.8 \times 10^{-7}$ mhos /m

$$f = 1.\text{MHz} = 1 \times 10^6$$

$$\mu = 4\pi \times 10^{-7}\,\text{H/m}$$

So, skin depth (depth of penetration)

$$\delta = \frac{1}{\sqrt{\pi f \mu \sigma}}$$

$$= \frac{1}{\sqrt{\pi \times 1 \times 10^6 \times 4\pi \times 10^7 \times 5.8 \times 10^7}}$$

$$= \frac{1}{2\pi \times 10^3 \sqrt{5.8}}$$

$$\delta = 6.608 \times 10^{-5}\,\text{m}$$

Example 2 (*a*). A plane wave having frequency of 1 MHz has an average Poynting vector of 1 Watt/m^2. If the medium is lossless with $\varepsilon_r = 2$ and $\mu_r = 3$, then find (*a*) the velocity of propagation (*b*) the wavelength (*c*) The impedance of the medium (*d*) the r.m.s of the electric fields.

Solution : Given $f = 1 \times 10^6$ Hz, $P = 1$ Watt / m^2 , $\mu_r = 3$ $\varepsilon_r = 2$

(*i*) The velocity of propagation,

$$\upsilon = \frac{1}{\sqrt{\mu\varepsilon}} = \frac{1}{\sqrt{\mu_0 \mu_r \varepsilon_r \varepsilon_r}} = \frac{3 \times 10^8}{\sqrt{3 \times 2}}$$

$$= \frac{3\times 10^8}{\sqrt{6}} = \frac{3\times 10^8}{\sqrt{6}} = 1.22 \times 10^8 \text{ m/sec}$$

(*ii*) $V = \lambda . f$

$$\lambda = \frac{\upsilon}{f} = \frac{1.22\times 10^8}{1\times 10^6} 1.22 \times 10^8 \text{ meters}$$

$$\lambda = 1.22 \times 10^2 \text{ meters}$$

(*iii*) $P = E . H$

$$= E . \frac{E}{\eta} = \frac{E^2}{\eta}$$

$$E^2 = \eta P$$

$$= P\sqrt{\frac{\mu_r \mu_0}{\varepsilon_r \varepsilon_0}} = P\sqrt{\frac{\mu_0}{\varepsilon_0}} . \sqrt{\frac{\mu_r}{\varepsilon_r}}$$

$$= P . 120\,\pi \sqrt{\frac{2}{3}}$$

$$= 1 \times 120\,\pi \sqrt{\frac{2}{3}}$$

$$E^2 = 120\,\pi \sqrt{\frac{2}{3}}$$

$$\left(\frac{E}{\sqrt{2}}\right)^2 = \frac{120\pi}{\sqrt{2}} . \frac{2}{\sqrt{3}}$$

$$E_{rms} = 14.75 \text{ V /m}$$

(*iv*) The impedance of the medium,

$$\eta = \frac{E^2}{P} = \frac{(\sqrt{2}E_{rms})^2}{P} = \frac{2(14.75)^2}{1}$$

$$\eta = 435.\ \Omega$$

Example 2 (*b*): In a loss-less dielectric medium, $\eta = 80\pi$, $\mu_r = 1$ Then find, ε_r, and ω.

Solution: For loss-less medium, $\sigma = 0$, $\alpha = 0$ and $\beta = 1$.

$$\eta = \frac{120\pi}{\sqrt{\varepsilon_r}}$$

$$\Rightarrow \quad e_r = \left(\frac{120\pi}{80\pi}\right)^2 = (1.5)^2 = 2.25$$

$$\beta = \omega\sqrt{\mu\varepsilon} = \frac{\omega}{c}\sqrt{\varepsilon_r} = \frac{\omega}{c} \times 1.5$$

$$\omega = \frac{\beta . c}{1.5} = \frac{3\times 10^8}{1.5} = 2\times 10^8 \text{ rad/sec}$$

Example 3. A plane wave in Free-space has average Poynting vector of 3 W/m^2. Find the average energy density.

Solution : Since $\quad P = E . H$

$$W = \frac{\mu}{2} H^2$$

So, $\frac{P}{W} = \frac{1}{\mu}\frac{E}{H} = \frac{1}{\mu} = \frac{1}{\sqrt{\mu\varepsilon}}$ = velocity of light

So, Average energy density = $\frac{\text{Poynting vector}}{\text{Average celocity}} = \frac{1}{3 \times 10^8}$

$$= 3.3 \times 10^{-9} \text{ J / m}$$

Example 4. Prove that for the plane wave $H = H_o e^{+yz} . a_H$ where a_H is the fixed unit vector.

Solution : $a_H = (a_H . a_x) a_x + (a_H . a_x) a_y + (a_H . a_x) a_z$

and hence $H = H_0 e^{\pm yz} [(a_H . a_x) a_x + (a_H . a_y) a_y + (a_H . a_z) a_z]$

$$= H_0 e^{\pm yz} [(a_H . a_x) a_x + H_0 e^{\pm yz} (aH . ay) a_y + H_0 e^{\pm yz} (a_H . a_z) a_z$$

Now, from the Maxwell's equation

$$\nabla . H = 0$$

$$\frac{\partial}{\partial x} H_x + \frac{\partial}{\partial y} H_y + \frac{\partial}{\partial z} H_z = 0$$

or $\frac{\partial}{\partial z} H_0 e^{\pm yz} (a_H . a_z) = 0$

$$rz H_0 e^{\pm yz} (a_H . a_z) = 0$$

This con be only true when

$a_H . a_z = 0$ Hence proved.

Example 5. $E = E_o \sin (\acute{\omega}t - \beta z) a_y$ is in free space. Calculate B and H.

Solution : From Maxwell's equation

$$\nabla \times E = -\frac{\partial B}{\partial t}$$

$$\begin{bmatrix} a_x & a_y & a_z \\ \frac{\partial}{\partial x} & \frac{\partial}{\partial y} & \frac{\partial}{\partial z} \\ 0 & E_0 \sin(\omega t - \beta z) & 0 \end{bmatrix} = -\frac{\partial B}{\partial t}$$

$$a_x \left[-\frac{\partial}{\partial z} E_0 \cdot \sin(\omega t - \beta z)\right] = -\frac{\partial \beta}{\partial t}$$

$$a_x [+ E_0 \cos (\acute{\omega}t - \beta z) . \beta] = -\frac{\partial \beta}{\partial t}$$

On integration of both side

$$\int E_0 \cos (\acute{\omega}t - \beta z) a_x . \beta \, dt = f - \frac{\partial \beta}{\partial t}$$

$$\frac{E_0 \beta}{W} . \sin (\omega t - \beta z) a_x . (-1) = - B$$

$$B = \left(-\frac{\beta E_0}{\omega}\right) \sin(\omega t - \beta z)\, a_x$$

and
$$B = \mu H$$

$$H = \frac{B}{\mu} = -\frac{\beta E_0}{\mu\omega} \sin(\omega t - \beta z)\, a_x$$

Example 6. In free space $H = H_0\, e^{j(\omega t + \beta z)}\, a_x$. Calculate E.

Solution :
$$\nabla \times H = -\frac{\partial D}{\partial t}$$

$$\begin{vmatrix} a_x & a_y & a_z \\ \frac{\partial}{\partial x} & \frac{\partial}{\partial y} & \frac{\partial}{\partial z} \\ H_0 e^j(\omega t + \beta z) & 0 & 0 \end{vmatrix} = \frac{\partial D}{\partial t}$$

$$a_y \cdot \frac{\partial}{\partial z} H_0\, e^{j(\omega t + \beta z)}\, j\beta = \frac{\partial D}{\partial t}$$

or
$$a_y H_0 \cdot e^{j(\omega t + \beta z)}\, j\beta = \frac{\partial D}{\partial t}$$

Integrating both side

$$\frac{a_y \cdot H_0 e^j(\omega t + \beta z)}{j\omega} . j\beta = D$$

$$D = \left(\frac{\beta . H_0}{\omega}\right) . e^{j(\omega t - \beta z)}\, a_y$$

So,
$$E = \frac{D}{\varepsilon} = \frac{\beta H_0}{\omega\varepsilon} . e^{j(\omega t - \beta z)}\, a_y$$

EXERCISES

1. Derive the expression of Maxwell's first two equation, *i.e.,*
$$\nabla \times H = \sigma E + \frac{\partial D}{\partial t}$$
$$\nabla \times E = -\frac{\partial B}{\partial t}$$
2. Derive Maxwell's last two equations, *i.e.,* third $\nabla . D = \rho$ and fourth $\nabla B = 0$ equation. Also explain the significant of these equations.
3. In free space, it is given that
$$E = 20\,\pi e^{j(106t + \beta z)}\, a_x \text{ V/m and } H = H_0\, e^{j(106t + \beta z)}\, a_x \text{ A/m}$$
Calculate β and H_0 $\beta > 0$
4. In a homogeneous non – conducting where $\varepsilon_r = 1$. It is given that
$$E = 30\,\pi . e^{j\,\omega t - 7/5y)}\, a_y \text{ V/m and } H = \frac{1}{2} e^{j(\omega t - 7/5z)}\, a_x \text{ A/m}$$
Calculate ε_r and ω .
5. In a homogeneous region where $\mu_r = 1$ and $a_r = 4$, it is given that

$$E = 10\,\pi e^{j(\omega t - \beta z)}\, a_x \text{ V/m}$$

$$B = \mu_0 H_0 e^{j(\omega t - \beta z)}\, a_x \text{ Tesla}$$

Calculate ω and H_0 if wave length is 1.50 meters.

6. If magnetic field H has only a_x component given by $H_x = 3x\,.\cos\beta + 5y \sin y$ and if the field is invariant with time, what is expression for current density J?

7. The electric field intensity associated with a plane wave travelling in a perfect dielectric medium is given by

$$E_x(z, t) = 5 \cos(2\pi\, 10^6\, t - 0.5\, \pi x) \text{ V/m}$$

Calculate the propagation velocity and also find expression for magnetic fields.

8. In free space $H(z, t) = 2\,.e^{j(2 \times 108\,t - \beta z)} a_z$ V / m.

Derive expression for $E(z, t)$ and determine direction of propagation.

9. In free space $H(z,t) = 0.\ 13 \cos(4 \times 10^8\,t - \beta z)\, a_z$ V / m

Find the expression for $E(z, t)$, β and λ

10. In free space, it is given that

$$E(z, t) = 10^3 \sin(6 \times 10^8\, t - \beta z)\, a_y \text{ V/m}$$

Sketch the wave at t = 0. and the time 't' when it is advanced $\frac{\lambda}{4}$ along z axis Also calculate β and λ

11. Show that skin depth for good conductor, δ can be given by ;

$$\delta = \frac{1}{\sqrt{\pi f \mu \sigma}} = \frac{1}{\alpha}$$

- "Education is the ability to listen to almost anything without losing your temper or your self-confidence"

Robert Frost

Chapter 6

ELECTROMAGNETIC WAVES AND FLOW OF POWER

6.1 INTRODUCTION

The electromagnetic waves have two components namely electric and magnetic fields both carries energy from one point another. The total energy are distributed in both fields and time varying. Power in electromagnetic wave is nothing but energy per unit time. If power is known then power density (p_d) at any point in free space is ratio of the total power radiated (P_t) to the area (A) over which it is distributed. That is

$$p_d = \frac{P_t}{A} = \frac{P_t}{4\pi r^2} \qquad ...(6.1)$$

where P_t = Total power radiates by electromagnetic wave in watts

A = Area = $4\pi\, r^2$ (m^2)

It is also seen that the power density is inversely proportional to the square of the distance $\left(i.e. \frac{1}{r^2}\right)$. *i.e*, if distance doubles, the power density is reduced by $\frac{1}{4}$ the (inverse square law).

Assuing no losses,the energy carried out by an EM wave is contained in both its electric (E) and magnetic (H) fields and each cubic meter of space contains the total energy (W). The energy W can be given as;

$$W = \frac{1}{2}\,(DE + BH) \text{ J/m}^2$$

$$W = \frac{1}{2}\,(\mu_0 E^2 + \mu_0 H^2) \text{ J/m}^2 \qquad ...(6.2)$$

If an electromagnetic wave has only E_z and H_z components;

Then $$W = \frac{1}{2}\,(\varepsilon_0\, E_z^2 + \mu_0 H_z^2) \text{ J/m}^2$$

Under no loss condition;

Energy in electric field = Energy in magnetic fields

$$\frac{1}{2}\,\varepsilon_0\, E^2 = \frac{1}{2}\mu_0\, H^2 \qquad ...(6.3)$$

or $$\frac{1}{2}\varepsilon_0 E_z^2 = \frac{1}{2}\mu_0 H_z^2$$

or from equation (6.3)

$$\frac{E}{H} = \sqrt{\frac{\mu_0}{\varepsilon_0}}\ 120\pi = 377\ \Omega$$

The ratio $\left(\frac{E}{H}\right)$ called free space impedance.

If was assume that wave travels through free space with the velocity of light then the energy S passing through each square meters of the plane perpendicular to the direction of propagation in each second is given by

$$S = c \times W = \frac{c}{2}\,(\varepsilon_0\, E_z^2 + \mu_0 H_z^2)$$

$$= \frac{c}{2} \times 2\varepsilon_0\, E_z^2 = c\varepsilon_0 E_2^2$$

and from $$c = \frac{1}{\sqrt{\mu_0 \varepsilon_0}}$$

$$S = \frac{1}{\sqrt{\mu_0\, \varepsilon_0}}\,\varepsilon_0 E_z^2$$

$$= \sqrt{\frac{\varepsilon_0}{\mu_0}}\, E_z^2$$

$$= \frac{E_z^2}{\sqrt{\frac{\mu_0}{\varepsilon_0}}}$$

$$S = \frac{E_z^2}{120\pi} \qquad ...(6.4)$$

Similarly in terms of magnetic fields (H_z)

$$S = \frac{H_z^2}{120\pi} \qquad ...(6.5)$$

The Poynting Theorem

The Poynting theorem helps to calculate the energy transfer from a transmitting source to a receiver during electromagnetic propagation. It is a simple relation between energy transfer and the amplitude of the electric and magnetic field strength of electromagnetic wave.

Since the time average of any two complex vectors is equal to the real part of the product of one complex vector multiplied by the complex conjugate of the other vector; hence the time average of the instantaneous Poynting vector in steady state in given by,

$$<P> = <E \times H> = \frac{1}{2}\, R_c\,(E \times H^*) \qquad ...(6.6)$$

Where the notation $<>$ stands for average and it is represected by a factor of $\frac{1}{2}$ appears in the equation (6.6). But from the Maxwell's equation,

$$\nabla \times H = J + \frac{\partial D}{\partial t} = J + \varepsilon.\frac{\partial E}{\partial t} \qquad ...(6.7a)$$

$$\nabla \times E = \frac{-\partial B}{\partial t} = -\,\varepsilon\,\frac{\partial H}{\partial t} \qquad ...(6.7b)$$

Again from equation (6.7*a*)

$$\Delta \times H = J + \varepsilon \frac{\partial E}{\partial t}$$

or $$J = \Delta \times H - \frac{\partial E}{\partial t} \quad ...(6.8)$$

When the multiply both side Eq. (6.8) by E we get the dimension of power as

$$J.E. = E(\Delta \times H) - \varepsilon E \frac{\partial E}{\partial t}$$

Using vector identity

$$A.\,(B \times C) = C.\,(A \times B) - B.\,(A \times C)$$

so $$E.J = H.\,(E \times \Delta) - \Delta.\,(E \times H) - \left(\frac{\varepsilon}{2}\right)\frac{\partial}{\partial t}E^2$$

$$= H.\,(\Delta \times E) - \Delta.\,(E \times H) - \left(\frac{\varepsilon}{2}\right)\frac{\partial}{\partial t}E^2$$

$$= H.\mu.\,\frac{\partial H}{\partial t} - \Delta\,(E \times H) - \left(\frac{\varepsilon}{2}\right)\frac{\partial E^2}{\partial t}$$

$$= -\frac{\mu}{2}\frac{\partial H^2}{\partial t} - \Delta.\,(E \times H) - \left(\frac{\varepsilon}{2}\right)\frac{\partial E^2}{\partial t}$$

$$(\text{E.J.}) = -\frac{\partial}{\partial t}\left(\frac{\mu}{2}H^2 + \frac{\varepsilon}{2}E^2\right) - \Delta(E \times H) \quad ...(6.9a)$$

Integrating over volume V

$$\int_V (E.J)\,dV = \int_V -\frac{\partial}{\partial t}\left(\frac{\varepsilon E^2}{2} + \frac{\mu H^2}{2}\right)dV - \int_V \Delta(E \times H)\,dV$$

But from Divergence theorem

$$\int_V (\Delta.\,A)\,dV = \int_S (A.dS)$$

$$\int_V \Delta(E \times H)\,dV = (E \times H).dS$$

so $$\int_V \Delta(E \times H)\,dV = \int_V -\frac{\partial}{\partial t}\left(\frac{\varepsilon E^2}{\partial t} + \frac{1}{2}\mu H^2\right)dV - (E \times H)dS$$

and $$\int_V (E.J)\,dV = \frac{\partial}{\partial t}\int_V \left(\frac{\varepsilon}{2}E^2 + \frac{\mu}{2}H^2\right)dV - (E \times H)ds \quad ...(6.9b)$$

The terms left side of the Eq. (6.9*b*)

$\int_V (EJ)\,dV$ = instantaneous power dissipateal in the volume V.

The first term on the right hand side of equation $\frac{\varepsilon}{2}E^2$ represents power stored energy density of electric field at steady state and $\frac{\mu H^2}{2}$ represents the stored energy density of magnetic field at steady state. The interpretation of the remaining term follows from the application of law of conservation of energy.

The rate of energy dissipated in the volume V must be equal to the rate at which the stored energy in (V) is decreasing plus the rate which energy is entering the volume V from outside then,

$(E \times H)\, dS$ is rate of energy flowing out word through the surface on closing the volume V. The integral of $E \times H$ over any closed surface gives the rate of energy flow through the surface.

Thus $$P = (E \times H) \text{ watts/m}^2 \quad ...(6.10)$$

The Poynting theorem gives $P = E \times H$ at any point which is a measure of the rate of energy flow per unit area at that point and under no loss condition.

$$\frac{1}{2}\mu_0 H^2 = \frac{1}{2}\mu_0 E^2$$

$$\frac{E}{H} = \sqrt{\frac{\mu_0}{\varepsilon_0}} = 120\,\pi$$

Power Flow in Coaxial Cable

The Poynting vector within the insulater of a coaxial cable is nearly parallel to the wire axis (assuming no fields outside the cable). So electric energy is flowing through the dielectric between the conductors. If a conductor has significant resistance, then near the surface of that conductor, the Poynting vector would be found toward impinging upon the conductor. Once the Poynting vector enters the conductor, it is bent to a direction that is almost perpendicular to the surface. This is consequences, the Poynting vector represents energy flow from the electromagnetic fields into the wire, producing resistive Joule heating in the coaxial cable.

Instantaneous and Average power

In general an *ac* power W is the product of instantaneous voltage V and instantaneous current I,

so $$\overline{W} = \overline{V}\,.\,\overline{I}$$

The quantities $\overline{V}$ and $\overline{I}$ may be represented in terms of phaser V and I as,

$$\overline{V} = Re\,(V e^{j\omega t}.\, e^{j\theta}) = Re\,[V e^{j(\omega t + \theta)}]$$

$$\overline{V} = |V|.\cos(\omega t + \theta_v)$$

and $$\overline{I} = |I|.\cos(\omega t + \theta_i)$$

So, $$|V|\cos(\omega t + \theta_i).\,|I|.\cos(\omega t + \theta_i) = \frac{|V||I|}{2}[2\cos(\omega t + \theta_v).\cos(\theta t + \theta_i)]$$

$$= \frac{|V|.|I|}{2}[\cos(\omega t + \theta_v - \omega t - \theta_i) + \cos(\omega t + \theta v + \omega t + \theta_i)]$$

$$\overline{W} = \frac{|V||I|}{.2}[\underbrace{\cos(\theta_v - \theta_i)}_{\text{I}} + \underbrace{\cos(2\omega t + \theta_v + \theta_i)}_{\text{II}}] \quad \text{(say)}$$

We observe that this consist of *a.c* average part (*I*) and an oscillatory part (*II*). Therefore the average instantaneous power

$$W_{av} = \frac{|V||I|}{2}\cos(\theta_v - \theta_i)$$

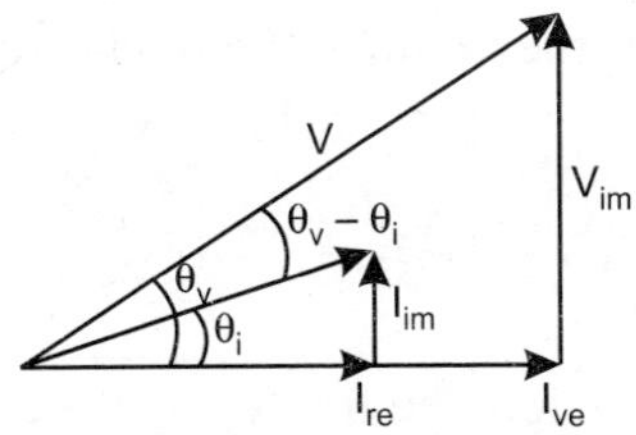

Fig. 6.1. Real and imaginary parts of V and I

if $\theta_v - \theta_i = \theta$

then $$W_{av} = \frac{|V||I|}{2}\cos\theta$$

Another useful quantity reactive power, which is also called VAR (reactive volt ampere) will be defined as;

$$W_{\text{react}} = \frac{|V||I|}{2}\sin\theta \quad ...(6.11)$$

That is W_{av} and W_{rect} are in phase and out of phase component of the volt ampre product. The presence of an out of phase or reactive component indicates that the instantaneous power changes sign.

Now complex power can be given as,

$$W = \frac{1}{2}VI^*$$

$$= \frac{1}{2}|V|e^{j\theta v}|I|e^{-j\theta}$$

$$= \frac{1}{2}|V||I|e^{+j\theta}$$

Hence, $$W = W_{av} + jW_{react} \quad ...(6.12)$$

The similar relations exist between Poynting vector, the electric fields strength and magnetic field strength.

$$P = E \times H$$

In phasor form the this relation can be given as

$$P = \frac{1}{2}(E \times H^*) \quad ...(6.13)$$

Therefore average power and reactive power can be given as;

$$P_{av} = \frac{1}{2}R_e(E \times H^*)$$

and reactive power

$$P_{\text{react}} = \frac{1}{2}I_m(E \times H^*)$$

and total power

$$P_T = P = P_{av} + P_{\text{react}} = \frac{1}{2}(E \times H^*) \quad ...(6.14)$$

Thus complex pointing vector gives as the average and reactive parts of the power flow per square meter.

Reflection and Refraction of plane waves

During electromagnetic wave propagation, the following phenomenons may occur;

(*i*) A portion of electromagnetic wave goes from one medium to other medium.

(*ii*) Some part of the electromagnetic wave get reflected back.

So, total incident wave has two component reflected and transmitted waves. Now electromagnetic wave may be incident normally or at any angle on the boundary surface.

(1) Normal-incidance reflection

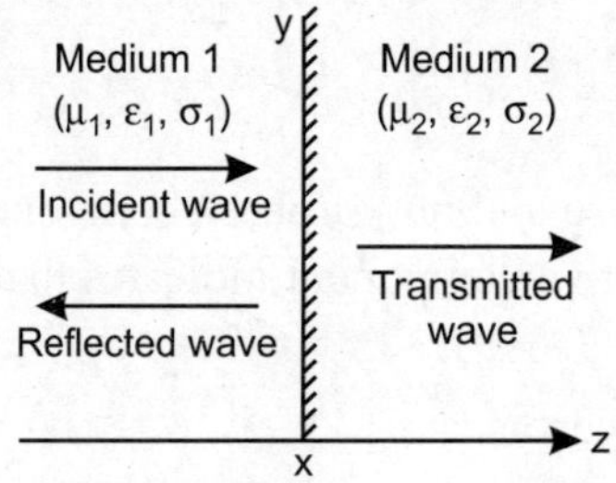

Fig. 6.2 Uniform plane wave reflection.

The simplest reflection case is that uniform plane wave normally incident on an uniform plane boundary between two dielectric melia with no surface charge density and current density. This situation is shown in Fig. 6.2

In medium 1 the field will be the sum of an incident wave and reflected wave

$$E_x = (1)\, E_0\, (e^{-j\beta_1 z} + \Gamma e^{j\beta_1 z}) \quad \text{...(6.15a)}$$

$$H_y(1) = \frac{E_0}{\eta_1} (e^{-j\beta_1 z} + \Gamma ej^{\beta z}) \quad \text{...(6.15b)}$$

where $\beta_1 = \omega \sqrt{\mu_1\, \varepsilon_1}$

$$\eta_1 = \sqrt{\frac{\mu_1}{\varepsilon_1}} = \frac{\eta_0}{\sqrt{\varepsilon_{r1}}} = \text{intrisic wave impeccant of medium 1}$$

$$\Gamma = \text{reflection coefficient}$$

In medium 2 there are transmitted waves

$$E_x\,(2) = E_0\, T\, e^{-j\beta_2 z} \quad \text{...(6.16a)}$$

$$H_y\,(2) = \frac{E_0}{\eta_2}\, Te^{-j\beta_2 z} \quad \text{...(6.16b)}$$

where $\beta_2 = \omega \sqrt{\mu_2 . \varepsilon_2}$

$$\eta_2 = \sqrt{\frac{\eta_2}{\varepsilon_2}} = \frac{\eta_0}{\sqrt{\varepsilon_{r2}}}$$

$$T = \text{transmission coefficient}$$

$$Z_2 = \left.\frac{E_x^{(1)}}{H_y^{(1)}}\right|_{z=0} = \left.\frac{E_x^{(2)}}{H_y^{(2)}}\right|_{z=0} \quad ...(6.16c)$$

Now
$$\frac{E_x^{(1)}}{H_y^{(1)}} = \frac{E_0\,(e^{-j\beta_2 z} + \Gamma e^{j\beta_2 z})}{\frac{E_0}{\eta_1}(e^{-j\beta_2 z} - \Gamma e^{j\beta_2 z})}$$

Now
$$\left.\frac{E_x^{(1)}}{H_y^{(1)}}\right|_{z=0} = \frac{E_0\,(e^{-0} + \Gamma e^{0})}{\frac{E_0}{\Gamma_1}(e^{-0} - \Gamma e^{0})}$$

$$= \eta_1 \frac{(1+\Gamma)}{(1-\Gamma)} \quad ...(6.17a)$$

and
$$\left.\frac{E_x^{(2)}}{H_y^{(2)}}\right|_{z=0} = \left.\frac{E_0\, Te^{-j\beta_2 z}}{\frac{E_0}{\eta_2}Te^{-j\beta 2z}}\right|_{z=0}$$

$$\frac{E_0\, Te^{-0}}{\frac{E_0}{\eta_2}Te^{-0}} = \eta_2 \quad ...(6.17b)$$

Now from equation (6.16c)

$$\frac{\eta_1(1+\Gamma)}{1-\Gamma} = \eta_2$$

$$\frac{1+\Gamma}{1-\Gamma} = \frac{\eta_2}{\eta_1}$$

$$\Gamma_1 + \eta_1\Gamma = \eta_2 - \eta_2\,\Gamma$$

$$\eta_1 - \eta_2 = -\Gamma_2\Gamma - \eta_1\Gamma$$

$$\eta_2 + \eta_1 = (\eta_1 - \eta_2)\,\Gamma$$

$$\Gamma = \frac{\eta_2 - \eta_1}{\eta_1 + \eta_2} = \left(\frac{\eta_2 - \eta_1}{\eta_2 + \eta_1}\right)$$

$$\Gamma = \left(\frac{\eta_2 - \eta_1}{\eta_2 + \eta_1}\right) \quad ...(6.18)$$

Again from the boundary condition the tangential components of the electric intensity are continuous across the interface $z = 0$ hence ;

$$E_x^{(1)} = E_0(e^{-j\beta_1 z} + \Gamma\, e^{+j\beta_1 z})$$

at
$$z = 0,\ E_x\,(1)\,E_0\,(e^{-0} + Te^{0}) = E_0(1 + T)$$

and
$$E_x\,(2) = E_0 Te^{-j\beta_2 z}$$

at
$$z = 0$$

$$E_x(2) = E_0 Te^{-0} = E_0\, T$$

so
$$E_0\,(1+\Gamma) = E_0 T \qquad (T = 1 + \Gamma)$$

Transmission coefficient = 1 + reflection coefficient, hence the transmission coefficient in terms of impedance in medium (2) and (1) can be given as;

$$T = 1 + \Gamma$$

$$= 1 + \frac{\eta_2 - \eta_1}{\eta_2 + \eta_1}$$

$$= \frac{\eta_2 + \eta_1 + \eta_2 - \eta_1}{\eta_2 + \eta_1}$$

$$T = \frac{2\eta_2}{\eta_2 + \eta_1} \quad ...(6.19)$$

If medium (1) is loss less ($\sigma = 0$), the standing wave ratio may be defined as;

$$SWR = \frac{\eta_2}{\eta_1}$$

The power density transmitted across the boundary is;

$$P_{tr} = \frac{1}{2}(E \times H^*).u_z$$

$$P_{tr} = \frac{E_0}{2\eta_1}(1 - T^2)$$

Then $$P_{tr} = P_{inc}(1 - T^2)$$

$$= P_{inc} - P_{inc} T^2$$

$$P_{tr} = P_{inc} - P_{ref}$$

$$P_{inc} = P_{tr} + P_{ref}$$

Where $$P_{ref} = P_{inc} T_2$$

Electromagnetic wave oblique incidence: When an electromagnetic wave is incident on any surface at an angle besides 90°, it is called oblique incidence,

***(a)* *E* is in two plane of incidence :** The plane of incidence is defined by the direction of propagation and the line normal to the boundary. Let a linearly polarized uniform wave with *E* laying in and *H* normal to the plane of incidence are impinging oblique on a boundary between two lossless dielectric materials as shown in Fig. 6.3.

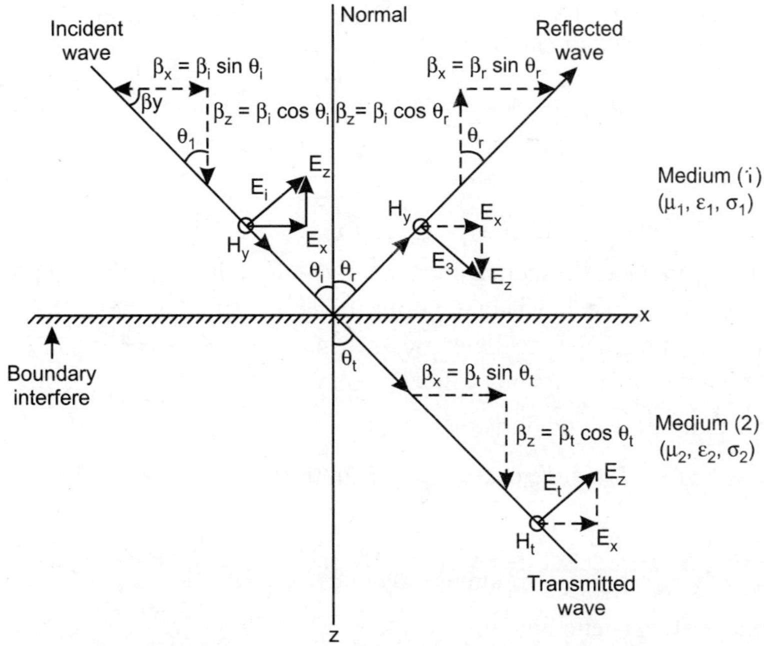

Fig. 6.3 Reflection and transmission of oblique incident.

whenever, an electromagnetic is wave incident obliquely on the boundary surface between two media, the polarization of the wave is vertical or horizontal depends on the direction of electric fields. As Fig. 6.3 shows, for lossless dielectric medium, the phase constants (propagation constant) of the two media in the *x* direction on the continuity of the tangential *E* and *H* on the boundary will be equal. Thus

$$\beta_i \sin\theta_i = \beta_r \sin\theta_r \quad \text{...(6.20a)}$$

$$\beta_i \sin\theta_i = \beta_t \sin\theta_t \quad \text{...(6.20b)}$$

Since, $\beta_i = \beta_r = \beta_t$

So, $\sin\theta_i = \sin\theta_r$, or $\theta_i = \theta_r$

Now from equation (6.20b)

$$\beta_i \sin\theta_i = \beta_t \sin\theta_t$$

$$\frac{\sin\theta i}{\sin\theta t} = \frac{\beta_t}{\beta_i}$$

or

$$\frac{\sin\theta t}{\sin\theta i} = \frac{\beta_1}{\beta_t} = \frac{\omega\sqrt{\omega_1\varepsilon_1}}{\omega\sqrt{\mu_2\varepsilon_2}} = \frac{\sqrt{\omega_1\varepsilon_1}}{\sqrt{\omega_2\varepsilon_2}} = \frac{v_2}{v_1} = \sqrt{\frac{\mu_1\varepsilon_1}{\mu_2\varepsilon_2}} \quad \text{...(6.21)}$$

since

$$v = \frac{1}{\sqrt{\mu\varepsilon}}$$

So $$v_1 = \frac{1}{\sqrt{\mu_1 \varepsilon_1}}$$

$$v_2 = \frac{1}{\sqrt{\mu_2 \varepsilon_2}}$$

So $$\frac{\sin \theta_t}{\sin \theta_i} = \frac{\beta_i}{\beta_t} = \frac{v_2}{v_1} = \sqrt{\frac{\mu_1 \varepsilon_1}{\mu_2 \varepsilon_2}} \qquad ...(6.22)$$

Where v_1 and v_2 are phase velocities in respective medium. This is well known Snell's law. In general all low loss dielectrics have equal permeability, *i.e.*, $\mu_1 = \mu_2 = \mu_0$. If the medium 2 is free space and medium is a non megnetic dielectric. the right hand side of equation (6.22) becomes $\sqrt{\varepsilon_r}$ which is called the index of refraction of the dielectirc.

The component of electric field intensity E are

$$E_x = E_0 \cos \theta_i \, e^{-j\beta_1 (x \sin \theta_i + z \cos \theta_i)} \qquad ...(6.23a)$$

$$E_y = 0 \qquad ...(6.23b)$$

$$E_z = -E_0 \sin \theta_i \, e^{-j\beta_1 (x \sin \theta_i + z \cos \theta_i)} \qquad ...(6.23c)$$

The component of magnetic field intensity are

$$H_x = 0$$

$$H_y = \frac{E_0}{\eta_1} . e^{-j\beta_1 (x \sin \theta_i + z \cos \theta_i)} \qquad ...(6.24)$$

$$H_z = 0$$

The wave impedance in the z direction is given by

$$Z_z = \frac{E_x}{H_y} = \eta \cos \theta \qquad ...(6.25)$$

The wave impedance can be expressed in terms of the reflection coefficient of the normal component at the boundary $z = 0$ in medium (1);

$$E_x(1) = E_0 \cos \theta_i \, [e^{-j\beta_1 x \cos \theta_i} + \Gamma \, e^{j\beta_1 z \cos \theta_r}] \qquad ...(6.26a)$$

$$H_y^{(1)} = \frac{E_0}{\eta_1} [e^{-j\beta_1 x \cos \theta_i} - \Gamma e^{j\beta_1 z \cos \theta_r}] \qquad ...(6.26b)$$

$$Z_z = \frac{E_x^{(1)}}{H_y^{(1)}} = \eta_1 \cos \theta_i \left(\frac{1+\Gamma}{1-\Gamma}\right) \qquad ...(6.27)$$

$$\frac{\eta_1 \cos \theta_t - \eta_1 \cos \theta_i}{\eta_2 \cos \theta_t + \eta_1 \cos \theta_i} = \frac{1+\Gamma-1+\Gamma}{1+\Gamma+1-\Gamma} = \frac{2\Gamma}{2}$$

$$\Gamma = \frac{\eta_2 \cos \theta_t - \eta_1 \cos \theta_i}{\eta_2 \cos \theta_t + \eta_1 \cos \theta_i} \qquad ...(6.28a)$$

Similarly transmission coefficient

$$T = 1 + T$$

$$T = \frac{2\eta_2 \cos \theta_t}{\eta_2 \cos \theta_t + \eta_1 \cos \theta_i} \qquad ...(6.28b)$$

The preceding two equation are known as Fresnel's formula for E in the plane of incident.

***H* is in the plane of incident:** If the magnetic field intengity H is in the plane of incidence, the components of H are,

$$H_x = H_0 \cos \theta_i \, e^{-j\beta_1 (x \sin \theta i + z \cos \theta i)} \quad ...(6.29a)$$

$$H_y = 0 \quad ...(6.29b)$$

$$H_z = H_0 . \sin \theta_i \, e^{-j\beta_1 (x \sin \theta i + z \cos \theta i)} \quad ...(6.29c)$$

And the electric field component E normal to the plane of incident are

$$E_x = 0 \quad ...(6.30a)$$

$$E_y = -\eta_1 H_0 \, e^{-j\beta_1 (x \sin \theta i + z \cos \theta i)} \quad ...(6.30b)$$

$$E_z = 0 \quad ...(6.30c)$$

And the wave impedance in the z direction can be given as;

$$Z_z = -\frac{E_y}{H_x} = \frac{\eta}{\cos \theta} = \eta \sec \theta \quad ...(6.31)$$

The Fresnel's formula for H in the plane of incidence are,

$$\Gamma = \frac{\eta_2 \sec \theta_t - \eta_1 \sec \theta_i}{\eta_2 \sec \theta_t + \eta_1 \sec \theta_i} \quad ...(6.32a)$$

and

$$T = 1 + \Gamma$$

$$T = \frac{2\eta_2 \sec \theta_t}{\eta_2 \sec \theta_t + \eta_1 \sec \theta_i} \quad ...(6.32b)$$

Surface resistance (R_s): At high frequency as frequence increases, from skin-effect it is cleared that δ decrease, *i.e*, most of current waves try to propagate near to the surface. The intrisic impedance of a good conductor for any instant can be given by,

$$\eta = \sqrt{\frac{Z}{Y}} = \sqrt{\frac{j\omega\mu}{(\sigma + j\omega\varepsilon)}} \quad ...(6.33)$$

for good conductor $\sigma >> \omega s$

$$= \sqrt{\frac{j\omega\mu}{\sigma\left(1 + \frac{j\omega\varepsilon}{\sigma}\right)}} \quad ...(6.34)$$

$$= \sqrt{\frac{j\omega\mu}{\sigma}} \left\{\text{since } \frac{\omega\varepsilon}{\sigma} <<< 1\right\}$$

$$= \sqrt{\frac{\omega\mu}{\sigma}} \left(\frac{1}{2} + \frac{j}{2}\right)$$

Since

$$\sqrt{j} = \left(\frac{1}{\sqrt{2}} + \frac{j}{\sqrt{2}}\right) \quad ...(6.35)$$

$$(1 + j) \sqrt{\frac{\omega H}{2\sigma}} = (1 + j) \text{ Rs}$$

where $\text{Rs} = \sqrt{\frac{\omega H}{2\sigma}}$ called skin resistance or surface resistance of the conductor. ...(6.36)

Wave Impedances : When an uniform electromagnetic wave propagate in a given medium, it offers some resistance to the wave which is called wave impedance. This wave impedance is also called intrinsic impedance.

For any other medium.

$$\eta = \frac{E_x}{H_y} = \sqrt{\frac{\mu}{\varepsilon}} = \sqrt{\frac{\mu_r \mu_0}{\varepsilon_r \varepsilon_0}} \quad ...(6.37)$$

$$= \sqrt{\frac{\mu_0}{\varepsilon_r \varepsilon_0}}$$

$$= \frac{1}{\sqrt{\varepsilon_r}} \cdot \sqrt{\frac{\mu_0}{\varepsilon_0}}$$

$$\eta = \frac{\eta_0}{\sqrt{\varepsilon_r}} \quad \text{where} \quad \eta_0 = 377\ \Omega \quad ...(6.38)$$

Power Loss in Plane Waves

Brewster Angle : Brewster angle is defined as the angle of incidence at which there is no reflection of electromagnetic waves, *i.e.*, reflection coefficient = 0.

Now from equation

$$\Gamma = \frac{\eta_2 \cos \theta_t - \eta_1 \cos \theta_i}{\eta_2 \cos \theta_t + \eta_t \cos \theta_i} \quad ...(6.39)$$

$$\eta_2 \cos \theta_t - \eta_1 \cos \theta_i = 0$$

$$\eta_2 \cos \eta_t = \eta_1 \cos \theta_i = 0$$

$$\eta_2 \cos \theta_t = \eta_1 \cos \theta_i$$

$$\eta_2 \cos \theta_t = \eta_1 \cos \theta_B \qquad \theta_B = \text{Brewster angle.} \quad ...(6.40)$$

$$\cos \theta_B = \frac{\eta_2}{\eta_1} \cos \theta_t$$

$$\cos \theta_B = \frac{\eta_2}{\eta_1} \sqrt{1 - \sin^2 \theta_t} \quad ...(6.41)$$

$$\frac{\sin \theta_t}{\sin \theta_i} = \sqrt{\frac{\mu_1 \varepsilon_1}{\mu_2 \varepsilon_2}}$$

$$\frac{\sin \theta_t}{\sin \theta_B} = \sqrt{\frac{\mu_1 \varepsilon_1}{\mu_2 \varepsilon_2}}$$

$$\sin \theta_t = \sqrt{\frac{\varepsilon_1 \varepsilon_1}{\mu_2 \varepsilon_2}} \quad ...(6.42)$$

$$\cos \theta_B = \frac{\eta_2}{\eta_1} \sqrt{1 - \sin^2 \theta_t}$$

$$\sqrt{1 - \sin^2 \theta_B} = \frac{\eta_2}{\eta_1} \sqrt{1 - \sin^2 \theta_t}$$

$$1 - \sin^2 \theta_B = \left(\frac{\eta_2}{\eta_1}\right)^2 (1 - \sin^2 \theta_t)$$

$$1 - \sin^2 \theta_B \left(\frac{\eta_2}{\eta_1}\right)^2 \left(1 - \frac{\mu_1 \varepsilon_1}{\mu_2 \varepsilon_2} \sin^2 \theta_B\right)$$

$$= \left(\frac{\sqrt{\frac{\mu_2}{\varepsilon_2}}}{\sqrt{\frac{\mu_1}{\varepsilon_1}}}\right) \left(1 - \frac{\mu_1 \varepsilon_1}{\mu_2 \varepsilon_2} \sin^2 \theta_B\right)$$

$$= \frac{\mu_2 \varepsilon_1}{\varepsilon_2 \mu_1} - \frac{\mu_2 \varepsilon_1}{\varepsilon_2 \mu_1} \times \frac{\mu_1 \varepsilon_1}{\mu_2 \varepsilon_2} \sin^2 \theta_B$$

$$1 - \sin^2 \theta_B = \frac{\mu_2 \varepsilon_1}{\mu_1 \varepsilon_1} - \frac{\varepsilon_1^2}{\varepsilon_2^2} \sin^2 \theta_B$$

$$\left(1 - \frac{\mu_2 \varepsilon_1}{\mu_1 \varepsilon_1}\right) = \sin^2 \theta_B \left[1 - \left(\frac{\varepsilon_1}{\varepsilon_2}\right)^2\right]$$

$$\sin^2 \theta_B = \frac{1 - \frac{\mu_2 \varepsilon_1}{\mu_1 \varepsilon_2}}{1 - \left(\frac{\varepsilon_1}{\varepsilon_2}\right)}$$

if $\mu_1 = \mu_2 = \mu_0$

Then
$$\sin^2 \theta_B = \frac{1 - \frac{\mu_1 \varepsilon_1}{\mu_0 \varepsilon_2}}{1 - \left(\frac{\varepsilon_1}{\varepsilon_2}\right)^2} = \frac{1 - \left(\frac{\varepsilon_1}{\varepsilon_2}\right)}{1 - \left(\frac{\varepsilon_1}{\varepsilon_2}\right)^2}$$

$$\sin^2 \theta_B = \frac{1 - \left(\frac{\varepsilon_1}{\varepsilon_2}\right)}{\left[1 + \left(\frac{\varepsilon_1}{\varepsilon_2}\right)\right]\left[1 - \left(\frac{\varepsilon_1}{\varepsilon_2}\right)\right]}$$

$$\sin^2 \theta_B = \frac{1}{1 - \left(\frac{\varepsilon_1}{\varepsilon_2}\right)}$$

$$\sin \theta_B = \sqrt{\frac{\varepsilon_2}{\varepsilon_1 + \varepsilon_2}} \quad \text{...(6.43}a\text{)}$$

$$\theta_B = \sin^{-1} \sqrt{\frac{\varepsilon_2}{\varepsilon_1 + \varepsilon_2}}$$

or
$$\theta_B = \tan^{-1} \sqrt{\frac{\varepsilon_2}{\varepsilon_1}} \quad \text{...(6.43}b\text{)}$$

θ_B is Brewster angle.

Polarization : The polarization of wave is defined as a direction of the electric field at a given point as function of time. The polarization of electromagnetic wave may be categories as;

1. Linear polarization
2. Circler polarization

3. Elliptical polarization

(1) Linear polarization : A wave is said to be linear polarization if electric field as function of time remains constant along a straight line at some point in the medium. It is of three types:

(*i*) Horizontal polarization

(*ii*) Vertical polarization

(*iii*) Theta polarization

If $E_y = 0$ and E_x is present and wave traveling in z-direction with $\vec{E}$ and $\vec{H}$ field lying in xy plane it is said to be x polarized or horizontal polarized and to be y polarized or vertical polarized. On the other hand if E_x and E_y both are present and are in phase, then the wave is said to the theta (θ) polarized.

(2) Circular polarization : A wave is said to be circular polarized when the electric field traces a circle. If E_x and E_y are in equal magnitude and 90° out of phase difference, the locus of the resultant E is a circle and the wave is circular polarization wave.

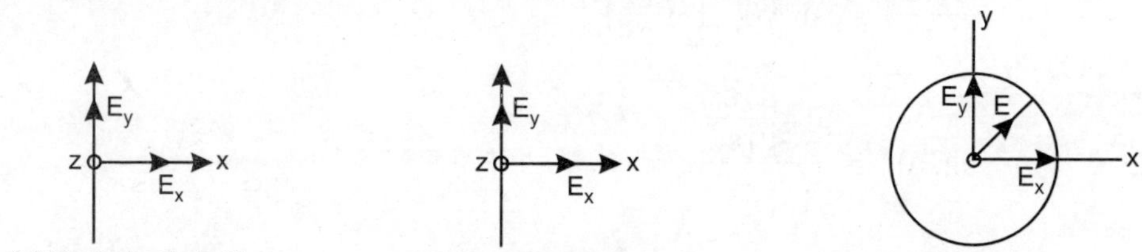

(a) Vertical polarization E_x - 0, Ey ≠ 0 (b) Horizontal polarization E_y = 0, Ex ≠ 0 (c) Circular polarization $E_x = E_y$

Fig. 6.4. Different Types of Palarization.

Let E of uniform plane wave propagation in z-direction be represented by,

$$E(z) = E_c\, e^{-j\beta z} \qquad ...(6.44a)$$

and in time varying form

$$E(z, t) = R_e\, \{E_c\, e^{-j\beta z}\, e^{j\omega t}\} \qquad ...(6.44b)$$

as wave moves in z-direction E and H lie in x-y plane. Here E_c is a complex vector, *i.e*, E_c can be written as

$$\vec{E}_c = E_1 + j\,E_2$$

where E_1 and E_2 are real vector at some point in space, say $z = 0$, then equation (6.44b) be comes;

$$E(0, t) = R_e\, [(E_1 + j\,E_2)e^{j\,\omega t}]$$

$$= E_1 \cos \omega t - E_2 \sin \omega t$$

as per the definition of circular polarization, the electric vector at $z = 0$, is expressed,

$$E_c = E_k a_x + E_k a_y$$

Hence, $$E(0, t) = (E_k \cos \omega t)\, a_y - (E_k \sin \omega t)\, a_y$$

$$E_x = E_k \cos \omega t$$

$$E_y = -E_k \sin \omega t$$

$$E_x^2 + E_y^2 = (E_k \cos \omega t)^2 + (-E_k \sin \omega t)^2$$

$$= E_k^2 (\cos^2 \omega t + \sin \omega t)$$

$$E_x^2 + E_y^2 = \text{E}K_k^2 \quad \text{...(6.45)}$$

The equation (6.45) represents a circle, and hence the wave is circular polarized wave.

(3) Elliptical polarization : If E_x and E_y are not equal in magnitude and differ by 90° phase, then, the Fig. 6.5 of resultant electric vector traces an ellipse. Then wave is said to be elliptically polarized wave.

Hence E_c can be written as,

$$E_c = a\,a_x + jba_y,$$

Hence $$E(0, t) = (a \cos \omega t)\, a_x - (b \sin \omega t)\, a_y$$

$$E_x = a \cos \omega t$$

$$E_y = -b \sin \omega t$$

$$\frac{E_x^2}{a^2} + \frac{E_y^2}{b^2} = 1$$

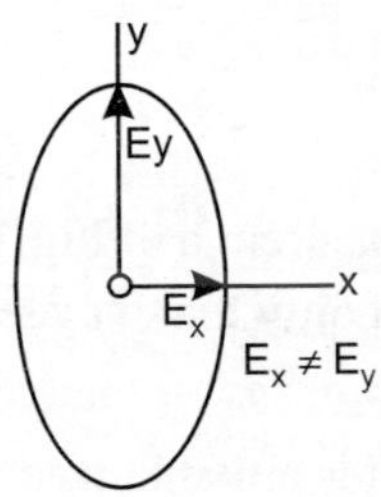

Fig. 6.5. Elliptical polarization.

This is the equation of ellipse and hence the wave is elliptically polarized.

Propagation of Electromagnetic Waves in Free Space

Electromagnetic wave being propagated in free space is divided into three (3) parts : (*i*) The ground wave, (*ii*) The sky wave (*iii*) Space waves. The ground wave is further divided into a direct wave and earth-reflected (indirect wave). Fig. 6.6 shows the wave compounds of eletromagnetic wave from a non-direction antenna to a receiving antenna.

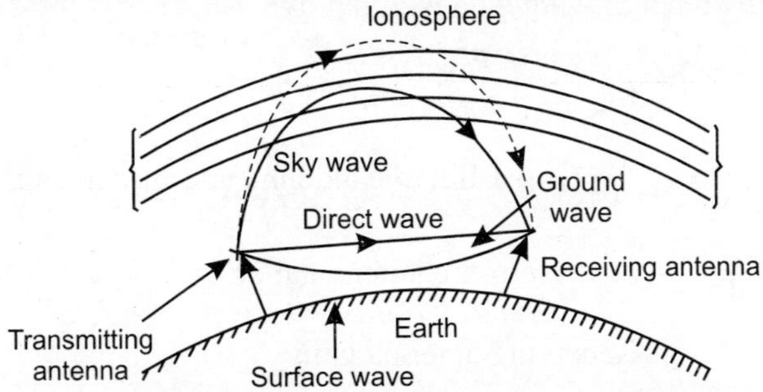

Fig. 6.6 Propagation of electromagnetic.

The ground wave of frequency up to 3MHz is also known as surface waves. Ground wave propagation depends on P_t, f and condition of earth. Sky wave mainly depends on angle of radiation P_t and strength of ionosphere, hence also known as ionosphere wave propagation. Space waves propagate to receiver with any reflation or refraction, following straight line, hence also known as LOS propagation.

The ionosphere is that region of the earth's atmosphere in which the constituent gases are ionized by solar radiation. The region intends from about 50km to 240 km and has different layers such as *C*,*D*,*E* and *F* layer in order of height.

The critical frequency for a particular layer can be given as,

$$f_c = 9\sqrt{N_{\max}} \qquad \text{...(6.46)}$$

where N_{max} is max electron density (electrons cubic per meter) of layer.

The sky wave reaches the receiving antenna after reflection from the ionosphere, although important in many communication systems, the sky wave need not be considered in most microwave and high frequency application; because a wavelength greater than about 4m, will not return back to the earth corresponding frequency f

$$f = \frac{c}{\lambda}$$

$$= \frac{3\times 10^8}{4} = 0.75 \times 10^8$$

$$= 75 \text{ MHz}$$

Energy radiated from the non-direction antenna of Fig. 6.6 strikes the earth at all points between the base of the antenna and the horizon, but only that wave which leaves the antenna in the direction shown reaches the receiver.

When power radiates from transmitting antenna, the power density carried by the spherical wave will decrease with distance as the energy in the wave spreads out over-increasing surface area as the wave progresses.

The power density is given by

$$P_d = \frac{P_t G_t}{4\pi R^2} \qquad \text{...(6.47)}$$

where P_t = Transmitting power in watts

G_t = Transmitting antenna gain

R = Distance between antenna and field point in meters.

The power received by receiving antenna is given by,

$$P_r = P_d A_c = \left(\frac{P_t G_t}{4\pi R^2}\right)\left(\frac{\lambda^2}{4\pi} G_r\right) \qquad \text{...(6.48)}$$

where $A_c = \frac{\lambda^2}{4\pi} G_r$ = Effective antenna aperture in square meters.

$\frac{\lambda^2}{4\pi} = A_q$ = Isotropic antenna aperture.

G_r = Receiving antenna gain

Fig. 6.7 shown the relationship of electromagnetic energy transmission in free space between two antennas.

If the receiving power is expressed in terms of decibels the equation (6.48) becomes

$$P_r = P_t + G_t + G_r + -20 \log\left(\frac{4\pi R}{\lambda}\right) \qquad ...(6.49)$$

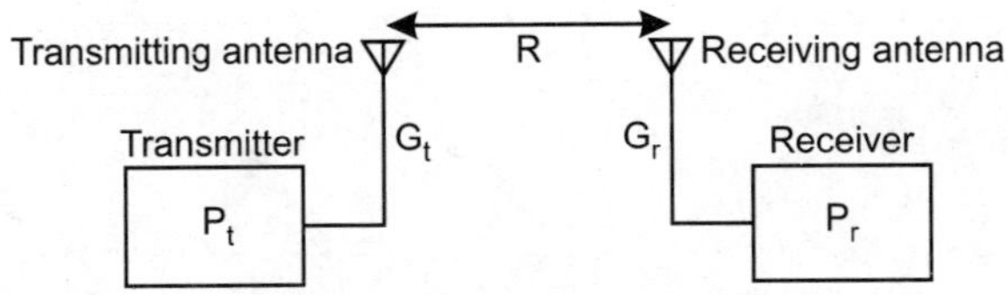

Fig. 6.7 Electromagnetic energy transmission between two antenna.

where P_t is *dBw*, G_r, G_t, are in dB

The term $20 \log\left(\frac{4\pi R}{\lambda}\right)$ is well known as frequency space attenuation in dB, and given by

$$L = 32.45 + 20 \log d\,(\text{km}) + 20 \log f\,(\text{MHZ})$$

SOLVED EXAMPLES

Example 1. Calculate the propagation constant γ for a material with $\mu_r = 1$, $\varepsilon_r = 8$ and $\sigma = 0.25$ ρ*S*/m; if the wave frequency is 1.6 MHz.

Solution : Given $\sigma = 0.25$ ρS/m

$$\mu_r = 1, \varepsilon_r = 8, f = 1.6 \times 10^6 \text{ Hz}$$

$$\frac{\sigma}{\omega\varepsilon} = \frac{0.25 \times 10^{-12}}{2\pi \times 1.6 \times 10^6 \times \varepsilon_0 \varepsilon_r}$$

$$= \frac{0.25 \times 10^{-12}}{2\pi \times 1.6 \times 10^6 \times \frac{1}{36\pi} \times 10^{-9} \times 8}$$

$$= 0.3515 \times 10^{-9} << 0$$

or
$$\frac{\sigma}{\omega\varepsilon} <<< 0$$

$$\sigma <<< 0$$

So it is perfect dielectric materials, and

$$\beta = \omega\sqrt{\mu\varepsilon}$$

$$= 2\pi f \sqrt{\mu_0 \mu_r \varepsilon_0 \varepsilon_r}$$

$$= 2\pi \times 1.6 \times 10^6 \sqrt{4\pi \times 10^{-7} \times 1 \times \frac{1}{36\pi} \times 10^{-2} \times 8}$$

$$\beta = 9.471 \times 10^{-2} \text{ rad/m}$$

so
$$\gamma = \alpha + j\beta$$

$$\gamma = 0 + j\,0.9471 \times 10^{-3} = j\,9.471 \times 10^{-2}$$

Example 2. Calculate the skin depth δ, γ and wave velocity v at a frequency of 1.6 MHz in alumenium where σ = 38.2 μ*s*/*m* and $\mu_r = 1$.

Solution : Since skin depth $\delta = \frac{1}{\alpha} = \frac{1}{\sqrt{\pi f \mu \sigma}} = \frac{1}{\sqrt{\pi \times 1.6 \times 10^6 \times \mu_0 \mu_r \sigma}}$

$$= \frac{1}{\sqrt{\pi \times 1.6 \times 10^6 \times 4\pi \times 10^{-7} \times 1 \times 38.2 \times 10^6}}$$

$$\delta = 64.409 \times 10^{-6} \text{ m}$$

$$\alpha = \beta = \frac{1}{\delta} = \frac{1}{64.409 \times 10^{-6}} = 1.5525 \times 10^4$$

$$\gamma = (\alpha + j\beta))$$

$$\gamma = (1.5525 + j\, 1.5525) \times 10^4$$

$$v = \frac{\omega}{\beta} = \frac{2\pi f}{\beta} = 2\pi f \delta$$

$$= 2\pi \times 1.6 \times 10^6 \times 64.409 \times 10^6 = 647.18 \text{ m/s}$$

Example 3. Assuming that the constitution parameters of alumenium are given by $\mu = \mu_0 = 4\pi \times 10^{-7}$ H/m and $\sigma = 3.54 \times 10^7$ mho/m calculate the frequency at which the aluminium has depth of penetration = 0.0001 cm.

Solution :

$$\delta = \frac{1}{\sqrt{\pi f \mu \sigma}}$$

$$f = \frac{1}{\delta^2 \pi \mu_0 \sigma} = \frac{1}{(0.0001 \times 10^{-2})^2 \times \pi \times 4\pi \times 10^{-7} \times 3.54 \times 10^7}$$

$$f = 71.625 \text{ MHz}$$

Example 4. Calculate the skin depth in copper at 10 GHz. Assume conductivity of copper is $\sigma = 5.8 \times 10^7$ mho/m and permeability equal to that of free space.

Solution :

$$\delta = \frac{1}{\sqrt{\pi f \mu \sigma}} = \frac{1}{\sqrt{\pi \times 10 \times 10^9 \times 4\pi \times 10^{-7} \times 5.8 \times 10^7}}$$

$$\delta = 6.348 \times 10^{-7} \text{ m}$$

Example 5. Find the intrinsic impedance η, the propagation constant γ, and the wave velocity v for a conduction medium in which $\sigma = 58 \times 10^6$ \$/m, $\mu_r = 1$, at a frequency $f = 100 \times 10^6$ Hz.

Solution : Since

$$\alpha = \beta = \frac{1}{\delta}$$

$$\delta = \frac{1}{\sqrt{\pi f \mu \sigma}} = \frac{1}{\sqrt{\pi f \mu_0 \mu_r \sigma}}$$

$$= \frac{1}{\sqrt{\pi \times 100 \times 10^6 \times 4\pi \times 10^{-7} \times 58 \times 10^6}}$$

$$= \frac{1}{151.24 \times 10^3} = \frac{1}{\alpha} = \frac{1}{\beta}$$

$$\alpha = \beta = 151.24 \times 10^3$$

$$\delta = \frac{1}{151.24 \times 10^3}\,\mu\text{m} = 6.612\ \mu\text{m}$$

$$v_p = \frac{\omega}{\beta} = \omega\delta = 2\pi f\delta$$

$$= 2\pi \times 100 \times 10^6 \times 6.612 \times 10^{-2} = 4152.336 \text{ m/s}$$

For good conductor

$$\gamma^2 = j\omega\mu_r\left(1 + \frac{j\omega\varepsilon}{\sigma}\right) = \sqrt{j\omega\mu\sigma}$$

$$= \sqrt{\omega\mu\sigma} < 45°$$

$$= \sqrt{2\pi f\mu_r,\mu_0\sigma} < 45°$$

$$= \sqrt{2\pi \times 100 \times 10^6 \times 4\pi \times 10^{-7} \times 1 \times 58 \times 10^6}$$

$$\gamma = 213.89 \times 10^3 < 45°$$

$$\eta = \frac{j\omega\mu}{\gamma} = \sqrt{\frac{j\omega\mu}{\sigma + j\omega\varepsilon}} = \sqrt{\frac{j\omega\mu}{\sigma}} = \sqrt{\frac{\omega\mu}{\sigma}} < 45°$$

$$= \sqrt{2\pi f\mu_0\mu_r/\sigma}$$

$$= \sqrt{\frac{2\pi \times 100 \times 10^6 \times 1 \times 4\pi \times 10^{-7}}{58 \times 10^6}}$$

$$\eta = 3.6877 \times 10^{-3} < 45°$$

Example 6: If x < 0 defines region 1 and $x > 0$ defines region 2. Region is characterized by $\mu_{r1} = 3.0$ and region 2 is characterized by $\mu_{r2} = 5.0$. If the electric field in region 1 is given by $H_1 = 4.0a_x + 1.5a_y + 3.6a_z$ A/m find $\overline{H}_2$ and H_2.

Solution : $H_1 = 4.0a_x + 1.5a_y - 3.0a_z$ A/m

for the regions given $H_{t1} = 1.5ay - 3.0a_z$ A/m

$$H_{tt1} = 4.0a_x$$

as $H_{t1} = H_{t_2}$ boundary condition $H_{t_2} = 1.5a_y = 3.0a_x$ A/m

$$H_{tta} = \frac{\mu_1}{\mu_2} = H_{n1} = \frac{3}{5}4.0a_x = 2.4a_x$$

so

$$\vec{H}_2 = \overline{H}_{t1} + \overline{H}_{tt2}$$

$$H_2 = 2.4a_x + 1.5a_y - 3.0a_z$$

The magnitude of H_2 is

$$H_2 = \sqrt{H_x^2 + H_y^2 + H_z^2} = \sqrt{(2.4)^2 + (1.5)^2 + (-3)^2} = 4.1243\text{A/m}$$

Example 7: If $H = \cos(10^8 t - \beta z)\, a_y$, A/m and $E = 377 \cos(10^8 t - \beta z)a_x$ V/m in free space, Find the frequency λ, phase β, and intrinsic impedance of the medium.

Solution :

$$\omega = 2\pi f = 10^8$$

$$f = \frac{10^8}{2\pi} = 15.9 \text{ MHz}$$

$$\beta = \frac{2\pi}{\lambda} = \frac{2\pi f}{\lambda f} = \frac{\omega}{v}$$

$$\lambda = \frac{v_0}{f} = \frac{3\times 10^8}{15.9\times 10^6} = 18.86 \text{ m}$$

$$\lambda = 18.86 \text{ m}$$

$$\beta = \frac{2\pi}{\lambda} = 0.3329 \text{ rad/m}$$

$$\eta_0 = \left|\frac{E}{H}\right| = 377\ \Omega$$

Example 8: When the intensity of the magnetic field in a plane wave is 2 A/m. (*a*) Determine the intensity of the electron field for the plane wave in free space. (*b*) Determine the electric field where wave propagates in medium which is characterized by,

$$\sigma = 0, \mu = \mu_0 \text{ and } \varepsilon = 4\varepsilon_0$$

Solution : (*a*) $\frac{E}{H} = \eta_0 = \sqrt{\frac{\mu_0}{\varepsilon_0}} = 120\pi$

$$H = 2 \text{ A/m.}$$

$$E = \eta_0 H = 120\pi \times 2\ \text{E} = 240\ \pi\text{V/m}$$

(*b*) $\sigma = 0, \varepsilon_r = 4, \mu_r = 1$

$$\eta_1 = \sqrt{\frac{\mu_0}{\varepsilon_0}} = \sqrt{\frac{\mu_0}{4\varepsilon_0}} = \frac{1}{2}\sqrt{\frac{\mu_0}{\varepsilon_0}} = \frac{120\pi}{2} = 60\pi$$

so $E = \eta H = 60\pi \times 2$

$$E = 120\pi \text{ V/m}$$

Example 9: The wavelength of x-direction plane wave in a loss-less medium is 0.25 m and the velocity of propagation is 1.5×10^{10} cm/s. The wave has electric fields with an amplitude equal 10 V/m. Find the frequency and permitivity of the medium, if $\mu = \mu_0$.

Solution : $v = 1.5 \times 10^{10} \text{ cm/sec} = 1.5 \times 10^8 \text{ m/sec}$

$$f = \frac{v}{\lambda} = \frac{1.5\times 10^8}{25\times 10^{-2}} = 0.06 \times 10^6 = 600 \text{ MHz.}$$

and we have $v_2 = \frac{1}{\sqrt{\mu\varepsilon}} = \frac{1}{\sqrt{\mu_0\varepsilon_0\varepsilon_r}} = \frac{v_0}{\sqrt{\varepsilon_r}} = 1.5 \times 10^8$

$$\sqrt{\varepsilon_r} = \frac{3\times 10^8}{1.5\times 10^8} = 2$$

$$\varepsilon_r = 1.141$$

Example 10: Identity frequency and phase constant when electriccc field of an EM wave is given $E = 5.0 \sin(10^8 t - 4.0x)a_z$. Also find λ.

Solution : $E = 5.0 \sin(10^8 t - 4.0x)a_z$

$$\omega = 10^8$$

$$2\pi f = 10^8$$

$$f = \frac{10^8}{2\pi} = 15.915 \text{ MHz}$$

$$\beta = 4$$

$$\lambda = \frac{v_0}{f} = \frac{3 \times 10^8}{15.915 \times 10^6} = 18.85 \text{m}$$

Example 11: Find the depth of penetration (δ) of an EM wave in copper at f = 60 Hz and 100 MHz for copper $\sigma = 5.8 \times 10^7$ mho/ m, $\mu_1 = 1$, $\varepsilon_r = 1$.

Solution : For copper f = 60 Hz

$$\frac{\sigma}{\omega\varepsilon_0} = \frac{5.8 \times 10^7}{2\pi \times 60 \times 8.854 \times 10^{-12}} = 174 \times 10^4 >> 1.$$

Therefore at f = 60 Hz the copper is a good conductor.

The depth of penetration

$$\delta = \frac{1}{a} = \sqrt{\frac{2}{\omega\mu\sigma}} = \sqrt{\frac{1}{\pi f \mu\sigma}} = \frac{1}{\sqrt{\pi \times 60 \times 4\pi \times 10^{-7} \times 5.8 \times 10^7}}$$

$$\delta = 8.53 \times 10^{-3} \text{ m.}$$

at f = 100 MHz $\frac{\sigma}{\omega\varepsilon_0} = \frac{5.8 \times 10^7}{2\pi \times 100 \times 10^6\ 8.8854 \times 10^{-12}} = 0.10425 \times 10^{11} = 10.425 \times 10^9 >> 1.$

Hence copper is good conductor at f = 100 MHz

so $$\delta = \frac{1}{\alpha} = \frac{1}{\sqrt{\pi f \mu\sigma}} = \frac{1}{\sqrt{\pi \times 100 \times 10^6 \times 4\pi \times 10^{-7} \times 5.8}}$$

$$= 6.608 \times 10^{-6} \text{ m}$$

EXERCISES

1. If $E = 2 \times \cos(10^8 t = 20\underline{x} + 40)a_z$, what is the phaser form of E.
2. If the electric and magnetic fields in free space are given by $E = 10 \cos(\omega t - \beta z)a_x$ and $H = \frac{10}{\eta} = \cos(\omega t - \beta z)ay$ find η.
3. What do you understand by Brewster angle show that $\theta_B = \tan^{-1}\sqrt{\frac{\varepsilon_2}{\varepsilon_1}}$.
4. Explain the polarization of electromagnetic waves.
5. What do you understand by skin effect? Show that $\delta = \left(\frac{1}{\sqrt{\pi f \mu\sigma}}\right)$.
6. Sea water has conductivity of 4 mhos/m and relative dielectric constant $\varepsilon_r = 50$ at a frequency of 4 GHz. Compute
   (*a*) The intrinsic impedance (*b*) The propagation constant
   (*c*) The phase velocity
7. Repeat Numerical 6 for dry sand ($\sigma = 2 \times 10^{-4}$ mho/m and $\varepsilon_r = 4$) and copper ($\sigma = 5.8 \times 10^7$ mhos/m).
8. Show that reflection coefficient of oblique incidence reflection when H is in the plane of incident can be given as

$$\Gamma = \frac{\eta_2 \sec\theta_t - \eta_1 \sec\theta_i}{\eta_2 \sec\theta_t + \eta_1 \sec\theta_i}$$

9. What will be transmission coefficient in Question 8?

10. Show that energy carried out by electric field in z direction in the free space is given as

$$S = \frac{E_z^2}{120\pi} \text{ J/m}^2 \quad \text{and} \quad \text{also } S = \frac{H_z^2}{120\pi} \text{J/m}^2$$

11. What do you understand by Polynting theorem? Explain and prove this theorem.

12. Explain the propagation of the plane electromagnetic wave in lossy dielectircs and find expression for γ, α, β η and phase velocity.

13. What do you understand by loss tangent? Show that it can be given by $\theta = \tan{-1}\left(\frac{\sigma}{\omega\varepsilon}\right)$

14. What do you understand by skin effect? Calculate depth of penetration of plane electromagnetic wave and surface resistance wave and surface resistance for good conductor.

15. What do you understand by skin effect? Calculate depth of penetration of plane electromagnetic wave and surface resistance wave and surface resistance for good conductor.

16. Prove boundary expression for conduction medium and obtain expression for the propagation constant γ and characteristics impedance (η) for good dielectric ($\alpha <<< \omega\varepsilon$) and good conductor ($\alpha >>> \omega\varepsilon$).

17. Calculate the value of α, β, γ, λ, V_p and η for a damp soil at $f = 10^6$ Hz, such that for damp soil, $\varepsilon_r = 12$, $\sigma = 2 \times 10^{-2}$ mho/m, $\mu_r = 1$.

18. Calculate α, β, λ, γ, V_p and η for rubber at $f = 10^6$ Hz. It is given that $\varepsilon_r = 3.5$, $\sigma = 10^{-6}$ mho, $\mu_r = 1$ rubber,

19. Find the velocity of plane wave in a loss less medium having a relative permeability of 4 and relative permeability of unity.

20. For a uniform plane wave in fresh lake water $\alpha = 10^{-3}$ mho/m, $\varepsilon_r = 80$, $\mu = \mu_0$, Calculate α, β, γ and for two frequencies 100 MHz and 20 KHz.

21. Calculate the phase velocity and the magnitude of the attenuation constant of the plane wave at a frequency 5 GHz in polythene. It is given that $\mu = \mu_0$, $\varepsilon_r = 2.3$ and $\sigma = 2.56 \times 10^{-4}$ mho/m

(**Hint :** Ans. $V_p = 1.9782 \times 108$ m/s, s = 318,03 $\times 10^{-4}$ neper/m.)

- "When people throw stones on you convert them into milestones"

Sachin Tendulkar

Chapter 7

TRANSMISSION LINES

7.1 INTRODUCTION

Transmission lines are defined as means to convey electrical signal or energy between two points. These two points may be transmitter and receiver or Source and destination, as shown in Fig. 7.1.

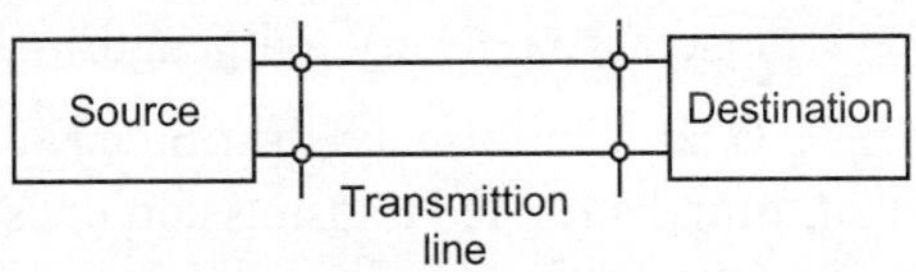

Fig. 7.1. Basic use of a Tx line

The source may be a hydro electric generator, a transmitter or an oscillator, however load *TV* receiver and oscilloscope. Transmission lines may be of following types:

(*i*) Parallel wire or twin wire
(*ii*) Coaxial cable
(*iii*) Waveguides
(*iv*) Striplines
(*v*) Microstrip line

The circuit theory are generally used usual to the analyse the properties of the *Tx* lines.

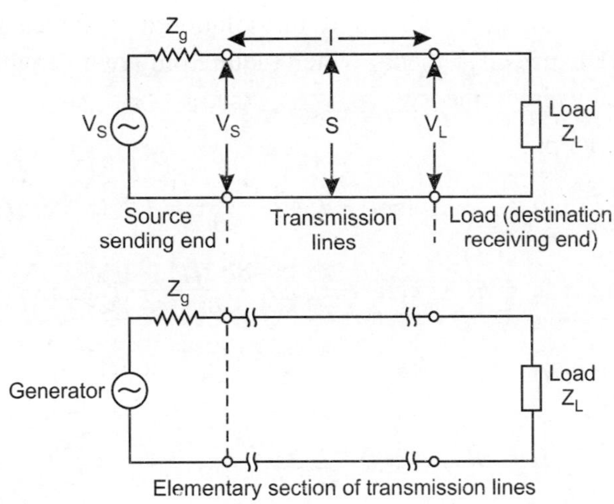

Fig. 7.2 Transmission line as a distributed circuit elements.

Transmission lines are passive components. Parallel wire is called balanced transmission lines and others are unbalanced transmission lines. Since, each conductor of transmission lines has a certain length and diameter it must have resistance *R* and since resistances vary with distance, Tx line is also called distributed elements.

$$R = \text{Resistance } (\Omega/\text{unit-length})$$

Since each conductor is closed to each other and separated by insulating material so it has some capacitance called distributed capacilance.

$$C = \text{Capacitance} = (\text{F/unit length})$$

Lastly since wires are separated by a medium known as dielectric medium that can not be perfect in its insulation and hence current leaks through it. This leakage of current through the dielectric is representey a short conductance G.

$$G = \text{Conductance (Mhos/unit-length)}$$

Conducting material has some inductive effect also

$$L = \text{Inductance (Henary/unit length)}$$

When all elements R, L,C, G are uniformaly distributed along, the entire length of the transmission lines it is called uniform distributed transmission lines (Fig. 7.2).

The equivalent ckt of transmission lines is shown in Fig. 7.3.

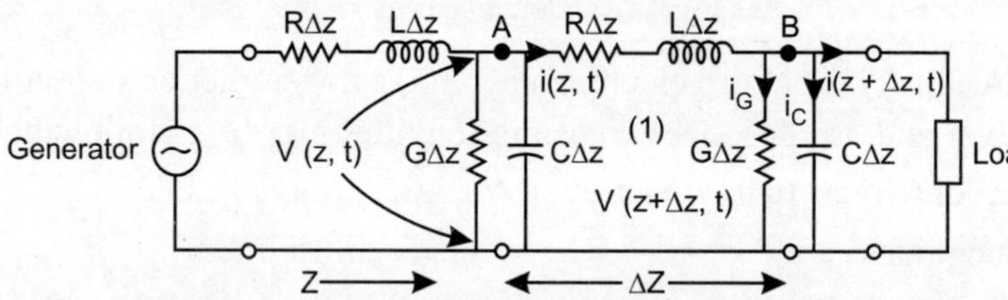

Fig. 7.3 Equivalent circuit of transmission line.

Transmission line can also be analysed by the solution of Maxwell's equation but it requires four variable (*x*, *y*, *z*, *t*), hence CKT theory which require only two variable *z*, *t* is prefered. So, we will analyze a Tx line by circuit theory.

By the KVL in (1) loops

$$V(z, t) = R\,\Delta z i\,(z, t) + L\,\Delta z\,\frac{\partial i\,(z, t)}{\partial t} + V\,(z + \delta z, t)$$

$$= R\,\Delta z\,i\,(z, t) + L\,\Delta z\,\frac{\partial i(z, t)}{\partial t} + \left[V\,(z, t) + \frac{\partial V\,(z, t)}{\partial z}\Delta z\right]$$

$$V\,(z, t) = R\,\Delta\,i\,(z, t) + L\,\Delta z.\,\frac{\partial i\,(z, t)}{\partial t} + V\,(z, t) + \frac{\partial V\,(z, t)}{\partial z}\,\Delta z$$

ignoring $V\,(z, t)$ $$0 = R\,\Delta z i\,(z, t) + L\,\Delta z\,\frac{\partial i\,(z, t)}{\partial t} + \frac{\partial V\,(z, t)}{\partial z}\,\Delta z \qquad (\Delta z \neq 0)$$

$$Ri + L\frac{di}{dt} + \frac{\partial V}{\partial z} = 0$$

$$\Rightarrow \qquad -\frac{\partial V}{\partial z} = Ri + L\frac{di}{dt} \qquad ...(7.1)$$

Now by KCL at node (B).

$$i(z, t) = i_G + i_C + i(z + \Delta z, t)$$

$$\Rightarrow \qquad i(z, t) = G\,\Delta z\, V(z + \Delta z, t) + C\,\Delta z\, \frac{\partial V(z + \Delta z, t)}{\partial t} + i(z + \Delta z, t)$$

$$i(z, t) = G\,\Delta z \left[V(z, t) + \frac{\partial V(z, t)}{\partial z}\Delta z\right] + C\Delta z \frac{\partial}{\partial t}$$

$$\left[V(z, t) + \frac{\partial V(z, t)\Delta z}{\partial z}\right] + \left[i(z, t) + \frac{\partial i(z, t)}{\partial z}\,\Delta z\right]$$

$$= G\Delta z\, V(z, t) + G\frac{\partial V(z, t)}{\partial z}(\Delta z)^2 + C\Delta z\, \frac{\partial V(z, t)}{\partial t}$$

$$+ C\Delta z \frac{\partial^2}{\partial t}\frac{V(z, t)}{\partial z}(\Delta z)^2 + i(z, t) + \frac{\partial i(z, t)}{\partial z}\,\Delta z$$

Omitting the $(\Delta z)^2$ (Since $\Delta z <<< 1$)

$$= G\Delta z\, V(z, t) + C\Delta z \frac{\partial V(z, t)}{\partial t} + \frac{\partial i(z, t)}{\partial z}\Delta z$$

$$0 = \left[GV(z, t) + \frac{C\Delta V(z, t)}{\partial t} + \frac{\partial i'(z, t)}{\partial z}\right]\sigma z \qquad \Delta z \neq 0$$

$$\Rightarrow GV(z, t) + \frac{C\partial V(z, t)}{\partial t} + \frac{\partial i(z, t)}{\partial z} = 0$$

$$\Rightarrow \qquad -\frac{\partial i}{\partial z} = GV + C\frac{\partial V}{\partial t} \qquad ...(7.2)$$

Now differentiating eqn. (7.1) w.r.t. z and (7.2) w.r.t t and combining the result gives;

$$\frac{\partial^2 V}{\partial z^2} = RGV + (RC + LG) \times \frac{\partial V}{\partial t} + LC\,\frac{\partial^2 V}{\partial t^2} \qquad ...(7.3)$$

Also differentiating 7.1 w.r.t. t and (7.2) w.r.t. z and combining the result gives

$$\frac{\partial^2 i}{\partial z^2} = RGi + (RG + LG) \times \frac{\partial i}{\partial t} + LC\frac{\partial^2 i}{\partial t^2} \qquad ...(7.4)$$

Let phasor form voltage (V) and current (I) are defined as;

$$V = V_0 e^{j\omega t} \text{ and } I = I_0 e^{j\omega t}$$

$$\Rightarrow \qquad \frac{\partial V}{\partial t} = V_0 j\omega e^{j\omega t} = j\omega V$$

$$\frac{\partial V}{\partial t} = j\omega V \qquad ...(7.5a)$$

Similarly,

$$\frac{\partial I}{\partial t} = j\omega I \qquad ...(7.5b)$$

and also (By second derivatives) $\dfrac{\partial^2 V}{\partial t^2} = -\omega^2 V$ ...(7.6)

$$\frac{\partial^2 I}{\partial t^2} = -\omega^2 I \qquad ...(7.7)$$

Eqn. (7.1), (7.2), (7.3) and (7.4) may be combined as;

$$-\frac{dV}{\partial z} = (R + L\,j\omega)I$$

or
$$-\frac{dV}{dz} = ZI \qquad ...(7.8)$$

$Z = (R + j\omega L)$ imbedance/unit length

Similarly
$$\frac{\partial I}{\partial z} = (G + j\omega C)\, V = YV$$

$$\frac{dI}{\partial z} = -YV \qquad ...(7.9)$$

Y = admittance mho/unit length

and from eqn. (7.3) and (7.4)

$$\frac{d^2 V}{dz^2} = \gamma^2 V \qquad ...(7.10a)$$

$$\frac{d^2 I}{dz^2} = \gamma^2 I \qquad ...(7.10b)$$

where,
$$\gamma = \text{called propagation constant}$$
$$= \alpha + i\beta = Z.Y$$
$$\alpha = \text{called attenuation constant N/m.}$$
$$\beta = \text{phase constant rad/ length}$$

Solution of eqn. (7.10a) and (7.10b) will be standard linear differential equation of the form;

$$V = V_+ e^{-\gamma z} + V_- e^{rz} \qquad ...(7.11a)$$

$$I = I_+ e^{-\gamma z} + I_- e^{\gamma z} \qquad ...(7.11b)$$

Hower for loss transmission lines,

$$R = G = 0 \qquad ...(7.12)$$

and at radio and microwave freq R <<< ωL and G << ωC

Expressions (7.1), (7.2), (7.3) and (7.4) are called general equations of transmission lines (Tx lines).

Impedance (Admittance) and Forward, (Backward) Traveling Waves

When a source (V, I) is applied at the input of transmission line. it propagates along the transmission lines. The voltage and current at any point of transmission line is the summation of transmitted and reflected waves due to impedance mismatch between transmission line and load. Transmitted voltage wave or current wave and reflected voltage or current wave can be represented as,

$$V = V\,\text{transmitted} + V\,\text{reflected}$$

$$= A.e^{-\gamma z} + Be^{+\gamma z}$$

$$V = V_+e^{-\gamma z} + V_-e^{\gamma z} \qquad ...(7.13a)$$

and forward, backward current

$$I = Ae^{-\gamma z} + Be^{\gamma z}$$

$$I = Ie^{-\gamma z} + I_-e^{\gamma z} \qquad ...(713b)$$

First terms of equation (7.13a) and (7.13b) called forward traveling wave and second term are called backward traveling wave. These two components can be given in Fig. 7.4.

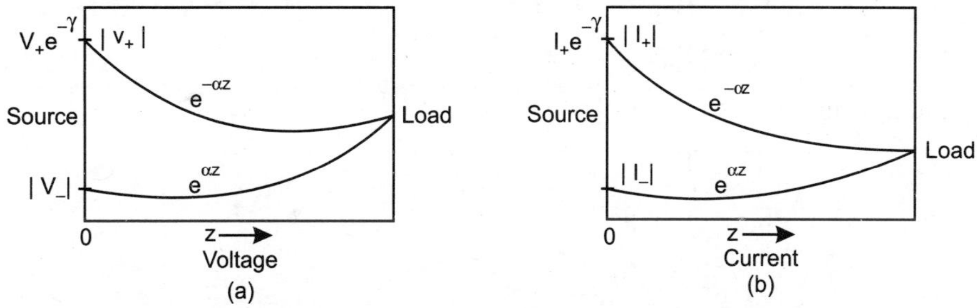

Fig. 7.4 Variation of voltage and current

The above two equations (7.13*a* & 7.13*b*), are the solutions of transmission line equations (7.3) and (7.4).

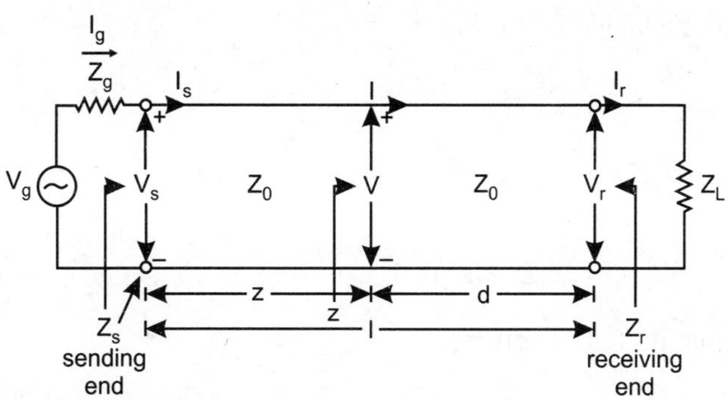

Fig. 7.5 Equivalent circuit of a Tx line with incident and ventral waves.

The impedance of a transmission line is a complex ratio of the voltage (phasor) and current phasor at that point. It is defined as,

$$Z \equiv \frac{V(z)}{I(z)} \qquad ...(7.14)$$

In general the voltage or current along a Tx line is the sum of the respective incident and reflected waves i.e., (Fig. 7.5)

$$V = V_{inc} + V_{ref} = V_+e^{-\gamma z} + V_-e^{\gamma z} \qquad ...(7.15a)$$

$$I = I_{inc} + I_{ref} = I_+e^{-\gamma z} + I_-e^{\gamma z} \qquad ...(7.15b)$$

If characteristic impedance of the line is Z_0;

then
$$I_+ = \frac{V_+}{Z_0}$$

$$I_- = \frac{-V_-}{Z_0}$$

So from eqn. (7.15*a*) and (7.15*b*), we get

$$V = V_+ e^{-\gamma z} + V_- e^{\gamma z} \quad ...(7.16a)$$

$$I = \left(\frac{V_+}{Z_0} e^{-\gamma z} - \frac{V_- e^{\gamma z}}{Z_0}\right) \quad ...(7.16b)$$

$$I = Y_0(V_+ e^{-\gamma z} - V_- e^{\gamma z})$$

At the sending end $Z = 0$, so $V = V_s$, $I = I_s$

$$V = V_+ + V_-$$

$$V_S = V_+ + V_- \quad ...(7.17)$$

$$I = \frac{(V_+ - V_-)}{Z_0}$$

$$I_s Z_0 = (V_+ - V_-) \quad ...(7.18)$$

$$V_+ + V_- = V_S = I_S Z_S$$

Now solving eqn. (7.17) and (7.18); $V_+ - V_- = I_S Z_0$

$$2V_+ = I_S(Z_S + Z_0)$$

$$V_+ = \frac{I_S(Z_S + Z_0)}{2} \quad ...(7.19)$$

$$V_- = \frac{I_S}{2}(Z_S - Z_0) \quad ...(7.20)$$

Replacing these value in (7.15b) gives;

$$V = \frac{I_S(Z_S + Z_0)e^{-\gamma z}}{2} + \frac{I_S(Z_S - Z_0)e^{+\gamma z}}{2} \quad ...(7.21a)$$

$$I = \frac{Z_S(Z_S + Z_0)e^{-\gamma z}}{2Z_0} - \frac{I_S(Z_S - Z_0)e^{-\gamma z}}{2Z_0} \quad ...(7.21b)$$

Dividing (7.21*a*) by (7.21*b*) yields;

$$\frac{V}{I} = \frac{\dfrac{I_S(Z_S + Z_0)e^{-\gamma z}}{2} + \dfrac{I_S(Z_S - Z_0)e^{\gamma z}}{2}}{\dfrac{I_S(Z_S + Z_0)e^{-\gamma z}}{2Z_0} - \dfrac{I_S(Z_S - Z_0)e^{+\gamma z}}{2Z_0}}$$

$$= \frac{(Z_S + Z_0)e^{-\gamma z} + (Z_S - Z_0)e^{\gamma z}}{(Z_S + Z_0)e^{-\gamma z} - (Z_S - Z_0)e^{\gamma z}}$$

$$= Z_0 \left[\frac{Z_S e^{-\gamma z} + Z_0 e^{-\gamma z} + Z_s e^{-\gamma z} - Z_0 e^{\gamma z}}{Z_S e^{-\gamma z} + Z_0 e^{-\gamma z} - Z_S e^{\gamma z} + Z_0 e^{\gamma z}} \right]$$

$$= Z_0 \left[\frac{Z_S (e^{\gamma z} + e^{-\gamma z}) Z_0 (e^{\gamma z} - e^{-\gamma z})}{Z_S (-e^{\gamma z} + e^{+\gamma z}) + Z_0 (e^{\gamma z} + e^{-\gamma z})} \right] \qquad ...(7.22)$$

But we know that

$$\frac{e^{j\theta} + e^{-j\theta}}{2} = \cos\theta$$

$$\frac{e^{j\theta} - e^{-j\theta}}{2j} = \sin\theta$$

and

$$\frac{e^{\theta} + e^{-\theta}}{2} = \cosh\theta$$

Therefore,

$$\frac{e^{\theta} - e^{-\theta}}{2} = \sinh\theta$$

$$\frac{V}{I} = Z_0 \left[\frac{Z_S 2 \cosh \gamma z + Z_0 2 \sin \gamma z}{-Z_S 2 \sinh \gamma z + Z_0 2 \cosh \gamma z} \right]$$

or

$$Z = Z_0 \left[\frac{Z_S \cosh \gamma z + Z_0 \sinh \gamma z}{Z_0 \cosh \gamma z - Z_S \sinh \gamma z} \right]$$

$\Rightarrow$

$$Z = Z_0 \left[\frac{Z_S + Z_0 \tanh \gamma z}{Z_0 - Z_S \tanh \gamma z} \right] \qquad ...(7.23)$$

Similarly, by taking load impedance or impedance at load side, replacing $Z_l = Z_s$, $z = l$

$$Z = Z_0 \left[\frac{Z_l + Z_0 \tanh \gamma l}{Z_0 - Z_l \tan \gamma l} \right]$$

Similarly line impedance at any point on receiving end will be

$$Z = Z_0 \frac{Z_l \cosh \gamma l + Z_0 \sinh \gamma l}{Z_0 \cosh \gamma l - Z_l \sinh \gamma l}$$

$$= Z_0 \left[\frac{Z_l + Z_0 \tanh \gamma l}{Z_0 - Z_l \tanh \gamma l} \right] \qquad ...(7.24)$$

If transmission line is loss less, *i.e.*,

$$\alpha = 0$$

$$\gamma = \alpha + j\beta = 0 + j\beta = j\beta$$

$$\sinh j\beta z = j \sin \beta z$$

$$\cosh j\beta z = \cos \beta z$$

$$Z = Z_0 \left[\frac{Z_S + Z_0\, j \tan \beta z}{Z_0 - Z_S j \tan \beta z} \right] \quad \text{...(7.25a)}$$

and

$$Z = Z_0 \left[\frac{Z_l + Z_0 j \tan \beta z}{Z_0 - jZ_l \tan \beta z} \right] \quad \text{...(7.25b)}$$

All these equations; (7.23), (7.24), (7.25*a*) and (7.25*b*) express impedance of the transmission lines.

Reflection and Transmission Coefficeints of Transmission Lines

Any travelling wave along transmission line contains two components one travelling in the positive direction and other in negative direction. If load impedance is equal to the line characteristics impedance, then the reflected wave does not exist. Fig. 7.6 shows a transmission line terminated with an impedance Z_l.

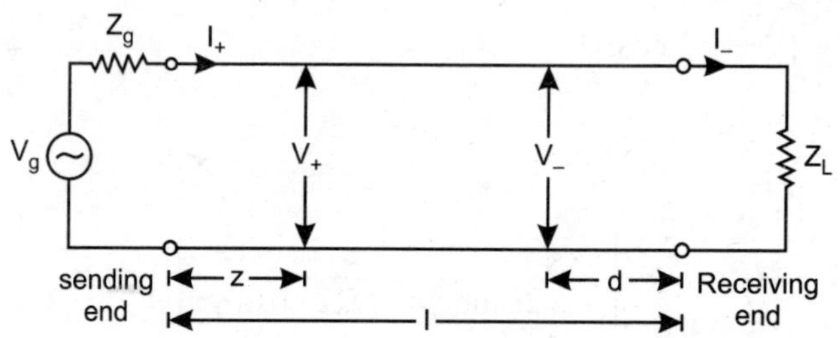

Fig. 7.6 Transmission line terminated with load impedance.

The reflection coefficient Γ can be defined as,

$$= \frac{\text{reflected voltage or current}}{\text{incident voltage or current}}$$

$$\Gamma = \frac{V_{ref}}{V_{inc}} = \frac{I_{ref}}{I_{inc}} \quad \text{...(7.26)}$$

Now, from equn. (7.16*a*) and (7.16*b*)

$$Z_l = \frac{1 + \dfrac{V_{\_} e^{\gamma e}}{V + e^{-\gamma e}}}{1 - \dfrac{V_{\_} e^{-\gamma e}}{V_{+} e^{-\gamma e}}}$$

$$Z_l = Z_0 \left[\frac{1+\Gamma}{1-\Gamma} \right] \quad \text{...(7.27)}$$

$$\frac{Z_l}{Z_0} = \frac{1+\Gamma}{1-\Gamma}$$

$$Z_l - Z_l \Gamma = Z_0 + Z_0 \Gamma$$

$$Z_l - Z_0 = \Gamma (Z_0 + Z_l)$$

$$\Gamma = \frac{Z_l - Z_0}{Z_l + Z_0} \quad ...(7.28)$$

Now if load and characteristic impedance is complex quantity, the value of Γ will also be complex and can be expressed as

$$\Gamma_l = |\Gamma_l| e^{-j\theta_l}. \quad ...(7.29)$$

where,

$$|\Gamma_l| = \text{Magnitude} \leq 1.$$

θ_l = Phase angle between incident and reflected voltages at receiving end.

The general solution of reflection coefficient at any point along the transmission line is given as;

$$\Gamma = \frac{V_+ e^{-\gamma z}}{V_- e^{-\gamma z}} \quad ...(7.30)$$

However, from the sending end,

$$\Gamma_l = \left|\frac{V_+}{V_-}\right| e^{2\gamma z}$$

$$\Gamma_l = |\Gamma_l| e^{2\gamma z} \quad ...(7.31)$$

Now the reflection coefficient at any point from the receiving end ($z + d = l$) or $z = l - d$ will be;

$$\Gamma_d = \frac{V + e^{-\gamma(l-d)}}{V_- e^{+\gamma(l-d)}} \quad ...(7.32)$$

$$= \left(\frac{V_r}{V_i}\right) \frac{e^{-\gamma e}\, e^{\gamma d}}{e^{-\gamma e}\, e^{-\gamma d}}$$

$$\Gamma_d = \Gamma_l e^{-2\gamma l} \quad ...(7.33)$$

Therefore, $0 \leq |\Gamma_l| \leq 1$

It has been proved that Γ_l will be zero, and there will be no reflection from the receiving end provided the termination impedance is equal the characteristic impedance of the line. A line terminated in its characteristic impedance Z_0 is called a proper terminated line, otherwise it is called an improper terminated line.

Transmission Coefficient: It is defined as

$$T = \frac{\text{transmitted voltage or current}}{\text{incident voltage or current}}$$

$$= \frac{V_{tr}}{V_{inc}} = \frac{I_{tr}}{I_{inc}} \quad ...(7.34)$$

Fig. 7.7 shows the transmission of power along a transmission line (where P_{inc} is the incident power, P_{ref} is the reflected power and P_{tr} is the transmitted power).

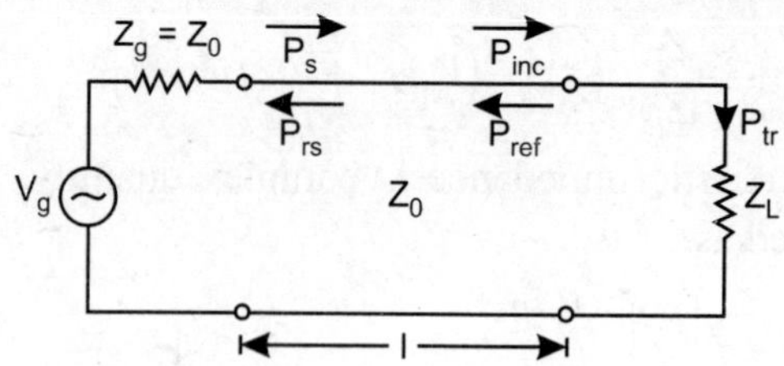

Fig. 7.7 Power along a transmission line.

Traveling waves at receiving end;

$$V_+e^{-\gamma l} + V_-e^{\gamma l} = V = V_{tr}e^{-\gamma l} \quad ...(7.35a)$$

$$I_+e^{-\gamma l} + I_-e^{\gamma l} = I = I_{rt}\,e^{-\gamma l} \quad ...(7.35b)$$

or
$$\frac{V+e^{-\gamma l}}{Z_0} - \frac{V_-e^{\gamma l}}{Z_0} = I = \frac{V_{tr}e^{-\gamma l}}{Z_l} \quad ...(7.36)$$

$$\frac{V_+e^{-\gamma l}}{Z_0} - \frac{V-e^{\gamma l}}{Z_0} = \frac{1}{Z_l}\,[V_{tr}e^{-\gamma l}] = \frac{1}{Z_l}[V_+\,e^{-\gamma l} + V_-e^{\gamma l}]$$

$$\Rightarrow \quad \frac{Z_l}{Z_0}[V_+e^{-\gamma l} - V_-e^{\gamma l}] = [V_+e^{-\gamma l} + V_-e^{\gamma l}]$$

$$\frac{Z_l}{Z_0} = \frac{V_+e^{-\gamma l} + V_-e^{\gamma l}}{V_+e^{-\gamma l} - V_-e^{\gamma l}}$$

$$\frac{Z_l}{Z_0} = \frac{1+\Gamma_l}{1-\Gamma_l}$$

or
$$\Gamma_l = \frac{Z_l - Z_0}{Z_l - Z_0} = \frac{V_-e^{\gamma l}}{V_+e^{-\gamma l}} \quad ...(7.37)$$

$$V_+e^{-\gamma l} + V_-e^{\gamma l} = V_{tr}\,e^{-\gamma l}$$

$$V_+e^{-\gamma l}\left[1 + \frac{V_-e^{\gamma l}}{V_+e^{\gamma l}}\right] = V_{tr}e{-}\gamma l$$

$$[I + \Gamma] = \frac{V_{tr}e^{-\gamma l}}{V_+e^{-\gamma l}} = T$$

Putting the value of Γ_l from eqn. (7.37), we get;

$$1 + \frac{Z_l - Z_0}{Z_l + Z_0} = T$$

$$T = \frac{2Z_l}{Z_l + Z_0} \quad ...(7.38)$$

Power carried by two wave;

$$P_{tr} = P_{inc} - P_{ref} = \frac{\left(V_+ e^{-\gamma l}\right)^2}{2Z_0} - \frac{\left(V_- e^{-\gamma l}\right)^2}{2Z_0}$$

$$P_{tr} = \frac{\left(V_{tr} e^{-\gamma l}\right)^2}{2Z_l}$$

by setting $P_{inc} = P_{tr}$ and using equation (7.37) and (7.38), we get

$$T^2 = \frac{Z_l}{Z_0}\,(1 - T_l^2) \qquad \text{...(7.39)}$$

Standing Wave

As mention a transmission line terminated in its characteristic impedance Z_0 is called properly terminated line otherwise it is called an improperly terminated line. In case transmission line is not correctly terminated then travelling electromagnetic wave has two component; forward travelling wave and backword travelling wave. The combination of these waves give rise to interference phenomenon and standing wave are generated along the transmission line, with definite maxima and minima of current and voltages along the line as shown in Fig. 7.8.

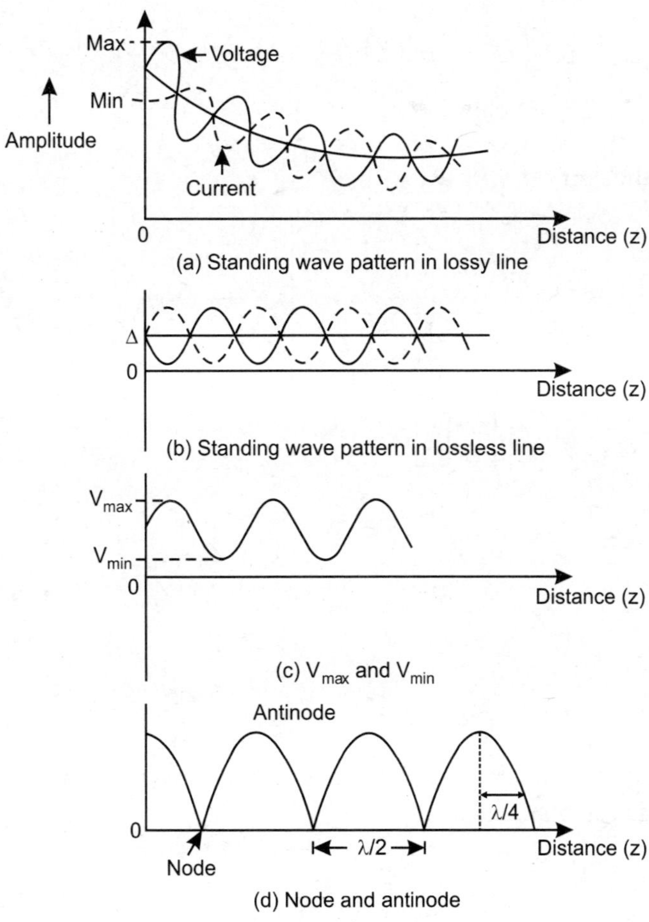

Fig. 7.8 SW pattern along a Tx line

In which distance, from Node to node = $\lambda/2$

Node to antinode = $\lambda/4$

Standing Wave Ratio (SWR): The ratio of maximum to minimum voltage or current on a line having standing waves, is defined by;

$$\text{SWR} = s = \left|\frac{V_{max}}{V_{min}}\right| = \left|\frac{I_{min}}{I_{min}}\right| \quad ...(7.40)$$

Standing waves ratio results from the fact that two travelling waves component adding in phase some points and substrating at other points. The distance between to successive maxima and minimum is $\frac{\lambda}{2}$.

Again, $V = V_+e^{-\gamma z} + V_-e^{\gamma z}$

$$= V_+e^{-\alpha z}\, e^{-j\beta z} + V_-e^{\alpha z}e^{j\beta z} \quad (\text{Since } \gamma = \alpha + j\beta)$$

The value V will be maixmum when $\beta z = n\pi$ where n = 0, 1, 2, 3, ...

$$V_{max} = V_+e^{-\alpha z} + V_-e^{\alpha z} \quad ...(7.41)$$

and V_{min} = occurs at $\beta z = (2n-1)\frac{\pi}{2}\, 1\ (n = 0, \pm 1, \pm 2, \pm 3,...)$

$$V_{min} = V_+e^{-\alpha z} - V_-e^{+\alpha z} \quad ...(7.42)$$

Standing wave ratio, $s = \dfrac{V_{max}}{V_{min}}$

$$\frac{V_+e^{-\alpha z} + V_-e^{\alpha z}}{V_+e^{-\alpha z} - V_-e^{\alpha z}} = \frac{V_+\, e^{-\alpha z}\left[1 + \dfrac{V_-e^{\alpha z}}{V_+e^{-\alpha z}}\right]}{V_+e^{-\alpha z}\left[1 - \dfrac{V_-e^{\alpha z}}{V+e^{\alpha z}}\right]}$$

$$s = \left[\frac{1+|\Gamma|}{1-|\Gamma|}\right]$$

$$s = \frac{1 + \dfrac{Z_l - Z_0}{Zl + Z_0}}{1 - \dfrac{Z_l - Z_0}{Z_l + Z_0}} = \frac{Z_l + Z_0 + Z_l - Z_0}{Z_l + Z_0 - Z_l + Z_0} = \frac{2Z_l}{2Z_0}$$

$$s = \frac{Z_l}{Z_0} \quad ...(7.43)$$

The general solution of standing wave consists of two travelling waves in opposite directions with general amplitude can be found as;

$$V = V_+e^{-\gamma z} + V_-e^{\gamma z}$$

and $I = I_+e^{-\gamma z} + I_-e^{+\gamma z}$.

Hence, $$V = V_+e^{-(\alpha + j\beta)z} + V_-e^{(\alpha z + j\beta)z} = V_+e^{-\alpha z}e^{-j\beta z} + V_-e^{\alpha z}e^{j\beta z}$$

$$= V_+e^{-\alpha z}[\cos\beta z - j\sin\beta z] + V_-e^{\alpha z}[\cos\beta z + j\sin\beta z]$$

$$V = [V_+e^{-\alpha z} + V_-e^{\alpha z}]\cos\beta z + j[V_+e^{-\alpha z} - V_-e^{\alpha z}]\sin\beta z \qquad ...(7.44)$$

Above equation (7.44) can be write as:

$$V_s = V_0e^{-j\phi} \qquad ...(7.45)$$

It is called equation of standing waves.

where $$V_0 = \sqrt{(V_+e^{-\alpha z} + V_-e^{\alpha z})^2\cos^2\beta z + (V_+e^{-\alpha z} - V_-e^{\alpha z})^2\sin^2\beta z}$$

and $$\phi = \tan^{-1}\left[\frac{V_+e^{-\alpha z} - V_-e^{\alpha z}}{V_+e^{-\alpha z} + V_-e^{\alpha z}}\right]$$

Similarly for current we get

$$I_S = I_0e^{-j\phi}$$

where $$I_0 = \sqrt{(I + e^{-\alpha z} + I_-e^{\alpha z})2\cos^2\beta z + (I_+e^{-\alpha z} - I_-e^{\alpha z})^2\sin^2\beta z}$$

and $$\phi = \tan{-1}\left[\frac{I_+e^{-\alpha z} - I_-e^{\alpha z}}{I_+e^{-\alpha z} + I_-e^{\alpha z}}\right]$$

$$V_{max} \text{ when } \beta z = n\pi \text{ when } n = 0, \pm 1, \pm 2, ...$$

$$V_{max} = (V_+e^{-\alpha z} + V_-e^{\alpha z}) \qquad ...(7.46)$$

$$V_{min} \text{ when } \beta z = (2n-1)\frac{\pi}{2}, n = 0, \pm 1, \pm 2, \pm 3...$$

Therefore SWR corresponding to voltage are defined by;

$$V_{min} = V_+e^{-\alpha z} - V-e^{\alpha z} \qquad ...(7.47)$$

$$\frac{V_{max}}{V_{min}} = \left[\frac{V_+e^{-\alpha z} + V_-e^{\alpha z}}{V_+e^{-\alpha z} - V_-e^{\alpha z}}\right] = \frac{V + e^{-\alpha z}}{V_+e^{-\alpha z}}\left[\frac{1+|\Gamma|}{1-|\Gamma|}\right] \qquad ...(7.48)$$

Similarly for current,

$$ISWR = \frac{I_{max}}{I_{min}} = \frac{I_+e^{-\alpha z} + I_-e^{\alpha z}}{I_+e^{-\alpha z} - I_-e^{\alpha z}} = \frac{I_+e^{-\alpha z}}{I_+e^{-\alpha z}}\left[\frac{1+|\Gamma|}{1-|\Gamma|}\right] \qquad ...(7.49)$$

There are three possible cases;

Case 1: When $V_- = 0, V_+ \neq 0$ reflected wave zero

$$V_0 = \sqrt{[(V_+e^{-\alpha})^2\cos^2\beta z + (V_+e^{-\alpha z})^2\sin^2\beta z]} \text{ and } V_s = V_+e^{\alpha z} \qquad ...(7.50)$$

Case 2: When $V_+ = 0\ V_- \neq 0$

$$V_0 = V_-e^{\alpha z} \qquad ...(7.51)$$

Case 3: When positive and negative waves have equal amplitudes *i.e.*

$$|V_+e^{-\alpha z}| = |V_-e^{+\alpha z}|$$

That is magnitude of the reflection coefficient is unity and wave is called pure standing wave, and V_S is given as;

$$V_S = 2V_+e^{-\alpha z}\cos\beta z \quad ...(7.52)$$

Similarly for current $I_S = -2Y_0V_+e^{-\alpha z}\sin\beta z$ ...(7.53)

i.e. Current and voltage standing waves are 90° out of phase along the line.

ISWR and *V*SWR may be identical in magnitude, but they will be out of phase (Fig. 7.9).

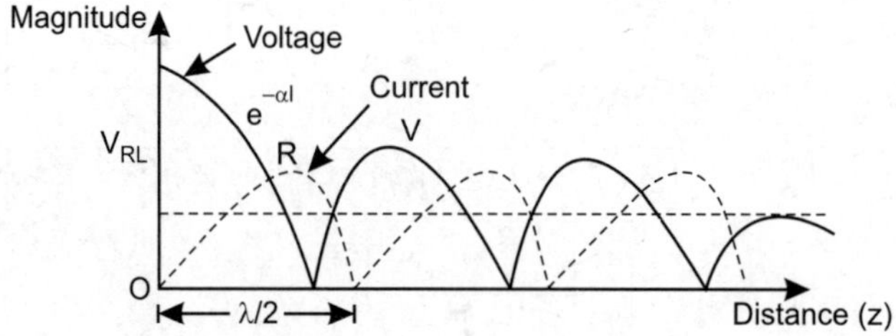

Fig. 7.9 Pure standing waves.

Smith Chart

Smith chart is a graphical representation on of transmission line. All the transmission parameter can be calculate with the help of this chart It is also known as impedance calculator, or as polar impedance diagram and consist two set of circles. The Smith chart consists of a plot of the normalized impedance or admittance with the angle and magnitude of a complex reflection coefficient in a unity circle. It is applicable for loss less as well as lossy line.

We know that $\Gamma = \dfrac{Z_l - Z_0}{Z_l + Z_0}$

$$\Gamma_l = \frac{Z_l - Z_0}{Z_l + Z_0} = \Gamma_r + \Gamma_i \text{(Say)} \quad ...(7.54)$$

Since $|\Gamma_l| \le 1$ so Γ_l must lie on or within unith circle with a radius of 1.

$$Z = R + jX$$

$$\frac{Z}{Z_0} = \frac{R + jX}{Z_0} = r + jx = z \text{ normalized impedance} \quad ...(7.55)$$

$$\Gamma_l = \frac{Z_l - Z_0}{Z_l + Z_0} = \frac{\dfrac{Z_l}{Z_0} - 1}{\dfrac{Z_l}{Z_0} + 1} = \frac{Z - 1}{Z + 1}$$

$$\Gamma_l Z + \Gamma_l = Z - 1$$

$$\Gamma_l Z - Z = \Gamma_l - 1 + Z(1 - \Gamma_l) = 1 - 1$$

$$Z = \frac{1 + \Gamma_l}{1 - \Gamma_l}$$

$$r + jx = \frac{1+\Gamma_l}{1-\Gamma_l} = \frac{1+T_r+T_i}{1-\Gamma_r - j\Gamma_i}$$

$$= \frac{(1+\Gamma_r + j\Gamma_i)(1-\Gamma_r - j\Gamma_i)}{(1-\Gamma_r - j\Gamma_i)(1-\Gamma_r + jT_i)} = \frac{(1+j\Gamma_i)^2 - \Gamma_r^2}{(1-\Gamma_r)^2 - (j\Gamma_i)^2}$$

$$r + jx = \frac{1-\Gamma_i^2 + 2j\Gamma_i - \Gamma_r^2}{(1-\Gamma_r)^2 + \Gamma_i^2} \qquad ...(7.56)$$

Comparing imaginary parts of the left hand side and right hand side of (7.56) gives;

$$r = \frac{1-\Gamma_i^2 - \Gamma_r^2}{(1-\Gamma_r)^2 + \Gamma_i^2} \qquad ...(7.57a)$$

$$x = \frac{2\Gamma_i}{(1-\Gamma_r)^2 + \Gamma_i^2} \qquad ...(7.57b)$$

$$(1-\Gamma_r)^2 + \Gamma_i^2 = \frac{2\Gamma_i}{x}$$

Now from equation (7.57*b*), $(1-\Gamma_r)^2 + \Gamma_i^2 - \dfrac{2\Gamma_i}{x} = 0$

$$(1-\Gamma_r)^2 + \Gamma_i^2 - \frac{2\Gamma_i}{x} + \frac{1}{x^2} = \frac{1}{x^2}$$

$$(1-\Gamma_r)^2 + \left(\Gamma_i - \frac{1}{x}\right)^2 = \frac{1}{x^2} \qquad ...(7.58a)$$

$(1-\Gamma r)^2 + \left(\Gamma_1 - \dfrac{1}{x}\right)^2 = \dfrac{1}{x^2}$ it is a equation of circle, with

Centre $\left(1, \dfrac{1}{x}\right)$ and radius $\dfrac{1}{x}$.

Now from eqn. (7.57*a*);

$$r = \frac{1-\Gamma_i^2 - \Gamma_r^2}{(1-\Gamma_r)^2 + \Gamma_i^2}$$

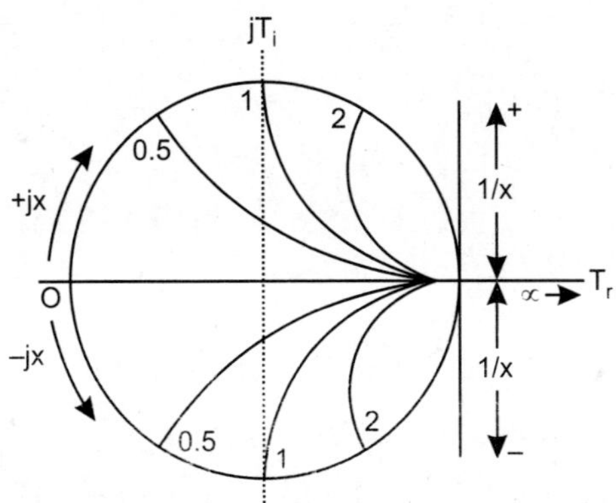

Fig. 7.10. Constant reactant reactance X circle.

or $$r(1-\Gamma_r)^2+\Gamma_i^2 = 1-\Gamma_r^2-\Gamma_r^2$$

$$\Rightarrow \quad r(1-\Gamma_r)^2+r\Gamma_i^2+\Gamma_r^2+\Gamma_r^2 = 1$$

$$r(1-\Gamma_r)^2+(r+1)\,\Gamma_x^2+\Gamma_r^2 = 1$$

$$r+r\Gamma_r^2-2r\Gamma_r^2+(r+1)\Gamma_i^2 = 1$$

$$(r+1)\Gamma_r^2+(r+1)\Gamma_i^2+r-2r\Gamma_r = 1$$

$$(r+1)\gamma_r^2+(r+1)T_i^2-2rT_r = 1-r$$

$$\Gamma_r^2+\Gamma_r^2-\frac{2r\Gamma_i}{1+r} = \frac{1-r}{1+r}$$

$$\Gamma_r^2+\Gamma_i^2-\frac{2r\Gamma_i}{1+r}+\left(\frac{r}{1+r}\right)^2-\left(\frac{r}{1+r}\right)^2 = \frac{1-r}{1+r}$$

$$\Gamma_r^2+\left(\Gamma_i-\frac{r}{1+r}\right)^2 = \frac{\dfrac{(1+r)}{(1+r)}+\dfrac{r^2}{(1+r)^2}}{\dfrac{(1-r)(1+r)+r^2}{(1+r)^2}} = \frac{1-r^2+r^2}{(1+r)^2} = \frac{1}{(1+r^2)}$$

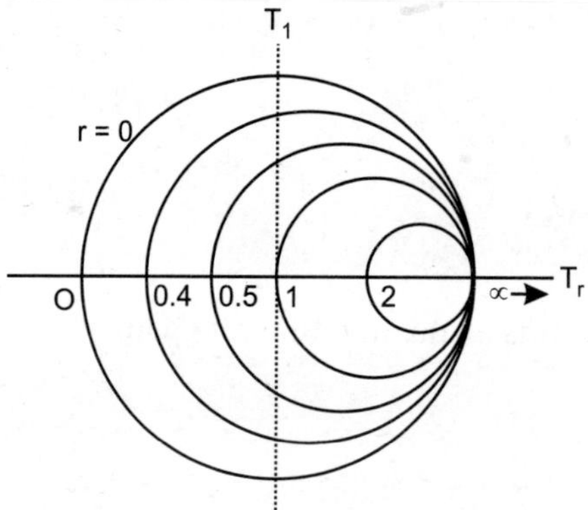

Fig. 7.11 Constant resistance circle.

$$\Gamma_r^2+\left(\Gamma_i-\frac{r}{1+r}\right)^2 = \frac{1}{(1+r)^2}$$

$$(\Gamma_r-0)^2+\left(\Gamma_i-\frac{r}{1+r}\right)^2 = \frac{1}{(1+r)^2} \qquad \text{...(7.58b)}$$

Which is again circle with

Centre $\left(0,\dfrac{r}{1+r}\right)$

and radius $\dfrac{1}{1+r}$. Equation (7.58*a*) and (7.58*b*) are called constant reactance and constant resistant circles respectively (Fig. 7.10 and 7.11).

When both the circle are combined it is called Smith chart and shown in Fig. 7.12.

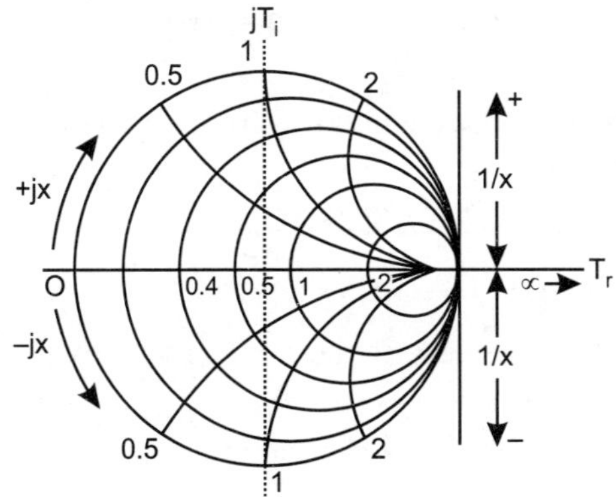

Fig. 7.12 Smith chart.

The upper half of constant reactance circle is $+jx$ and lower half is $-jx$. The distance around the Smith chart is $\lambda/2$.

Applications of Smith Charts:

(*i*) Calculations of admittance.

(*ii*) Calculation of impedance.

(*iii*) Measurement of SWR

(*iv*) Calculation of the length of short transmission line to give a required capactive and inductive reactances.

Characteristics of the Smith Chart

The characteristics of the Smith chart can be summerized as follows:

1. The constant r and constant x loci form two families of othogonal circles in the chart.
2. The constant r and constant x circles all pass through the point ($T_r = 1$, $T_i = 0$).
3. The upper half of the diagram represents $+jx$.
4. The lower half of the diagram represents $-jx$.
5. For admittance the constant r circles become constant g circle and the constant x circle becomes constant susceptance b circles.
6. The distance around Smith chart is one half wavelength ($\lambda/2$).
7. At a point of $Z_{min} = 1/\rho$ there is a V_{min} on the line.
8. At point of $Z_{max} = \rho$ there is a V_{max} on the line.
9. The horizontal radius to the right of the chart centre corresponds to V_{min}, I_{max}, Z_{min} and ρ (SWR).
10. The horizontal radius to the left of the chart centre corresponds to V_{max}, I_{min}, Z_{max} and $1/\rho$ (SWR).
11. Since the normalized admittance y is a reciprocal of the normalized impedance z, the corresponding quantities in the admittance chart 180° out of phase with those in the impedance chart.

12. The normalized impedance and admittance is repeated for every half wavelength of distance.
13. The distances are given in wavelength toward the generator and also toward the load.

There are two categories of *Tx* line;

(*a*) Lossless Transmission Line (*R* = *G* = 0)

A transmission line is said to lossless if the conducivity of the line is infinite ($\sigma = \infty$) and dielectric mediam separating them is lossless ($\sigma = 0$).

Since $$\mathrm{R} = \frac{1}{\pi a\, \delta\, \sigma_c}$$ for two wire Tx line.

$$= \frac{1}{\pi a\, \delta . \infty} = \frac{1}{\infty} = 0$$

and $$G = \frac{2\pi\sigma}{\cosh - (d/2a)} = \frac{2\pi \times 0}{\cosh - (d/2a)} = 0$$

So $$R = G = 0$$

This is necessary condition for a line to be lossless. Thus for such a line performance parameters can be given as

$$\alpha + j\beta = r = \sqrt{ZY} = \sqrt{(R + j\omega L)(G + j\omega C)}$$

$$= \sqrt{(0 + j\omega L)\ (0 + j\omega C)} = \sqrt{(j\omega L)(j\omega C)} = j\omega\sqrt{LC}$$

$$\alpha + j\beta = 0 + j\omega\sqrt{LC}$$

$$\alpha = 0$$

$$\beta = \omega\sqrt{LC} \qquad ...(7.59a)$$

and $$v = \frac{\omega}{\beta} = \frac{\omega}{\omega\sqrt{LC}} = \frac{1}{\sqrt{LC}} = \lambda f \qquad ...(7.59b)$$

and characteristic impedance

$$Z_0 = \sqrt{\frac{Z}{Y}} = \sqrt{\frac{j\omega L + R}{G + j\omega C}} = \sqrt{\frac{j\omega L}{j\omega C}} = \sqrt{\frac{L}{C}}$$

(*b*) Distortionless T_x Line $\left(\frac{R}{L} = \frac{G}{C}\right)$

A distortionless line is one in which the attenuation constant (α) is frequency independent while phase constant β is linearly dependent on frequency. Since $\gamma = \alpha + j\beta = \sqrt{(R + j\omega L)(G + j\omega C)}$ a distortionless line results if the line parameters are such as $\left(\frac{R}{L} = \frac{G}{C}\right)$.

Thus for a distortionless line,

$$\gamma = \sqrt{RG\left(1 + \frac{j\omega L}{R}\left(1 + \frac{j\omega C}{G}\right)\right)} = \sqrt{RG}\left(1 + \frac{j\omega C}{G}\right) = \alpha + \mathrm{j}\beta$$

Comparing both sides;

$$\alpha = \sqrt{RG}, \quad \beta = \omega\sqrt{LC} \qquad ...(7.60a)$$

$$Z_0 = \sqrt{\frac{R + j\omega L}{(G + j\omega C)}} = \sqrt{\frac{R\left(1 + \frac{j\omega C}{R}\right)}{G\left(1 + \frac{j\omega C}{L}\right)}} = \sqrt{\frac{R}{G}} = \sqrt{\frac{L}{C}} = R_0 + jX_0$$

or $$R_0 = \sqrt{\frac{R}{G}} = \sqrt{\frac{L}{C}} \text{ and } X_0 = 0 \qquad ...(7.60b)$$

$$v = \frac{\omega}{\beta} = \frac{1}{\sqrt{LC}} = f\lambda \qquad ...(7.60c)$$

That is for distortionless transmision line:

(*i*) The phase velocity is independent of frequency because the phase constant β linearly depends on frequency.

(*ii*) v and Z_0 remains the same as for lossless line.

(*iii*) A lossless line is also a distortionless line but distortionless line is not lossless line.

Therefore we can also summarize.

Conditions	**Propagation Constant**	**Characteristic Impedance**
	$\gamma = \alpha + j\beta$	$Z_0 = R_0 + X_0$
General	$\sqrt{(R + j\omega L)(G + j\omega C)}$	$\sqrt{\frac{R + j\omega L}{G + j\omega C}}$
Lossless	$0 + j\omega\sqrt{LC}$	$\sqrt{\frac{L}{C}} + j0$
Distortion line	$\sqrt{RG} + j\omega\sqrt{LC}$	$\sqrt{\frac{L}{C}} + j0$

EXAMPLES

Example 1. An air line has characteristic impedance of Z_0 and phase constant of 3 radian at 300 MHz; calculate the inductance and capacitance per meter of the line.

Solution: Air can be considered as lossless line since α = 0.

So, $$R = G = 0$$

$$Z_0 = \sqrt{\frac{L}{C}}$$

$$\beta = \omega\sqrt{LC}$$

$$\frac{Z}{\beta} = \frac{\sqrt{\frac{L}{C}}}{\omega\sqrt{LC}}.$$

$$\frac{Z_0}{\beta} = \frac{1}{\omega C}$$

$$\frac{Z_0}{3} = \frac{1}{2\pi fC} = \frac{1}{2\pi \times 100 \times 10^6 \times C}$$

$$C = \frac{1}{2\pi \times 100 \times 10^6} \times \frac{3}{Z_0} = 68.2 \text{ PF/m}$$

So,
$$L = Z_0^2 C = (Z_0)^2 \times 68.2 \times 10^{-12}$$

$$L = 334.2 \text{ mH/m}$$

Example 2. A distortionless line has $Z_0 = 60\ \Omega$, $\alpha = 20$ mHp/m, $u = 0.6c$ where $c = 3 \times 10^8$ m/sec.

Find R, L, C, G and [illegible] at [illegible]0 MHz.

Solution: For distortionless line

$$\frac{R}{L} = \frac{G}{C} \Rightarrow RC = LG$$

$$Z_0 = \sqrt{\frac{L}{C}}$$

$$\alpha = \sqrt{RG} = R\sqrt{\frac{C}{L}} = \frac{R}{Z_0}$$

$$R = \alpha Z_0$$

$$v = \frac{\omega}{\beta} = \frac{1}{\sqrt{LC}}$$

So,
$$R = \alpha Z_0 = 20 \times 10^{-3} \times 60 = 1200 \times 10^{-3} = 1.2\ \Omega/\text{m}$$

$$L = \frac{Z_0}{4} = \frac{60}{0.6(3 \times 10^8)} = 333 \text{ mH/m}$$

$$G = \frac{\alpha^2}{R} = \frac{400 \times 10^{-6}}{1.2} = 333\ \mu\text{s/m}$$

$$uZ_0 = \frac{1}{C}$$

$$C = \frac{1}{uZ_0} = \frac{1}{0.6(3 \times 10^8) \times 60} = 9259 \text{ PF/m}$$

$$\gamma = \frac{u}{f} = \frac{0.6(3 \times 10^8)}{10^8} = 1.5 \text{ m}$$

Example 3. A loss-less Tx line of 60m length operating at 2MHz. If line is terminatal with a load $Z_l = (60 + j\,40)\Omega$ assuming $Z_0 = 50\ \Omega$, $u = 0.3c$, find: Γ, 5 and Z is.

Solution: We know that

(*a*)
$$\Gamma = \frac{Z_l - Z_0}{Z_L + Z_0} = \frac{60 + j40 - 50}{60 + j40 + 50} = \frac{10 + j40}{110 + j40}$$

$$= 0.353 < 56°$$

(*b*) $$\text{VSWR } (s) = \frac{1+|\Gamma|}{1-|\Gamma|} = \frac{1+0.353}{1-0.353} = 2.091$$

(*c*) $$\beta_l = \frac{\omega l}{u} = \frac{2 \times 3.14 \times (2 \times 10^6) \times 60}{0.3 \times 3 \times 10^8} = 120°$$

$$Z_{in} = Z_0 \left[\frac{Z_L + jZ_0 \tan \beta l}{Z_0 + jZ_L \tan \beta l} \right]$$

$$= 50 \left[\frac{60 + j40 + j \times 50 \tan 120°}{50 + j(60 + j40) \tan 120°} \right]$$

$$= 24.011 < 3.22°$$

$$= (23.94 + j\, 1.35)\Omega$$

EXERCISES

1. What do you understand by balanced and unbalanced transmission line? Derive the expression transmission line equation.
2. Describe the Smith charts. List out the applications of the Smith chart.
3. Prove that, $Z_l = Z_0 \dfrac{Z_l + jZ_0 \tanh \beta l}{Z_0 + jZ_l \tan \beta l}$
4. What do you understand by standing wave and standing wave ratio? Write the expressions for VSWR and ISWR.
5. Defined the following terms and their physical significance with reference to a transmission line: (*i*) Characteristic impedance (*ii*) Phase velocity (*iii*) Phase constant (*iv*) VSWR (*v*) ISWR
6. Show that a lossy transmission line of length l has an input impedance $Z_{sc} = Z_0 \tanh \gamma l$ when shorted and $Z_{oc} = Z_0 \cot \gamma l$, when open circuited.
7. Measurement on a terminated transmission line gave the following results. VSWR = 3.2, location of first V_{min} = 0.237 from load characteristic impedance = 50Ω. Calculate the terminating impedance.
8. A 50 Ω transmission line connect a signal by 300 kΩ to a load of 75 Ω. If the load power is 25 mW, determine (*a*) VSWR (*b*) Position of first $V_{\min}$, $V_{\max}$ (*c*) $V_{\min}$ and $V_{\max}$ (*d*) Impedance of $V_{\min}$ and $V_{\max}$.
9. The short circuit and open circuit impedance of a 10 km long open wire transmission line are Z_{sc} =3030 < 25° Ω and Z_{oc} = 260 < 30° at frequency 1.5 kHz. Calculate the characteristics impedance.
10. A 50 Ω coaxial cable feeds a (75 +*j*25) Ω dipole antenna. Find the standing wave ratio.

- Learning never exhausts the mind.

Leonardo da Vinci

Chapter 8

BROADBAND IMPEDANCE MATCHING

8.1 INTRODUCTION

The term impedance matching often comes across in the various field of technology; *particularly in communication.* The idea of impedance matching could be understand very easily from Fig. 8.1. In which a source of field intensity (Eg.) is connectal to a load though two wire Tx line.

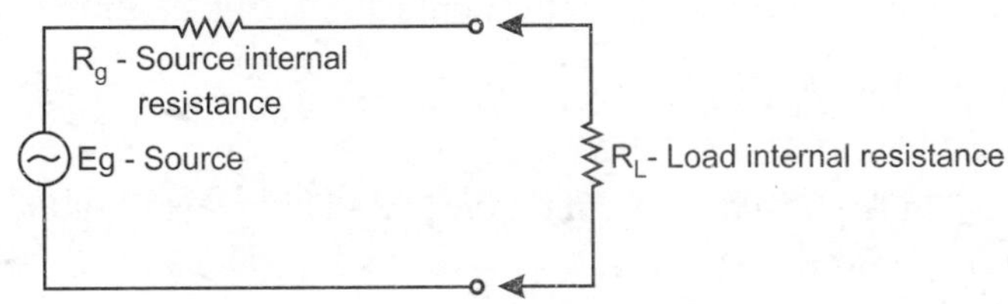

Fig. 8.1 Source driving a load resistance R_L and its own internal resistance is R_g.

Source always has its own internal resistance, so wherever a load is connected to its output terminal, some of output power is dissipated as a heat, unless we maintain the source temperature close to absolute zero (–273°C). Survey on "How the power fed to load varies as the load resistance is varied" shows that the amount of power reaches maximum when the load resistance is same as / matches to the source resistance, otherwise its falls away in either cases (higher or lower). Thus, the idea of matching/adjusting the load and source resistances is nothing but the matching the source and load impedances. In general practice,we used guiding device (Transmission line or coaxial cable) to connect source and load (Antenna) to transmitter or receiver (Fig. 8.2).

In which Z_A = Antenna impedance.

Z_{in} = Impedance of matching network,

Z_0 = Characteristic impedance of Tx line

Z_l = Impedance of Tx/Rx.

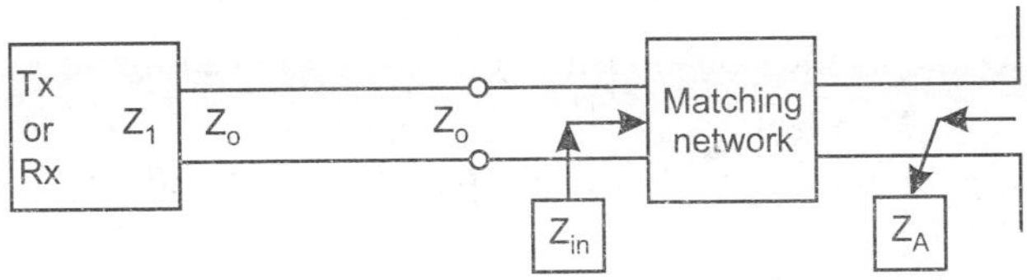

Fig. 8.2 Set-up of typical transmitter/receiver matching.

Here, the co-axial cable behaves as a transmission line at RF and as a result it has its own characteristic impedance. This simply means that because of the inductance to capacitance ratio (I/C) of the cable, RF energy tends to move along it, with a particular ratio between the electric and magnetic fields (i. e., voltage to current). In most cases,when the energy reaches at the end of the cable, it is highly desired to transfer it's maximum portion into load (in the case of transmitter) or the input RF stage (in the case of receiver). In case of transmitter, this offers the highest power efficiency, however for a receiver it gives the best noise performance. This is only possible when the characteristic impedance of cable is exactly same as the antenna impedance. If it is not so, portion of power reflected back, setting up standing waves across the cable which may damage the cable.

Some time there may be advantages of having mismatching (transmitter impedance is much lower than the cable) to minimize power loss in the final stage and also ensure that it RF waves are reflected back from the antenna and, most of it's portion is bounced right back up again. As well as the matching in any consummation system is concern, it is important to make effective use of all available power from the transmitter in the transmitting case and from the antenna in the receiving case. In general there are two basic considerations to be focused for impedance matching.

1. Impedance matching between the antenna and transmission line.
2. The excitation of the current distribution on the antenna.

Generally, we refer two cases; Low VSWR and high VSWR. For many applications extremely low VSWR is luxury and net necessity. If at the transmission line of high quality (low loss),reflected wave represents low – dissipative losses, a VSWR 2 : 1 leads to 89% power transmission. In case, VSWR is very high the power will travel back forth along the T_x line. In addition if line is lossy and long, dissipative losses may be sound/significant. It is found that at location closer to earth's surface impedance ranges 30 to 130 Ω. Another disadvantage of high VSWR, that very high voltages will be generated on the conductors at certain points along a transmission line which are called' hot spots and may cause Arcing. If the frequency of the transmitter changes due to severe input impedance mismatching; this is called frequency pulling. Because, as mismatching changes Γ_L, which results in frequency changes. That is frequency pulling is specified as the maximum deviation from the usual operating frequency, giving some worst case load (unknown). In addition, if output frequency changes with supply power, it is termed as frequency pushing. It is expressed in Hz/V, and can be either positive or negative value. One can minimize the frequency pulling by using matching network between source and load.

The most common matching network is an quarter wave transformer which is a quarter wave – length long transmission line of characteristic impedance $\sqrt{Z_0 Z_A}$ .The other matching networks/ devices are: Lumped elements, sub turners, and various other types of transformers. A UHF and microwave frequencies turning devices such as stub turners and irises are used to transform the real part of the impedance to that of the transmission line as well as tuning out the reactive component. However, at lower frequencies, reactive tuning is accomplished with variable capacitors and coils, this is because electrical dimensions are small compared to operating wave – length.

Disadvantages of Matching Networks

Despite of various advantages of a matching circuits there are few disadvantages of using matching networks too. For example if a matching network is designed to obtain a perfect matching it will be achieved at a single frequency, hence usually it will be narrow – banded. On the other hand if the matching network is designed to be broad banded, it will usually not yield a perfect matching at all frequencies over the band or perhaps at any other frequency over the band. Advantages and disadvantages of individual types of matching will be described in the sequent sections.

Impedance Matching for Maximum power Transfer

In actual sense, impedance matching is the control of impedance for the purpose of achieving maximum power/ minimum reflection. A variable load impedance connected to a source will receive the maximum possible power from a source, when it is adjusted to equal the complex conjugate of the impedance of source.

i.e., maximum power – transfer results only when the impedance is purely resistive, in other word the load and source impedances are then matched on a conjugate– impedance basis. Whereas, fixed source impedance and fixed load impedance may be coupled for maximum power transfer by a matching network (Fig. 8.3). In this matching technique, network transforms the source– impedance to complex conjugate of the load impedance, and *i.e.*, a conjugate – impedance match ing occurs at the input to the network as well as at the output of the network. Transmission lines used in antenna system commonly have characteristic impedances that is purely resistive. If the line has finite attenuation, maximum power transfer takes place between source and load, only when the following conditions are satisfied.

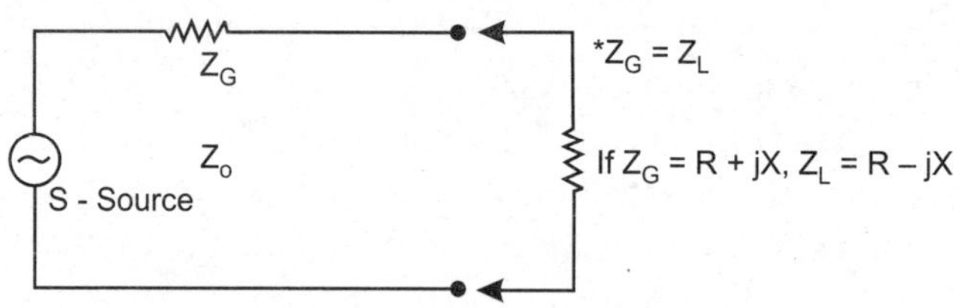

Fig. 8.3 Source and load on complex - conjugate basis.

(*a*) The source is loaded by the characteristic impedance and *provides maximum power delivery from the source.*

(*b*) The line is terminated in the characteristic impedance and provides minimum power dissipation in the line by eliminating the reflected wave on the line. In case of mismatching the line will be subjected to two waves; the direct and reflected waves propagating on opposite directions creating new wave pattern called standing wave patterns of voltages or currents.

If a system matched both on a conjugate – impedance as well as on a characteristic – impedance basis, it is termed as ideal matched condition. A measure of mismatch at the load terminal is called the voltage reflection coefficient (ρ)

$$\rho = \frac{Z_L - Z_o}{Z_L + Z_0} \quad ...(8.1)$$

The power at the load is thus reduced from the maximum available power by the ratio of $(1 - \rho^2)$. The reflection coefficient is a vector quantity related to impedance ratio of two mismatch impedances. The magnitude of reflation coefficient should not exceed unity. The reduction in transmission is called a reflection or transmission loss and expressed in dB. The combination of reflected and incident waves forms standing wave with definite maxima and minima of voltage or current on the line (Fig. 8.4).

The ratio of the maximum and minimum magnitudes of current or voltage on the Tx line having standing wave is called the standing wave ratio (CSWR |VSWR), denoted by s.

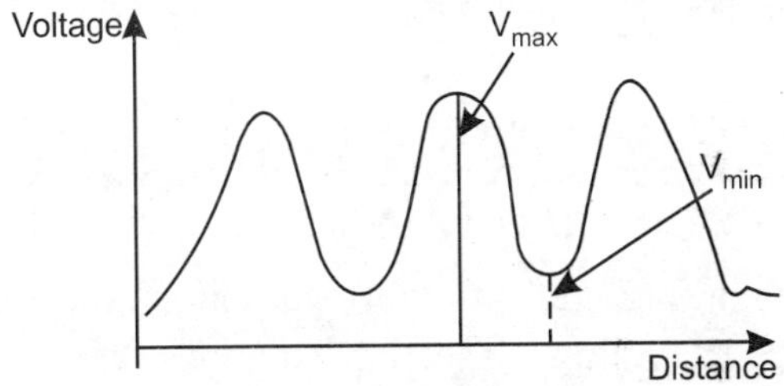

Fig. 8.4 VSWR pattern.

Therefore,

$$s = \text{VSWR} = \left|\frac{V_{\max}}{V_{\min}}\right| = \left|\frac{I_{\min}}{I_{\max}}\right| \quad ...(8.2)$$

i.e., VSWR is always greater than 1, usually its theoretical values lies between 1 to ∞. When its value is one, means the line is terminated by its Z_0 and there is no reflection of waves. (i.e. no loss of power).

If V_r is r.ms. value of reflected wave and V_i is r.m.s. value of incident wave, then

$$|V_{\max}| = |V_i| + |V_r|$$

$$|V_{\min}| = |V_i| - |V_r|$$

$$\text{VSWR} = \frac{V_{\max}}{V_{\min}}\frac{|V_i| + |V_r|}{|V_i| - |V_r|} = \frac{1 + \left|\frac{V_r}{V_i}\right|}{1 - \left|\frac{V_r}{V_i}\right|}$$

$$\text{VSWR} = \frac{1 + |\rho|}{1 - |\rho|}$$

where $\rho = \frac{V_r}{V_i}$ = voltage reflection coefficient.

$$= \frac{VSWR - 1}{VSWR + 1} \quad ...(8.3a)$$

The refection (or return) loss in given by;

$$\text{Return loss} = -10 \log |\rho| (\text{dB}) \quad \text{...(8.3b)}$$

The line impedance at voltage maximum and minimum points are purely real, and also have maximum and minimum values respectivally. They are expressed as follows:

$$Z_{in} = Z_{max} = \frac{V_{max}}{I_{min}} = \frac{V_i\,(1+|\Gamma|)}{I_i\,(1-|\Gamma|)} = Z_0 \,.\, s \quad \text{...(8.3c)}$$

and

$$Z_{in} = Z_{in} = \frac{V_{min}}{I_{max}} = \frac{V_i\,(1-|\Gamma|)}{I_i\,(1-|\Gamma|)} = \frac{Z_0}{s} \quad \text{...(8.3d)}$$

The presence of the standing waves increase the maximum voltage and current on the Tx line for the same delivered power by the ratio $\sqrt{S}$. The efficiency of the line may be expressed as follows:

$$\eta = \frac{S_2 - \dfrac{1}{S_2}}{S_1 - \dfrac{1}{S_1}} \quad \text{...(8.4)}$$

$$= \frac{4S_1}{10^{0.1A}(S_1+1)^2 - 10^{-0.1A}(S_1-1)^2} \quad \text{...(8.5)}$$

Where the subscripts (1) and (2) refer to the load and source ends VSWR of line, and A is normal line attenuation in dB. The following section examines several matching systems that might be very useful in a number of systems in different situations.

8.2 IMPEDANCE MATCHING TECHNIQUES

Quarter - wave Transformer Matching

Quarter wave transfer is nothing but a section of loss less transmission line of length $\frac{\lambda}{4}$. The useful quarter – wave transformer results from the general line transformer, where the electrical length is equal to $(\beta l) = \frac{\lambda}{2}$. Referring to Fig. (8.3) for any terminating impedance $Z_L = R_L + jL$, the voltage and currents on Tx can be expressed as;

$$V_s = V_r \cos \beta l + j\, I_r Z_o \sin \beta l \quad \text{...(8.6a)}$$

$$I_s = I_r \cos \beta l + j\,(V_r / Z_o) \sin \beta l \quad \text{...(8.6b)}$$

In which

V_s and I_s = voltage and current at sending / source end.

V_r and I_r = voltage and current at receiving / load end.

l = electrical length of the line.

β = phase constant.

Equations (8.6a and 8.6b) show that a loss less T_x line can be used as voltage and current transformers with secondary values, V_r and I_r at the load end and the primary values V_s and I_s at the sending end by selecting the proper values of l and Z_0. If a line of electrical length $\frac{\lambda}{4}$, i.e., $\beta l = \frac{\pi}{2}$

having characteristic impedance Z_0 is terminated by an impedance Z_L, the input impedance of the line can be given by;

$$Z_{in} = Z_o\left[\frac{Z_L + jZ_0 \tan\beta\ell}{Z_0 + j\tan\beta\ell}\right] = \frac{Z_0^2}{Z_L} \qquad ...(8.7)$$

If this Z_{in} matched to the source of impedance Z_s, *i. e.*, $Z_{in} = Z_s$, then

$$Z_o = \sqrt{Z_S.Z_L} \quad (Z_s \pm Z_L) \qquad ...(8.8)$$

Therefore, a source of impedance Z_s can be matched to a load (if $Z_s \pm Z_L$) by inserting in between a transmission line section of length $l = \lambda/4$ where characteristic impedance is $\sqrt{Z_S Z_L}$. This section of *T* line called quarter wave matching transformer (Fig. 8.5), and matching is QWT matching. Using impedance transferring properties of Tx line, the QWT provides a matched impedance at its input by properly transferring the load impedance.

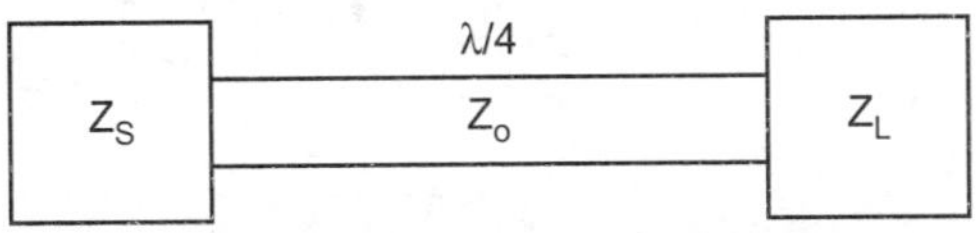

Fig. 8.5 Quarter wave transformer matching.

From the equation (8.7) it is clear that, the input impedance in inversely proportional to the load impedance. The phase angle of input impedance (Z_{in}) is the negative of that of the load impedance.

Therefore QWT can also be used, for example to transform an inductive low impedance to a capacitive high impedance or *vice – versa.*

Quarter wave co axial line

In a case of mismatched co axial line, a quarter wave transformer formed by a simple sleeve is generally used to tune out the reflection. The sleeve forms an enlargement of the inner conductor, outer side. The characteristic impedance of the line is reduced over the quarter wave length by the sleeve to a value of

$$\frac{Z_0}{\sqrt{S}} \qquad ...(8.9)$$

where S is the initial value of SWR of the line. The load end of the sleeve is positioned at a voltage minima a on the standing wave pattern . The impedance here looking toward the load is $\left(\frac{Z_0}{S}\right)$. The transformer transforms this impedance to Z_0, so that the line on the source end is perfectly matched.

The single quarter wave transformer is best suitable if the value of both impedances to be matched is closer. If the Z_s and Z_L are widely different better matching can be obtained by using multi – section QWT, each having characteristic impedance equal to the square root of the product of those on both the sides,*i.e.*,

$$Z_1 = \sqrt{Z_s \cdot Z_2}$$

$$Z_2 = \sqrt{Z_1 \cdot Z_3}$$

and ....

$$Z_n = \sqrt{Z_n - 1 \cdot Z_n + 1} \qquad ...(8.10)$$

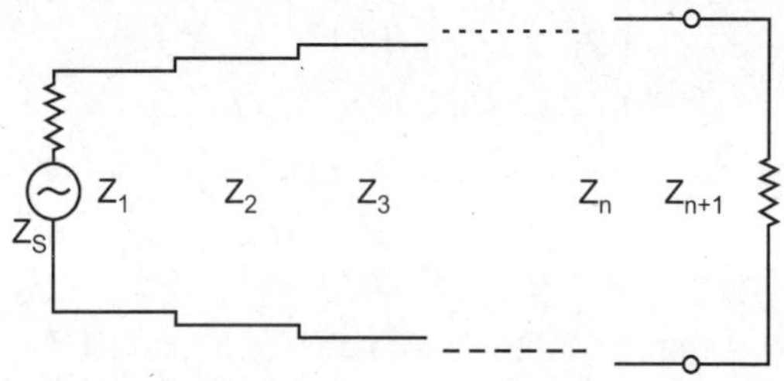

Fig. 8.6 Multi–section quarter wave transformer

The quarter wave – transformer provides a perfect matching at the single designed frequency and at all the odd harmonics for which the transformer length becomes equal to an odd multiple of $\lambda/4$. That is, QWT is narrow band width matching device. The bandwidth of a single section QWT is determined by considering the electrical length $\beta l = \theta$, at any other frequency $f \neq f_0$ and then the input reflection coefficient (Γ). The bandwidth is given by

$$\Delta f = 2(f_0 - f_m) = 2f_0(1 - 2\theta_m/\pi) \qquad ...(8.11a)$$

where f_0= Centre frequency and

$$\theta_m = \cos^{-1}\left[\frac{2|\Gamma_m|\sqrt{Z_s \cdot Z_L}}{(Z_L - Z_S\sqrt{1-|\Gamma_m|})^2}\right] \qquad ...(8.11b)$$

where $|\Gamma_m|$ is the maximum allowable limit of reflection coefficient around the mid band frequency to, where $|\Gamma| = 0$, and $\theta = \frac{\pi}{2}$ or $f = f_0$.

Transmission line stub matching

Once again from the expression of input impedance (Z_{in}) of loss less line;

$$Z_{IN} = Z_0\left[\frac{Z_L + jZ_0 \tan \beta\ell}{Z_L + jZ_L \tan \beta\ell}\right] \qquad ...(8.12)$$

Therefore, for a short-circutied line, $Z_L = 0$, So

$$Z_{in} = Z_{sc} = jZ_0 \tan \beta l \qquad ...(8.13a)$$

and for an open-circuited line, $Z_L = \infty$, so that

$$Z_{in} = Z_{ac} = -jZ_0 \cot \beta l \qquad ...(8.13b)$$

In order to achieve perfect matching between source and load, the characteristic impedance (Z_0) and its etectric length can be rearranged in terms of source and load impedances. Therefore from equation (8.12)

$$Z_0 = \left[\frac{R_L|Z_S|^2 - R_S(Z_L)^2}{R_S - R_L}\right]^{\frac{1}{2}} \qquad ...(8.14)$$

$$\beta = \tan^{-1}\left[\frac{Z_0(R_L - R_S)}{R_L \cdot X_S - R_S \cdot X_L}\right] \qquad ...(8.15)$$

For the positive characteristic impedance (Z_0), the following condition must satisfy;

$$\frac{1}{2} > \frac{\log\left|\frac{Z_L}{Z_S}\right|}{\log\left|\frac{R_L}{R_S}\right|} < \infty \qquad ...(8.16)$$

Stub Matching :

Thus, from previous section,it is cleared that matching can be obtained by inserting a section of Tx line between source and load of specific characteristic impedance (Z_0) and electrical length βl. It is also possible to achieve matching by connecting a section of open / short-circuited Tx line, known as stub in shunt arm with the main line at some point / points; called sub matching. The stub matching has following advantages;

(*a*) The length and Z_0 of the line is remains unaltered

(*b*) Adjustable susceptances are added in shunt with the line

The very commonly used stub matching is single stub matching which is shown in Fig. (8.7).

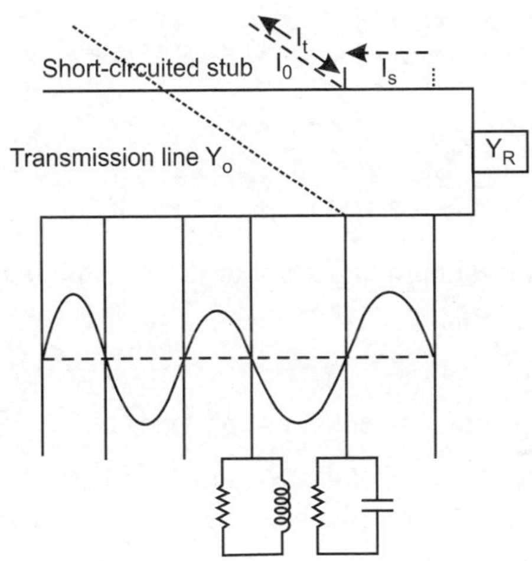

Fig. 8.7 (*a*) A single stub matching.

In Fig. (8.7*a*), a transmission line of characteristic admittance (y_0) is terminated with a pure inductances Y_R. A short – circuited stub of length 'l_s' has been connected in parallel with the main line at distance 'l_t' from the terminated end. From the theoretical analysis, it is found that matching between source and load occurs when the values of 'l_s', *i.e.*,location of stub and the length of the stub 'l_t' are defined as follows:

$$\ell_s = \frac{\lambda}{2\pi}\tan^{-1}\left[\frac{Z_R}{Z_0}\right]^{1/2} \text{ and } \ell_t = \frac{\lambda}{2\pi}\tan^{-1}\left[\frac{\sqrt{Z_R \cdot Z_0}}{Z_0}\right] \qquad ...(8.17)$$

where λ is operating wave – length and Z_R and Z_0 are the impedances.

Equation (8.17) indicates that a single stub matching is a narrow band matching system,because both 'l_s' and 'l_t' involve λ. That is the change of λ (or operating frequency)will result in change of length as well as position of stub,which is not appreciable. So it is possible in open wave line only not is co–axial line, hence single stub matching may become inaccurate in practice. The short comings of single stub matching can be overcome by cascaded quarter wave transformers.

Double Stub Matching

As we have seen in single stub matching, the stub is placed on the Tx line at a specified point. Location of these point varies with Z_R and operating frequency f, which causes some difficulties as the specified point may fall at an undesirable location. To avoid such limitations, double stubs are preferred, where impedance matching is achieved by inserting two stubs at specified locations along transmission line as shown Fig 8.7b. The distance between the two stubs is fixed such as $\lambda/16$, $\lambda/8$, $3\lambda/16$, $3\lambda/8$ or even closer and the lengths of the two stubs are adjusted to match the load.

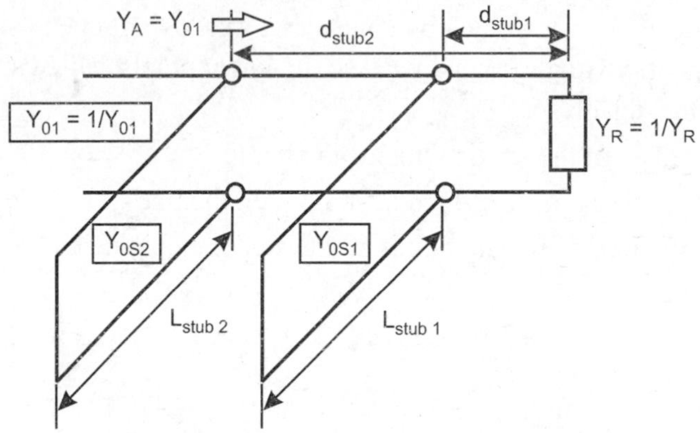

Fig. 8.7 (b) Double stud matching

That is there are two main design parameters for double stub matching:

- Length of the first stub line L_{stub1}
- Length of the second stub line L_{stub2}

While distance between location is kept constant and typically it is $\lambda/8$.

In the double stub configuration, the stubs are inserted at predetermined locations. In this way, if the load impedance is changed, we simply have to replace the stubs with another set of different length. Because the lengths l_1 and l_2 can be easily altered this matching network is physically tunable.

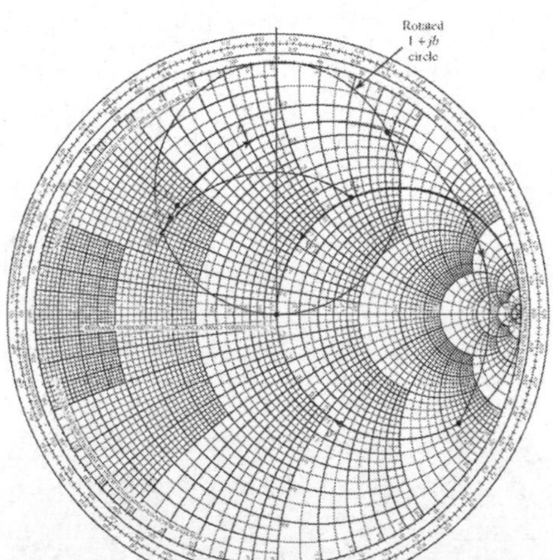

Fig. 8.7 (c) Solution to Smith chart for the double-stub tuners.

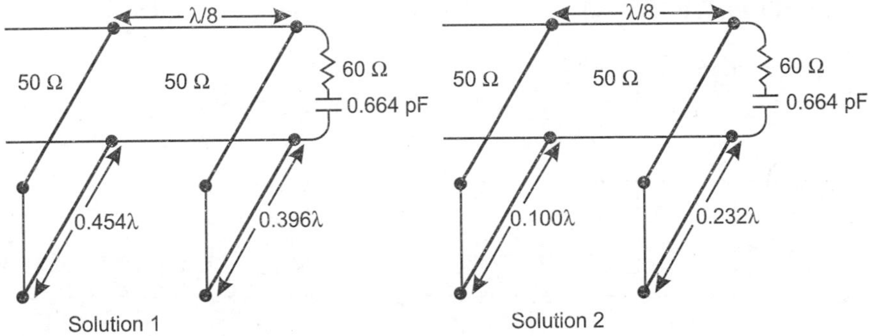

Fig. 8.7 (*d*) The two double-stub tuning solutions

The drawback of double stub tuning is that a certain range of load admittances cannot be matched once the stub locations are fixed. Design of a double-stub shunt tuner can be best understood by following example;A double-stub shunt tuner has to match a load impedance $Z_L = 60 - j80\ \Omega$ to a 50 Ω line. The stubs are to be short-circuited stubs and spaced λ/8 apart. If this load consists of a series resistor and capacitor, and that the match frequency is 3 GHz, find possible soultions of the marching and find actual length and distance between stubs.

The normalized load admittance is $y_L = (60 - j80)/50 = 0.3 + j0.4$, which is plotted on the Smith chart of Fig. 8.7*c*. Next, we construct the rotated $1 + jb$ conductance circle, by moving every point on the g = 1 circle λ/8 toward the load. We then find the susceptance of the first stub, which can be one of two possible values: $b_1 = 1.314$, or $b = -0.1\ 14$. We now transform through the (λ/8) section of line by rotating along a constant radius (SWR) circle λ/8 toward the generator. This brings the two solutions to the following points: $y_2 = 1 - j3.38$, or $y = 1 + j\ 1.38$. Then the susceptance of the second stub should be $b_2 = 3.38$, or $b = -1.38$. The lengths of the short-circuited stubs are then found ($\ell_1 = 0.396\lambda$, $\ell_2 = 0.454\lambda$), or ($\ell_2 = 0.232\lambda$, $\ell_2 = 0.100\lambda$). Since operating frequency is 3 GHz, λ = 10 cm, ($\ell_1 = 3.96$, $\ell_2 = 4.54$), or ($\ell_1 = 2.32$, $\ell_2 = 1.00$,), and λ/8 = 1/8 = 0.125 cm.

This is complete two solutions for the double-stub tuner design. Now if the resistor-capacitor load $Z_L = 60 - j80\ \Omega$ at $f = 3$ GHz, then $R = 60\ \Omega$ and $C = 0.995$ pF. The two tuning circuits are then shown in Fig. 8.7*d*. Note that the first solution has a much narrower bandwidth than the second (primed) solution, due to the fact that both stubs for the first solution are somewhat longer (and closer to λ/2) than the stubs of the second solution.

Cascaded Quarter Wave Transformers

In this case a number of quarter wave T_x line sections are arranged in cascade fashion to achieve great improvement in wide band performance over a single transformer. The characteristic impedance of the successive sections are proportioned to divide the overall transformation systematically. There are two common types of cascade quarter wave transformers namely

(*i*) Binomial transformer

(*ii*) Chebyshev transformer

The terms used in cascade quarter wave transformers are described in Table 8.1

Table 8.1 : Design specifications for cascade transformers.

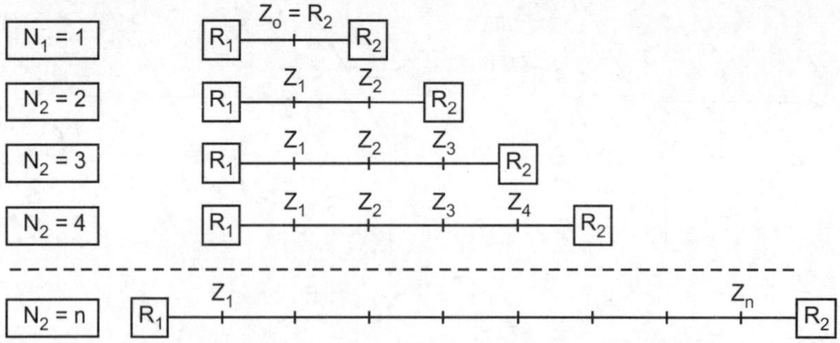

$$Z_0 = R_2$$

in which

$$R_1\,R_2 = \text{Terminating impedances } (R_1 < R_2)$$

$$R_0 = \sqrt{R_1 R_2} \text{ mean impedance}$$

Each section of a quarter wave transformer operates at f_0.

$$f_0 = \text{Design frequency} = \frac{f_u + f_l}{2}$$

where f_u = upper frequency limit, f_l = lower frequency limit.

Binomial Transformer

The binomial transformer distribution gives almost maximally flat performance and the logarithms of the impedance ratios of the steps between sections are made to be in the ratio of the trinomial coefficients. Table 8.2 may be used to determine the characteristic impedance Z_n of nth section in an N– section binomial transformer as a function of (R_2/R_1). The ratio:

$$\frac{\log(Z_n/R_1)}{\log(R_2/R_1)} = \frac{\log\left(\frac{R_2}{R_1}\times\frac{Z_n}{R_2}\right)}{\log\left(\frac{R_2}{R_1}\right)} = 1 - \frac{\log\left(\frac{R_2}{Z_n}\right)}{\log\left(\frac{R_2}{R_1}\right)} \qquad \text{...(8.18)}$$

The input SWR of an N-section binomial transformer is given by

$$S = 1 + (\cos\theta)^N \ln\left(\frac{R_2}{R_1}\right) \qquad \text{...(8.19)}$$

where θ is the electrical length of each section. This expression is applicable only in the case of small steps, zero – discontinuity capacitance and the sections of equal length.

Typically for $N = 2$ *i.e.*, the two – section transformer has maximally flat bandwidth curve for all transformation ratios, however for other values of N (i.e., multi section binomial transformer, except two sections), the performance approximates the maximally flat wave for transformation ratios near unity.

Table 8.2 : Design Ratios for Binomial Transformer

$N=1$ \ $n=1$	1	2	3	4	5	6	7
1	1/2	–					
2	1/4	3/4	–				
3	1/8	4/8	7/8	–			
4	1/16	5/16	11/16	15/16	–		
5	1/32	6/32	16/32	26/32	31/32	–	
6	1/64	7/64	22/64	42/65	57/64	63/64	–
7	1/13	8/128	29/128	64/128	99/128	120/12	127/12

If we choose the reflection coefficient of the multi-section matching transformer such that;

$$\Gamma(\theta) = A(1 + e^{j2\theta})N$$
$$= A[e^{-j\theta} + e^{-j\theta})^N]$$
$$= A(2e^{j\theta}\cos\theta)^N \qquad ...(8.20)$$

Then the reflection coefficient $\Gamma(\theta)$ has the following properties;

$$\frac{dn|\Gamma(\theta)|}{d\theta^n} = 0 \text{ at } \theta = \frac{\pi}{2}, n = 1, 2, 3,.....n$$

Which is defined as the maximally flat response for the reflection coefficient at the centre frequency f_0, $\theta = \frac{\beta\ell}{2} = \frac{\pi}{2}$ or length $\ell = \lambda / 4$. The characteristic impedance of the individual Transformer sections must be determined in such a way that the maximally flat response is achieved

The general form of the reflection coefficient for the N– section matching transformer can be easily approximated and written in terms of the binomial series as follows:

$$\Gamma(\theta) = \Gamma_0 + \Gamma_1 e^{-2j\theta} + \Gamma_2 e^{-j4\theta} ...+ \Gamma_N e^{-2jN\theta}$$
$$= A\sum_{n=0}^{N} C_n e^{-2jn\theta} \qquad ...(8.21)$$

where 'A' is the general amplitude coefficient and C_n is binomial coefficients, which is given by;

$$C_n = \frac{N}{n!(N-n)!} \qquad ...(8.22)$$

Equating coefficient for the binomial series and the reflection coefficient approximation gives;

$$\Gamma_n = AC_n$$

The amplitude coefficient A may be determined by taking the limit of the reflection coefficient equations as the frequency decreases to zero. In that case;

$$\theta = \beta\ell = \frac{2\pi l}{\lambda} < 0$$

and each section of the matching transformer is of zero length. This gives

$$\Gamma(0) = 2^{N} A = \frac{Z_L - Z_0}{Z_L + Z_0}$$

or
$$A = \frac{Z_L - Z_0}{Z_L + Z_0} . 2^{-N} \qquad ...(8.23)$$

The characteristic impedance of the adjacent sections of the binomial transformer are related by

$$\Gamma_n = AC_n = \frac{Z_{n+1} - Z_n}{Z_{n+1} + Z_n}$$

Which yields

$$Z_{n+1} = Z_n \frac{1 + AC_n}{1 - AC_n} \qquad ...(8.24)$$

So, to determine the individual characteristic impedance of the multi–section binomial matching transformer, if this formula is used then significant errors caused from the input of the matching section to the load,may be introduced. The reason behind that the theory of small reflections was applied to obtain the binomial constants though an exact formula was used to determine the amplitude constant. However a more accurate formula results when the amplitude coefficient A is also determined in terms of small reflections. The local reflection coefficients for the connections of the individual *T*x line sections may be defined as;

$$\Gamma_n = \frac{Z_{n+1} - Z_n}{Z_{n+1} + Z_n} = \frac{Z_r - 1}{Z_r + 1} \qquad ...(8.25)$$

where $Z_r = \frac{Z_{n+1}}{Z_n}$, and according to the theory of small reflections since the discontinuities in the characteristic impedance at the junctions of the multi–section matching are small, the value of Z_r may approximated to one. The reflection coefficients may then be expressed in terms of the natural log of this quotient, because

$$\ln x = 2\frac{x-1}{x+1} \text{ for } x = 1$$

Therefore
$$\ln(Z_r) \approx 2 \cdot \frac{Z_r + 1}{Z_r + 1}$$

$$= 2 \cdot \Gamma_n = 2 \cdot A \cdot C_n = \frac{2C_n}{2^n} \frac{Z_L - Z_0}{Z_L + Z_0} \approx \frac{C_n}{2^N} \ln\left(\frac{Z_L}{z_0}\right) \qquad ...(8.26)$$

Solving this equation for Z_{n+1} in terms of Z_n yields;

$$Z_n + 1 = Z_n \exp\left[\frac{C_n}{2_n} \ln\left(\frac{Z_L}{Z_0}\right)\right] \qquad ...(8.27)$$

We can also show the impedance Z_{n+1} is exactly equal to Z_L,as follows.

$$Z_1 = Z_0 \exp\left[\frac{C_0}{2_N} \ln\left(\frac{Z_L}{Z_0}\right)\right]$$

$$Z_2 = . Z_1 \exp\left[\frac{C_0}{2_N} \ln\left(\frac{Z_L}{Z_0}\right)\right]$$

$$= Z_0 \exp\left[\frac{C_0}{2_N} \ln\left(\frac{Z_L}{Z_0}\right)\right] \exp\left[\frac{C_1}{2_N} \ln\left(\frac{Z_L}{Z_0}\right)\right]$$

....

$$Z_{n+1} = Z_n \exp\left[\frac{C_n}{2_N}\ln\left(\frac{Z_L}{Z_0}\right)\right]$$

$$= Z_0 \exp\left[\frac{C_0}{2_N}\ln\left(\frac{Z_L}{Z_0}\right)\right] \exp\left[\frac{C_1}{2_N}\ln\left(\frac{Z_L}{Z_0}\right)\right] \ldots \exp\left[\frac{C_n}{2_N}\ln\left(\frac{Z_L}{Z_0}\right)\right]$$

Therefore
$$Z_{n+1} = Z_0 \exp\left[\frac{\sum_{k=1}^{n} C_k}{2^N}\ln\left(\frac{Z_L}{Z_0}\right)\right] \quad \ldots(8.28a)$$

The binomial series has the special property that $\sum_{k=1}^{n} C_k = 2^N$,so that

$$Z_{n+1} = Z_0 \exp\left[\frac{\sum_{k=1}^{n} C_k}{2^N}\ln\left(\frac{Z_L}{Z_0}\right)\right] = Z_0 \exp\left[\ln\left(\frac{Z_L}{Z_0}\right)\right] = Z_L \quad \ldots(8.28b)$$

Bandwidth

Since the angle θ for the binomial multi–section matching transformer relates the frequency as follows :

$$\theta = \beta l = \frac{2\pi f}{\upsilon_p}\frac{\upsilon_p}{4f_0} = \frac{\pi}{2}\frac{f}{f_0}$$

We get $f = \left(\frac{2f_0}{\pi}\right)\theta$, provided each section of the transformer is λ/4 long.

Thus, the bandwidth of the transformer may be defined in terms of the angle θ. If Γ_m is a magnitude of maximum allowable reflection coefficient, then the bandwidth Δf can be expressed as

$$\Delta f = f_2 - f_1 = \frac{2f_0}{\pi}(\theta_2 - \theta_1) \quad \ldots(8.29a)$$

where f_1 and f_2 are defined as the lower and upper frequencies respectively, whereas angles θ_1 and θ_2 corresponds to the frequencies f_1 and f_2 . And they are found by solving the expressions for the overall reflection coefficient evaluated at the maximum allowable value.

$$|\Gamma(\theta) = \Gamma_m = 2^N |A|| e^{-jN\theta} | \cos^N \theta$$

Solving this equation yields;

$$\theta_{1,2} = \cos^{-1}\left[\frac{1}{2}\left(\frac{\Gamma_m}{|A|}\right)^{\frac{1}{N}}\right]$$

The angle θ_1 is the solution to the this equation which is less than $\theta_0 = \pi/2$. However due to symmetry, the angles θ_1 and θ_2 are equally spaced about θ_0. Therefore,the bandwidth may be re,-written as;

$$\Delta f = f_2 - f_1 = \frac{2f_0}{\pi}(\theta_2 - \theta 1) = \frac{2f_0}{\pi}[2(\theta_0 - \theta_1)]$$

$$= \frac{4f_0}{\pi}\left\{\frac{\pi}{2} - \cos^{-1}\left[\frac{1}{2}\left(\frac{\Gamma_m}{|A|}\right)^{\frac{1}{N}}\right]\right\}$$

$$= \frac{2f_0}{\pi}\left\{\frac{\pi}{2} - \cos^{-1}\left[\frac{1}{2}\left(\frac{\Gamma_m}{|A|}\right)^{\frac{1}{N}}\right]\right\} \quad \ldots(8.29b)$$

Chebyshev Transformer

This matching transformer is based on Chebyshev polynomial of mth degree $T_N(x)$ and defined by

$$T_0(x) = 1$$

$$T_1(x) = x$$

$$T_2(x) = 2x^2 - 1$$

or in general

$$T_{m+1}(x) = 2xT_m(x) - T_{m-1}(x) \quad ...(8.30)$$

Input SWE of the Chebyshev transformer is given;

$$S = 1 + \ln\left(\frac{R_2}{R_1}\right)\frac{T_N\left(\frac{\cos\theta}{\cos\theta_l}\right)}{T_N\left(\frac{1}{\cos\theta_l}\right)} \quad ...(8.31)$$

where 'θ' is the electrical length of each section and θ_l length of section at frequency 'f'.

Chebyshev transformer provides larger bandwidth than a binomial multi–section matching transformer for the given number of Tx line sections. The increase of Cheysgev transformer results at the cost of increased ripple over the pass band of the matching network. However, one may still designate some maximum allowable reflection coefficient for the design of the Chebyshev transformer.

Properties of Chebyshev Polynomials

(*a*) Even order Chebyshev polynomials are even function.

(*b*) Odd order Chebyshev polynomials are odd function.

(*c*) The magnitude of any Chebyshev polynomials is unity or less and in the range of $-1 \leq x \leq 1$.

(*d*) $T_n(1) = 1$ for all Chebyshev polynomials.

(*e*) All the roots (zeroes) of the Chebyshev polynomials lie within the range of $-1 \leq x \leq 1$.

Therefore using the properties of Chebyshev polynomials one can design matching networks with a refection coefficient at or below some prescribed level over a wide band width.

Substring $x = \cos\theta$, the Chebyshev polynomials may be written as follows

$$T_m(\cos\theta) = \cos n\theta \quad ...(8.32)$$

As given the maximum magnitude of each Chebyshev polynomials should be unity within the pass band, the maximum magnitude of the reflection coefficient in the pass band is

$$T_m = |A|$$

where A is a constant and determined by taking the limit as θ approaches 0 and equal to

$$A = \frac{\ln(Z_L/Z_0)}{2T_N(\sec\theta_m)} \quad ...(8.33)$$

where, $$\sec\theta_m = \cos\left[\frac{1}{N}\cos^{-1}\left(\frac{\ln(Z_L/Z_0)}{2\sqrt{m}}\right)\right] \quad ...(8.34)$$

After determining the reflection coefficients (Γ_n), the characteristic impedances are found as follows;

$$Z_{n+1} = Z_n e^{2\Gamma n} \quad ...(8.35)$$

The reflection coefficient of an N–section matching transformer may be give as

$$\Gamma(\theta) = 2e^{-jN\theta}[\Gamma_0 \cos N\theta + \Gamma_1 \cos(N-2)\theta + ... \Gamma_n \cos(N-2n)\theta]$$

$$= A \cdot e^{-jN\theta} \cdot \Gamma_N \cdot (\sec\theta_m, \cos\theta)$$

The resulting fractional bandwidth for the Chebyshev multi – section transformer then found to be;

$$\frac{\Delta f}{f_0} = 2 - \frac{4\theta_m}{\pi}$$

or $$\Delta f = 2f_0\left(\frac{1-2\theta_m}{\pi}\right) \quad ...(8.36)$$

Which is similar to band width of multi-section transformer.

8.3 BALUN TRANSFORMER MATCHING

Balun stands for balance to unbalance matching transformer. It is a form of a quarter wave–transformer matching balanced line (parallel wire line) to an unbalanced transmission circuit, like antenna and co-axial cable). This problem arises when an ungrounded antenna is being fed by co-axial cable (Fig. 12.8*a*)

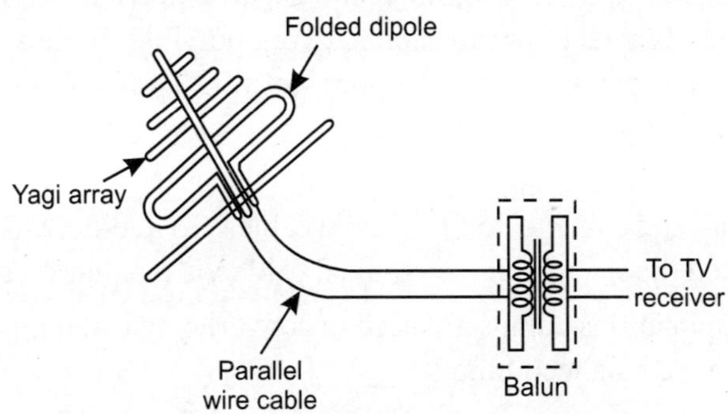

Fig. 8.8 (*a*) Use of balun in the circuit.

The cable from the yagi – array antenna offers reasonably high output impedance, which does not match with the input impedance of the TV receiver. Therefore, a transformer is required that matches the output impedance to a parallel wire cable with input impedance of the TV set. Hence this transformer is termed as balun and has got few turns of winding on a ferrite core.

A balance line consists of two parallel conductors separated by a fixed distance apart by some low loss material where as an unbalanced line is a co-axial cable which consists of a wire inside a tabular conductor. With the circuit diagram shown in Fig. (8.8*b*), one can make a balun that will transform impedance at a 4 : 1 ratio, with $R_2 = 4.R_1$. Thus a 220 Ω folded dipole feed point wi be transferred to 55 Ω unbalanced, and this types of balun is often included inside antenna tuners

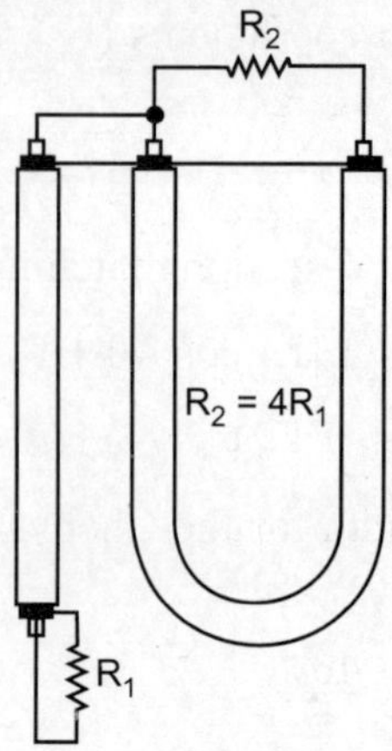

Fig. 8.8 (*b*) Co–axial balun transformer.

The length of the balun section of co-axial cable is given by

$$L_n = \frac{492V}{F_{MHz}}$$

where L_n is the length is feet.

V is velocity factor of the co-axial cable.

F_{MHz} is operating frequency in MHz.

A balun's main function is to achieve compatibility between systems, and as such, find expansive applications in modern communications,particularly in realizing frequency conversio mixer to make cellular and data transmission network possible. Baluns are useful everywhere suc as in RADARS, in satellites, in every telephone network and probably in most wireless networ modern / routers used in homes.

The baluns used for home television antennas have a 4 : 1 impedance ratio,to match th standard 300 Ω twin– lead cable to 75 Ω co-axial cables. To avoid EMC problems, it is a good ide to connect a centre fed dipole antenna to co-axial cable via a balun. In addition

(*a*) In audio applications baluns are used to convert between high impedance unbalanced an low impedance balanced limes.

(*b*) In power line communications, baluns are used in coupling signals on to a power line.

(*c*) In electronic communications, baluns are used to convert TWIN–X cables to category cables and back.

Some common types of baluns are shown in Fig . 8.9

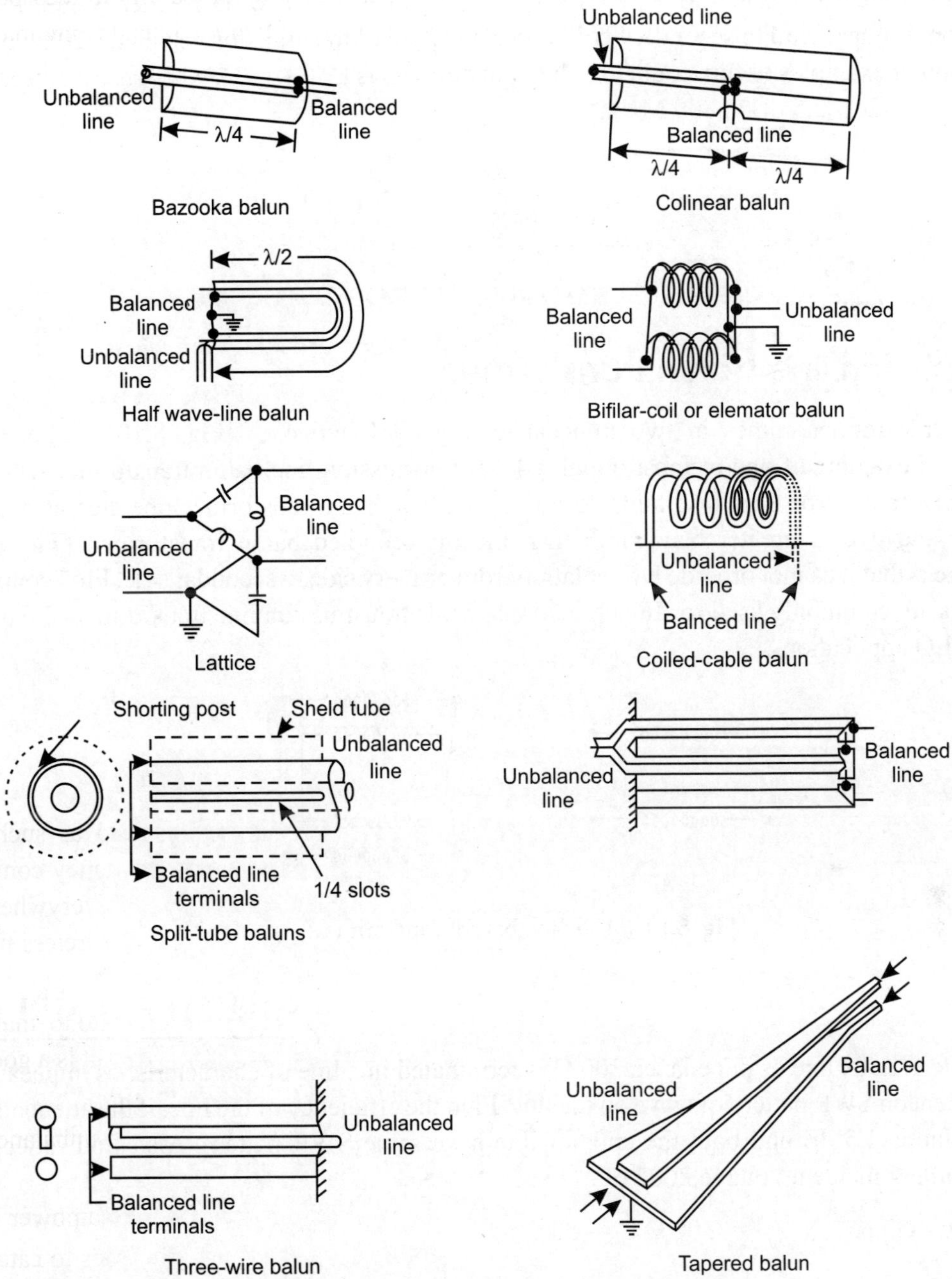

Fig. 8.9 Various types of baluns

12.4 RF BALUN TRANSFORMERS

Basically there two types of RF balun tranformers;

Flax coupled balun transformer

It can provide a wide range of impedance ratios; (1 : 1), (4 : 1), (9 : 1) and (16 :1), which are most coemmonly used. It has a low – loss and good balance over a wide frequency range. DC isolation from primary coil to secondary coil is also a feature of this transformer (Fig.8.10). One of the most popular examples of flux coupled balun transformers is ETC 41 – 2, which has an insertain loss of approximately 2.5 dB at 800 MHz.

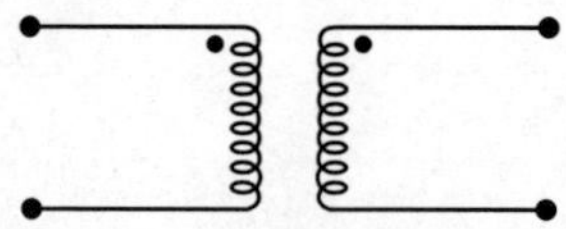

Fig. 8.10 (*a*) Flux coupled balun transformer

Transmission line balun transformers

This balun transformer comes in two impedance ratios 1:1 and 4 : 1(Fig. 8.10*b*). The ETC 1–4–2–3 is an example of one of most popular 4 : 1 transmission line balun transformer with an optimum performance of 1dB insertaion loss at 900 MHz. This transformer operates at higher frequencies and over a greater bandwidth than the flux coupled balun transformer. Only one disadvantage is that it cannot provide DC isolation from primary side to secondar side. Flux coupled transformers are commonly used for IF applications, while transmission line transformers are used for RF and LO applications.

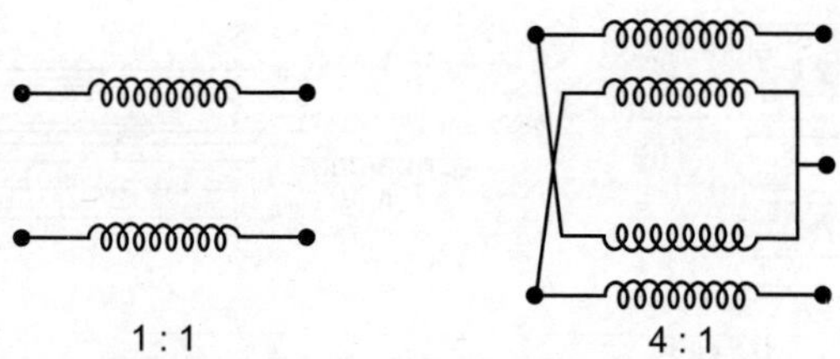

Fig. 8.10 (*b*) Tx line balun transformer.

SOLVED EXAMPLES

Example 1. A Tx line of impedance 300 Ω is terminated in a line of characteristics impedance 50 Ω Comment on SWR pattern of wave on the line. Find the efficiency of the line, if the attenuation across the line is 1.5 dB and both the ends of line have same SWR pattern. Also find reflected voltage assuming incident voltage 200 volt.

Solution : Given $Z_L = 300\ \Omega, Z_0 = 50\ \Omega$

$$\Gamma = \frac{Z_L - Z_0}{Z_L + Z_0}$$

$$= \frac{300-50}{300+50} = \frac{250}{350} = \frac{5}{7} = 0.71$$

$$\Gamma = 0.71$$

$$\text{VSWR} = \frac{1+\Gamma}{1-\Gamma} = \frac{1+0.71}{1-0.71} = \frac{1.71}{0.29} = 5.869$$

$$\Gamma = \frac{V_r}{V_i} \Rightarrow Vr = 0.71 \times 200 = 14.2 \text{ volts}$$

Efficiency of the line is (from 8.5);

$$\eta = \frac{4S_1}{10^{0.1A}(S_1+1)^2 - 10^{-0.1A}(S_1-1)^2}$$

$$= \frac{4 \times 5.896}{10^{0.15}(5.8+1)^2 - 10^{-0.15}(5.8-1)^2} = \frac{23.584}{48.97} = 0.48$$

$$\eta = 48\ \%$$

Example 2. A quarter wave transformer is used to match a 120 Ω to a 50 Ω air filled line at 10 GHz in the dominant mode considering both rectangular and co-axial lines cases. Find

(*i*) Length and impedance of transformer, if a = 2.268 cm.

(*ii*) Frequency band of operation for co–axial line over which the VSWR remain 1.5.

(*iii*) Position of V_{max} and V_{max}

(*iv*) Impedances at V_{max} and V_{max} and values of V_{max} and V_{max} if the load power is 100 mW.

Solution : Given $Z_L = 50\Omega$, $Z_s = 120\Omega$, $f_0 = 10\ GHz$

Hence, $\lambda_0 = \frac{c}{f_0} = \frac{30}{10} = 3cm$

(*i*) For rectangular $\frac{\lambda}{4}$ transformer

$$\lambda_g = \frac{\lambda_0}{\left[1 - \left(\frac{\lambda}{2a}\right)^2\right]^{\frac{1}{2}}} = \frac{3}{\left[1 - \left(\frac{3}{2 \times 2.286}\right)^2\right]^{\frac{1}{2}}} \quad \text{or} \quad \lambda_g = 4.0\ cm.$$

Hence length of rectangular quarter wave transformer is $\frac{\lambda_g}{4} = 1cm$ And co-axial quarter wave transformer

$$\frac{\lambda_g}{4} = \frac{\lambda_0}{4} = \frac{3}{4} = 0.75 \text{ cm.}$$

The characteristic impedanco of transformer section is;

$$Z_0 = \sqrt{Z_s Z_L} = \sqrt{50 \times 120} = 77.5\Omega$$

(*ii*) VSWR =1.5

$$\Gamma = \frac{S-1}{S+1} = \frac{1.5-1}{1.5+1} = \frac{5}{2.5} = 0.2$$

$$\frac{\Delta f}{f_0} = 2 - \frac{4}{\pi}\cos^{-1}\left[\frac{2.\Gamma_m\sqrt{Z_s \cdot Z_s}}{(Z_L - Z_s)\sqrt{(1-|\Gamma_m|^2)}}\right]$$

$$= 2 - \frac{4}{180}\cos^{-1}\left[\frac{2 \times 0.2 \times 77.5}{70 \times \sqrt{1-0.04}}\right]$$

$$= 2 - \frac{4}{180}\cos^{-1}\left[\frac{31}{68.585}\right]$$

$$= 2 - \frac{4}{180} \times 63.12$$

$$\Delta f = 0.598 \times 10 = 5.98 \text{ GHz}$$

(*iii*) Since there is impedance difference between source load impedance, standing wave occurring and hence voltage and current maxims and minims, At the voltage maximum / current minimum

$$Z_{in} = Z_{max} = \frac{V_{mix}}{I_{min}} = \frac{E_i + E_r}{I_i + I_r} = \frac{E_i(1+|\Gamma|)}{I_i(1-|\Gamma|)} = Z_0.s$$

Similarly, at voltage minimum / current maximum

$$Z_{in} = Z_{max} = \frac{V_{min}}{I_{max}} = \frac{E_i - E_r}{I_i + I_r} = \frac{E_i(1-|\Gamma|)}{I_i(1-|\Gamma|)} = \frac{Z_0}{s}$$

Since, $Z_L > Z_0$ first V_{max} is located at load and first V_{min} is located at a distance λ/4 i.e.θ.75 cm.

(*iv*) Load power $PL = \frac{V^2 max}{Z_L} = 100\ MW$

$$V_{max} \sqrt{100 \times 10^{-3} \times 120} = \sqrt{12} \text{ Volt.}$$

$$V_{min} = \frac{V_{max}}{s} = \frac{\sqrt{12}}{1.5} = 2.30 \text{ volt.}$$

At $V_{max}, Z_{in} = Z_0 \times s = 50 \times 1.5 = 75\Omega.$

At $V_{min}, Z_{in} = \frac{Z_0}{s} = \frac{50}{1.5} = 33.3\ 3\ \Omega.$

Returnloss = – log (Γ) = 20 log (0.2) = 13.97 dB.

Example 3. A pure resistive Tx line having $Z_0 = 350\ \Omega$ is terminated by a load resistance 50Ω. What type of stub will be required to provide impedance matching when stub is placed nearest to load? Find length and position of stub, if operating frequency is 30 MHz.

Solution : No doubt the short – circuit will be used because its radiation loss is less as the ends are closed and its effective length may be varied by means of a shorting bar which normally takes the shape of shorting plugs. Given

$$\lambda_0 = \frac{300}{30} = 10\ cm,\ Z_R = 50\Omega \text{ and } Z0 = 350\Omega$$

Location of the stub form load is given by;

$$l_s = \frac{\lambda}{2\pi} \tan^{-1} \sqrt{\frac{Z_R}{Z_0}} = \frac{10}{2 \times 3.14} \tan^{-1} \left[\sqrt{\frac{50}{350}}\right]$$

$$= \frac{10}{6.28} \times 8.130 = 12.95\ m$$

Length of the stub is given by;

$$l_t = \frac{\lambda}{2\pi} \tan^{-1} \left[\frac{\sqrt{Z_0 Z_R}}{Z_R - Z_0}\right] = \frac{10}{2 \times 3.14} \tan^{-1} \left[\sqrt{\frac{50 \times 350}{300}}\right]$$

$$= \frac{10}{6.28} \times 23.795 = 37.89\ m$$

Example 4. Design an quarter wave transformer to match an antenna array of impedance 300Ω, fed by a two wire transmission lines. The spacing between the wires is four times the diameter of Tx lines. Assume spacing between matching line is 10% lesser than the main line.

Solution : The characteristic impedance of feeder can be calculated by

$$Z_f = 276 \quad 10 \log\left(\frac{S}{r}\right), \quad \text{As } d = 2r,\ S = 4d = 4 \times 2r = 8$$

$$= 276 \quad 10 \log_{10}(8) = 525\Omega.$$

Therefore the Characteristic impedance of the quarter wave line is;

$$Z_0 = \sqrt{Z_R \times Z_f} = \sqrt{300 \times 525} = 396.86\ \Omega$$

For the convenient point of view, the spacing between matching should be same as that of main line, therefore, if 'r_1' is radius of matching line then

$$396.86 = 276 \quad \log\left(\frac{S}{2r_1}\right)$$

$$1.437 = \log 10\left(\frac{S}{2r_1}\right) \Rightarrow \text{then} \quad \frac{S}{r_1} = 54.81$$

In case the radius of feeder line wire is 0.1 cm, then spacing between lines, s = 80 × 0.1 = 8 *cm*

Therefore, radius of feeder wire is

$$r_1 = \frac{54.81}{80} = 0.68\ cm \qquad \therefore d_1 = 1.36\ cm$$

Since,spacing between feeder line radius is 10% lesser *i.e.*, 72 *cm*

$$r_1 = \frac{54.81}{72}\ 0.76\ cm \qquad \therefore d_1 = 1.52\ cm$$

Therefore, full matching set – up will be

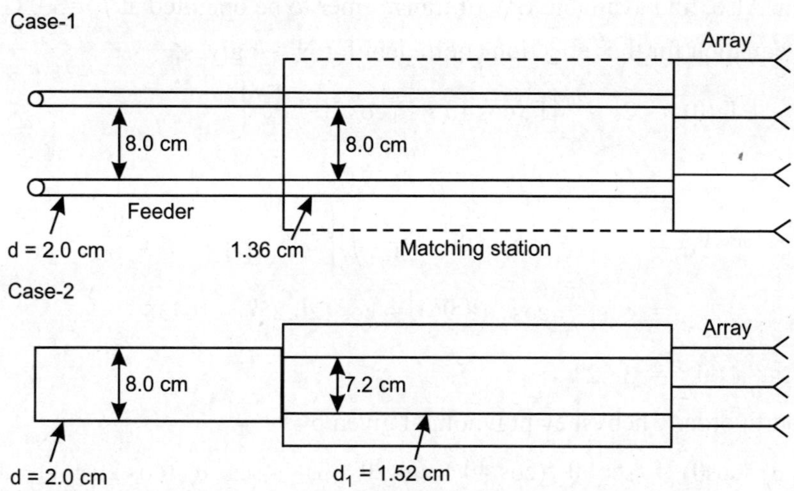

Fig. 8.12

Example 5. Calculate the impedances of a 4-section binomial matching transformer to match a 300 Ω line and also determine the resulting percentage bandwidth for $\Gamma_m = 0.1$.

Solution : The binomial coefficients are given by $C_n \frac{N!}{n!(N-n)!}$

Hence $\quad C_0 = \frac{4!}{0!4!} = 1 = C_4\ C_1 = \frac{4!}{0!3!} = 4 = C_3$ and $C_2\ \frac{4!}{2!2!} = 6$

Therefore from (8.28) impedances of the 4 transformer sections are;

$$Z_1 = Z_0 \exp\left[\frac{C_0}{2^4}\ln\left(\frac{Z_L}{Z_0}\right)\right] = 50 \exp\left[\frac{1}{16}\ln(6)\right] = 55.9\ \Omega$$

$$Z_2 = Z_0 \exp\left[\frac{C_0 + C_1}{2^4}\ln\left(\frac{Z_L}{Z_0}\right)\right] = 50 \exp\left[\frac{5}{16}\ln(6)\right] = 87.5\ \Omega$$

$$Z_3 = Z_0 \exp\left[\frac{C_0 + C_1}{2^4}\ln\left(\frac{Z_L}{Z_0}\right)\right] = 50 \exp\left[\frac{11}{16}\ln(6)\right] = 171.4\ \Omega$$

$$Z_4 = Z_0 \exp\left[\frac{C_0 + C_1 + C_2 + C_3}{2^4}\ln\left(\frac{Z_L}{Z_0}\right)\right] = 50 \exp\left[\frac{15}{16}\ln(6)\right]$$

$$= 268.25\ \Omega$$

The percentage bandwidth is given by :

$$\frac{\Delta f}{f_0} = 2 - \frac{4}{\pi}\cos^{-1}\left[\left(\frac{1}{2}\frac{\Gamma_m}{|A|}\right)^{\frac{1}{N}}\right]$$

In which

$$A = 2^{-4}\frac{Z_L - Z_0}{Z_L + Z_0} = \frac{1}{16}\ \frac{1}{16}\frac{300-50}{300+50} = \frac{5}{112}$$

$$\frac{\Delta f}{f_0} = 2 - \frac{4}{x}\cos^{-1}\left[\frac{1}{2}\left(\frac{(0.1)(112)}{5}\right)^{\frac{1}{4}}\right]$$

$$= 2 - \frac{4}{\pi}\cos^{-1}(0.6117) = 0.84 \text{ (or 84\%)}$$

Example 6. Design a 4 section Chabyshev matching transformer to match a 300 Ω load to 50 Ω line for Γ = 0.1. Also, find available BW of transformer to be operated at f_0 =3.0 GHz

Solution : The equation for the reflection coefficient for N = 4 gives;

$$\Gamma(\theta) = 2e^{-j4\theta}\left[\Gamma_0\cos 4\theta + \Gamma_1\cos 2\theta + \frac{\Gamma_2}{2}\right]$$

$$= A.e^{-j4\theta}.\Gamma 4.(\sec\theta_m.\cos\theta)$$

Angle

$$\sec\theta_m = \cos\left[\frac{1}{N}\cos^{-1}\left(\ln\left(\frac{Z_L/Z_0}{2\Gamma}\right)\right)\right] = \cos\left[\frac{1}{4}\cos^{-1}\left(\frac{\ln(6)}{0.2}\right)\right]$$

$$= \cos\left[\frac{1}{N}\cos^{-1}(8.96)\right] = \cos(20.25^0) = 0.938$$

$$\theta_m = 38.2^0$$

Inserting the fourth order Chebyshev polynomial given by

$$\Gamma_4(\sec\theta_m.\cos\theta) = \sec^4\theta_m(\cos 4\theta + \cos 2\theta + 3) - 4\sec^2\theta_m(\cos 2\theta + 1) + 1$$

into the reflection coefficient of cosine gives

$$\Gamma(\theta) = 2e^{-j4\theta}\left[\Gamma_0\cos 4\theta + \Gamma_1\cos 2\theta + \frac{\Gamma_2}{2}\right]$$

$$= Ae^{-j4\theta}\ \sec^4\theta_m(\cos 4\theta + \cos 2\theta + 3) - 4\sec^2\theta_m(\cos 2\theta + 1) + 1$$

Equation coefficient of cosine gives

$$2\Gamma_0 = A\sec^4\theta_m \Rightarrow \Gamma_0 = \frac{A}{2}\sec^4\theta_m = \frac{|\Gamma_m|}{2}\sec^4\theta_m$$

$$= \frac{0.1}{2} \sec^4 (38.2^0) = 0.131$$

and $$2\Gamma_0 = 4A.(\sec^4 \theta_m - 4 \sec^2 \theta_m)$$

$$= 2 \times 0.\ 1[3\sec^4 (38.2) - 4 \sec^2 (38.2) + 1] = 0.\ 201$$

$$\Gamma_2 = A.\ (3 \sec^4 \theta_m - 4\sec^2 \theta_m + 1)$$

$$= 0.1[3 \sec^4 (38.2) - 4 \sec^2 (38.2) + 1] = 0.239$$

However, other terms and found symmetry *ie*.

$$\Gamma_4 = \Gamma_0\ , \Gamma_3 = \Gamma_1, \ldots\ldots.$$

Therefore resulting impedances are;

$$Z_{n+1} = Z_n e^{2\Gamma_n}$$

$$Z_1 = Z_0\ e^{2\Gamma 0} = 50 \times e^{0.262} = 50 \times 1.3 = 65\ \Omega$$

$$Z_2 = Z_1\ e^{2\Gamma 1} = 65 \times e^{0.401} = 65 \times 1.49 = 97.12\ \Omega$$

Similarly, $$Z_3 = Z_2\ e^{2\Gamma 2} = 156.\ 7\Omega$$

$$Z_4 = Z_3\ e^{2\Gamma 3} = 234.\ 0\Omega$$

And bandwidth is

$$\Delta f = 2 \times f_0 \left(1 - \frac{2\theta_m}{\pi}\right)$$

$$= 2 \times 3.0\left(1 - \frac{2 \times 38.2 \times \pi}{180 \times \pi}\right)$$

$$= 6.\ 0 \times 0.\ 576 = 3.456 \text{ GHz}$$

Available bandwidth Δf = 0. 456 GHz = 456 MHz

Example 7. Design a co–axial balun transformer for velocity factor 0. 176, to be operated at 30 MHz. What would be its length if '*f*' increases by 20%

Solution : (*i*) We know that

$$L_n = \frac{492V}{f_{MHzs}} = \frac{492 \times 0.176}{30} = 2.87$$

20% increase means 30 + 30 × 0.20 = 36 MHz

$$L'_n = \frac{492 \times 0.176}{36} = 2.4 \text{ feet.}$$

Example 8. Design a two section symmetrical binomial transformer to match the 100 Ω impedance to a 50 Ω air filled line at 30 GHz. And also find the B/ W in order to obtained reflection coefficient (Γ = 0. 11).

Solution : Given number of section, $N = 2$, so number of steps required,

$$n = N + 1 = 3. \text{ and } Z_s = 100\ \Omega\ Z_L = 50\ \Omega.$$

$$\Gamma = 0.\ 11, \text{ and } \lambda_0 = \frac{c}{f} = \frac{3 \times 10^{10}}{3 \times 10^9} = 10.0\ cm.$$

$$\text{Therefore} \quad \Gamma_1 + \Gamma_2 + \Gamma_3 = \frac{1}{2} \ln\left(\frac{100}{50}\right) = 0.347$$

We know that for symmetrical transformer, $\Gamma_1 = \Gamma_3$, $\Gamma_2 = NT_1 = 2\Gamma_1$

$$\Gamma_1 + 2\Gamma_1 + \Gamma_1 = 0.347$$

$$\Gamma_1 = 0.0868$$

$$\Gamma_2 = 2 \times 0.\ 0868 = 0.\ 174$$

Hence,
$$Z_2 = Z_L\, e^{2\Gamma 1} = 50 \times e^{0.174} = 59.\ 473\ \Omega$$

$$Z_3 = Z_2\, e^{2\Gamma 2} = 59.473 \times e^{0.348} = 84.227\ \Omega$$

Length of the section is $\frac{\lambda_g}{4} = \frac{\lambda_0}{4} = \frac{10}{4} = 2.5\ cm.$

$$|\Gamma_m| = \Gamma_1 .\ 2^{n-1} \cos^{n-1}\theta_1 = 0.0868 \times 4 \times \cos^2\theta_1$$

$$\Rightarrow \quad \theta_1 = \cos{-1}\left[\frac{0.11}{0.08868 \times 4}\right]^{\frac{1}{2}} = 55.\ 745^0$$

$$\theta_2 = 180 - \theta_1 = 180 - 55.\ 745° = 124.254°$$

Therefore
$$\theta = \theta_2 - \theta_1 = 68.\ 779°$$

Therefore,the ratio (r) of the wavelength at the band edges frequencies f_1 and f_2 is equal to

$$r = \frac{\lambda_{g1}}{\lambda_{g2}} = \frac{\theta_2}{\theta_1} = \frac{124.254}{55.745} = 2.228$$

Bandwidth
$$\Delta f = f_0 \times \frac{2\theta}{\pi} = \frac{3 \times 2 \times 68.779}{180^0} = 2.293 \text{ GHz}$$

Example 9. Design a two section Chebychev transformer to match a 110 Ω load a 50 Ω line, at a frequency of 9 GHz. Find operating bandwidth, if maximum allowable VSWR is 1.20.

Solution : No . of Section = 2,no of steps = 3,

$$Z_s = Z_4 = 110\ \Omega,\ Z_L = 50\ \Omega$$

$$|\Gamma_m| = \frac{VSWR-1}{VSWR+1} = \frac{0.2}{2.2} = 0.091$$

Total reflection coefficient for symmetrical transform (n Old)

$$\Gamma = |\Gamma_m| T_{n-1}(x) = 0.091 \qquad T_2(x) = 2[\Gamma_1 \cos 2\theta] + \Gamma_2$$

$$0.091[2x^2 - 1] = 2[\Gamma_1 (2\cos^2\theta - 1)] + \sqrt{2}$$

$$\text{Therefore,} \quad 4\Gamma_1 \cos^2\theta_1 = 0.182$$

$$\Gamma_1 = \frac{0.0455}{\cos^2\theta_1} \qquad \text{Equation coefficient of } (x_2)$$

Similarly, equation coefficient of 'x' gives;

$$\Gamma_2 = 2\Gamma_1 - 0.091$$

Now,
$$\frac{Z_S - Z_L}{Z_S + Z_L} = \Gamma_m .T_{n-1}(\sec\theta_1)$$

$$= \frac{110-50}{110+50} = 0.091.\ T_2(\sec\theta_1) = \frac{60}{170} = 0.091\,(2\sec^2\theta_1 - 1)$$

$$\Rightarrow \quad \sec\theta_1 = 1.56 \cos\theta_1 = 0.64 \text{ or } \theta_1 = \cos^{-1}(0.64) = 50.\ 13^0$$

$$\Gamma_1 = 0.1107 \quad \text{and} \quad \Gamma_2 = 2 \times 0.1107 - 0.091 = 0.1304$$

Now,

$$Z_1 = \text{impedance of } 1^{st} \text{ section transformer}$$

$$\left(\frac{1+\Gamma_1}{1-\Gamma_1}\right) \times Z_L = \left(\frac{1+0.01107}{1-0.1107}\right) \times 50 = \left(\frac{1.1107}{0.8893}\right) \times 50 = 62.45\ \Omega$$

$$Z_2 = \text{impedance of } 1^{st} \text{ section transformer}$$

$$= \left(\frac{1+\sqrt{2}}{1-\sqrt{2}}\right) \times Z_1 = \left(\frac{1+0.1304}{1-0.1304}\right) \times 62.45 = 81.\ 18\ \Omega$$

The

$$B/W = \Delta f = \frac{4}{\pi}\left(\frac{\pi}{2} - \theta_1\right) \times f_0 = \frac{4}{180^0}(90° - 50.\ 13°) \times 9 = 7.974 \text{ GHz}$$

Ans.

Example 10. Describe Ferranti The Effect with suitable example.

Solution: The Ferranti effect is an increase in voltage occurring at the receiving end of a long transmission line, above the sending end voltage. This effect occurs when the line is energized, and line is open circuited or very lightly loaded at the receiving end. This effect is due to the voltage drop across the line inductance (due to charging current) being in phase with the sending end voltages. As this voltage drop affects the sending end voltage, the receiving end voltage becomes greater. Therefore, both line inductance and capacitance are responsible for this phenomenon. The relative voltage rise is proportional to the square of the line length. The Ferranti Effect will be more pronounced for longer Tx line and for the higher voltage applied.

Let us consider a Tx line is terminated with resistive load 30 Ω, as shown in Fig 8.12 (*a*).

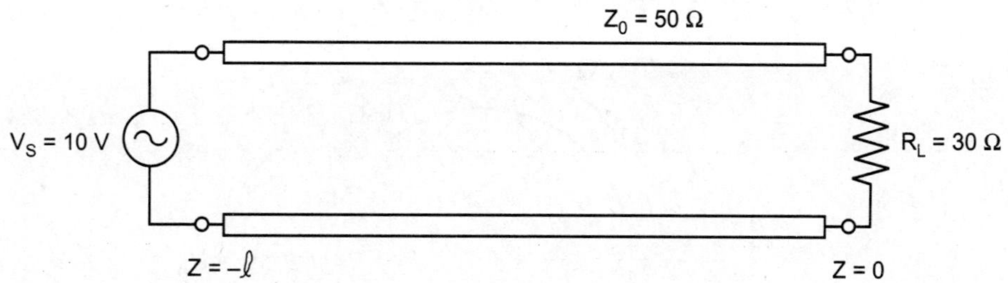

Fig 8.12 (*a*) Geometry for example 8.10

We can find the VSWR, as follows;

Reflection coefficient

$$\Gamma = \frac{30-50}{30+50} = -0.25$$

$$\text{VSWR} = \frac{1+0.25}{1-0.25} = 1.67$$

The voltage at $Z = -l$ is always fixed to be V_s. Hence, we can see that $|V(z)|$ on parts of the transmission line can be longer than V_s If l is chosen so that V_s is at V_{min} then

$$V_{max} = V_{min} \times \text{VSWR} = 10 \times 1.67 = 16.7 \text{ volt}$$

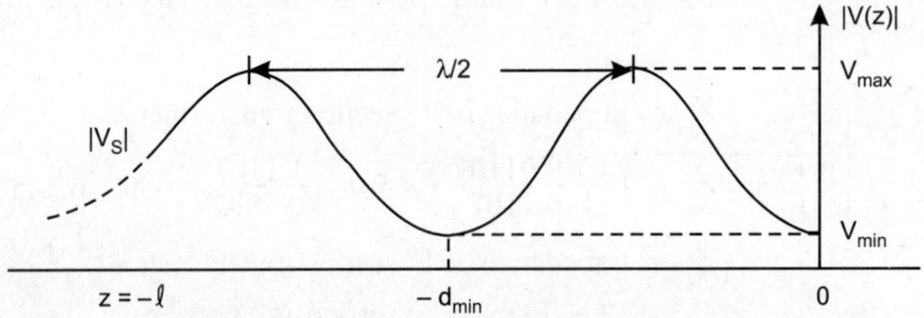

Fig. 8.12 (*b*) VSWR pattern for example 8.10

This amplification of voltage on a line is only Ferranti's effect. If the VSWR on the line is very high, V_{max} can be so high that it reaches the breakdown voltage of the line. So, this is important for one to be cautious in designing transmission line circuits.

Example 11. If a load impedance $Z_L = (100 + j85)\ \Omega$ has to be matched a Tx line of $Z_o = 100\ \Omega$. Using single stub matching, find the minimum d and l, that will reduce the VSWR of the main line to 1. If matching load consists of serious resistance and capacitance, then find its values at assume operating frequency is 3 GHz.

Solutions:

First let find that the normalized load $Z_{nL} = (2 + j1.7)$ as shown on the Smith Chart (Fig 8.13). Since there are parallel connections, it is more convenient to calculate admittances, using $Y_{nL} = (1/Z_{nL})$. When we move toward the generator, $Y_n(z)$ traces out a locus on the Smith Chart. It intersects the $G = 1$ circle after moving through 0.216λ. Therefore, $d = 0.216\ \lambda$.

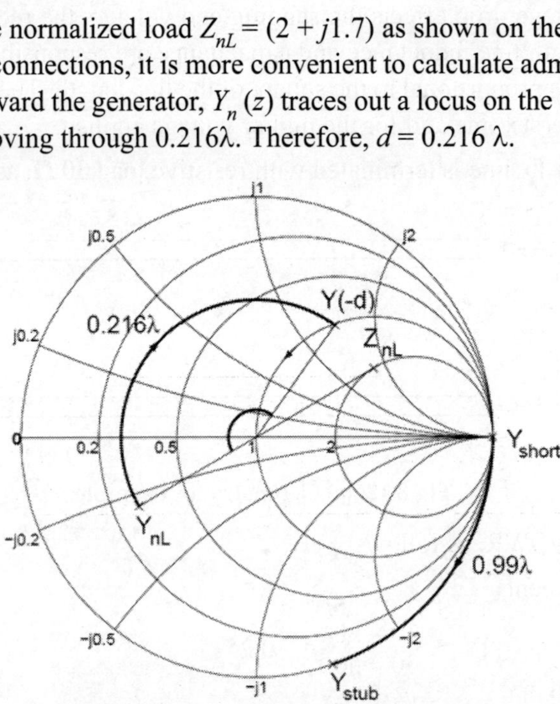

Fig. 8.13 (*a*) Smith chart for example 8.11

Now, $Y_n(-d) = 1 + j1.4$. Hence, $Y_{n\ stub} = -j1.4$. From the Smith Chart, we observe that the admittance for a short is infinity, and is at the right end of the Smith Chart. To get a $Y_{n\ stub} = -j1.4$, we have to move toward the generator for 0.99λ. Hence, $l = 0.99\lambda$. As frequency is 3 GHz, $\lambda = 10$ cm, therefore position of stub (d) and length of stub (l) will be 2.16 cm and 9.9 cm respectively. Since load impedance is $(100 + j\ 85) = (R + j\omega C)$, hence $R = 100\ \Omega$ and $C = 0.624$ pF, as shown in Fig.8.13 (b).

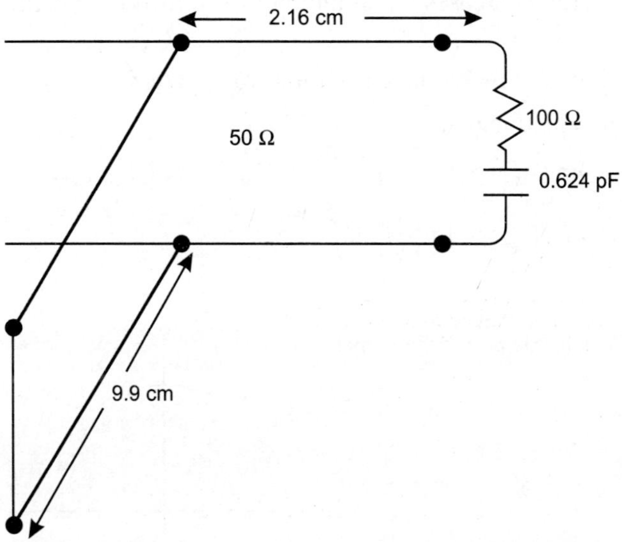

Fig. 8.13 (b) Single stub matching for example 8.11.

Example 12. Describe voltage standing wave pattern in different types of Tx line circuit

Solutions: Standing Wave Patterns

n practical applications it is very convenient to plot the magnitude of phasor voltage and phasor current along the transmission line. These are the standing wave patterns:

$$\text{Loss-less line} \begin{cases} |V(d)| = |V^+|.|(1+\Gamma(d))| \\ |I(d)| = \left|\dfrac{V^+}{Z_0}\right|.|(1-\Gamma(d))| \end{cases}$$

$$\text{Lossy line} \begin{cases} |V(d)| = |V^+ e^{\alpha d}|.|(1+\Gamma(d))| \\ |I(d)| = \left|\dfrac{V^+ e^{\alpha d}}{Z_0}\right|.|(1-\Gamma(d))| \end{cases}$$

The voltage standing wave pattern provides immediate information on the transmission line circuit

- If the load is matched to the transmission line ($Z_R = Z_0$) the voltage standing wave pattern is flat, with value $|V+|$.
- If the load is real and $Z_R > Z_0$, the voltage standing wave pattern starts with a maximum at the load.

- If the load is real and $Z_R < Z_0$, the voltage standing wave pattern starts with a minimum at the load.
- If the load is complex and $\text{Im}(Z_R) > 0$ (inductive reactance), the voltage standing wave pattern initially increases when moving from load to generator and reaches a maximum first.
- If the load is complex and $\text{Im}(Z_R) < 0$ (capacitive reactance), the voltage standing wave pattern initially decreases when moving from load to generator and reaches a minimum first.

Maxima and minima of the voltage standing wave pattern.

- Load with inductive reactance

$$\text{Im}(Z_R) > 0$$

$$\Rightarrow \quad \text{Im}(\Gamma_R) = \text{Im}\left(\frac{Z_R - Z_0}{Z_R + Z_0}\right) > 0$$

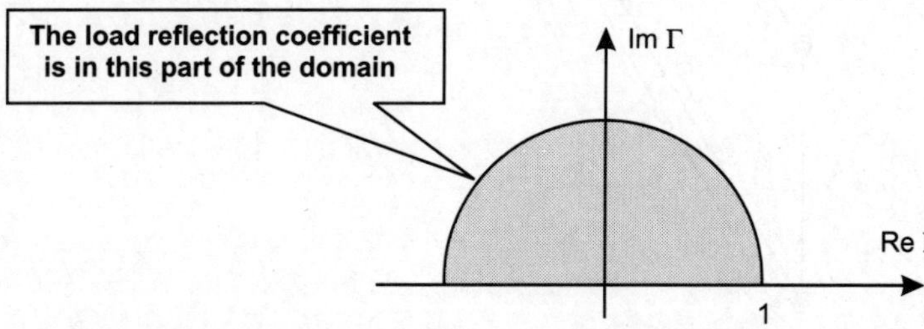

The first maximum of the voltage standing wave pattern is closest to the load, at location

$$\angle\Gamma(d) = \phi - 2\beta\, d_{\max} = 0$$

$$\Rightarrow \quad d_{\max} = \frac{\phi}{4\pi}\lambda$$

- Load with capacitive reactance

$$\text{Im}(Z_R) < 0$$

$$\Rightarrow \quad \text{Im}(\Gamma_R) = \text{Im}\left(\frac{Z_R - Z_0}{Z_R + Z_0}\right) < 0$$

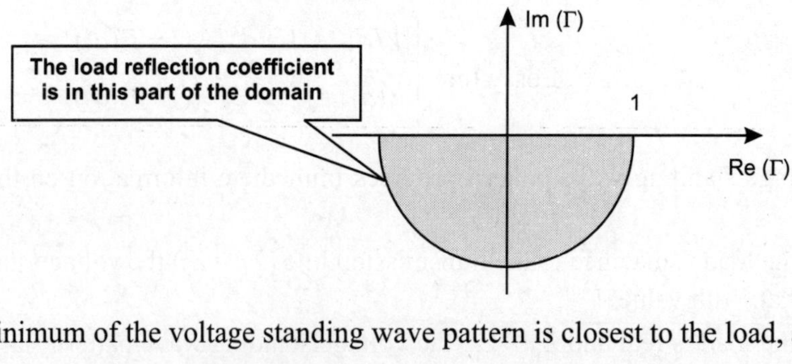

The first minimum of the voltage standing wave pattern is closest to the load, at location

$$\angle\Gamma(d) = |\phi - 2\beta\, d_{\max}| = \pi$$

$$\Rightarrow \quad d_{\max} = \frac{|\pi - \phi|}{4\pi}\lambda$$

A measurement of the voltage standing wave pattern provides the locations of the first voltage maximum and of the first voltage minimum with respect tot he load. The ratio of the voltage magnitude at these points gives directly the voltage standing wave ratio (VSWR). This information is sufficient to determine the load impedance Z_R, if the characteristic impedance of the transmission line Z_0 is known.

EXERCISES

1. A Tx line of impedance 250 Ω is terminated with a line of characteristic impedance 50 ohms, both the ends of line have same SWR pattern at a frequency of 6 GHz. Find the efficiency of the line if the attenuation across the line is 2.0 dB. Also find VSWR and reflected voltage,if incident voltage is 180 Volt.
2. A quarter wave transformer is used to match a 150 Ω to a 50 air filled line at 9 GHz in the dominant mode, considering both rectangular and co– axial lines cases. Find
   (*i*) Length and impedance of transformer, if $a = 2.4$ cm.
   (*ii*) Frequency band of operation for co– axial line over which the reflection coefficient remains 0.5
   (*iii*) Impedance at V_{min} and V_{max} and values of V_{max} and V_{min}, if the load power is 200 mW..
3. A Tx line of characteristics impedance 300 Ω, is terminated by a load resistance 50 Ω using stub placed nearest to the load. Find length and position of sub at operating wavelength 6 cm.
4. Design a (λ/4) wave transformer to match a two wire Tx line feel antenna array of impedance 250 ohms. Estimate design parameters; if the spacing between the wires is 1.5 times the diameter fo Tx lines and spacing between matching line is 15% lesser than the main line.
5. What are the impedances of a 3–section binomial matching transformer to provide matching between a 250 Ω load and a 50 Ω line ? Also estimate the percentage bandwidth for $\Gamma_m = 0.15$.
6. Design a two section Chebyshev matching transformer to match a 250 Ω load to a 50 Ω line for $\Gamma = 0.12$ Also find available BW of transformer to be operated at $f_0 = 2.0$ GHz.
7. Design a two section Chebychev transformer to match a 100 Ω load to a 50 Ω line,at a frequency of 10 GHz. Find operating bandwidth, if maximum allowable VSWR is 1.5.
8. Find the length of a co–axial balun transformer for velocity factor 0.176, to be operated at 35 MHz.
9. Design a single section symmetrical binomial transformer to match the 150 Ω impedance to a 50 Ω air filled line at 25 GHz. And also find the B/W in order to obtained reflection coefficient $(\Gamma = 0.1)$.
10. Describe the working printable of double stub matching. How it is better than the single stub matching?

- If you cannot do great things, do small things in a great way.

Napoleon Hill

Chapter 9

WAVEGUIDES

9.1 INTRODUCTION

Wave-guide is a hollow metallic structure of uniform cross section and used for transmitting and receiving electromagnetic waves by successive reflection from the inner walls of guide. Waveguides are preferred at microwave frequency because they are much less lossy at the microwave frequency and handle greater power transmission. Wave guide is used at microwave frequency ranging (3 – 300) GHz.

This is because above 3 GHz frequency transmission line shows higher loss, since at higher frequency solid dielectrics need to support the conductor and conductor itself shows much losses. A wave-duide acts as high-pass filter, because it can be operated only above certain frequency known as cut-off frequency.

It may also noted that no TEM (Transverse electromagnetic wave) wave can exist in the waveguide but TE and TM waves can. Induced current into waveguide wall gives rise to power losses and to minimize these losses the inner surface of the waveguide is usually coated with either gold or silver to improve the conductivity and minimize losses in the waveguide. Generally it is manufactured with copper. Applications of waveguide are as capacitor, inductor, resonant circuit etc. Wave-guide can have any arbitrary shape of uniform cross section, but most common geometry of wave-guide is either rectangular or circular.

Comparision with Transmission Lines

Similarities:

1. Wave travelling in waveguide has phase velocity (v_p) and hence will be attenuated as in a transmission line.
2. When the EMW (Electromagnetic wave) reaches the end the of the waveguide it get reflected in similar fashion as in transmission line.
3. Any irreguality in the waveguide produces reflection just like an irregularity in the transmission line.
4. Reflected wave in waveguide can be eliminated by proper matching in waveguide as transmission lines.

5. When both incident and reflected waves are present in waveguide a standing wave pattern results as in transmission lines.

Dissimilarities:

1. Waveguide has a cut-off-value of the frequency which only depends on the dimension of the waveguide. Only wave having frequencies greater than cut-off frequency f_c will propagates. Hence waveguide behave as high pass filter, whereas in 2-wire transmission lines all the frequencies (f = 0 to any value) can pass through.
2. Waveguide is one conductor transmission line system. The whole body of the waveguide behaves as ground and the wave propagats through multiple reflection from the wall of waveguide, whereas conventional transmission line is two wire transmission line system and wave propagated due to potential gradiant.
3. In waveguide wave has two velocity called group and phase velocities having different value than velocity of wave in free space whereas in transmission line the wave propagates about same velocity as in free space.
4. Mechanical simplicity and operating (miximum) frequency (300 GHz) is higher than transmission line.
5. Power handling ability in wage guide is about ten times higher than transmission line.
6. In wave guide the wave has wave impedance which is function of frequency as

$$Z_g = \frac{Z_0}{\sqrt{1-\left(\frac{\lambda}{\lambda_c}\right)^2}} = \frac{Z_0}{\sqrt{1-\left(\frac{f_c}{f}\right)^2}} \qquad \left(Since\ f \propto \frac{1}{\lambda}\right)$$

where

Z_g = Waveguide impedance

Z_0 = Free space impedance

λ_c = Cut-off wavelength

f_c = Cut-off frequency

whereas in a loss-less transmission line the impedance is called characteristic impedance Z_T

$$Z_T = \sqrt{\frac{Z}{Y}} = \sqrt{\frac{R+j\omega L}{G+j\omega C}} = \sqrt{\frac{L}{C}}$$

where

Z = Impedance, Ω/m

Y = Admittance, mho/m

is the function of transmission line parameters.

7. The system of wave propagation in wageguide is in accordance with field theory (E and H) whereas in transmission line in accordance with circuit theory (V and I) hence, internal conductor is not required in the wageguide.

8. Wageguide generally used for indoor applications whereas a transmission line for outdoor applications.

Propagation in TE and TM Mode

The electromagnetic wave inside a wageguide can have infinite number of patterns which are called modes. We know that in electromagnetic wave electric and magnetic fields are always perpendicular to each other. And at the surface of conductor the electric fields cannot have a component parallel to the surface. This indicates that electric field must always be perpendicular to the surface of the conductor.

The magnetic field on the other hand is always parallel to the surface of the conductor and cannot have component perpendicular to the surface.

There are two kinds of mode in a wageguide. In the first type, the electric field is always perpendicular to the direction of propagation and is called the Transverse Electric or TE wave. In the second type, the magnetic field is always perpendicular to the direction of propagation and is called the Transverse Magnetic or TM wave. Let us assume the wave is propagated in z-direction and the E_z and H_z will be as follows;

for TE mode $E_Z = 0$ $H_Z \neq 0$

for TM mode $E_Z \neq 0$ $H_Z = 0$

for TEM mode $E_Z = 0$ $H_Z = 0$

Where E_z and H_z are *z*-componerts of an EM wave propagating in a medium having componerts (E_x, E_y, E_z) and (H_x, H_y, H_z).

Rectangular Wave Guide

A hollow pipe with any short of cross-section could be used as waveguide, but the simplest cross section are prefered and rectangular shape is most simplest one, hence called rectangular wave-guide (Fig. 9.1).

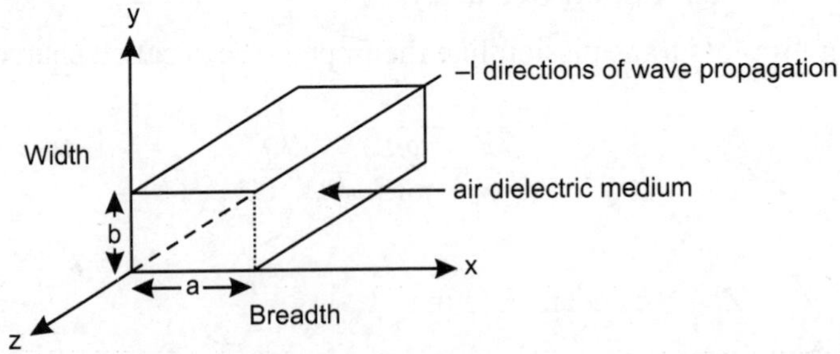

Fig. 9.1. Basic geometry of a rectangular wave-guide

As electromagnetic wave propagating in *z* direction, the general wave equation can be derived from Helmholtz equation, as

$$\nabla^2 E = \gamma^2 E$$

$$\nabla^2 H = \gamma^2 H$$

where $\gamma = \sqrt{j\omega\mu(\sigma + j\omega\varepsilon)}$; propagation constant

Since wave is propagating in z direction, then for TE module and TM mode;

$$\nabla^2 E_z = \gamma^2 E_z \text{ (for TM mode } H_z = 0) \quad \text{...(9.1 a)}$$

$$\nabla^2 H_z = \gamma^2 H_z \text{ (for TE mode } E_z = 0) \quad \text{...(9.1 b)}$$

The equation (9.1 a) can be expanded as (TM mode)

$$\left(\frac{\partial^2}{\partial x^2} + \frac{\partial^2}{\partial y^2} + \frac{\partial^2}{\partial z^2}\right) E_z = \gamma^2 E_z$$

$$\left(\frac{\partial^2 E_z}{\partial x^2} + \frac{\partial^2 E_z}{\partial z^2} + \frac{\partial^2 E_z}{\partial z^2}\right) = j\omega\mu(\sigma + j\omega\varepsilon) E_z$$

For loss-less medium $\sigma = 0$,

$$\frac{\partial^2 E_z}{\partial x^2} + \frac{\partial^2 E_z}{\partial y^2} + \frac{\partial^2 E_z}{\partial z^2} = -\omega^2 \mu\varepsilon E_z \quad \text{...(9.2a)}$$

Since, wave propagating in is z direction so, E-field can be given as;

$$E_z = E_{z_0} . e^{-\gamma z}$$

$$\frac{\partial E_z}{\partial z} = E_{z_0} = (-\gamma e^{-\gamma z}) = -\gamma E_z$$

and

$$\frac{\partial^2 E_z}{\partial z^2} = -\gamma \frac{\partial E_z}{\partial z}$$

$$= -\gamma(-\gamma e^{-\gamma} E_z) = \gamma^2 E_z$$

Putting this value in eq. (9.2a), gives

$$\text{So } \frac{\partial^2 E_z}{\partial x^2} + \frac{\partial^2 E_z}{\partial y^2} + \gamma^2 E_z = -\omega^2 \mu\varepsilon E_z$$

$$\frac{\partial^2 E_z}{\partial x^2} + \frac{\partial^2 E_z}{\partial y^2} = -(\gamma^2 + \omega^2\mu\varepsilon) E_z \quad \text{...(9.2b)}$$

Let $(\gamma^2 + \omega^2\mu\varepsilon) = h^2$, constant then (9.2$b$) can be written as,

$$\frac{\partial^2 E_z}{\partial x^2} + \frac{\partial^2 E_z}{\partial y^2} + h^2 E_z = 0 \quad \text{...(9.3)}$$

Similarly, for TE mode the wave equation can be written as;

$$\frac{\partial^2 H_z}{\partial x^2} + \frac{\partial^2 H_z}{\partial y^2} + h^2 H_z = 0 \text{ (TE Mode)} \quad \text{...(9.4)}$$

Solving eqn. (9.3) and (9.4) we can calculate E_x, E_y and H_x, H_y components for TE and TM modes. From Maxwell's first equation

$$\nabla \times H = J + \frac{\partial D}{\partial t}$$

$$\nabla \times H = \sigma E + \frac{\varepsilon \partial E}{\partial t} \quad ...(9.5)$$

Since, $J = \sigma E$ and $D = \varepsilon E$

for loss less medium $\sigma = 0$ so,

$$\nabla \times H = j\omega\varepsilon E$$

$$\left(\frac{\partial}{\partial x} a_x + \frac{\partial}{\partial z} a_y + \frac{\partial}{\partial z} a_z\right) \times (H_x a_x + H_y a_y + H_z a_z) = j\omega\varepsilon\,(E_x a_x + E_y a_y + E_z a_z)$$

or
$$\begin{bmatrix} a_x & a_y & a_z \\ \frac{\partial}{\partial x} & \frac{\partial}{\partial y} & \frac{\partial}{\partial z} \\ H_x & H_y & H_z \end{bmatrix} = (j\omega\varepsilon E_x)\, a_x + (j\omega\varepsilon E_y) a_y + (j\omega\varepsilon E_z)\, a_z$$

but
$$\frac{\partial}{\partial z} = -\gamma \text{ (an operator)}$$

$$\begin{bmatrix} a_x & a_y & a_z \\ \frac{\partial}{\partial x} & \frac{\partial}{\partial y} & -\gamma \\ H_z & H_y & H_z \end{bmatrix} = (j\omega\varepsilon E_x)\, a_x + (j\omega\varepsilon E_y) a_y + (j\omega\varepsilon E_z) a_z$$

$$\left(\frac{\partial H_z}{\partial y} + \gamma H_y\right) a_x - \left(\frac{\partial H_z}{\partial x} + \gamma H_z\right) a_y + \left(\frac{\partial H_y}{\partial x} - \frac{\partial}{\partial y} H_x\right) a_z$$
$$= (j\omega\varepsilon\, E_x)\, a_x + (j\omega\varepsilon\, E_y)\, a_y + (j\omega\varepsilon\, E_z) a_z$$

Comparing individual components, gives;

$$\frac{\partial H_z}{\partial y} + \gamma H_y = j\omega\varepsilon\, E_x \quad ...(9.6a)$$

$$\frac{\partial H_z}{\partial x} + \gamma H_x = -\, j\omega\varepsilon\, E_y \quad ...(9.6b)$$

$$\frac{\partial H_y}{\partial x} - \frac{\partial H_x}{\partial y} = j\omega\varepsilon\, E_z \quad ...(9.6c)$$

Similarly, from Maxwell's second equation

$$\frac{\partial E_z}{\partial y} + \gamma E_y = -j\omega\mu\, H_x \quad ...(9.7a)$$

$$\frac{\partial E_z}{\partial x} + \gamma E_x = j\omega\mu\, H_y \quad ...(9.7b)$$

$$\frac{\partial E_y}{\partial x} - \frac{\partial E_x}{\partial y} = -j\omega\mu H_z \quad ...(9.7c)$$

Combing equation (9.6*a*) and (9.7*b*) gives;

$$\frac{\partial H_z}{\partial y} + \gamma H_y = j\omega\varepsilon\, E_x$$

$$E_x = \frac{1}{j\omega\varepsilon}\left[\frac{\partial H_z}{\partial y} + \gamma H_y\right]$$

$$= \frac{1}{j\omega\varepsilon}\left[\frac{\partial H_z}{\partial y} + \frac{\gamma}{j\omega\mu}\left(\frac{\partial E_z}{\partial x} + \gamma E_z\right)\right]$$

$$= \frac{1}{j\omega\varepsilon}\frac{\partial H_z}{\partial y} - \frac{\gamma}{\omega 2\mu\varepsilon}\frac{\partial E_z}{\partial x} - \frac{\gamma^2 \varepsilon E_x}{\omega^2 \mu\varepsilon}$$

$$E_x = -\frac{\gamma}{h^2}\cdot\frac{\partial E_z}{\partial x} - \frac{j\omega\mu}{h^2}\cdot\frac{\partial H_z}{\partial y} \quad ...(9.8a)$$

$$E_y = -\frac{\gamma}{h^2}\cdot\frac{\partial E_z}{\partial y} + \frac{j\omega\mu}{h^2}\cdot\frac{\partial H_z}{\partial x} \quad ...(9.8b)$$

Similarly, for magnetic fields

$$H_x = -\frac{\gamma}{h^2}\cdot\frac{\partial H_z}{\partial x} + \frac{j\omega\varepsilon}{h^2}\cdot\frac{\partial E_z}{\partial y} \quad ...(9.8c)$$

$$H_y = -\frac{\gamma}{h^2}\cdot\frac{\partial H_z}{\partial y} - \frac{j\omega\varepsilon}{h^2}\cdot\frac{\partial E_z}{\partial x} \quad ...(9.8d)$$

The above equations (9.8*a* – 9.8 *d*) represents the general field components. There are three cases.

(*i*) Propagation of TEM mode:

If we substitute

$E_z = H_z = 0$ in eqn (9.8*a* – 9.8*d*)

Then, $E_x, E_y, H_x, H_y = 0$, that is TEM wave cannot exist inside the waveguide.

(*ii*) TM Mode propagation: In rectangular waveguide, for TM mode,

$$H_z = 0 \; E_z \neq 0$$

Therefore from (9.3)

$$\frac{\partial^2 E_z}{\partial x^2} + \frac{\partial^2 E_z}{\partial y^2} + h^2 E_z = 0 \quad ...(9.9a)$$

This is partial differential eqn. which can be solved by separation of variable method,

Let, $E_z = X.Y$ ...(9.9*b*)

where X is the only function of x, and Y is the only function of y.

X, Y are independent variable.

So, putting the values of E_z in eqn. (9.9.*b*);

$$\frac{\partial^2}{\partial x^2}X.Y + \frac{\partial^2}{\partial y^2}X.Y + h^2 XY = 0$$

dividing both sides by XY, $\frac{Y}{XY}\frac{\partial^2 X}{\partial x^2} + \frac{X}{XY}\frac{\partial^2 Y}{\partial y^2} + h^2 = 0$

$\frac{1}{X}\cdot\frac{\partial^2 X}{\partial x^2} + \frac{1}{Y}\frac{\partial^2 Y}{\partial y^2} + h^2 = 0$, which indicates that there are three terms and they are (*i*) Term only function of x,

(*ii*) Term only function of y only,

(*iii*) Term is constant.

So each individual term must be equal to zero or constant separately, since each individual term can not be equal to zero.

So, they must be constant separately, as

$$\frac{1}{X}\frac{\partial^2 X}{\partial x^2} = -k_x^2 \quad ...(9.10a)$$

$$\frac{1}{Y}\frac{\partial^2 Y}{\partial y^2} = -k_y^2 \quad ...(9.10b)$$

Hence, $$-k_x^2 - k_y^2 + h^2 = 0$$

or $$h^2 = \sqrt{k_x^2 + k_y^2} \quad ...(9.11)$$

and also, $$\frac{1}{X}\frac{\partial^2 X}{\partial x^2} = -k_x{}^2 \quad \text{or} \quad \frac{\partial^2 X}{\partial x^2} + Xk_x^2 = 0 \quad ...(9.12)$$

The solution of eqn. (9.12) can be given as;

$$X = (A \cos k_x x + B \sin k_x x)$$

Similarly, from equation (9.10b), $Y = (C \cos k_y y + D \sin k_y y)$

So $$E_z = X \cdot Y$$

$$E_z = (A \cos k_x x + B \sin k_x x).\ (C \cos k_y y + D \sin k_y y) \quad ...(9.13)$$

Now using boundary conditions, we have to find the values of constants A, B, C, and D.

Since surface of the rectangular waveguide acts as a short circuit (as ground) for electric field, then $E_z = 0$. (Fig. 9.2), hence out coming waves will be transverse in nature. So, the boundary condition will be;

1st boundary condition: (Bottom plane)

i.e., $$E_z = 0 \text{ at } y = 0 \text{ and } x \to 0 \text{ to } a$$

hence $x \to 0$ to a, means 'x' varying between 0 at a.

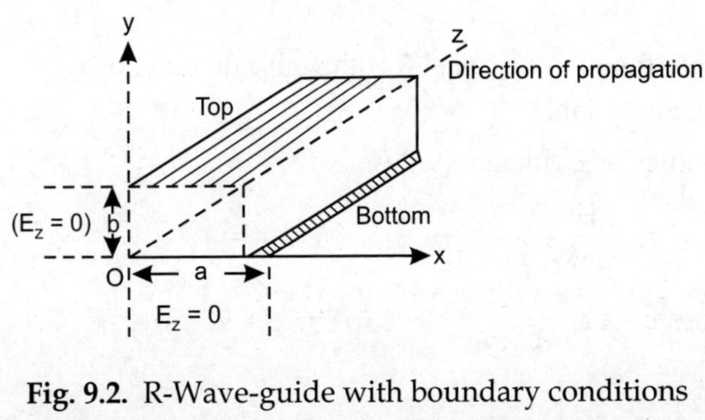

Fig. 9.2. R-Wave-guide with boundary conditions

So, $$E_z = (A \cos k_x x + B \sin k_x x).\ (C \cos k_y y + D \sin k_y y)$$

$$0 = (A \cos k_x x + B \sin k_x x)(C \cos k_y (0) + D \sin k_y (0))$$

$$0 = (A \cos k_x x + B \sin k_x x)(C)$$

Since, $(A \cos k_x x + B \sin k_x x) \neq 0$

So, $C = 0$

So modified eqn for E_z

$$E_z = (A \cos k_x x + B \sin k_x x).\ (0) \sin k_y y = A \cos k_x x + B \sin k_x . x$$

2nd boundary condition (top plane)

$$E_z = 0 \text{ at } y = b \text{ and } x = 0 \rightarrow 0 \text{ to a}$$

$$E_z = (A \cos k_x x + B \sin k_x x) D \sin k_y y$$

or $= (A \cos k_x x + B \sin K_x x).\ D \sin k_y b = 0$

Since, $A. \cos k_x x + B \sin k_x x \neq 0$

and $D \neq 0$

So, third term must be equal to zero,

$$\sin k_y b = 0$$

$$k_y b = \pm n\pi$$

$$k_y = \pm \frac{n\pi}{b} \qquad ...(9.14)$$

where $n = 0, 1, 2, 3, 4$, called half wave variation of electric and magnetic fields, taking only positive:

$$k_y = \frac{n\pi}{b}$$

So, again modified equation after putting therse values,

$$E_z = (A \cos k_x x + B \sin k_x x) D \sin \left(\frac{n\pi}{b}\right) y$$

3rd boundary conditions (left side plane):

$$E_z = 0 \text{ at } x = 0 \quad y \rightarrow 0 \text{ to } b$$

$$E_z = (A \cos k_x x + B \sin k_x x) D \sin \left(\frac{n\pi}{b}\right) y$$

$$0 = (A \cos (0) + B \sin (0)) D \sin \left(\frac{n\pi}{b}\right) y$$

$$0 = A.\ D \sin \left(\frac{n\pi}{b}\right) y$$

Since, $D. \sin\left(\frac{n\pi}{b}\right) y \neq 0$ otherwise $E_z = 0$

So, $A = 0$

Modified equation $\quad E_z = (B.\sin k_x x)\, C \sin\left(\frac{n\pi}{b}\right)y$

4th boundary condition:

$$E_z = 0 \text{ at } x = a \quad y \to 0 \text{ to } b$$

$$0 = (B \sin k_x a)\left(D \sin\left(\frac{n\pi}{b}\right)y\right)$$

$B \neq 0 \qquad D \sin\left(\frac{n\pi}{b}\right)y \neq 0$

So

$$\sin k_x a = 0$$

$$k_x a = 0$$

$$k_x = \pm \frac{m\pi}{a}$$

or

$$k_x = \frac{m\pi}{a} \qquad \text{(taking + ve value only)}$$

$$E_z = B.D.\sin\left(\frac{m\pi}{a}\right)x \,.\, \sin\left(\frac{n\pi}{b}\right)y$$

$$E_z = E_0 \sin\left(\frac{m\pi}{a}\right)x \,.\, \sin\left(\frac{n\pi}{b}\right)y \qquad \text{...(9.15)}$$

Now, from general field equation (9.8*a*) to (9.8*d*), we get:

$$E_x = -\frac{\gamma}{h^2}.\frac{\partial E_z}{\partial x} - \frac{j\omega\mu}{h^2}.\frac{\partial H_z}{\partial y}$$

$$H_z = 0$$

$$= -\frac{\partial}{h^2}\frac{\partial}{\partial x}(E_z)$$

$$= -\frac{\gamma}{h^2}E_0.\cos\left(\frac{m\pi}{a}\right)x \sin\left(\frac{n\pi}{b}\right)y$$

$$E_x = E_{0x}\cos\left(\frac{m\pi}{a}\right)x \sin\left(\frac{n\pi}{b}\right)y$$

However, if time-variation factor ($e^{j\omega t}$) is also intruded in the variation of field components, then corresponding expressions becomes:

$$E_Z = E_{0Z}\sin\left(\frac{m\pi}{a}\right)x.\cos\left(\frac{n\pi}{b}\right)y.e^{j\omega t}.e^{-\gamma z}$$

$$E_y = E_{0y}\sin\left(\frac{m\pi}{a}\right)x.\cos\left(\frac{n\pi}{b}\right)y.e^{j\omega t}.e^{-\gamma z}$$

$$E_x = E_{0x}\cos\left(\frac{m\pi}{a}\right)x.\text{in}\left(\frac{m\pi}{a}\right)xy.e^{j\omega t}.e^{-\gamma z}$$

And

$$H_z = -\frac{\gamma}{h^2}\frac{\partial H_z}{\partial x} + \frac{j\omega\varepsilon}{h^2}\frac{\partial E_z}{\partial y}$$

$$= \frac{j\omega\varepsilon}{h^2}.\frac{\partial}{\partial y}(E_z)$$

and magnetic fields $H_z = H_{0x} . \sin\frac{m\pi}{a}x . \sin\left(\frac{n\pi}{b}\right)y . e^{j\omega t} . e^{-\gamma z}$

$$H_y = H_{0y} \cos\frac{m\neq}{a}x . \cos\left(\frac{n\neq}{b}\right)y . e^{j\omega t} . e^{-\gamma z}$$

Cut off frequency of a wavequide, we know that

$$\gamma^2 + \omega^2\mu\varepsilon = h^2 = k_x^2 + k_y^2$$

$$\gamma^2 + \omega^2\mu\varepsilon = \left(\frac{m\pi}{a}\right)^2 + \left(\frac{n\pi}{b}\right)^2$$

$$\gamma = \sqrt{\left(\frac{m\pi}{a}\right)^2 + \left(\frac{n\pi}{b}\right)^2 - \omega^2\mu\varepsilon}$$

at lower frequency

$$\left(\frac{m\pi}{a}\right)^2 + \left(\frac{n\pi}{b}\right)^2 > \omega^2\mu\varepsilon$$

and $\gamma = \alpha$, real and positive

Since, $\gamma = \alpha + i\beta$

$\gamma = \alpha$, means the wave is get attenuated and hence it cannot propagates.

Now when one increases the frequency, at a frequency known as cut-off angular frequency ω_c, γ becomes zero, *i.e.*,

$$\left(\frac{m\pi}{a}\right)^2 + \left(\frac{n\pi}{b}\right)^2 = \omega_c^2\mu\varepsilon$$

where ω_c is cut off angular frequency and given by

$$\omega_c = \frac{1}{\sqrt{\mu\varepsilon}}\sqrt{\left(\frac{m\pi}{a}\right)^2 + \left(\frac{n\pi}{b}\right)^2}$$

(Since $c = \frac{1}{\sqrt{\mu\varepsilon}}$ the velocity of light of electromagnetic wave)

Hence, cut-off frequency, $f_c = \frac{1}{2\pi\sqrt{\mu\varepsilon}}\sqrt{\left(\frac{m\pi}{a}\right)^2 + \left(\frac{n\pi}{b}\right)^2}$

$$= \frac{1}{2\pi}\left(\frac{1}{\mu\varepsilon}\right) . \sqrt{\left(\frac{m\pi}{a}\right)^2 + \left(\frac{n\pi}{b}\right)^2}$$

$$f_c = \frac{c}{2\pi}\sqrt{\left(\frac{m\pi}{2}\right)^2 + \left(\frac{n\pi}{b}\right)^2}$$

and corresponding cutt-off wavelength;

$$\lambda_c = \frac{c}{f_c} = \frac{2ab}{\sqrt{(mb)^2 + (na)^2}}$$

Waveguide wavelength (λ_g): The distance travelled by the wave in wave-guide undergo a phase shift of 2π radian, is called wavelength of the electromagnetic wave and it is shown in Fig. 9.3

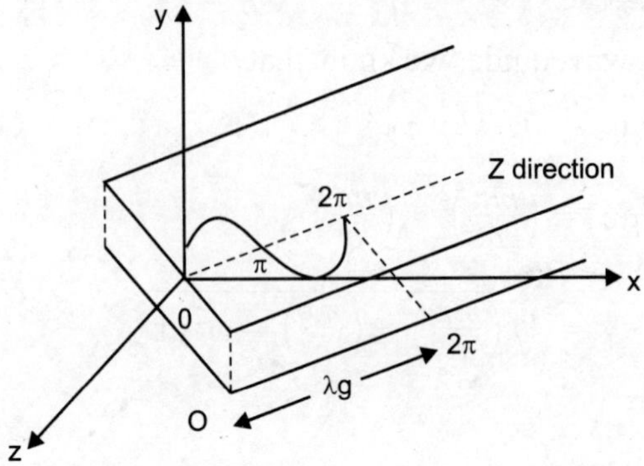

Fig. 9.3. Distance λ_g in a wave-guide

It is related with phase constant β as, $\lambda_g = \dfrac{2\pi}{\beta}$

Wavelength in waveguide (λ_g) different than wavelength in free space. (λ_0).

Phase velocity: The rate at which the wave changes its phase in terms of the guide wavelength is called the phase velocity, i.e.,

$$\text{Phase velocity,} \quad v_p = \frac{\lambda_g}{\text{unit time}}$$

$$= \frac{\lambda_g}{T} = \lambda_g f \qquad (T = \text{time period, } \lambda = \text{wavelength, } f = \text{frequency})$$

$$= \frac{\lambda_g f}{2\pi}(2\pi) \qquad \text{(by multiply } 2\pi \text{ in numerator and denominator)}$$

$$= \frac{(2\pi f)\lambda_g}{2\pi}$$

$$= \frac{\omega\lambda_g}{2\pi} = \frac{\omega}{\beta}, \quad \text{as } \beta = \frac{\lambda_g}{2\pi}$$

$$v_p = \frac{\omega}{\beta} \qquad \text{...(9.16)}$$

Phase velocity in terms of waveguide length and cut off wave length and free space wavelength can be derived as:

$$h^2 = k_x^2 + k_y^2.$$

$$\gamma^2 + \omega^2\mu\varepsilon = k_x^2 + k_y^2$$

$$= \left(\frac{m\pi}{a}\right)^2 + \left(\frac{n\pi}{b}\right)^2$$

$$\gamma^2 = \left(\frac{m\pi}{a}\right)^2 + \left(\frac{n\pi}{b}\right)^2 - \omega^2\mu\varepsilon$$

$$(\alpha + j\beta)^2 = \left(\frac{m\pi}{a}\right)^2 + \left(\frac{n\pi}{b}\right)^2 - \omega^2\mu\varepsilon$$

(For propagation of wave in waveguide attenuation $\alpha = 0$)

$$(j\beta)^2 = \left(\frac{m\pi}{a}\right)^2 + \left(\frac{n\pi}{b}\right)^2 - \omega^2\mu\varepsilon.$$

at $\omega = \omega_c$, $\beta = 0$,

$$\Rightarrow \qquad \omega_c^2\mu\varepsilon = \left(\frac{m\pi}{a}\right)^2 + \left(\frac{n\pi}{b}\right)^2$$

So,

$$-\beta^2 = \omega\varepsilon^2\,\mu\varepsilon - \omega^2\mu\varepsilon$$

$$\beta^2 = \omega^2\mu\varepsilon - \omega_c^2\mu\varepsilon$$

$$\beta = \sqrt{\omega^2\,\mu\varepsilon - \omega_c^2\,\mu\varepsilon}$$

$$\beta = \omega^2\,\sqrt{\mu\varepsilon}\sqrt{1-\left(\frac{\omega_c}{\omega}\right)^2}$$

or

$$\frac{\beta}{\omega} = \frac{1}{c}\sqrt{1-\left(\frac{f_c}{f}\right)^2}$$

$$\frac{\omega}{\beta} = v_p = \frac{c}{\sqrt{1-\left(\frac{f_c}{f}\right)^2}}$$

or in terms of cut off wave length:

$$v_p = \frac{c}{\sqrt{1-\left(\frac{\lambda}{\lambda_c}\right)^2}} \qquad \text{...(9.17}a\text{)}$$

That is v_p is greater than velocity of light, and it is the velocity with which EM wave changes its phase.

Group velocity: It is the velocity at which EMW propagate in waveguide and defined as a rate of change of angular velocity with phase.

$$v_g = \frac{d\omega}{d\beta} \qquad \text{...(9.17}b\text{)}$$

We know that

$$\beta = \sqrt{(\omega^2\,\mu\varepsilon - \omega_c^2\,\mu\varepsilon)}$$

$$\frac{d\beta}{d\omega} = \frac{2\omega\mu\varepsilon}{2\sqrt{\omega^2\,\mu\varepsilon - \omega_c^2\,\mu\varepsilon}}$$

$$\frac{d\beta}{d\omega} = \frac{\omega\mu\varepsilon}{\sqrt{\omega^2\,\mu\varepsilon - \omega_c^2\,\mu\varepsilon}} = \frac{\sqrt{\mu\varepsilon}}{\sqrt{1-\left(\frac{\omega_c}{\omega}\right)^2}}$$

$$\frac{d\beta}{d\omega} = \frac{1}{c\sqrt{1-\left(\frac{\omega_c}{\omega}\right)^2}}$$

$$\frac{d\omega}{d\beta} = c\sqrt{1-\left(\frac{\omega_c}{\omega}\right)^2}$$

$$v_g = c\sqrt{1-\left(\frac{f_c}{f}\right)^2} \text{ or } v_g = c\sqrt{1-\left(\frac{\lambda}{\lambda_c}\right)^2} \quad ...(9.18)$$

This velocity is lower than velocity of light. The maximum value of group velocity is found when

$$\lambda = \lambda_c$$

$$v_g = c$$

Relationship between phase velocity v_p and group velocity v_g:

If we multiply equations (9.17*a*) and (9.18), wet get:

$$v_p \times v_g = \frac{c}{\sqrt{1-\left(\frac{\lambda}{\lambda_c}\right)}} \times c\sqrt{1-\left(\frac{\lambda}{\lambda}\right)^2} = c^2 \times \frac{\sqrt{1-\left(\frac{\lambda}{\lambda}\right)^2}}{\sqrt{1-\left(\frac{\lambda}{\lambda_c}\right)^2}}$$

$$v_p.v_g = c^2$$

If wave propagates in free space then $v_p = v_g$

So, $$v_p.\, v_g = c^2$$

$$v_p^2 = c^2$$

$$v_p = c$$

So $$v_p = c_g = c$$

i.e., both velocities are equal in free space or in vacuum.

Relationship between wavelengths: λ, λ_c and λ_g

Since, $$v_p.\, v_g = c^2$$

$$v_g.\, \frac{\omega}{\beta} = c^2$$

$$V_g.\left(\frac{\omega}{\frac{2\pi}{\lambda_g}}\right) = c^2 \qquad \left(\textit{Since } \beta = \frac{2\pi}{\lambda}\right)$$

$$\frac{2\pi f}{2\pi/\lambda_g} c\sqrt{1-\left(\frac{\lambda}{\lambda_c}\right)^2} = c^2$$

$$\lambda_g\sqrt{1-\left(\frac{\lambda}{\lambda_c}\right)^2} = \left(\frac{c}{f}\right)$$

$$\lambda_g\sqrt{1-\left(\frac{\lambda}{\lambda_c}\right)^2} = \lambda$$

$$\sqrt{1-\left(\frac{\lambda}{\lambda_c}\right)^2} = \frac{\lambda}{\lambda_g}$$

$$1-\left(\frac{\lambda}{\lambda_c}\right)^2 = \left(\frac{\lambda}{\lambda_g}\right)^2$$

$$1 = \left(\frac{\lambda}{\lambda_g}\right)^2 + \left(\frac{\lambda}{\lambda_c}\right)^2$$

$$\frac{1}{\lambda^2} = \frac{1}{\lambda_g^2} + \frac{1}{\lambda_c^2}$$

or

$$\lambda_g = \frac{\lambda}{\sqrt{1-\left(\frac{\lambda}{\lambda_c}\right)^2}} \qquad ...(9.19)$$

Wave impedance Z_z in TM and TE Modes

Wave impedance is the ratio of the strength of electric field in one transverse direction to the strength of the magnetic field along the other transverse direction as shown. Fig. 9.4.

$$Z_z = \frac{E_x}{H_y} = -\frac{E_y}{H_x}$$

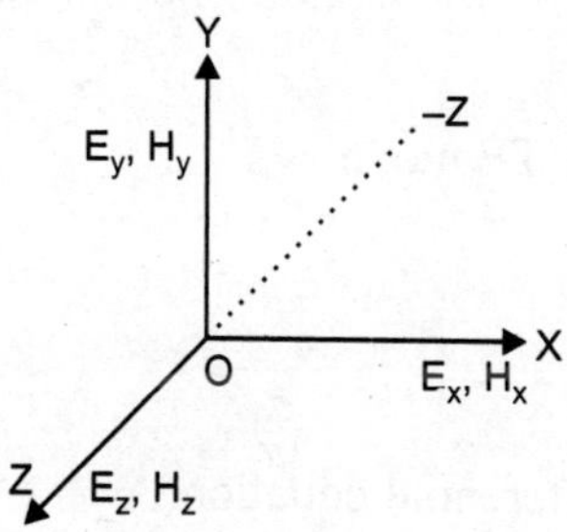

Fig. 9.4 Fidd components of EM waves

or

$$Z_z = \frac{E}{H} = \frac{\sqrt{E_x^2 + E_y^2}}{\sqrt{H_x^2 + H_y^2}}$$

So,

$$Z_{TM} = \frac{E_x}{H_y} = \frac{-\frac{\gamma}{h^2}\cdot\frac{\partial E_z}{\partial x} - \frac{j\omega\mu}{h^2}\cdot\frac{\partial H_z}{\partial y}}{-\frac{\gamma}{h^2}\frac{\partial H_z}{2y} - \frac{j\omega\varepsilon}{h^2}\frac{\partial E_z}{\partial x}}$$

But for TM mode $H_z = 0$, so

$$\frac{-\frac{\gamma}{h^2}\cdot\frac{\partial E_z}{\partial x}}{-\frac{j\omega\varepsilon}{h^2}\cdot\frac{\partial E_z}{\partial x}}$$

$$\frac{\gamma}{h^2} \times \frac{h^2}{j\omega\varepsilon} = \frac{\gamma}{j\omega\varepsilon} = \frac{\alpha + j\beta}{j\omega\varepsilon}$$

$$= \frac{j\beta}{j\omega\varepsilon} = \frac{\beta}{\omega\varepsilon}$$ (Since for propagation in waveguide, $\alpha = 0$)

Hence, $$Z_{TM} = \frac{\beta}{\omega\varepsilon} = \frac{\sqrt{\omega^2\mu\varepsilon - \omega_c^2\mu\varepsilon}}{\omega\varepsilon}$$

$$= \sqrt{\frac{\omega^2\mu\varepsilon}{\omega^2\varepsilon^2} - \frac{\omega_c^2\mu\varepsilon}{\omega^2\varepsilon^2}}$$

$$= \sqrt{\frac{\mu}{\varepsilon} - \left(\frac{\omega_c}{\omega}\right)^2\frac{\mu}{\varepsilon}} = \sqrt{\frac{\mu}{\varepsilon}}.\sqrt{1-\left(\frac{\omega_c}{\omega}\right)^2}$$

$$Z_{TM} = \eta\sqrt{1-\left(\frac{\omega_c}{\omega}\right)^2} = \eta\sqrt{1-\left(\frac{\lambda}{\lambda_c}\right)^2} \quad ...(9.20)$$

where $\eta = \sqrt{\frac{\mu}{E}}$ and called intrinsic impedance of the medium of wave guide.

***(iii)* TE mode Propagation**

This mode is also an transverse mode in which electric field are perpendicular to the direction of propagation. If the electromagnetic wave propagation is in z-direction, then for TE modes;

$$E_z = 0$$

Since, in order to energy flow in waveguide one component must be in the direction of propagation. So,

$$H_z \neq 0 \text{ for TE mode}$$

So, from equation (9.4), we get

$$\frac{\partial^2 H_z}{\partial x^2} + \frac{\partial^2 H_z}{\partial y^2} + h^2 H_z = 0 \qquad (\text{Since } h^2 = \gamma^2 + \omega^2\mu\varepsilon) \quad ...(9.21)$$

This equation is called partial differential equation.

Let the above equations has solution

$$H_z = X.Y \quad ...(9.22)$$

where, X is the only function of x. and Y is the only function of y.

So, putting the value of H_z from equation (9.22) in (9.21), yield ds.

$$\frac{\partial^2 X.Y}{\partial x^2} + \frac{\partial^2 XY}{\partial y^2} + h^2 X.Y = 0$$

Deviding by $X.Y$ in both side

$$\frac{1}{X.Y}.\frac{\partial^2 X.Y}{\partial x^2} + \frac{1}{XY}\frac{\partial^2 X.Y}{\partial y^2} + h^2 = 0$$

or $$\frac{Y}{X.Y}\frac{\partial^2 X}{\partial x^2} + \frac{X}{XY}\frac{\partial^2 Y}{\partial y^2} + h^2 = 0$$

$$\underset{(i)}{\frac{1}{X}.\frac{\partial^2 X}{\partial x^2}} + \underset{(ii)}{\frac{1}{Y}.\frac{\partial^2 Y}{\partial y^2}} + \underset{(iii)}{h^2} = 0 \quad ...(9.23)$$

The equation (9.23) has three terms and each one are independent to other and summation of each is equal to zero, and third term is constant and positive so first two term must be negative and constant.

Let $$\frac{1}{X}\cdot\frac{\partial^2 X}{\partial x^2} = -k_x^2 \quad ...(924a)$$

and $$\frac{1}{Y}\cdot\frac{\partial^2 Y}{\partial y^2} = -k_y^2 \quad ...(2.24b)$$

So, $$\frac{1}{X}\frac{\partial^2 X}{\partial x^2} + \frac{1}{Y}\frac{\partial^2 Y}{\partial y^2} + h^2 = 0$$

$$-k_x^2 - k_y^2 + h^2 = 0$$

or $$h^2 = k_x^2 + k_y^2 \quad ...(9.25)$$

Now, the eqn. (9.24a) and (9.24b) can be solved by separation of variable method, hence X and Y can be given as,

$$X = [A \cos (k_x .x) + B \sin (k_x x)]$$

$$Y = [C \cos (k_y y) + D \sin (k_y y)]$$

So, $$H_z = (A \cos k_x x + B \sin k_x x).\,(C \cos k_y y + D \sin k_y y) \quad ...(9.26)$$

Here, A, B, C and D are constant and equation (9.26) can be solved with the boundary conditions as in case of TM mode.

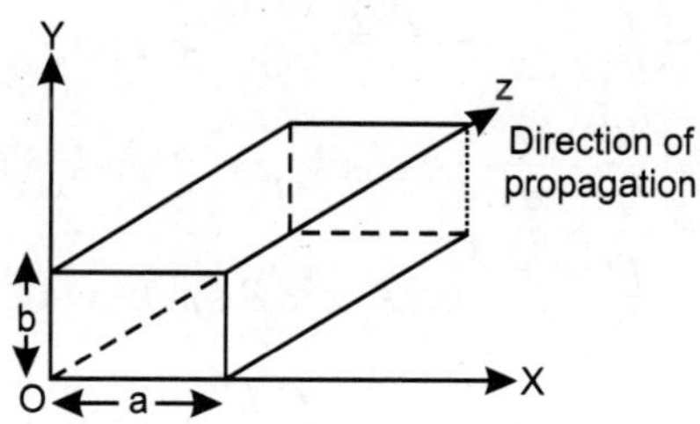

Fig. 9.5 Simple geometry of R-wave-guide

Since, $E_z = 0$ for TE modes but E_x and E_y components exists.

$E_x = 0$ at top and bottom wall of the waveguides

$E_y = 0$ at left and right side of the wall.

1st Boundary Condition

$$E_x = 0\text{, at bottom walls}$$

or $$E_x = 0\text{, at } y = 0 \text{ and } x \to 0 \text{ to a}$$

Now, from wave equation;

$$E_x = -\frac{\gamma}{h^2}\cdot\frac{\partial E_x}{\partial x} - \frac{j\omega\mu}{h^2}\cdot\frac{\partial H_z}{\partial y}$$

$$= -j\omega\mu.\,\frac{\partial H_z}{\partial y} = 0$$

$$\frac{\partial H_z}{\partial y} = 0$$

$$\frac{\partial}{\partial y}(A \cos k_x x + B \sin k_x x).(C \cos k_y y + D \sin k_y y) = 0$$

$$(A \cos k_x x + B \sin k_x x)\frac{\partial}{\partial y}(C \cos k_y y + D \sin k_y y) = 0$$

$$(A \cos k_x x + B \sin k_x x)(-C k_y \sin k_y y + D k_y \cos k_y y) = 0$$

putting $y = 0$

$$(A \cos k_x x + B \sin k_x x).(-C k_y .\sin k_y (0) + D k_y \cos k_y (0)) = 0$$

$$(A \cos k_x x + B \sin k_x x)(+ D k_y) = 0$$

Since $(A \cos k_x x + B \sin k_x x) \neq 0$

So, $Dk_y = 0$ or $D = 0$

So, $$H_z = (A \cos k_x x + B \sin k_x x)(C \cos k_y y) \quad \text{...(9.27)}$$

2nd Boundary Condition

Now, at top wall of waveguide

$$E_x = 0 \text{ at } y = b \quad x \to 0 \text{ to } a$$

$$E_x = -\frac{\gamma}{h^2}\cdot\frac{\partial E_z}{\partial x} - \frac{j\omega\mu}{h^2}\cdot\frac{\partial H_z}{\partial y}$$

$$= -\frac{j\omega\mu}{h^2}\frac{\partial}{\partial y}(A \cos k_x x + B \sin k_x x)(C \cos k_y y)$$

$$= -\frac{j\omega\mu}{h^2}.(A \cos k_x x + B \sin k_x x)\, C.\frac{\partial}{\partial y}\cos k_y y$$

$$= -\frac{j\omega\mu}{h^2}.(A \cos k_x x + B \sin k_x x)(k_y - \sin k_y y)$$

Now, putting boundary condition.

$$0 = +\frac{j\omega\mu}{h^2}.(A \cos k_x x + B \sin k_x x)\, C k_y \sin k_y b$$

Since $(A \cos k_x x + B \sin k_x x) \neq 0$

So, $$\sin k_y b = 0$$

$$k_y b = \pm n\pi$$

$$k_y = \pm\frac{n\pi}{b} \quad \text{...(9.28)}$$

So, $$H_z = (A \cos k_x x + B \sin k_x x)\, C \cos\frac{n\pi}{b}y$$

3rd Boundary Condition

$$E_y = 0 \text{ at } x = 0 \quad y = 0 \text{ to b}$$

$$E_y = -\frac{\gamma}{h^2}\frac{\partial E_z}{\partial y} + \frac{j\omega\mu}{h^2}\cdot\frac{\partial H_z}{\partial x}$$

$$= \frac{j\omega\mu}{h^2}\cdot\frac{\partial}{\partial x}\cdot(A\cos k_x x + B\sin k_x x)\, C.\cos\left(\frac{n\pi}{b}\right)y$$

$$= \frac{j\omega\mu}{h^2}(-A\,k_x\sin k_x x + B\,k_y\cos k_x x).\, C\cos\left(\frac{n\pi}{b}\right)y$$

$$0 = \frac{j\omega\mu}{h^2}(-A\,k_x\sin k_x(0) + Bk_y\cos k_y(0))\, C.\cos\left(\frac{n\pi}{b}\right)y$$

$$= \frac{j\omega\mu}{h^2}(B\,k_y).\, C\cos\left(\frac{n\pi}{b}\right)y$$

Since, $\cos\frac{n\pi}{b}y \neq 0$

So, $B\,k_y = 0$ and $B = 0$

$$B = 0$$

$$H_z = (\text{A}\cos k_x x)\, C.\cos\left(\frac{n\pi}{b}\right)y$$

$$H_z = A.\text{C}\cos(k_x x).\cos\left(\frac{n\pi}{b}\right)y$$

4th Boundary Condition

$$E_y = 0 \text{ at } x = 0 \quad y = 0 \text{ to b}$$

$$E_y = -\frac{\gamma}{h^2}\cdot\frac{\partial E_z}{\partial y} + \frac{j\omega H}{h^2}\cdot\frac{\partial H_z}{\partial x}$$

$$= \frac{j\omega\mu}{h^2}\cdot\frac{\partial}{\partial x}(\text{A.C})\cos k_x x\cos\left(\frac{n\pi}{b}\right)y$$

$$= \frac{j\omega\mu}{h^2}.(AC)\,k_x\sin k_x x\cos\left(\frac{n\pi}{b}\right)y$$

$$0 = \frac{j\omega\mu}{h^2}(AC).k_x.\sin k_x a\cos\left(\frac{n\pi}{b}\right)y$$

So, $\sin k_x a = 0$ (other terms n of equal to zero)

$$k_x a = \pm m\pi$$

$$k_x = \frac{m\pi}{a} \qquad \text{...(9.29)}$$

So, $H_z = (A.C).\cos\left(\frac{m\pi}{a}\right).\cos\left(\frac{n\pi}{b}\right)y$

$$H_z = H_{0z}.\cos\left(\frac{n\pi}{a}\right).\cos\left(\frac{n\pi}{b}\right)y$$

Now, from H_z components all electric and magnetic field components can be obtained. as follows:

$$E_x = -\frac{\gamma}{h^2}\cdot\frac{\partial E_z}{\partial x} - \frac{j\omega\mu}{h^2}\frac{\partial H_z}{\partial y}$$

$$= 0 - \frac{j\omega\mu}{h^2}\frac{\partial}{\partial y}\cdot H_{0z}\cos\left(\frac{m\pi}{a}\right)x.\cos\left(\frac{n\pi}{b}\right)y$$

$$= +\frac{j\omega\mu}{h^2}H_{0z}\left(\frac{n\pi}{b}\right)\cos\left(\frac{m\pi}{a}\right)x.\sin\left(\frac{n\pi}{b}\right)y$$

Let $\frac{j\omega\mu}{h^2}H_{0z}.\left(\frac{n\pi}{b}\right) = E_{0x}$

$$E_x = E_{0x}.\cos\left(\frac{m\pi}{a}\right)x.\sin\left(\frac{n\pi}{b}\right)y$$

and similarly, $E_y = -\frac{\gamma}{h^2}\cdot\frac{\partial E_z}{\partial y} + \frac{j\omega\mu}{h^2}\cdot\frac{\partial H_z}{\partial x}$

$$= 0 + \frac{j\omega\mu}{h^2}\cdot\frac{\partial}{\partial x}\cdot H_{0z}\cdot\cos\left(\frac{m\pi}{a}\right)x.\cos\left(\frac{n\pi}{b}\right)y$$

$$= \frac{j\omega\mu}{h^2}H_{0z}\cdot\sin\left(\frac{m\pi}{a}\right)x.\cos\left(\frac{n\pi}{b}\right)y\left(\frac{m\pi}{a}\right)y$$

$$= \left(\frac{j\omega\mu}{h^2}\right)H_{0z}\cdot\left(\frac{m\pi}{a}\right).\sin\left(\frac{m\pi}{a}\right)a\,x.\cos\left(\frac{n\pi}{b}\right)y$$

$$E_y = E_{0y}\sin\left(\frac{n\pi}{a}\right)\cos\left(\frac{n\pi}{b}\right)y \qquad ...(9.30)$$

Similarly from the H_x and H_y :

$$H_x = -\frac{\gamma}{h^2}\frac{\partial H_z}{\partial x} + \frac{j\omega\varepsilon}{h^2}\cdot\frac{\partial E_z}{\partial y} \text{ and } H_y = -\frac{\gamma}{h^2}\frac{\partial H_z}{\partial y} - \frac{j\omega\varepsilon}{h^2}\frac{\partial E_z}{\partial x}$$

On simplification:

$$H_x = H_{0x}\sin\left(\frac{m\pi}{a}\right).\cos\left(\frac{n\pi}{b}\right)y \qquad ...(9.31a)$$

$$H_y = H_{0x}\cos\left(\frac{m\pi}{a}\right).\sin\left(\frac{n\pi}{b}\right)y \qquad ...(9.31b)$$

These are general equation for the electric and magnetic filed component in x and y directions. Again if time-variation ($e^{j\omega t}$) of the fields are added with exponential variation ($e^{-\gamma z}$) ;

$$E = E_{0z}\,e^{j\omega t}.\,e^{-\gamma z} = E_0\,e^{j(\omega t-\gamma z)}$$

and $H = H_{0z}\,e^{j(\omega t-\gamma z)}$.

Then for TE mode, modified field component are found to be as follow;

$$E_z = 0 \qquad ...(9.32a)$$

$$H_z = H_0 z.\cos\left(\frac{m\pi}{a}\right)x.\cos\left(\frac{n\pi}{b}\right)y.e^{j\omega t}.e^{-\gamma z} \qquad ...(9.32b)$$

$$E_x = E_{0x}.\cos\left(\frac{m\pi}{a}\right)x.\sin\left(\frac{n\pi}{b}\right)y.e^{j\omega t}.e^{-\gamma z} \qquad ...(9.32c)$$

$$E_y = E_{0y}.\sin\left(\frac{m\pi}{a}\right)x.\cos\left(\frac{n\pi}{b}\right)y.e^{j\omega t}.e^{-\gamma z} \qquad ...(9.32d)$$

$$H_x = H_{0x}.\sin\left(\frac{m\pi}{a}\right)x.\cos\left(\frac{n\pi}{b}\right)y.e^{j\omega t}.e^{-\gamma z} \quad ...(9.32e)$$

$$H_y = H_{0y}.\cos\left(\frac{m\pi}{a}\right)x.\sin\left(\frac{n\pi}{b}\right)y.e^{j\omega t}.e^{-\gamma z} \quad ...(9.32f)$$

Now, different mode configuration can be given as

(*a*) TE_{mn} if $m = 0, n = 0$

$TE_{0,0}$ a field component E_x, E_y, H_x, H_y, H_z will be zero hence TE_{00} mode does not exist.

(*b*) Now, if $m = 0$, $n = 1$, for TE_{01},

$$E_y = 0, H_x = 0 \text{ and } E_x \text{ and } H_y \text{ exist}$$

(*c*) Similarly, TE_{10}, $m = 1, n = 0$

$$E_x = 0, H_{y=0} \text{ and } E_y \text{ and } H_x \text{ exist.}$$

In other values of *m* and *n*, electromagnetic wave do exist in waveguide.

Dominant Modes

The mode corresponding to which waveguide has lowest cut-off frequency called dominant modes or has maximum cut-off wavelengths.

Since,
$$\lambda_c = \frac{2}{\sqrt{\left(\frac{m}{a}\right)^2 + \left(\frac{n}{b}\right)^2}}$$

or
$$\lambda_{cmn} = \frac{2}{\sqrt{\left(\frac{m}{a}\right)^2 + \left(\frac{n}{b}\right)^2}} \quad ...(9.33a)$$

where '*m*' and '*n*' represent number of mods in wave-guide.

(*i*) Now, if $m = 0, n = 0$

$$\lambda_{c00} = \frac{2}{\sqrt{\left(\frac{0}{a}\right)^2 + \left(\frac{0}{b}\right)^2}} = \frac{2}{0} = \infty \text{ which is not valid.}$$

(*ii*) For $m = 0, n = 1$, TE_{01}

$$\lambda_{c01} = \frac{2}{\sqrt{\left(\frac{m}{a}\right)^2 + \left(\frac{n}{b}\right)^2}}. = \frac{2}{\sqrt{\left(\frac{0}{a}\right)^2 + \left(\frac{0}{b}\right)^2}} = \frac{2}{\frac{1}{b}} = 2\text{b}$$

$$\lambda_{c01} = 2b$$

(*iii*) For $m = 1, n = 0$, TE_{10}

$$\lambda_{c1,0} = \frac{2}{\sqrt{\left(\frac{1}{a}\right)^2 + \left(\frac{0}{b}\right)^2}} = 2a$$

Since, $a > b$ for standard waveguide, however in optimum condition, $a = 2b$.

So,
$$\lambda_{c1,0} > \lambda_{c01}$$

TE_{10} called dominant mode, the other modes are lossy or leaky, they does not propagate upto the destination.

Note: All the parameter for TE modes (v_p, v_g, f_c, λ_c) are same as in case of TM mode except the wave impedance.

For TE mode Z_{TE}:

$$\text{ZTE} = \frac{E_y}{H_x} = \left(\frac{-\frac{\gamma}{h^2}\frac{\partial E_z}{\partial x} - \frac{j\omega\mu}{h^2}\frac{\partial H_z}{\partial y}}{-\frac{\gamma}{h^2}\frac{\partial H_z}{\partial y} - \frac{j\omega}{h^2}\frac{\partial E_z}{\partial x}} \right)$$

$$\text{E2} = 0\text{ZTE} = \frac{j\omega H}{\gamma} = \frac{j\omega H}{\alpha + i\beta} = \frac{j\omega\mu}{j\beta} = \left(\frac{\omega\mu}{\beta}\right) = \frac{\omega\mu}{\sqrt{\omega^2\mu\varepsilon - \omega_c^2\mu\varepsilon}}$$

$$\text{ZTE} = \frac{\eta}{\sqrt{1-\left(\frac{\omega_c}{\omega}\right)^2}} = \frac{\eta}{\sqrt{1-\left(\frac{\lambda}{\lambda_c}\right)^2}} \quad ...(9.33b)$$

This is Wave impedance in the TE mode is always greater than in free space.

Circular Waveguide

A circular waveguide (W/G) has circular cross sectional area and simple in construction. There are other similar waveguides such as ellipitical (W/g) and recentant W/g etc. as shown in Fig. 9.6.

It is used an alternative to the R-wave-guide. Like other wave-guides, it is constructed from a single, enclosure conductor, and support TE and TM modes. It is single degree (W/g) that is only radius (a) is taken as design parameter. Like R-wave-guide C-wave guide also acts as high-pass filter.

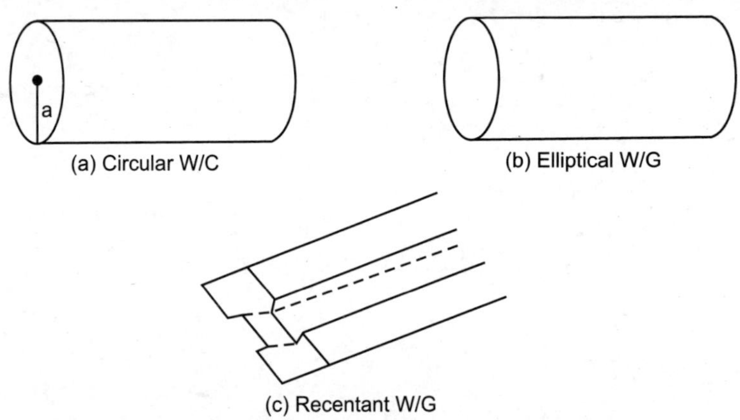

Fig. 9.6. Geometry of C-waveguides and its derivatives

Solution of wave equation in circular (W/G). (TE mode)

In order to analysis a circular wave-guide, let us consider basis geometry of it, as shown in

Fig. 9.7, where '*a*' is radius of wave-guide oriented along *z*-axis.

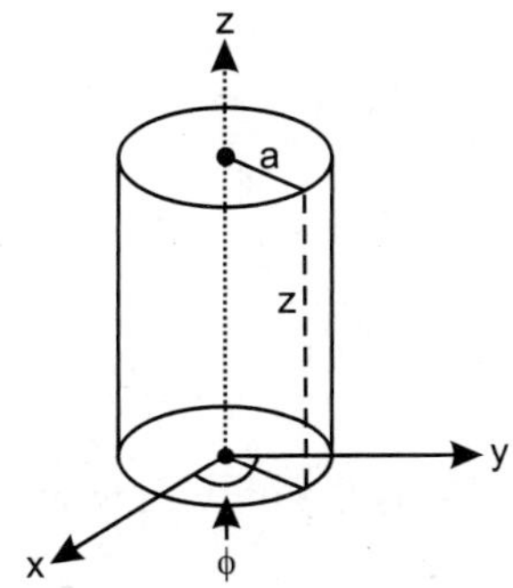

Fig. 9.7. C wave-guide oriented along *z*-axis.

The scaler helmholtz in cylindrical coordinates is given by

$$\frac{1}{r}\cdot\frac{\partial}{\partial r}\left(r.\frac{\partial\psi}{\partial r}\right)+\frac{1}{r^2}\frac{\partial^2\psi}{\partial\phi^2}+\frac{\partial^2\psi}{\partial z^2}=r_1^2\psi \quad \text{...(9.34a)}$$

Now separation of variable method can be used to should equation (9.34*a*)

$$\psi = R(r).\,\phi(\psi)\,Z(z) \quad \text{...(9.34b)}$$

Where $R(r)$ = a function of the radius only.

$\phi(\psi)$ = a functional the ϕ coordinates only

$Z(z)$ = a function of the *z* only.

Now putting the value of eqn. (9.34*b*) in to (9.34*a*) geves:

$$\frac{1}{r}\cdot\frac{\partial}{\partial r}\,(r.\,\frac{\partial}{\partial r}R.\,\phi.\,z)+\frac{1}{r^2}\frac{\partial^2}{\partial\phi^2}\,(R.\,\phi.\,z)+\frac{\partial^2}{\partial z^2}(R.\,\phi.\,z)-r_1^2\,(R.\,\phi.\,z)=0$$

or

$$\frac{\phi.Z}{r}\frac{\partial}{\partial r}\,(rR)+\frac{R.z}{r^2}\frac{\partial^2\phi}{\partial\phi^2}+\left(R\,\phi\,\frac{\partial^2 z}{\partial z^2}\right)-r_1^2\,(R.\,\phi.\,z)=0$$

Dividing by $R\phi z$ in both sides:

$$\frac{1}{r.R}\frac{\partial(rR)}{\partial r}+\frac{1}{r^2\phi}\cdot\frac{\partial^2\phi}{\partial\phi^2}+\frac{1}{Z}\times\frac{\partial^2 Z}{\partial z^2}-r_1^2=0 \quad \text{...(9.35)}$$

Since sum of all three independent term is zero so all individual terms will be equal to constant. Let third term is equal to

$$\frac{1}{Z}\frac{\partial^2 Z}{\partial z^2} = \gamma_g^2$$

or

$$\frac{\partial^2 Z}{\partial z^2} = \gamma_g^2 Z \quad \text{...(9.36)}$$

Hence the solution of equation (9.36) is,

$$Z = Ae^{-\gamma_g z} + \mathrm{B}e^{\gamma_g z} \quad \text{...(9.37)}$$

where γ_g = is propagation constant in the waveguide. Now if we insert the value of $\frac{1}{Z}\cdot\frac{\partial^2 Z}{\partial z^2} = \gamma_g^2$ in equation (9.35), it yields: (Assuming $r_1 - \gamma$).

$$\frac{1}{r}\,\text{R}.\,\frac{\partial}{\partial r}\left(r.\frac{dR}{dr}\right)+\frac{1}{r^2.\phi}\cdot\frac{\partial^2\phi}{\partial\phi^2}+\gamma_g^2-\gamma^2=0$$

Multiplying by r^2 in above equation, gives

$$\frac{\gamma}{R}\frac{\partial}{\partial r}\left(r\frac{dR}{dr}\right)+\frac{1}{\phi}\frac{\partial^2\phi}{\partial\varphi^2}-(\gamma^2-\gamma_{g\cdot}^2)\,r^2=0$$

$$\frac{r}{R}\frac{d}{dr}\left(r\frac{dR}{dr}\right)+\frac{1}{\varphi}\cdot\frac{\partial^2\varphi}{\partial\phi^2}-(\gamma^2-\gamma_g^2)\,r^2=0 \qquad \text{...(9.38)}$$

The second term in above eqn. (9.38) is the only function of ϕ. Then

$$\frac{1}{\phi}\frac{\partial^2\phi}{\partial\phi^2} = -n^2 \qquad \text{...(9.39)}$$

Solution of eqn. is (9.39) of the form:

$$\phi = A_{n}.\cos n\phi + B_n\,.\sin n\phi \qquad \text{...(9.40)}$$

Replacing second term from equation (9.39), equation (9.35) gives:

$$\frac{r}{R}\cdot\frac{\partial}{\partial r}\left(r.\frac{\partial R}{\partial r}\right)-n^2-(\gamma^2-\gamma_g^2)\,r^2=0 \qquad (\text{As } r_1-\gamma)$$

Let $\gamma^2-\gamma_g^2 = -k_c^2$

or $\gamma^2+k_c^2 = \gamma_g^2$ ...(9.41a)

$$\frac{r}{R}\cdot\frac{d}{dr}\left(r.\frac{dR}{dr}\right)-n^2+(k_cr)^2=0$$

$$\frac{r}{R}\frac{d}{dr}\left(r.\frac{dR}{dr}\right)+(k_cr)^2-n^2=0$$

Multiplying by R, then

$$r\frac{d}{dr}\left(r.\frac{dR}{dr}\right)+R\,[(k_cr)^2-n^2]=0 \qquad \text{...(9.41}b\text{)}$$

The eqn. (9.41b) is called characteristics equation of Bessel's functions.

For lossless waveguide, the characteristics equation (from 9.41a), we get;

$$k_c^2+(\alpha+j\beta)^2 = (\alpha+j\beta_g)^2$$

$$k_c^2-\beta^2 = -\beta_g^2 \qquad (\text{for loss less } \alpha=0)$$

So $k_c^2-\omega^2\mu\varepsilon = -\beta_g^2$

$$\beta_g \pm \sqrt{\omega^2\mu\varepsilon-k_c^2} \qquad \text{...(9.42)}$$

The solution of Bessel's equations (9.41 b) will be

$$R = C_nJ_n(k_c\gamma)+D_nN_n(k_c\gamma) \qquad \text{...(9.43)}$$

Where $J_n(k_c\gamma)$ Bessel function n^{th} order of first kind

$N_n(k_c\gamma)$ = Bessel's function n^{th} order of second type

Hence the total solution of Bessel's equation can be given as:

$$\psi = C_n J_n (k_c r) + D_n N_n (k_c r). (A_n. \cos n\phi + B_n. \cos n\phi) e^{\pm j\beta_g z} \quad ...(9.44)$$

The $J_n (k_c r)$ and a argument of $J_n (k_c r)$ plot can be given as shown in Fig. 9.8.

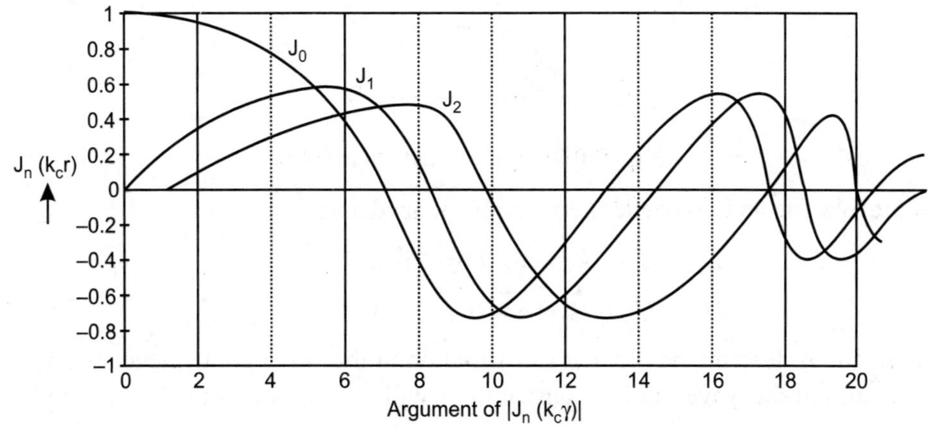

Fig. 9.8 Plot for $J_n (k_c\gamma)$

Second term $N_n (k_c\gamma)$ can also be plotted as:

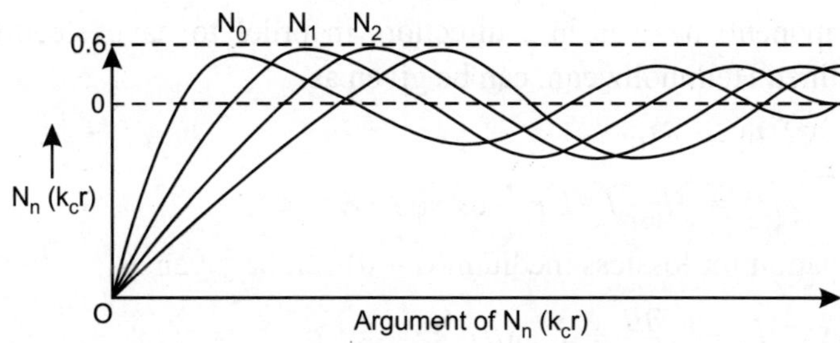

Fig. 9.9. Plot for $N_n (k_c\gamma)$

At $r = 0$, $k_c r = 0$, then the function $N_n (k_c r)$ approach to the infinity, so $D_n = 0$.

This mean for r = 0 at z axis, the field must be finite and the second term of equation (9.44) can be written as,

$$A_n \sin (n\phi) + B_n. \cos (n\phi) = \frac{A_n}{\sqrt{A_n^2 + B_n^2}} \sin(n\phi) + \frac{B_n}{\sqrt{A_n^2 + B_n^2}} \cos (n\phi)$$

$$= \sqrt{A_n^2 + B_n^2}. (\sin \theta. \sin n\phi + \cos \theta \cos n\phi) \text{ (using Fig. 9.10)}$$

$$= \sqrt{A_n^2 + B_n^2} \cos (n\phi - \theta) \approx F_n .\cos n\phi \text{ (say)}$$

So the final solution of the Helm holtz equation can be given as

$$\psi = F_n J_n (k_c r) \cos (n\phi)\, e^{-j\beta_g z}$$

or

$$\psi = \psi_0 J_n (k_c r) \cos n\phi. e^{-j\beta_g z} \quad ...(9.45)$$

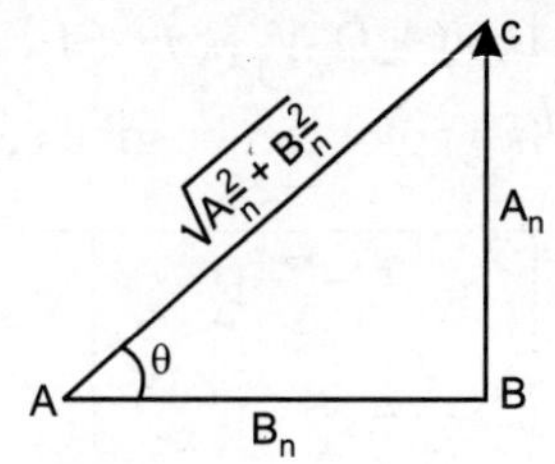

Fig. 9.10 Representation An and Bn.

Therefore, electric and magnetic fields can be defined as

$$E = E_0 J_n (k_c r) \cos (n\phi)\, e^{-j\beta_g z} \quad ...(9.46)$$

$$H = H_0 J_n (k_c r) \cos (n\, \phi) .e^{-j\beta_g z} \quad ...(9.47)$$

These equations describe be the fields distribution in the wave guide, hence useful to analyse the characteristics of the wave-guide. As usual there will be two modes in C-wave-guide.

(*i*) **The TE modes in cylinderical or circular weaveguide:** We assume that wave is propagating in z direction, the mode is characterized by TE_{np} for TE mode;

$$E_z = 0 \text{ and } H_z \neq 0 \quad ...(9.48)$$

It means one component must be in z direction, in order to have electromagnetic energy transmission. The modified Helmholtz eqn. can be given as,

$$\nabla^2 H_z = \gamma^2 Hz$$

and $$H_z = H_{0z} .J_n (k_c r). \cos n\phi e^{-j\beta_g z} \quad ...(9.49)$$

The Maxwell's equation for lossless medium ($\sigma = 0$) can be given as;

$$\nabla \times E = -\frac{\partial B}{\partial t} = -j\omega\mu H \quad ...(9.50)$$

$$\nabla \times H = -\frac{\partial D}{\partial t} = j\omega\varepsilon E \quad ...(9.51)$$

From eqn. (9.50);

$$\nabla \times \text{E} = -j\omega\mu H$$

$$\left(\frac{\partial}{\partial r} a_r + \frac{1}{r} \cdot \frac{\partial}{\partial \phi} + \frac{\partial}{\partial z} a_z\right) \times (E_r a_r + E_\phi\, \mu_\phi + E_z a_z)$$

$$= -j\omega\mu(H_r a_r + H_\phi a_\phi + H_z a_z)$$

$$\begin{vmatrix} a_r & a_z & a_z \\ \frac{\partial}{\partial r} & \frac{1}{r}\frac{\partial}{\partial \phi} & \frac{\partial}{\partial z} \\ E_r & E_\phi & E_2 \end{vmatrix} = -j\omega\mu\, (H_r a_r + H_\phi a_\phi + H_z a_z)$$

$$\left(\frac{1}{r} \cdot \frac{\partial E_2}{\partial \phi} - \frac{\partial}{\partial z} E_\phi\right) + \left(\frac{\partial}{\partial z} E_r - \frac{\partial}{\partial r} E_z\right) \text{a}_\phi + \left(\frac{\partial}{\partial r} E_\phi - \frac{1}{r} \frac{\partial}{\partial \phi} E_r\right) a_z$$

$$= -j\omega\mu\, (H_r a_r + H_\phi a_\phi + H_z a_z)$$

Comparing each component from right and left gives

$$\frac{1}{r}.\frac{\partial E_z}{\partial \phi} - \frac{\partial E_\phi}{\partial z} = -j\omega\mu H_r \quad ...(9.52a)$$

$$\frac{\partial}{\partial z}E_r - \frac{\partial E_z}{\partial r} = -j\omega\mu H_\phi \quad ...(9.52b)$$

$$\frac{\partial}{\partial r}E_\phi - \frac{1}{r}\frac{\partial E_r}{\partial \phi} = -j\omega\mu Hz$$

or
$$\frac{1}{r}\frac{\partial}{\partial r}(rE_\phi) - \frac{1}{r}.\frac{\partial E_r}{\partial \phi} = -j\omega\mu H_z \quad ...(9.52c)$$

Similarly from equation (9.51);

$$\nabla \times H = j\omega\varepsilon E$$

$$\left(\frac{\partial}{\partial r}a_r + \frac{1}{r}.\frac{\partial}{\partial \phi}a_\phi + \frac{\partial}{\partial z}a_z\right) \times (H_r a_r + H_\phi a_\phi + H_z a_z)$$

$$= j\omega\varepsilon\,(E_r a_r + E_\phi a_\phi + E_z a_z)$$

$$\frac{1}{r}\begin{vmatrix} a_r & ra_\phi & a_z \\ \frac{\partial}{\partial r} & \frac{\partial}{\partial \phi} & \frac{\partial}{\partial z} \\ E_r & rH_\phi & H_z \end{vmatrix} = -j\omega\varepsilon\,(E_r a_r + E_\phi a_\phi + E_z a_z)$$

On solving this matrix and comparing both sides gives;

$$\frac{1}{r}\frac{\partial H_z}{\partial \phi} - \frac{\partial H_\phi}{\partial z} = j\omega\varepsilon E_r \quad ...(9.53a)$$

$$\frac{\partial H_r}{\partial z} - \frac{\partial H_z}{\partial r} = -j\omega\varepsilon E_\phi \quad ...(9.53b)$$

$$\frac{1}{r}\frac{\partial}{r}(rH\phi) - \frac{1}{r}\frac{\partial H_r}{\partial \phi} = -j\omega\varepsilon E_z \quad ...(9.53c)$$

Since $\frac{\partial}{\partial z} = -\gamma_g$

and $\frac{\partial^2}{\partial z^2} = \gamma_g^2$

So $\frac{\partial}{\partial z} = -(\alpha + j\beta_g) = -j\beta_g$ (for lossless medium $\alpha = 0$)

and $E_z = 0$ for *TE* mode.

Putting these two value in eqn. (9.52*a*) to (9.52*c*);

$$\frac{1}{r}.\frac{\partial}{\partial \phi}(0) - \frac{\partial E}{\partial z}\phi = +j\omega\mu H_r$$

$$-\frac{\partial}{\partial z}E_\phi = -j\omega\mu H_r$$

$$-(-j\beta_g)\,E_\phi = -j\omega\mu H_r$$

$$H_r = \frac{j\beta_g E_\phi}{-j\omega\mu} = -\left(\frac{\beta_g}{\omega\mu}\right)E_\phi \quad ...(9.54a)$$

From eqn. (9.52*b*);

$$\frac{\partial E_r}{\partial z} - \frac{\partial E_z}{\partial r} = -j\omega\mu H_\phi$$

$$-j\beta_g E_r = -j\omega\mu H_\phi$$

$$E_r = \left(\frac{\omega\mu}{\beta_g}\right) H_\phi \quad \text{...(9.54 b)}$$

From eq. (9.52*c*);

$$\frac{1}{r}\cdot\frac{\partial}{\partial r}(r\,E_\phi) - \frac{1}{r}\frac{\partial}{\partial\phi}(E_r) = -j\omega\mu H_z$$

$$\frac{1}{r}\cdot\frac{\partial}{\partial r}\left(\frac{r.\omega\mu}{\beta_g}H_r\right) - \frac{1}{r}\cdot\frac{\partial}{\partial\phi}\left(\frac{\omega\mu}{\beta_g}\right)H_\phi = -\,\mathrm{j}\omega\mu H_z$$

$$\frac{1}{r}\cdot\frac{\partial}{\partial r}\left(r.\frac{\omega\mu}{\beta_g}.H_r\right) - \frac{1}{r}\frac{\partial}{\partial\phi}\left(\frac{\omega\mu}{\beta_g}\right).\,H_\phi = -j\omega\mu H_z$$

$$\frac{1}{\beta_g r}\left[\frac{\partial}{\partial r}(r.\,H_r) - \frac{\partial}{\partial\phi}(H_\phi)\right] = -jH_z$$

Similarly from eqn. (9.53a), (9.53b) and (9.53c), we get

$$\frac{1}{r}\cdot\frac{\partial H_z}{\partial\phi} - \frac{\partial H_\phi}{\partial_z} = j\omega\varepsilon E_r$$

$$\mathrm{j}\omega\varepsilon E_r = \frac{1}{r}\frac{\partial H_z}{\partial\phi} - \frac{\partial}{\partial z}\left(\frac{\beta_g E_r}{j\omega\mu}\right)$$

$$= \frac{1}{r}\frac{\partial H_z}{\partial\phi} - \frac{\beta}{\omega\mu} - j\beta_g E_r$$

$$= \frac{1}{r}\frac{\partial H_z}{\partial\phi} + \frac{j\beta_g^2}{\omega\mu}E_r$$

$$\left(j\omega\varepsilon - \frac{j\beta_g^2}{\omega\mu}\right)E_r = \frac{1}{r}\cdot\frac{\partial H_z}{\partial\phi}$$

$$\frac{\left(j^2\omega^2\mu\varepsilon - 2\beta_g^2\right)E_r}{j\omega\mu} = \frac{1}{r}\cdot\frac{\partial H_z}{\partial\phi}$$

$$\left(-\frac{k_c^2}{j\omega\mu}\right)E_r = \frac{1}{r}\cdot\frac{\partial H_z}{\partial\phi}$$

$$E_r = -\frac{j\omega\mu}{k_c^2}\frac{1}{r}\cdot\frac{\partial H_z}{\partial\phi} \quad \text{...(9.55}a\text{)}$$

Similarly, $$E_\phi = \frac{j\omega\mu}{k_c^2}\cdot\frac{\partial H_z}{\partial r} \text{ and } E_z = 0 \quad \text{...(9.55}b\text{)}$$

Similarly H components can be also found as;

$$H_r = \frac{-j\beta_g}{k_c^2}\frac{\partial H_z}{\partial r} \quad \text{...(9.56}a\text{)}$$

$$H_\phi = \frac{-j\beta_g}{k_c^2}\frac{1}{r}\cdot\frac{\partial H_z}{\partial \phi} \quad ...(9.56b)$$

and $$H_z = H_{0z}.J_n(k_c r).\cos(n\phi).e^{-j\beta_g^z} \quad ...(9.56c)$$

Boundary conditions

The ϕ component of electric field at $r = a$ or magnetic field components at $r = a$ must be vanish so,

$$E_\phi = 0 \text{ at } r = a$$

or $$H_r = 0 \text{ at } r = a$$

$$E_\phi = \frac{j\omega\mu}{k_c^2}\cdot\frac{\partial H_z}{\partial r}$$

$$\Rightarrow \quad \frac{j\omega\mu}{k_c^2}\cdot\frac{\partial}{\partial r}[H_{0z}J_n(k_c r).\cos n\phi e^{-j\beta_g^z}] = 0$$

$$\frac{j\omega\mu}{k_c^2}\frac{\partial}{\partial r}[H_{0z}J_n(k_c a)\cos n\phi.e^{-j\beta_g^z}] = 0 \text{ at } r = a$$

$$\frac{j\omega\mu}{k_c^2}H_{0z}.\cos(n\phi)e^{-j\beta_g^z}.\frac{\partial}{\partial r}J_n(k_c a) = 0$$

$$\Rightarrow \quad \frac{j\omega\mu}{k_c^2}H_{0z}\cos(n\phi)e^{-j\beta_g^z}.j_n^1(k_c a) = 0 \quad ...(9.57a)$$

Since $$\frac{j\omega\mu}{k_c^2}H_{0z}.\cos n\psi\, e^{-j\beta_g^z} \neq 0$$

So, $$\frac{\partial}{\partial r}J_n'(k_c a) = 0 \quad ...(9.57b)$$

Since $J_n(k_c r)$ is oscillatory function, so derivalues of $J_n'(k_c a)$ will also be oscillatory function. So, it has infinite number of value, for which equation (9.57a) is valid.

The permissible value of $(k_c a)$ can be given as;

Let $$k_c a = \chi'_{np} \text{ or } k_c = \frac{\chi'_{np}}{a}$$

Now $$E_r = \frac{-j\omega\mu}{k_c^2}\cdot\frac{1}{r}\cdot\frac{\partial H_z}{\partial \phi}$$

$$= -\frac{j\omega\mu}{k_c^2}\frac{1}{r}\cdot\frac{\partial}{\partial \phi}\cdot H_{0z}.J_n(k_c r).\cos n\phi e^{-j\beta_g^z}$$

$$= -\frac{j\omega H}{k_c^2}\frac{1}{r}H_{0z}.J_n(k_c r)\, n\sin(n\phi)\, e^{-\partial\beta_g^z}$$

$$E_r = E_{0r}.J_n\left(\frac{\chi'_{np}}{a}r\right)\sin(n\phi)e^{-j\beta_g^z} \quad ...(9.57c)$$

$$E_\phi = \frac{j\omega\mu}{k_c^2}\cdot\frac{\partial H_z}{\partial r} = \frac{j\omega\mu}{k_c^2}\cdot\frac{\partial}{\partial \gamma} H_{0z} J_n\left(\frac{\chi'_{np}}{a}r\right)\cos n\phi e^{-j\beta_g z}$$

$$E_\phi = E_{0\phi}\cdot J_n'\left(\frac{x'_{np}}{a}r\right)\cos n\phi e^{-j\beta_g z} \qquad ...(9.57d)$$

Since $\frac{E}{H} = Z_g$

or $Z_g = \frac{E_r}{H_\phi} = -\frac{H_\phi}{H_r}$

$$H_\phi = \frac{E_r}{Z_g} = \frac{E_{0z}}{Z_g}.J_n\left(\frac{\chi'_{np}}{a}r\right)\sin n\phi . e^{-j\beta_g^z}$$

$$H_r = -\frac{E_\phi}{Z_g} = -\frac{E_{0\phi}}{Z_g}J_n\left(\frac{k_{np}}{a}r\right).\cos(n\phi)\, e^{-j\beta_g^z}$$

Here $n = 0, 1, 2, 3,$ and $p = 1, 2, 3, 4, 5, ...$

where n represents number of full cycles of field variation in one revolution through 2π radian of ϕ, and p represents number of zero of E_ϕ that is $J_n'\left(\frac{\chi'_{np}}{a}r\right)$ along the radial of a waveguide.

Now different parameter related with circular waveguide can be derived as follows;

The propagation constant ($-j\beta_g$):

Since $\gamma^2 + k_c^2 = \gamma_g^2$

$$\gamma_g = \pm\sqrt{\gamma^2 + k_c^2}$$

or $$\alpha_g + j\beta_g = \pm\sqrt{(\alpha + j\beta)^2 + \left(\frac{\chi'_{np}}{a}\right)^2}$$

For loss-less propagation $\alpha_g = 0, \Rightarrow 0 + (j\beta_g)^2 = (j\beta)^2 + \left(\frac{\chi'_{np}}{a}\right)^2$

$$-\beta_g^2 = -\omega^2\mu\varepsilon + \left(\frac{\chi'_{np}}{a}\right)^2$$

$$\beta_g^2 = \omega^2\mu\varepsilon - \left(\frac{\chi'_{np}}{a}\right)^2$$

$$\beta_g = \pm\sqrt{\omega^2\mu\varepsilon - \left(\frac{\chi'_{np}}{a}\right)^2} \qquad ...(9.58a)$$

The cut off Wave Number

The cut of wave number of a mode is that value for which the mode propagation constant k_c vanishes. Hence.

$$k_c = \frac{\chi'_{np}}{a} = \omega_c\sqrt{\mu\varepsilon}$$

The cutoff frequency

The cut off frequency for *TE* mode in circular waveguide can be given as

$$k_c = \frac{\chi'_{np}}{a} = \omega_c\sqrt{\mu\varepsilon}$$

i.e., it is the frequency at which $\beta_g = 0$, there is no phase shift.

So
$$\omega_c\sqrt{\mu\varepsilon} = \frac{\chi'_{np}}{a}$$

$$\omega_c = \frac{\chi'_{np}}{a\sqrt{\mu\varepsilon}}$$

$$2\pi f_c = \frac{\chi'_{np}}{a\sqrt{\mu\varepsilon}}$$

$$f_c = \frac{\chi'_{np}}{2\pi a\sqrt{\mu\varepsilon}}$$

or
$$f_c = \frac{\chi'_{np}}{2\pi a}\frac{1}{\sqrt{\mu\varepsilon}}$$

$$= \frac{\chi'_{np}}{2\pi a}\frac{1}{\sqrt{\mu_0\mu_r\varepsilon_0\varepsilon_r}}$$

$$f_c = \frac{\chi'_{np}}{2\pi a}\frac{1}{\sqrt{\mu_0\varepsilon_0}}\frac{1}{\sqrt{\mu_\gamma\varepsilon_r}} = \frac{\chi'_{np}.c}{2\pi a\sqrt{\mu_\gamma\varepsilon r}} = \frac{\chi'_{np}}{2a}.v_p \quad ...(9.58b)$$

Where v_p is called phase velocity in free medium or unbounded medium.

Waveguide wave-length λ_g;

Since
$$\beta_g = \sqrt{\omega^2\mu\varepsilon - \left(\frac{\chi'_{np}}{a}\right)^2}$$

$$\beta_g = \sqrt{\omega^2\mu\varepsilon - \omega_c^2\mu\varepsilon}$$

$$\beta_g = \omega\sqrt{\mu\varepsilon}\sqrt{1-\left(\frac{\omega_c}{\omega}\right)^2}$$

$$\lambda_g = \frac{2\pi}{\beta_g} = \frac{2\pi}{\omega\sqrt{\mu\varepsilon}}\sqrt{1-\left(\frac{\omega_c}{\omega}\right)^2}$$

$$= \frac{c}{f}\cdot\sqrt{1-\left(\frac{f_c}{f}\right)^2}$$

$$\lambda_g = \frac{\lambda}{\sqrt{1-\left(\frac{f_c}{f}\right)^2}} \quad ...(9.58c)$$

Wave impedance

$$Z_g = \frac{\omega\mu}{\beta_g} = \frac{\omega\mu}{\sqrt{\omega^2\mu\varepsilon - \omega c^2\mu\varepsilon}}$$

$$= \frac{\omega\mu}{\omega\sqrt{\mu\varepsilon}\sqrt{1-\left(\frac{\omega_c}{\omega}\right)^2}}$$

$$= \frac{\sqrt{\frac{\mu}{\varepsilon}}}{\sqrt{1-\left(\frac{\lambda}{\lambda_c}\right)^2}} = \frac{\eta}{\sqrt{1-\left(\frac{\lambda}{\lambda_c}\right)^2}}$$

$$Z_g = \frac{\eta}{\sqrt{1-\left(\frac{\lambda}{\lambda_c}\right)^2}} \quad \left(\text{since } \eta = \sqrt{\frac{\mu}{\varepsilon}}\right) \quad ...(9.58d)$$

That is Z_g is greater than free space impedance, $\eta = \sqrt{\frac{\mu_0}{\varepsilon_0}} = 120\,\pi$.

Dominant mode and degenerate mode

From the expression of cut-off frequency (f_c) it is observed that, f_c is lowest for wave-guide with $m = 1$ and $n = 0$ (i.e. TE_{10} mode). Since for TM_{mn} modes, $m \neq 0$ and $n \neq 0$, lowest order mode is TE_{10} only, hence it is termed as Dominant mode. Higher order mods, those have same 'f_c' are known as degenerate mode, For example, all the possible TE_{mn} and TM_{mn} modes for $m \neq 0$ and $n \neq 0$, are always degenerate modes. (e.g. TE_{12}, TE_{21} and TM_{12}, TM_{21}).

TM mode in circular W/G

The TM mode in circular waveguide is characterized by,

$$H_z = 0 \text{ and } E_z \neq 0$$

so the wave equation can be given as

$$\nabla^2 E = \gamma^2 E$$

In particular $\quad \nabla^2 E_z = \gamma^2 E_z$

Since from (9.46), $\quad E_z = E_{0z} \cdot J_n(k_c r) \cdot \cos(n\phi) \cdot e^{-j\beta_g z} \quad ...(9.59a)$

Boundary conditions

Since electric field vanishes at r = a, hence

$$E_z = 0 \text{ at } r = a$$

$$E_z = E_{0z}.\ J_n(k_c r)\cos(n\phi).e^{-j\beta_g^z}$$

$$0 = E_{0z}J_n(k_c r)\cos(n\phi).\ e^{-j\beta_g^z}$$

$$0 = (E_{0z}\cos(n\phi).e^{-j\beta_g^z}).\ J_n(k_c a)$$

$$J_n(k_c a) = 0$$

Since $E_{0z}\cos n\phi.e^{-j\beta_g^z} \neq 0$

so $J_n(k_c a) = 0$

$$k_c a = \chi'_{np}$$

$$k_c = \frac{\chi_{np}}{a};$$ where k_c is called wave number.

Again from Maxwell's lst and 2nd equations:

$$\nabla \times E = -j\omega\mu H \quad ...(9.59b)$$

$$\nabla \times H = j\omega\varepsilon E \quad ...(9.59c)$$

So, from (9.59*a*);

$$\left(\frac{\partial}{\partial r}a_r + \frac{1}{r}\frac{\partial}{\partial\phi}a_\phi + \frac{\partial}{\partial z}a_z\right) \times (E_r a_r + E_\phi a_\phi + E_z a_z)$$

$$= -j\omega\mu(H_r a_r + H_\phi a_\phi + H_z a_z)$$

$$= \frac{1}{r}\begin{vmatrix} a_r & ra_\phi & a_z \\ \frac{\partial}{\partial r} & \frac{\partial}{\partial \phi} & \frac{\partial}{\partial z} \\ E_r & rE_\phi & E_z \end{vmatrix} = -j\omega\mu(H_r a_r + H_\phi a_\phi + H_z a_z)$$

After solving and putting $\frac{\partial}{\partial z} = -j\beta_g$ and $H_z = 0$, As earlier)

We get

$$E_r = \frac{-j\beta_g}{k_c^2}.\frac{\partial E_z}{\partial r} \quad ...(9.60a)$$

$$E_\phi = \frac{-j\beta_g}{k_c^2}.\frac{1}{r}\frac{\partial E_z}{\partial \phi} \quad ...(9.60b)$$

$$E_z = E_{0z}J_n(k_c r)\cos n\phi e^{-j}\beta_g z \quad ...(9.60c)$$

and

$$H_r = \frac{j\omega\varepsilon}{k_c^2}.\frac{1}{r}\frac{\partial E_z}{\partial \phi} \quad ...(9.61a)$$

$$H_\phi = \frac{j\omega\varepsilon}{k_c^2}.\frac{\partial E_z}{\partial r} \quad ...(9.61b)$$

Now if replacing the value of E_z from (9.60*c*) to equation (9.60*a*) and adjusting the parameters we get;

$$E_r = -\frac{j\beta_g}{k_c^2}\frac{\partial}{\partial r}[E_{0z}]_n (k_c r)\cos(n\phi)\, e^{-j\beta_g z}]$$

$$= -\frac{j\beta_g}{k_c^2} E_{0z} . J_n^1 (k_c r)\cos(n\phi)\, e^{-j\beta_g Z}$$

or $$= E_{0r} J_n^1 (k_c r)\cos(n\phi)\, e^{-j\beta_g Z}$$

where $$E_{0r} = -\left(\frac{j\beta_g}{k_c^2}\right) E_{0z}$$

$$E_r = E_{0r} . J_n'\left(\frac{\chi_{np}}{a} r\right)\cos(n\phi) e^{-j\beta_g z} \quad ...(9.62a)$$

$$E_\phi = E_{0\phi} J'_n{}'\left(\frac{\chi_{np}}{a} r\right)\sin(n\phi) e^{-j\beta_g z} \quad ...(9.62b)$$

$$E_z = E_{0z} J_n'\left(\frac{\chi_{np}}{a} r\right) . \cos(n\phi)\, e^{-j\beta_g z} \quad ...(9.62c)$$

Similarly, $$H_r = \frac{E_{0r}}{Z_g} . J_n'\left(\frac{\chi_{np}}{a} r\right) . \cos(n\phi) e^{-j\beta_g z} \quad ...(6.63a)$$

$$H_\phi = \frac{E_{0\phi}}{Z_g} . J_n'\left(\frac{\chi_{np}}{a} r\right) . \sin(n\phi) e^{-j\beta_g z} \quad ...(9.63b)$$

$$H_z = 0 \quad ...(9.63c)$$

That is almost all characteristic equation for TM mode is similar to TE mode, but some are different that can be calculated as follow:

Propagation constant γ_g and β_g:

Since $$k_c a = \chi_{np}$$

$$k_c = \frac{\chi_{np}}{a}$$

and $$\gamma_g^2 = k_c^2 + \gamma^2$$

$$\gamma_g = \sqrt{k_c^2 + \gamma^2}$$

$$\gamma_g = \sqrt{\left(\frac{\chi_{np}}{a}\right)^2 + \gamma^2} \quad ...(9.64a)$$

Now for wave propagation in wave guide attenuation constant must be equal zero *i.e.* $\alpha_g = 0$ and $\alpha = 0$

So $$\gamma_g^2 = k_c^2 + \gamma^2$$

$$(\alpha_g + i\beta_g)^2 = k_c^2 + (\alpha + j\beta)^2$$

$$-\beta_g^2 = k_c^2 - \beta^2$$

or
$$\beta_g = \sqrt{\beta^2 - k_c^2}$$

$$\beta_g = \sqrt{\omega^2 \mu\varepsilon - k_c^2}$$

$$\beta_g = \sqrt{\omega^2 \mu\varepsilon - \left(\frac{\chi_{np}}{a}\right)^2} \quad \text{...(9.64b)}$$

Cut off number and cut off frequency

When operating frequency is low, $\omega^2\mu\varepsilon - \left(\frac{\chi_{np}}{a}\right)^2 < 0$.

Then β_g will be imaginary and wave will not propagate in wave guide because as frequency increase one situation arise where;

$$\omega^2\mu\varepsilon - \left(\frac{\chi_{np}}{a}\right)^2 = 0$$

Then $\omega \rightarrow \omega_c$, and called cut off frequency. It is defined as minimum frequency at which wave just start propagation into the wave guide.

Therefore, $\omega^2\mu\varepsilon - \frac{\chi_{np}}{a} = 0$

$$\frac{\chi_{np}}{a} = \omega_c^2\mu\varepsilon = k_c$$

$$\omega_c = \frac{\chi_{np}}{a.\sqrt{\mu\varepsilon}}$$

or
$$f_c = \frac{\chi_{np}}{2\pi a\sqrt{\mu\varepsilon}} = \left(\frac{\chi'_{np} \cdot v_p}{2\pi a}\right) \quad \text{...(9.64c)}$$

where u_p is phase-velocity of wave inside the wave-guide.

Phase velocity

Earlier phase velocity has been defined mathematically as follows:

$$v_p = \frac{\omega}{\beta} = \frac{\omega}{\beta_g}$$

$$= \frac{\omega}{\sqrt{\omega^2\mu\varepsilon - \omega_c^2\mu\varepsilon}}$$

$$= \frac{\omega}{\omega\sqrt{\mu\varepsilon}\sqrt{1-\left(\frac{\omega_c}{\omega}\right)^2}}$$

$$= \frac{c}{\sqrt{1-\left(\frac{f_c}{f}\right)^2}}$$

(since $\omega_c = 2\pi f_c$ and $\omega = 2\pi f$)

$$v_p = \frac{c}{\sqrt{1-\left(\frac{\lambda}{\lambda_c}\right)^2}} \quad ...(9.64d)$$

Therefore it is clear that;

$$v_p > c \text{ (c is velocity of wave in free-space).}$$

Group velocity v_g

Mathematically group velocity can be expressed

$$v_g = \frac{d\omega}{d\beta_g}$$

Since $$\beta_g = \sqrt{\omega^2 \mu\varepsilon - \omega_c^2 \mu\varepsilon}$$

$$\frac{d\beta_g}{d\omega} = \frac{1}{2}(\omega^2\mu\varepsilon - \omega_c^2\mu\varepsilon)^{-1/2}\, 2\omega\mu\varepsilon$$

$$\frac{d\beta_g}{d\omega} = \frac{\omega\mu\varepsilon}{\sqrt{\omega^2 \mu\varepsilon - \omega_c^2 \mu\varepsilon}}$$

$$= \frac{\omega\mu\varepsilon}{\omega\sqrt{\mu\varepsilon}\sqrt{1-\left(\frac{\omega_c}{\omega}\right)^2}}$$

$$= \frac{\sqrt{\mu\varepsilon}}{\sqrt{1-\left(\frac{\omega_c}{\omega}\right)^2}}$$

So $$\frac{d\omega}{d\beta_\delta} = \frac{\sqrt{1-\left(\frac{\omega_c}{\omega}\right)^2}}{\sqrt{\mu\varepsilon}} = c.\sqrt{1-\left(\frac{f_c}{f}\right)^2}$$

$$v_g = c\sqrt{1-\left(\frac{f_c}{f}\right)^2}$$

or $$v_g = c\sqrt{1-\left(\frac{\lambda}{\lambda_c}\right)^2} \quad ...(9.65a)$$

Waveguide wavelength λ_g;

Since $$\lambda_g = \frac{2\pi}{\beta_g}$$

$$= \frac{2\pi}{\sqrt{\omega^2 \mu\varepsilon - \omega_c^2 \mu\varepsilon}}$$

$$= \frac{2\pi}{\omega\sqrt{\mu\varepsilon}\sqrt{1-\left(\frac{f_c}{f}\right)^2}}$$

$$= \frac{2\pi}{2\pi f\sqrt{\mu\varepsilon}\sqrt{1-\left(\frac{\lambda}{\lambda_c}\right)^2}} \qquad (\text{since } \lambda \propto \frac{1}{f})$$

$$= \frac{1}{f\sqrt{\mu\varepsilon}\sqrt{1-\left(\frac{\lambda}{\lambda_c}\right)^2}}$$

$$\lambda_g = \frac{c}{f\sqrt{1-\left(\frac{\lambda}{\lambda_c}\right)^2}} = \frac{\lambda}{\sqrt{1-\left(\frac{\lambda}{\lambda_c}\right)^2}} \qquad ...(9.65b)$$

Waveguide impedance

For TM modes $\quad Z_g = \frac{\beta_g}{\omega\varepsilon} = \eta\sqrt{1-\left(\frac{f_c}{f}\right)^2}$

$$Z_g = \eta\sqrt{1-\left(\frac{\lambda}{\lambda_c}\right)^2} \qquad ...(9.65c)$$

Then for TEM mode

$$Z_g = \frac{\omega\mu}{\beta_g} = \frac{\eta}{\sqrt{1-\left(\frac{\lambda}{\lambda_c}\right)^2}}$$

Where $\eta = \sqrt{\frac{\mu_0}{\varepsilon_0}}$ is intrinsic impedance of free- space. = 120 π

Dominant and degenerate mode

The mode which has lowest cut off frequency called dominant mode or vice-versa the mode having highest cut off waveguide wave length is called dominant mode.

It can be given as,

$$\omega_c\sqrt{\mu\varepsilon} = \frac{\chi'_{np}}{a} \text{ for } TE \text{ mode} \qquad ...(9.66a)$$

and

$$\omega_c\sqrt{\mu\varepsilon} = \frac{\chi'_{np}}{a} \text{ for } TM \text{ mode} \qquad ...(9.66b)$$

Again from (9.66*a*), $\omega_c\sqrt{\mu\varepsilon} = \frac{X'_{np}}{a}$

$$\Rightarrow \qquad f_c = \frac{1}{2\pi a}\frac{\chi'_{np}}{\sqrt{\mu\varepsilon}}$$

$$\Rightarrow \qquad \lambda_c = \frac{2\pi a\sqrt{\mu\varepsilon}}{\chi'_{np}\cdot c}$$

$$\lambda_c = \frac{2\pi a}{\chi'_{np}} \qquad \left(\text{As } c = \frac{1}{\sqrt{\mu\varepsilon}}\right)$$

That is λ_c will be maximum for minimum value of χ_{np}', ($\approx$ 1.841)

So, $$\lambda_c = \frac{2\pi a}{1.841}, \qquad ...(9.67a)$$

where χ'_{np} = 1.841 for $m = 1$ $n = 1$, so, TE_{11} called dominent mode.

For TM mode

$$\lambda_c = \frac{2\pi a}{\chi'_{np}}$$

χ'_{np} is minimum for $m = 1$ and $n = 0$, hence

$$\lambda_c = \frac{2\pi a}{2.405}, \text{ so } TM_{10} \text{ called dominant mode.} \qquad ...(9.67b)$$

The degenerate mode has same cut-off frequency for different value of '*m*' and '*n*' where '*m*' and '*n*' show different modes.

Since at $\chi'_{op} = \chi_{1p}$ all the TM_{op} and TM_{1p} modes are degenerated in uniform circular wave guide. where p = 1, 2, 3. These modes are TM_{01}, TE_{21}, TE_{01} and TM_{11}.

Attenuation in Wave Guide

In waveguide analysis it has been assumed that the wall is perfect conductor and the dielectric medium within is loss less. Thus for frequencies greater than the cut off frequency ω_c, the propagation constant γ is purely imaginary,

$$\gamma = \sqrt{\left(\frac{m\pi}{a}\right)^2 + \left(\frac{n\pi}{b}\right)^2 - \omega^2\mu\varepsilon} \qquad ...(9.67c)$$

and the wave propagates untainted (*i.e.* without loos of energy) However, in practical waveguide, losses are present mainly because of infinite conductivity of conducting walls, *i.e.* attenuation occurs.

Attenuation can be calculating by calculating power loss of transmitted power using Poynting Theorem

$$P_{tr} = \oint \overline{P}\,.\,ds = \oint \frac{1}{2}(E \times H^*)\,ds$$

Hence for lossless dielectrics the average power flow through rectangular waveguide. will be

$$P_{tr} \frac{1}{2Z_g}\int_a |E|^2\,da = \frac{Z_g}{2}\int_a |H|^2 da$$

where $$Z_g = \frac{E_x}{H_y} = -\frac{E_y}{H_x} \quad \text{and} \quad a \rightarrow \text{Area of cross-section}$$

and $$|E|^2 = |E_x|^2 + |E_y|^2$$

$$|H|^2 = |H_x|^2 + |H_y|^2$$

For TE_{mm} mode

$$P_{tr} = \frac{\sqrt{1-\left(\frac{f_c}{f}\right)^2}}{2\eta} \int_0^a \int_0^b |E_x|^2 + |E_y|^2 \, dx \, .dy \qquad ...(9.68a)$$

for TM modes

$$P_{tr} = \frac{1}{2\eta\sqrt{1-\left(\frac{f_c}{f}\right)^2}} \int_0^a \int_0^b |E_x|^2 + |E_y|^2 \, dx \, dy \qquad ...(9.68b)$$

Total average power transmittal across the cross-section of R.Wave guide will be

$$P_{ove} = \int_{x=0}^{a} \int_{y=0}^{b} \frac{|E_x|^2 + |E_y|^2}{2\eta_g} \, dx. \, dy \qquad ...(9.68c)$$

where $\eta_g = \eta_{TE}$ for TE mode and $\eta_g = \eta_{TM}$ for TM mode.

There are two types of power loss in rectangular W/G

(*i*) Power loss in dielectric

(*ii*) Power loss in guide walls

Power loss in dielectric causes dielectric attenuation in low loss dielectric ($\sigma <<< \mu\varepsilon$). The propagation constant can be given as,

$$\gamma = \sqrt{Z.Y}; \quad \text{and} \quad \sigma << \omega\varepsilon$$

$$= \sqrt{j\omega\mu(\sigma + j\omega\varepsilon)}$$

$$= \sqrt{j\omega\mu j\omega\mu\left(1 - \frac{j\sigma}{\omega a}\right)}$$

$$= j\omega\sqrt{\mu\varepsilon}\left(1 - \frac{j\sigma}{2\omega\varepsilon}\right) + ...$$

$$\frac{\omega}{2\omega\varepsilon} \sigma\sqrt{\mu\varepsilon}$$

$$\alpha + j\beta = \frac{\sigma}{2}\sqrt{\frac{\mu}{\varepsilon}}$$

CASE-I $\quad \alpha = \frac{\sigma}{2}\sqrt{\frac{\mu}{\varepsilon}} = \frac{\sigma}{2}\eta = \frac{\eta\sigma}{2}$

So in rectangular waveguide.; ($\alpha \to \alpha_g$ and $\eta \to \eta_g$):

$$\alpha_g = \frac{\sigma}{2}\eta\sqrt{1-\left(\frac{f_c}{f}\right)^2} \text{ for } TM \text{ mode} \qquad ...(9.69a)$$

$$\alpha_g = \frac{\sigma}{2}\frac{\eta}{\sqrt{\left(1-\frac{f_c}{f}\right)^2}} \text{ for } TM \text{ mode} \qquad ...(9.69b)$$

Now power loss in guided wall will be:

$$|E| = |E_{0z}|.e^{-\alpha_g z}$$

$$|H| = |H_{0z}|.e^{-\alpha_g z}$$

where E_{0z} and H_{0z}, electric and magnetic field intensities at z. It is seen that average power flow decreases propertinary $e^{-2\alpha_g z}$.

$$P_{tr} = (P_{tr} + P_{loss})e^{-2\alpha_g z}$$

$$e^{2\alpha_g z} = \left(\frac{P_{tr} + P_l}{P_{tr}}\right) = \left(1 + \frac{P_l}{P_{tr}}\right)$$

$$1 + 2\alpha_g z = \left(\frac{P_{loss}}{P_{tr}} + 1\right) \text{ (if } 2\alpha_g z << 1)$$

$$\alpha_g = \frac{P_{loss}}{2zP_{tr}} = \left(\frac{P_l/z}{2P_{tr}}\right)$$

$$\Rightarrow \qquad \alpha_g = \frac{\text{Power loss per unit length}}{2 \times \text{Power transmittal}} \qquad ...(9.70a)$$

That is the attenuation constant of the guided wall is equal to the ratio of the power loss/unit length to twice the power transmitted through the guide.

Since the electric and magnetic field intensities established with respect to skin depth while wave propagats into the walls, it is better to defined the surface resistance, R_s as:

$$R_s = \frac{\rho}{\delta} = \frac{1}{\sigma\delta} = \frac{\sqrt{\pi f \mu}}{\sigma} \ \Omega/\text{square} \qquad ...(9.70b)$$

Here ρ = resistivity of the conducting wall (ohm meters)

σ = conductivity mho per meter and δ = skin depth in meters

CASE-II

Second loss can be calculated as

$$P_z = \frac{R_s}{2}\int_s |H_t|^2\, ds \text{ Watt/length} \qquad ...(9.70c)$$

H_t tangential component of magnetic fields.

$$\alpha_g = \frac{R_s \int_s |H_t|^2 ds}{2z_g \int_a |H|^2 da}$$

And α can be given as

$$\alpha = \frac{Rs}{b\eta_0}\frac{k_0}{\beta}\left[1 + 2\left(\frac{b}{a}\right)\left(\frac{f_c}{f}\right)^2\right] \qquad ...(9.70d)$$

A typical variation of α with frequency is found to be as shown in Fig. 9.11.

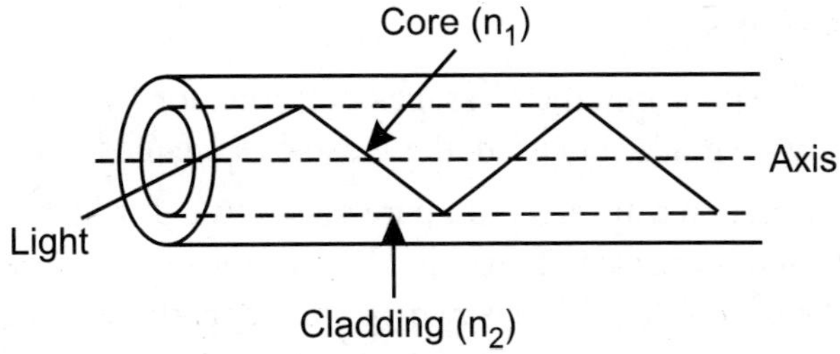

Fig. 9.11 Attenuation Vs frequency

Therefore may be moted that attenuation becomes infinite as f_c decrease to a minimum when frequency is increased and then increases gradually and monotimically as frequency is increased for then.

Impossibility of TEM Mode in W/G

Since in rectangular waveguide

$$E_x = -\frac{r}{h^2}\cdot\frac{\partial E}{\partial x} - \frac{j\omega\mu}{h^2}\cdot\frac{\partial H_z}{\partial y}, E_y = -\frac{\gamma}{h^2}\cdot\frac{\partial E_z}{\partial y} + \frac{j\omega\mu}{h^2}\frac{\partial H_z}{\partial x} \quad ...(9.71a)$$

$$H_x = -\frac{\gamma}{h^2}\cdot\frac{\partial H_z}{\partial x} + \frac{j\omega\varepsilon}{h^2}\frac{\partial E_z}{\partial y}, H_y = -\frac{\gamma}{h^2}\cdot\frac{\partial H_z}{\partial y} - \frac{j\omega\varepsilon}{h^2}\frac{\partial E_z}{\partial x} \quad ...(9.71b)$$

In all these four equation if we put $E_z = 0 = H_z$ (TEM) that is transverse electro-magnetic wave, all components of E and H becomes zero *i.e.*, no electromagnetic wave exist in wave guide when $E_z = 0, H_z = 0$.

Similarly for a circular waveguide it can be shown that

$$E_r = -\frac{j\omega\mu}{k_c^2}\cdot\frac{1}{r}\cdot\frac{\partial H_a}{\partial\phi} \quad ...(9.72a)$$

$$E_\phi = \frac{j\omega\mu}{k_c^2}\cdot\frac{\partial H_z}{\partial r} \quad ...(9.72b)$$

$$E_z = 0 \quad ...(9.72c)$$

$$H_r = -\frac{r}{k_c^2}\cdot\frac{\partial H_z}{\partial r} \quad ...(9.73a)$$

$$H_\phi = -\frac{r}{k_c^2}\cdot\frac{\partial H_z}{\partial\phi} \quad ...(9.73b)$$

$$H_z = H_{0z} J_n(k_c r)\cos n\phi e^{-j\beta_g Z} \quad ...(9.73c)$$

When $E_z = 0$, $H_z = 0$ are placed then all field component becomes zero. So, TEM mode does not exist in circular waveguide. In fact TEM mode can not exist in a single conductor waveguide because there is no axial component of either E or H, It only exist in two wire transmission line.

Optical wave guides

Optical wave guides used to trans mitt the message or information by optical means (light). For this purpose special type of wave guides are used, these cables are called optical fibre.

Optical fibre is transparent plastic and ultra pure glass through which optical signal transmitted inside the optical fibre. Light signals are transmitted inside the optical fibre are transmitted through total internal reflation. Optical fibre cable has two parts Inner part has higher refractive index called core and output has slightly lower refractive index called cladding. (Fig. 9.12). Fibres have its three types (*i*) Single mode fibre (*ii*) Step index multi-mode fibre (*iii*) Graded index multi-mode fibre.

Applications areas of fibre incudes; (*i*) Communications (*ii*) Military services (*iii*) Food and beverage industry (*iv*) Study of soil samples (*v*) Spectroscopy (*vi*) Plastic industry.

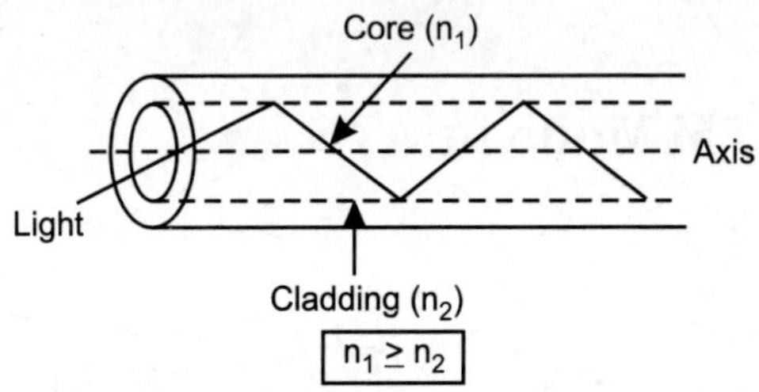

Fig. 9.12 Basic geometry of a optical fibre

SOLVED EXAMPLES

Example 1. An air filled rectangular W/G having dimension 7 cm × 3.5 cm operated in the dominant mode TE_{10}. as shown in 9.13.

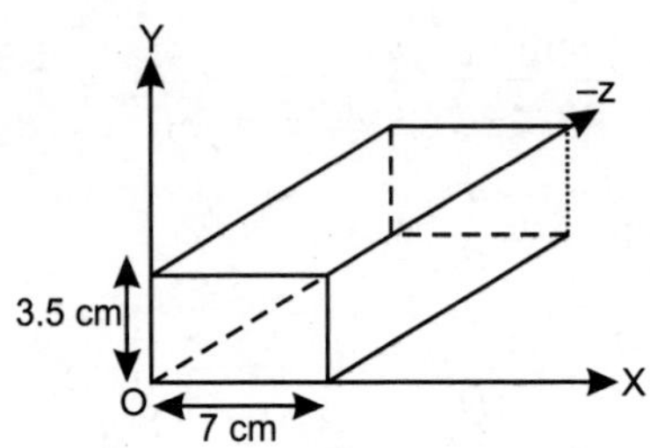

Fig. 9.13 Geometry of example 1.

Calculator following parameters at $f = 3.5$ GHz

(*i*) The cut off frequency f_c

(*ii*) The cut-off wave length λ_c

(*iii*) Phase and group velocities

(*iv*) The guided wave length

Solution: (*i*) Since, $f_c = \frac{c}{2}\sqrt{\left(\frac{m}{a}\right)^2 + \left(\frac{n}{b}\right)^2}$

for dominant dode $m = 1, n = 0$

$$f_c = \frac{c}{2}\sqrt{\left(\frac{1}{a}\right)+\left(\frac{0}{b}\right)}$$

$$f_c = \frac{c}{2a} = \frac{3.0\times10^8}{2\times7\times10^{-2}} = \frac{3\times10^{10}}{14} = 2.14 \times 10^9 \text{ Hz}$$

$$= 2.14 \text{ GHz}$$

(*ii*) Now cut off wavelength can be given as

$$c = f_c \times \lambda_c$$

$$\lambda_c = \frac{c}{f_c} = \frac{c}{\frac{c}{2}\sqrt{\left(\frac{m}{a}\right)^2+\left(\frac{h}{b}\right)^2}}$$

$$\lambda_c = \frac{2}{\sqrt{\left(\frac{m}{a}\right)^2+\left(\frac{n}{b}\right)^2}}$$

For dominant mode $m = 1$, $n = 0$

$$\lambda_c = \frac{2}{\sqrt{\left(\frac{1}{a}\right)^2+\left(\frac{0}{b}\right)^2}} = 2a$$

$$\lambda_c = 2 \times 7 = 14 \text{ cm}$$

(*iii*) Phase velocity

$$v_p = \frac{c}{\sqrt{1-\left(\frac{\lambda}{\lambda_c}\right)^2}} = \frac{c}{\sqrt{1-\left(\frac{f_c}{f}\right)^2}}$$

$$= \frac{3\times10^8}{\sqrt{1-\left(\frac{2.14}{3.5}\right)^2}}$$

$$v_p = 3.78 \times 10^8 \text{ m/sec}$$

Group velocity $$v_g = c\sqrt{1-\left(\frac{f}{f_c}\right)^2} = c\sqrt{1-\left(\frac{f_c}{f}\right)^2}$$

$$= 3 \times 108\sqrt{1-\left(\frac{2.14}{3.5}\right)^2}$$

$$v_g = 2.15 \times 10^8 \text{ m/sec}$$

(*iv*) The guided wavelength can be given as,

$$\lambda_g = \frac{\lambda}{\sqrt{1-\left(\frac{\lambda}{\lambda_c}\right)^2}} = \frac{\lambda}{\sqrt{1-\left(\frac{f_c}{f}\right)^2}}$$

$$= \frac{\dfrac{3 \times 10^8}{3.5 \times 10^9}}{\sqrt{1 - \left(\dfrac{3.5 \times 10^9}{2.14 \times 10^9}\right)^2}}$$

$$\lambda_g = 10.8 \text{ cm}$$

Example 2. A rectangular W/G is filled with a dielectric material of relative permitivity $\varepsilon_r = 9$. It has inside dimension of 7cm × 3.5 cm and operated in the dominant mode. Calculate the following parameter

(*i*) Cut off frequency and cut off wavelength

(*ii*) Phase and group velocity at 2 GHz

(*iii*) Guided wavelength at the same operating frequency.

Solution: Since cut off frequency $f_c = \frac{c}{f}\sqrt{\left(\frac{m}{a}\right)^2 + \left(\frac{n}{b}\right)^2}$

(*i*) But for deminant mode TE_{10}, $m = 1, n = 0$

$$f_c = \frac{c}{2}\sqrt{\left(\frac{1}{a}\right)^2 + \left(\frac{0}{b}\right)^2} = \frac{c}{2a}$$

$$\text{fc} = \frac{3 \times 10^8}{2 \times 7 \times 10^{-2}} = \frac{3}{14} \times 10^{10}$$

$$= 0.214 \times 10^{10}$$

$$= 2.14 \times 10^9$$

$$= 2.14 \text{ GHz}$$

and cutt off wavelength $\lambda_c = 2a$

$$= 2 \times 7 = 14 \text{ cm}$$

(*ii*) Phase velocity $\quad v_p = c.\sqrt{1 - \left(\frac{\lambda}{\lambda_c}\right)^2}$

but $\quad \lambda = \frac{c}{\sqrt{\varepsilon_r}\, f}$

$$= \frac{3 \times 10^8}{\sqrt{9} \times 2 \times 109} = \frac{3 \times 10^8}{3 \times 2 \times 10^9}$$

$$= \frac{10^{-1}}{2} = 0.5 \times 10^{-1}$$

$$= 5 \text{ cm}$$

$$v_p = 3 \times 1 \sqrt{1 - \left(\frac{5}{14}\right)^2} = 3.2 \times 10^8 \text{ m/sec}$$

and $\quad v_g = \frac{c}{\sqrt{1 - \left(\frac{\lambda}{\lambda_c}\right)^2}} = 1.07 \times 10^8$ m/sec

Example 3. A TE_{10} mode of 10 GHz is propagated in an air filled rectangular wave guide. The magnetic field in z direction is given by,

$$H_z = H_{0z} \cos\left(\frac{\pi x}{\sqrt{6}}\right).\ \cos\left(\frac{\pi y}{b}\right).\ \text{A/m}$$

The phase constant is β = 1.0475 rad/cm. The quantities x and y expressed in centimetre and $a = b = \sqrt{6}$ are also in cm. Determine the cut off frequency f_c. Phase velocity v_p. guided wavelength λ_g and magnetic field increasing in y direction.

Solution: Since general expression of magnetic field in z direction is given by

$$H_z = H_{0x} \cos\left(\frac{m\pi}{a}\right)x \cos\left(\frac{n\pi}{b}\right)y$$

By comparing

$$H_z = H_{0x} \cos \frac{\pi x}{\sqrt{6}}.\ \cos\frac{\pi y}{\sqrt{6}}$$

$$m = 1, n = 1, a = \sqrt{6}, b = \sqrt{6}$$

So cut off frequency

$$f_c = \frac{c}{2}\sqrt{\left(\frac{m}{a}\right)^2 + \left(\frac{n}{b}\right)^2}$$

$$= \frac{3\times 10^8}{2}\sqrt{\left(\frac{1}{\sqrt{6}}\right)^2 + \left(\frac{1}{\sqrt{6}}\right)^2}$$

$$= \frac{3\times 10^8}{2}\sqrt{\frac{1}{6}+\frac{1}{6}}$$

$$= \frac{3\times 10^8}{3}\frac{\sqrt{1}}{3}$$

$$= \frac{3\times 10^8}{2} \times \frac{1}{\sqrt{3}} = \frac{3\times 10^8}{2\times 1.73} = 8.66 \text{ GHz}$$

$$v_P = \frac{\omega}{\beta_g} = \frac{2\pi_f}{\beta_g} = \frac{2\pi\times 10\times 10^9}{1.0475\times 10^2} = 5.99 \times 10^8 \text{ m/sec}$$

$$\text{Phase velocity} = v_p = \frac{c}{c\sqrt{1-\left(\frac{f_c}{f}\right)^2}} = \frac{3\times 10^8}{\sqrt{1-\left(\frac{8.60}{10}\right)^2}}$$

$$= 3 \times 108 / \sqrt{1-0.74996} = \frac{3\times 10^8}{0.500043998}$$

$$= 5 \times 10^8 \times 0.000009398 = 1.5 \times 10^8 \text{ m/sec}$$

$$\lambda_g = \frac{\lambda}{\sqrt{1-\left(\frac{f_c}{f}\right)^2}}$$

$$= \frac{\frac{3\times 10^8}{10\times 10^9}}{\sqrt{1-\left(\frac{8.66}{10}\right)^2}}$$

$$= \frac{3\times10^8\times10^{-10}}{\sqrt{1-0.712336}} = 3\times10^{-2}\times1.86$$

$$= 5.58 \text{ cm}$$

The magnetic field intensity in *y*-direction is given as

$$H_y = \left(-\frac{\gamma}{h^2}\right)\frac{\partial H_z}{\partial y} - \frac{j\omega\varepsilon}{h^2}\cdot\frac{\partial E_z}{\partial x}$$

for *TE* mode $E_z = 0$

$$H_y = -\left(\frac{\gamma}{h^2}\right)\cdot\frac{\partial H_z}{\partial y} = \left(-\frac{\gamma}{h^2}\right)\cdot\frac{\partial}{\partial y}\left[H_{0z}\cos\frac{\pi x}{\sqrt{6}}\cos\frac{\pi y}{\sqrt{6}}\right]$$

$$= \left(-\frac{\gamma}{h^2}\right)H_{0z}\cos\frac{\pi\eta}{\sqrt{6}}\cdot\sin\frac{\pi y}{\sqrt{6}}\times\frac{\pi}{\sqrt{6}}$$

$$= \left(-\frac{\gamma}{h^2}\right)H_{0z}\frac{\pi}{\sqrt{6}}\cdot\frac{\cos\pi x}{\sqrt{6}}\cdot\sin\sqrt{\frac{\pi y}{\sqrt{6}}}\text{ A/m}$$

$$H_y = H_{oy}\cos\frac{\pi x}{\sqrt{6}}\sin\frac{\pi y}{\sqrt{6}} \Rightarrow H_y = H_{0y}\cos\frac{\pi x}{\sqrt{6}}\cdot\sin\frac{\pi y}{\sqrt{6}}$$

where $H_{0y} = \left(-\frac{\gamma}{h^2}\frac{\pi}{\sqrt{6}}H_{oz}\right)$

Example 4. A rectangular waveguide is designed to propagate in the dominant mode TE_{10} at a frequency of 5 GHz. The cut off frequency of the signal frequency is 0.8 time operating frequency. The ratio of height to width is 2 the find the dimensions: '*a*' and '*b*'.

Solution: Given frequency $f = 5$ GHz

so f_c = cut off frequency = 0.8 f

$$= 0.8\times5 = 4.0 \text{ GHz}$$

As $\frac{a}{b} = 2$,

$$f_c = \frac{c}{2}\sqrt{\left(\frac{m}{a}\right)^2+\left(\frac{n}{b}\right)^2}$$

$$4\times109 = \frac{3\times10^8}{2}\sqrt{\left(\frac{1}{a}\right)^2+\left(\frac{0}{b}\right)^2}$$

$$4\times10^9 = \frac{3\times10^8}{2}\sqrt{\frac{1}{a}} = \frac{3\times10^8}{2a}$$

$$a = \frac{3\times10^8}{4\times2\times10^9} = \frac{3}{8}\times10^{-1}$$

$$= 0.375\times10^{-1} = 3.75 \text{ cm}$$

$$\frac{a}{b} = 2$$

$$b = \frac{a}{2} = \frac{3.75}{2} = 1.875 \text{ cm}$$

$$\beta = \omega\sqrt{\mu\varepsilon}\sqrt{1-\left(\frac{f_c}{f}\right)^2} = \frac{2\times3.14\times5\times10^9}{3\times10^{10}}\left[1-\left(\frac{4}{5}\right)^2\right]^{1/2}$$

$$= 1.046 \times 0.6 = 0.06 \text{ rad/m}$$

Example 5. The inner readies of a circular wave-guide in 2 cm, the find its cut-off frequency, and band width for operation in its dominant mode only.

Solution: We know that

$$f_c = \frac{1.841}{2\pi}\left(\frac{c}{a}\right)$$

$$= \frac{1.841 \times 3. \times 10^{10}}{2 \times 3.14 \times 2} = 4.398 \text{ GHz}$$

Bandwidth $= f_c$ of next higher mode of $TM_{01} - f_c$ of TE_{11}

$$= \frac{2.405}{2\pi}\left(\frac{c}{a}\right) - 4.398$$

$$= \frac{2.405 \times 3 \times 10^{10}}{2 \times 3.14 \times 2} - 4.398$$

$$= 5.745 - 4.398$$

$$= 1.35 \text{ GHz}$$

Example 6. Calculate the cut-off frequency for first four modes of WR 284 wave-guide, having dimension: $a = 2.840$ mils and $b = 1.340$ mils. Also find Z_g for TE mode, if operating frequency is 2.2 GHz. Find the band width to be operated in dominant mode only.

Solution: We know that,

$$f_c = \frac{c}{2}\left[\left(\frac{m}{a}\right)^2 + \left(\frac{n}{b}\right)^2\right]^{1/2}$$

Scal $\quad$ 1 inch = 1000 mils

1 inch = 2.54 cm.

Hence, $\quad a = 7.214$ cm and $b = 3.404$ cm

where, $m, n = 0, 1, 2, ...3....$

Combinations of modes will be: TE_{10}, TE_{01}, TE_{11} and TE_{02} and TE_{20}, and corresponding frequency will be:

$$f_c(TE_{10}) = \frac{c}{2a} = \frac{3 \times 10^{10}}{2 \times 7.241} = 2.0793 \times 10^9 = 2.079 \text{ GHz}$$

$$f_c(TE_{01}) = \frac{c}{2b} = \frac{3 \times 10^{10}}{2 \times 3.404} = 4.407 \times 10^9 = 4.407 \text{ GHz}$$

$$f_c(TE_{11}) = \frac{c}{a}\left[\left(\frac{1}{a}^2\right) + \left(\frac{1}{b}\right)^2\right]^{1/2}$$

$$= 1.5 \times 10^{10}\left[\frac{1}{(7.214)^2} + \frac{1}{(3.404)^2}\right] = 4.87 \text{ GHz}$$

$$f_c(TE_{20}) = \frac{c}{a} = \frac{3 \times 10^{10}}{7.214} = 4.158 \times 10^9 = 4.158 \text{ GHz}$$

$$f_c(TE_{02}) = \frac{c}{b} = \frac{3 \times 10^{10}}{3.404} = 8.813 \times 10^9 = 8.813 \text{ GHz}$$

Therefore, lowest cut-off frequency of guide is 2.079 GHz.

Then $$\lambda_c = 2a = 2 \times 7.214 = 14.428 \text{ cm}$$

$$\lambda = \frac{3 \times 10^{10}}{2.2 \times 109} = 13.64 \text{ cm}$$

Band width $$\Delta f = 4.158 - 0.079 = 2.079 \text{ GHz}$$

$$Z_g = \frac{\eta}{\sqrt{1 - \left(\frac{\lambda}{\lambda\varepsilon}\right)^2}} \frac{377}{\sqrt{1 - \left(\frac{13.64}{14.428}\right)^2}} = \frac{377}{0.325} = 1160\ \Omega$$

Since operating frequency is 2.2 GHz, so only TE_{10} mode or $f_c = 2.09$ GHz will be supported in the guide.

Example 7. Find the wave-impedance of a 30 cm long short-circuital wave-guide (WR90) operating at 8.5 GHz.

Solution: For WR 90 wave-guide, 'a' and 'b' is found to be

$$a = 2.286 \text{ cm} \quad \text{and} \quad b = 1.143 \text{ cm.}$$

So, $$f_c(= TE_{10}) = \frac{c}{2a} = \frac{3 \times 10^{10}}{2 \times 2.286} = 6.56 \text{ GHz}$$

Since wave-guide is shorted at $l = 30$ cm. Its

$$Z_{\text{in}} = jZ_{10}^{TE} \tan(\beta l)$$

where $$\beta = \text{propagation constant}$$

$$= \frac{2\pi f}{c}\sqrt{1 - \left(\frac{f_c}{f}\right)^2}$$

$$= \frac{2 \times 3.14 \times 8.5 \times 10^9}{3 \times 10}\left[1 - \left(\frac{6.56}{8.5}\right)^2\right]^{11/2}$$

$$= 177.933 \times 0.623 = 110.85 \text{ rad/m}$$

$$Z_{10}^{TE} = \frac{\eta}{\sqrt{1 - \left(\frac{f_c}{f}\right)^2}} = \frac{377}{\sqrt{1 - \left(\frac{6.56}{8.5}\right)^2}} = \frac{377}{0.623} = 605.14\ \Omega$$

Hence, $$Z_{\text{in}} = (605.14) \tan(110.85 \times 0.3)$$

$$= j\,396\ \Omega$$

Phase-velocity, $$u_p = \frac{c}{\sqrt{1 - \left(\frac{f_c}{f}\right)^2}} = \frac{3 \times 10^8}{0.623} = 4.82 \times 10^8 \text{ m/s}$$

Group-velocity $$u_g = c\left[1 - \left(\frac{f_c}{f}\right)^2\right]^{1/2}$$

$$= 3 \times 10^8 \times 0.623 = 1.87 \times 10^8 \text{ m/s}$$

Example 8. A dielectric ($\varepsilon_r = 2.2$) R-wave-guide (copper) with $a = 2.28$ cm and $b = 1.02$ cm is operated at 10 GHz in dominant mode. Find total at attenecation loss if its length is 50 cm (TE_{10} mode)

Solution: Surface resistance $R_s = \sqrt{\dfrac{\pi f \mu}{\sigma}}$

and
$$f_c = \frac{c}{2a} = \frac{30}{2 \times 28} = 6.58 \text{ GHz}$$

$$= \sqrt{\frac{3.14 \times 10 \times 10^9 \times 4 \times 3.14 \times 10^{-7}}{5.8 \times 107}}$$

$$= 2.6 \times 10^{-2}\ \Omega$$

$$\alpha_c = \frac{R_s\left[1 + \left(\frac{2b}{a}\right)\left(\frac{f_c}{f}\right)^2\right]}{b\,\eta_0 \frac{1}{\sqrt{\varepsilon_r}}\left[1 - \left(\frac{f_c}{f}\right)^2\right]^{1/2}}$$

$$= \frac{2.6 \times 10^{-2}\left[1 + \left(\frac{2 \times 1.02}{2.28}\right)\left(\frac{6.58}{10}\right)^2\right]}{1.02 \times 377 \frac{1}{\sqrt{2.2}}\left[1 - \left(\frac{6.58}{10}\right)^2\right]^{1/2}}$$

$$= \frac{4.13 \times 10^{-2}}{195.23} = 2.115 \times 10^{-4} \text{ Np/m.}$$

$$\alpha_c \text{ (total)} = \alpha_c \times l = 2.115 \times 50 \times 10^{-4}$$

$$= 105.75 \times 10^{-4} \text{ Np/m}$$

$$= 105.75 \times 8.686 = 918.55 \times 10^{-4} \text{ dB}$$

$$\alpha_c \text{ (Total)} = 0.092 \text{ dB}$$

Example 9. Derive the expression for time-average power flow in R-wave-guide in a loss-less wave-guide for its dominant mode.

Solution: We know that, for TE_{10} mode:

$$P_{10} = \frac{Z_g(TE_{10})}{2}\int_b^a \int_0^b |H_x|^2 .\, dx.\, dy$$

where
$$H_z = \left(\frac{j\gamma}{\omega\mu}\right) E_0 \sin\left(\frac{\pi x}{a}\right) e^{-rz}$$

Hence
$$P_{10} = \frac{Z_g(TE_{10})}{2}\left|\frac{\gamma E_0}{\omega\mu} e^{-rz}\right|^2 b \int_0^a \sin^2\left(\frac{\pi x}{a}\right).\, dx.\, dy$$

$$= \frac{Z_y(TE_{10})}{4}\left|\frac{r E_0}{\omega\mu} e^{-rz}\right|^2 a.b$$

Since
$$Z_g(TE_{10}) = \frac{\omega\mu}{\gamma}$$

$$P_{10} = \frac{\omega\mu}{4\gamma}\left|\frac{\gamma E_0}{\omega\mu} e^{-\gamma z}\right|^2 (a.b)$$

Again for loss-less medium ($\alpha = 0$)

$$\gamma = j\beta = j.\frac{2\pi}{\lambda_g}$$

$$P_{10} = \frac{\omega\mu}{4\beta}\left(\frac{\beta E_0}{\omega\mu}\right)^2 a.b = \frac{\beta}{4\omega\mu} a.b\, E_0^2$$

or $$P_{10} = \frac{1}{4\eta} F.E_0 A$$

where $$A = a.b$$

$$F = \left[1-\left(\frac{f_c}{f}\right)^2\right]^{1/2}$$

Also find it value for $a = 2.2$ cm, $b = 1.02$ cm, $f = 7.5$ and $E_0 = 2.15$ kV.

$$f_c = \frac{c}{2a} = \frac{3\times10^{10}}{2\times2.2} = 6.82 \text{ GHz}$$

$$F = \left[1-\left(\frac{6.82}{7.5}\right)^2\right]^{1/2} = 0.42$$

So, $$P_{10} = \frac{1}{4\times377}\times0.42\times2.15\times10^3\times2.2\times1.02$$

$$= \frac{2026.332}{1508} = 1.343 \text{ W}$$

Also, find

$$\lambda_g = \frac{\lambda}{F} = \frac{4}{0.42} = 9.52 \text{ cm}$$

$$Z_g = 337\times0.42 = 158.34\ \Omega$$

and $$u_p = \frac{3\times10^8}{0.42} = 7.14\times10^8 \text{ m/s}$$

Example 10. Define Dominant Mode, and draw it field pattern

Solution: The dominant mode of any waveguide is that of the lowest cutoff frequency. Now, for a rectangular guide, the coordinate system may always be oriented to make $a \geq b$. Dominant mode of a rectangular guide is invariably TE_{10}, with

$$H_{x10} = j\left(\frac{2a}{\lambda_{10}}\right)H_{10}\sin\frac{\pi x}{a} \text{ and } E_{y10} = -\eta_{10}H_{x10} = -j\eta_0\left(\frac{2a}{\lambda_0}\right)H_{10}\sin\frac{\pi x}{a}$$

Both $|E_y|$ and $|H_z|$ vary as sin ($\pi x/a$). This is indicated in Fig. 9.14 by drawing the lines of E close together near $x = (a/2)$ and far apart near $x = 0$ and $x = a$. The lines of H are shown evenly spaced because there is no variation with y. This same line-density convention is used to indicate the local value of $|E| = |E_y|$ in Fig. 9.15 (a) and of $|H| = \sqrt{H_x^2 + H_z^2}$ in Fig. 9.15 (b). Since the lines of H are closed curves (div $H = 0$); the H field may be considered as circulating one about the perpendicular displacement current density J_d.

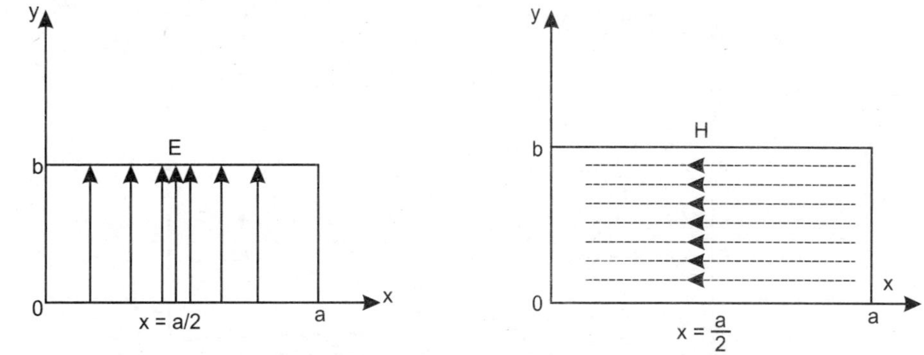

Fig. 9.14. Transverse cross section

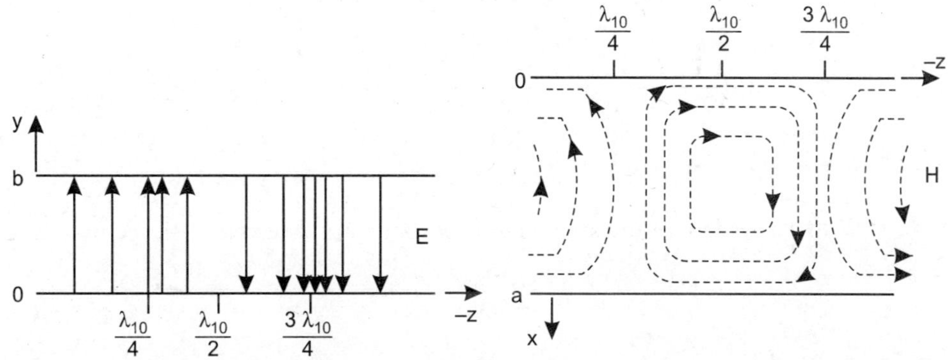

(a) x = a/2 (b) y = cont

Fig. 9.15. Longitudinal cross sections

Example 11. Sketch the variation of wave impedances, attenuation constant and wave velocities with frequency.

Solution: These variations are shown in Fig. 9.16 (*a* and *b*).

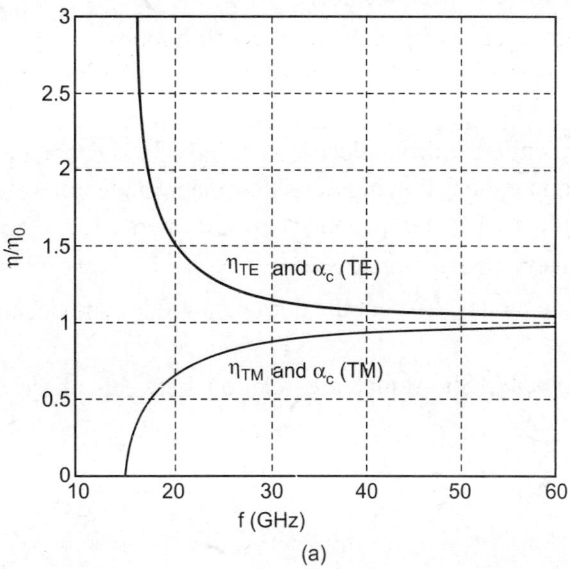

(a)

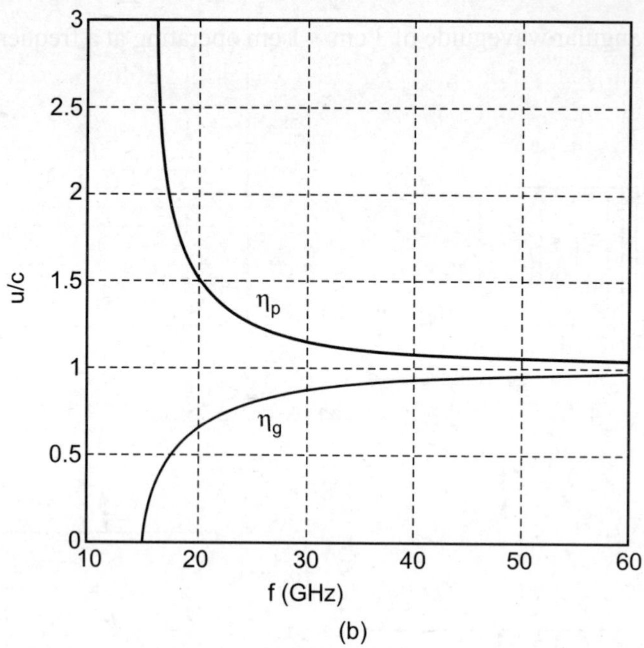

Fig. 9.16. (*a*) Wave impedances and attenuation constant (*b*) Wave velocities versus frequency

EXERCISES

1. Derive the expressions for attenuation and phase constants of a circular waveguide.
2. Derive the expressions of TE field components of a R-waveguide
3. Find all the possible modes that can propagate in a rectangular waveguide of dimensions: 5.2 and 2.5 cm, at operating frequency of 5 GHz.
4. A C-waveguide is filled with a dielectric material of $\varepsilon_r = 7$, and its inner radius is 5 cm. Then compute the following in its dominant mode: group velocity, phase constant, guide wavelength and wave impedance (Z_{TM})
5. A rectangular air filled copper waveguide with dimensions $a = 5.5$ cm, $b = 2.5$ cm and $l = 30.48°$ cm, operating at $f = 10.5$ GHz in its dominant mode. Determine its total attenuation. ($\sigma = 5.8 \times 10^7$ s/m)
6. A R-waveguide is filled with a dielectric material of $\varepsilon_r = 6$, and dimensions are 7.5 cm and 4 cm. Then compute the following in its dominant mode: phase velocity, phase constant, guide wavelength and wave impedance (Z_{TE}), if operating $f = 2.5$ GHz.
7. Describe dominant and degenerate modes.
8. Define waveguide modes. List out the dominant modes and degenerate modes of R and C waveguides.
9. Define wave-velocity. Show that the relation between phase and group velocities is $v_p = \dfrac{v_g}{[1-(\lambda/2a)^2]}$

10. Given a rectangular waveguide of 3 cm × 1 cm operating at a frequency of 8 GHz in TE_{10} mode. Estimate maximum power handling capability of the waveguide, if the maximum potential gradient of the signal is 3.2 kV/cm.

11. Find breakdown field intensity of an air filled R-waveguide of dimensions 2.25 cm × 1.0 cm for the dominant mode at $f = 9.375$ GHz.

12. Which one is the dominant mode: TM_{00}, TM_{01}, TM_{01}, TM_{10} and TM_{11}, justify.

13. Design an air filled R-waveguide to be operated at 5 GHz and modify it to be operated at 10 GHz such that design frequency must be in the middle of operating band. If $\sigma = 5.8 \times 10^7$ S/m then calculate dimensions (a, b), a_c, and P_T assuming that E_m is one half of the dielectric strength of air.

- Happiness can exist only in acceptance.

George Orwell

Chapter **10**

PRINCIPLE OF RADIATION AND ANTENNAS

10.1 INTRODUCTION

The word radiation means "*Ray of light*" and used, in general, as a term to cover all forms of energy flow through space from one point to another. Radiation may be in form of a spray of subatomic particles or in form of EM waves, which include light itself as well as radio-waves. A group of electric charges is source of EM field, while if the electric charges are time varying it produce EM radiation. On the other hand basic principle of physics state that accelerating charges radiate energy i.e. radiation comes from time-varying currents. That is radiation is phenomenon that is related with time-varying current, so there is no radiation for non-time varying currents. Further as the frequency of current increases radiation also increases. Therefore the radiation of EM waves from the current sources can be predicted using Maxwell's equations under time varying conditions. This chapter will first describe basic theory of radiation, then antenna's parameters.

10.2 VECTOR & RETARDED VECTOR POTENTIALS

In the EM theory, it is difficult to deal with electric and magnetic fields directly except for simple plane waves. Instead one must define the *Scalar and Vector potential functions* related to electric and magnetic fields and find their solutions using Maxwell equations and hence finally electric and magnetic fields. Scalar potential is just same as voltage. For example 10 V battery has a scalar potentials difference of 10 V between its terminals. Electric field, in static conditions, in general is given by the spatial rate of change of scalar potential, and directed towards lower region of the potential however vector potential is related with magnetism. In static situations, magnetic field is related with the rate of change of the vector poetical in various directions. If the potentials are changing with time as in radiation, the relation between fields and potentials is complicated. But in either case, the electric and magnetic fields are proportional to the rates of change of the potential in space and time. If there is a moving point charge somewhere on the surface of conductor, then scalar potential at some other point will be directly proportional to value of charge, and inversely proportional to the distance of the charge (Fig 10. a). The vector potential is related exactly same way to current and distance.

Therefore vector and scalar potentials they may be expressed

$$V(r) = \frac{1}{4\pi\varepsilon}\int \frac{\rho(r')}{R} dV' \quad \text{and} \quad A(r) = \frac{\mu}{4\pi}\int \frac{J(r')}{R} dV' \qquad ...(10.1)$$

Advanced theories such as relativity and quantum mechanics use the vector potential and scalar potential instead of electric and magnetic fields

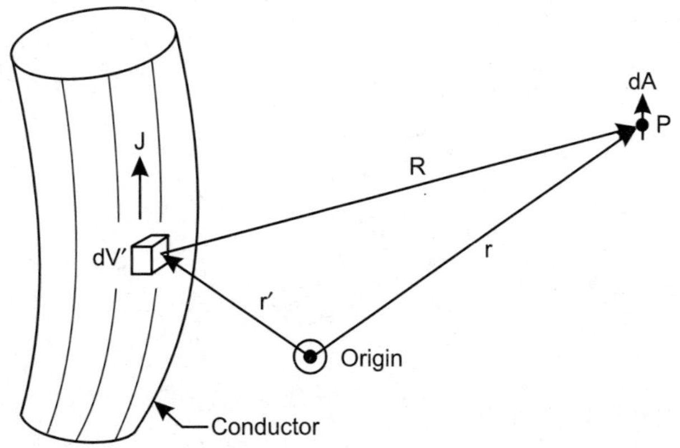

Fig 10.1 (*a*) Determination of scalar and vector potentials

We know that the Maxwell's equations for the time-varying fields are;

$$\nabla \times H = J + J_c = \sigma E + \varepsilon \frac{\partial E}{\partial t} \qquad ...(10.2a)$$

$$\nabla \times H = \frac{-\partial B}{\partial t} = -\mu \frac{\partial H}{\partial t} \qquad ...(10.2b)$$

$$\nabla . E = \frac{\rho}{\varepsilon} \qquad ...(10.2c)$$

$$\nabla . B = 0 \qquad ...(10.2d)$$

Where 'J' and ρ are source current density and charge current density respectively.

Again from vector identity for a vector A;

$$\nabla . \nabla \times A = 0 \qquad ...(10.3a)$$

The comparison of (10.2*d*) and (10.3), yields;

$$B = \nabla \times A$$

$$\Rightarrow \quad H = \frac{1}{\mu} \nabla \times A \qquad ...(10.3b)$$

Then 'A' is termed as *magnetic vector potential*. Substituting (10.3*b*) into (10.2*b*) gives;

$$\nabla \times E = -\mu \left[\frac{\partial}{\partial t}(\nabla \times A)\right] = -\nabla \times \frac{\partial A}{\partial t}$$

$$\Rightarrow \quad \nabla \times \left(E + \frac{\partial A}{\partial t}\right) = 0 \qquad ...(10.4a)$$

Again $\nabla \times (\nabla V)$ from for all values of V, (where V is scalar function).

$$E + \frac{\partial A}{\partial t} = \nabla V \Rightarrow E = \frac{\partial A}{\partial t} - \nabla V \qquad ...(10.4b)$$

Then 'V' is called *electric scalar potential.* The (–ve) sign introduced in equation (10.4b) is basically to make it consistent with electrostatic theory, where as if the fields are non-time varying (i.e. $\frac{\partial A}{\partial t} = 0$), electric field ($E$) should be equal to ($-\nabla V$).

Hence substituting the value of 'E' into (10.2c) yields;

$$\nabla.\left(-\frac{\partial A}{\partial t} - \nabla V\right) = \frac{\rho}{\varepsilon}$$

$$\nabla.\left(\frac{\partial A}{\partial t}\right) + \nabla^2 V = -\frac{\rho}{\varepsilon} \quad ...(10.5a)$$

Similarly, substituting the value of 'H' from (10.3b) into (10.2a) yields;

$$\frac{1}{\mu}\nabla \times \nabla \times A = J + \varepsilon\frac{\partial}{\partial t}\left(-\frac{\partial A}{\partial t} - \nabla V\right)$$

Using $\quad \nabla \times \nabla \times \text{A} = \nabla(\nabla.A) - \nabla^2 A$

We get

$$\nabla(\nabla.A) - \nabla^2 A = \mu J + \mu\varepsilon\frac{\partial A}{\partial t}\left(-\frac{\partial A}{\partial t} - \nabla V\right) \quad ...(10.5b)$$

Equations (10.5a) and (10.5b) are complicated as they are coupled with unknown functions 'A' and 'V'. However, if divergence of A is set equal to ($-\mu\,\varepsilon\,V$), these equations become uncoupled with 'A' and 'V', and reduces to;

$$\nabla^2 A - \mu\varepsilon\frac{\partial^2 A}{\partial t^2} = \mu J \quad ...(10.6a)$$

$$\nabla^2 V - \mu\varepsilon\frac{\partial^2 V}{\partial t^2} = -\frac{\rho}{\varepsilon} \quad ...(10.6b)$$

And particular choice taken in these equations i.e.

$$\nabla.A = -\mu\varepsilon\dot{V} \quad ...(10.7)$$

is known as *Lorentz gauge condition*. It is clear that if the source are charge free, equations (10.6 a & b) reduces to general *wave equations*. Also for no-time varying potentials (i.e. $\frac{\partial A}{\partial t} = 0$ and $\frac{\partial V}{\partial t} = 0$), they reduces to static equations;

$$\nabla^2 A = \mu J \text{ and } \nabla^2 V = -\frac{\rho}{\varepsilon} \quad ...(10.8)$$

i.e. selection of divergence of A equal to ($-\mu\,\varepsilon\,\dot{V}$) is indeed correct. The close observation of (10.6 a & b) also reveals that electric scalar potential reduces to the charge/charge density, whereas magnetic vector potential to the current/current density only. As ∇^2 operator is scalar, for a given current density, the direction of vector potential everywhere in the space is same and along the direction of current. Whereas same behavior is not found for electric and magnetic fields, hence the use of scalar and vector potentials is preferred for solving the Maxwell's equations with sources.

Since source of the EM fields are current and charge distributions and they are time dependent, therefore scalar and vector potentials may be express in the time-variables form as follows;

$$V(r,t) = \frac{1}{4\pi\varepsilon}\int\frac{\rho(r'\,t)}{R}dV' \quad \text{and} \quad A(r,t) = \frac{\mu}{4\pi}\int\frac{J(r'\,t)}{R}dV' \quad ...(10.9)$$

In which. $R = |r - r'|$ Since electromagnetic messages travel at the speed of light, equation (10.9)

are not instantaneous states of the source distribution that matter, but rather it is condition at some earlier time, *called reference time*, when the message left. Since the message must travel a distance R to reach the destination, the delay (R/v) occurs. Therefore, potential functions modified to;

$$V(r, t) = \frac{1}{4\pi\varepsilon}\int \frac{\rho(r', t - R/v)}{R} dV'$$

and

$$A(r, t) = \frac{\mu}{4\pi}\int \frac{J(r', t - R/v)}{R} dV' \quad ...(10.10)$$

Because the integrals in (10.10) are evaluated at the retarded time, they are called *Retarded potentials,* and very useful to describe radiation of EM waves from the finite conductors/sources, in terms of E and H– fields.

Concept of radiation

As already mentioned accelerating charges and changing currents produce EM waves i.e. radiation. Once established, the EM waves in free-space propagate out to infinity, carrying energy with them that is radiation is irreversible flow of energy away from the source. In particular, if a spherical source radiating energy is tropically, the total power passing out through this surface will be integral of Poynting vector;

$$P(r) = \int S.dA = \frac{v}{4\pi}\int (E \times B)dA \quad ...(10.11)$$

That is power radiated in the limit of this quantity |P(r)| as 'r' goes to infinity i.e.

$$P_{rad} = P(r)_{Limit\ r \to \infty}$$

That is the energy per unit time that is transported to infinity, and never returned back. Since area of sphere is $4\pi r^2$. Hence so far radiations occur the Poynting vector S should decrease at large distance 'r' no faster than ($1/r^2$). If we combine Coulomb's and Biot-Savert laws for E and B, both have the same 'r' dependence behavior, then found that for static configuration, Poynting vector varies as; $S \to \frac{1}{r^4}$. Which means static source almost do not radiate. Therefore study of radiation involves picking out the parts of E and H that fall off like (1/r) at large distance from the source, constructing from them the term ($1/r^2$) in S and integrating over a large spherical surface under limit $r \to \infty$. In order to describe principle of radiation, further, the radiation mechanism of certain wire antennas are considered in the following section.

10.9 TYPES OF ANTENNAS

Based on performance and frequency of operation, antennas are divided into four groups-

(*i*) Electrically small Antennas

(*ii*) Resonant Antennas

(*iii*) Broad band Antennas

(*iv*) Aperture Antennas

Electrically small antennas are much less than operating wavelength. They are simple in structure and their properties are not sensitive to construction details. These antennas are used for VHF frequencies applications. The vertical monopole used for AM reception on cars is a best example. It is about 0.002 λ long and has nearly Omni directional pattern in horizontal plane. *Resonant antennas* are often used were simple structure with good input impedance over a narrow band of frequencies is needed. It has a broad main beam and low or moderate gain.

The λ/2 dipole, patch antenna and Yagi antenna are the popular examples.

There are many applications that require antenna which could operates over a wide frequency range, such antenna is known as *broadband antenna.* They have acceptable performance over a 2:1 bandwidth ratio of upper to lower operating frequency. Examples are log-periodic dipole, Yagi and spiral antennas. Since only portion of a broadband antenna is responsible for radiation at a given frequency, the gain is low. But it may be an advantage to have gain that is nearly constant, although low.

Aperture antenna is usually several wave lengths long in one or more dimensions. The pattern usually has a narrow main beam leading to high gain. Bandwidth is moderate as large as 2:1. Most useful aperture antenna is a horn antenna; it acts as a "funnel" directing waves into the connected wave-guide. Geometrical configuration and properties of above antennas are given listed in Fig. 10.9.

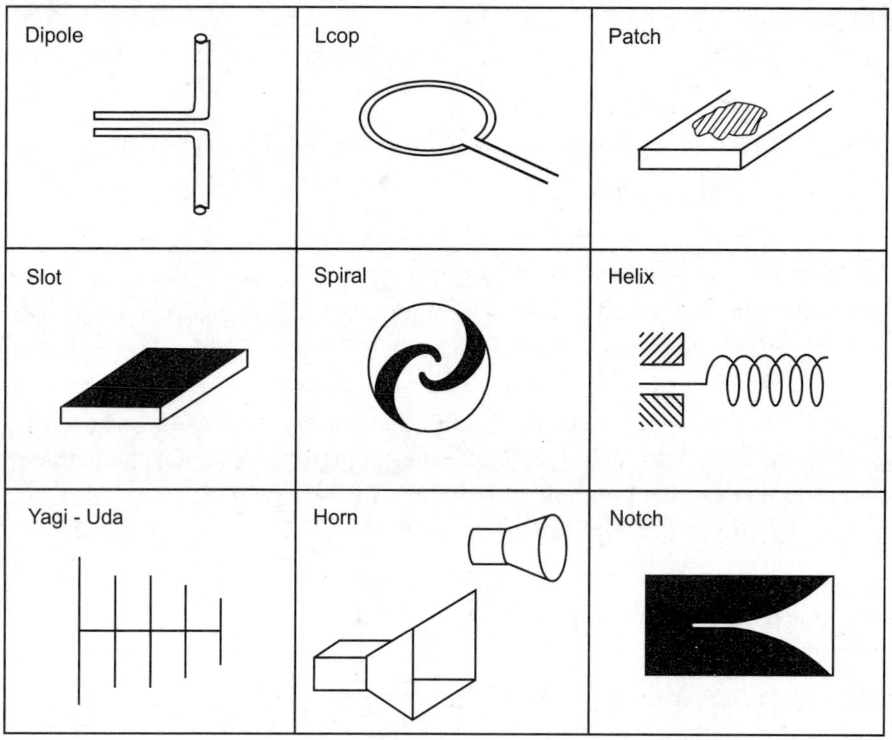

Fig. 10.9. (b) Various antenna configurations

10.3 HERTZIAN DIPOLE

In general, an infinitesimal current carrying element called as *Hertzian* dipole. Hertzian dipole is very useful to calculate the field of a large wire antenna, by considering it as a combination of a large number of Hertzian dipoles connected in series. In order to describe it , let us consider a Hertzian dipole of length dl (<< λ), which is located at the origin of a 3 D co-ordinate system, as shown in Fig 10.2.

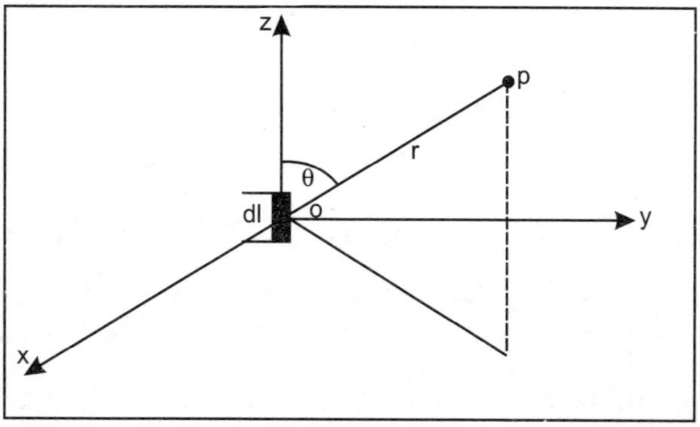

Fig. 10.2. Hertizian dipole in 3D co-ordinate system

As the diameter of dipole is very small than its length it is omitted in analysis. If the dipole carries a uniform constant current $I = I_0 \cos \omega t$, the retarded magnetic vector potential $\boldsymbol{A}$ at a far-field point P due to dipole is given by;

$$A = \frac{\mu [I] dl}{4\pi r} \hat{a}_z \qquad \text{...(10.12)}$$

Where $\hat{a}_z$ a unit vector as dipole is situated along the z-axis. $[I]$ is retarded current and corresponding retarded current density **[J]** is defined as

$$J = J_m e^{j(\omega t - \beta r)} \text{ A/m}^2$$

Since potential vector A is acting in z–direction, it will have only z-component i.e. A_z. Thus, we may write A in the phaser form as follows

$$A_z = \frac{\mu I_0 dl}{4\pi r} e^{-j\beta r}$$

Transforming this vector from Cartesian to Spherical co-ordinate and considering $A_x = 0$, $A_y = 0$ and $A_z = A_z$, gives, $A_r = A_z \cos \theta$, $A_\theta = -\sin \theta$ and $A\phi = 0$

We know that $\quad B = \mu H = \nabla \times \vec{A}$ (By definition)

Therefore the components of magnetic field (H) are found to be;

$$H_r = H_\theta = 0$$

$$H_\phi = \frac{I_0 \, dl}{4\pi} \sin \theta \left[\frac{j\beta}{r} + \frac{1}{r^2} \right] e^{-j\beta r} \qquad \text{...(10.13)}$$

Similarly by using Maxwell's equation,

$$\nabla \times H = \varepsilon \frac{\partial E}{\partial t} = j\varepsilon \, \omega \, E$$

We find electric field components and they are;

$$E_r = \frac{\eta I_0 dl}{2\pi} \cos\theta \left[\frac{1}{r^2} - \frac{j}{\beta r^3}\right] e^{-j\beta r} \quad ...(10.14a)$$

$$E_\theta = \frac{\eta I_0 dl}{4\pi} \sin\theta \left[\frac{j\beta}{r} + \frac{1}{r^2} - \frac{j}{\beta r^3}\right] e^{-j\beta r} \quad ...(10.14b)$$

and $$E_\phi = 0 \quad ...(10.14c)$$

Where $\eta = \sqrt{\frac{\mu}{\varepsilon}}$ and defined as characteristic impedance of medium surrounding the dipole. The close observation of expressions for magnetic and electric fields given in equations (10.13) and (10.14) reveals that there are terms varying as $\frac{1}{r^3}, \frac{1}{r^2}$ and $\frac{1}{r}$. The terms vary with $\left(\frac{1}{r^3}\right)$, known as electrostatic or simply electric field. Since it corresponds to the field of an electric dipole and dominates over other terms in a region very close to Hertzian dipole/Conductor. Both equations have components varying as $\frac{1}{r^2}$ and they are known as near field or induction field, and it is predictable from the Biot–Savart's low. Induction field is predominant at point near to dipole only and represents the energy stored in the magnetic field surrounding the current element. This energy is alternatively stored in the field and returned to the source current element during each half–cycle. As well as radiation is concerned this field is of little important and omitted in radiation field calculations.

The term varies as $\frac{1}{r}$ is known as *radiation/ distant or far-field*, because it is the only term that remains up to the far-zone. This field accounts for the radiation of EM waves from the conductor/ dipole. The radiation fields are of great significance at large distance from conductor/dipole. Since the components of the magnetic field are produced by the alternating electric fields and the electric field components arise from the alternating magnetic field and vice-versa; the flow of current establishes the local induction fields, whereas the radiation fields exist as sequence of the changing induction field. Just close to the conductor the magnetic field is in phase with the current; whereas the electric field varies in phase with the change on either side of the conductor. In far-field region, the magnetic and electric fields thus have a phase–difference of $\frac{\pi}{2}$ radian and are at the right angles to each other in space i.e. E_θ and H_ϕ are in phase in the far field region. Here we are concern mainly with far-field varying as $\frac{1}{r}$ i.e.

$$H_\phi = \frac{I_0 d\ell}{4\pi} \sin\theta [j\beta e^{-j\beta r}] \quad ...(10.15a)$$

and $$E_\theta = \frac{\eta I_0 d\ell}{4\pi} \sin\theta [j\beta e^{-j\beta r}] \quad ...(10.15b)$$

That is $\frac{E_\theta}{H_\phi} = \eta$ which is same as in the case of plane wave radiation.

The time–average power density, thus obtained

$$P_{av} = \frac{1}{2} R_e (E_\theta \times H_\phi) = \frac{1}{2} R_e (E_\theta . H_\phi^* a_r) = \frac{1}{2} \eta |H_\phi|^2 a_r$$

Therefore, the average radiated power is

$$P_{\text{av}} = \int P_{av}.dx = \int_0^{2\pi}\int_0^{\pi} \frac{(I_0 \beta dl \sin\theta)^2}{32\pi^2 r^2}\eta.r^2 \sin\theta\, d\theta\, d\phi$$

$$= \eta\frac{I_0^2 \beta^2 d\ell^2}{32\,\pi^2} 2\pi \int_0^{\pi} \sin^3\theta\, d\theta = \eta\,\frac{I_0^2\,\pi^2}{3}\left[\frac{d\ell}{\lambda}\right]^2$$

If medium surrounding the dipole is air $\eta = 120\pi$, we get

$$P_{\text{rad}} = \frac{I_0^2\,\pi \times 120\,\pi}{3}\left[\frac{d\ell}{\lambda}\right]^2 = 40\pi^2\,[d\ell_\lambda]^2\, I_0^2$$

This power may be considered as the power dissipated in a fictitious resistance R_{rad} by current I, which is equal to $I_0 \cos \omega t$.

That is $\quad P_{\text{rad}} = I_{rms}^2\, R_{rad} = \frac{1}{2}\, I_0^2\, R_{rad}$

$$\Rightarrow \quad R_{\text{rad}} = \frac{2 \times 40\,\pi^2 \left(\frac{d\ell}{\lambda}\right)^2 I_0^2}{I_0^2} = 80\,\pi^2 \left(\frac{d\ell}{\lambda}\right)^2 \qquad ...(10.15c)$$

Thus, antenna required large amount of radiation resistance to radiate large amount of power. As we have considered Hertzian dipole is very small, the radiation resistance R_{rad} is very small and no feeding lines of such low impedance are available to match the dipole. Also, since there is uniform current, so current be non-zero at the end points of the dipole, which is practically impossible because the surrounding medium is not conducting. These are limitations with Hertzian dipole.

10.4 HALF WAVE DIPOLE AND MONOPOLE ANTENNA

A half wave dipole is constructed from a conducting element which may wire or metal tube an electrical half wavelength long, and fed in the centre where the impedance falls to its lowest. This is because at the centre point the current is a maximum and the voltage is a minimum. This is much easier to feed as high RF voltages can present many problems for feeders and matching units. In this way, the antenna consists of the feeder connected to two quarter wavelength elements in line with each other. Half wave–dipole can be considered as the series combination of large numbers of Hertzian dipoles. As name indicates, it is dipole antenna of length of half-wave length i.e. $\lambda/2$ where λ is operating wave-length, hence fields due to $\lambda/2$-dipole antenna can be estimated easily by integrating the field of infinitesimal dipole under specified limit i.e. $\pm\,\lambda/4$. The basic half wave dipole antenna fed with two wire transmission line and its current distribution is shown in Fig 10.3 (a). Whereas the geometry of field calculation due to dipole, where an infinitesimal length $dl = dz$ of the dipole is located at distance z from the centre, is shown in Fig 10.3(b). The magnetic vector potential at observation point P due to '$dl = dz$' carrying current $I = I_0 \cos \omega t$ is given by

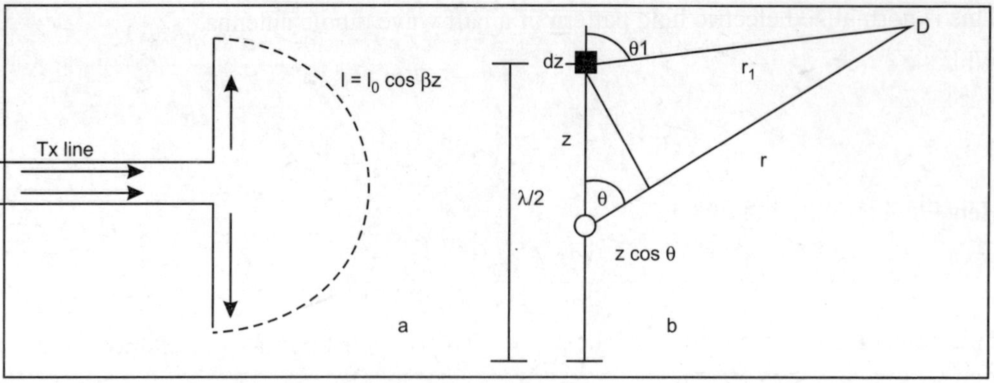

Fig 10.3 (*a*) Half-wave dipole and its current distribution (*b*) Geometry for field calculation

$$dA_z = \frac{\mu\, I_0 \cos \beta\, z.\, dz}{4\pi\, r_1} e^{-j\,\beta\, r_1} \quad \text{...(10.16)}$$

Where r_1 = distance of element from the point of observation P and $= r - z \cos\theta$.

As the values of r and r_1 are very large, one can replace r_1 by r. i.e. $r_1 \approx r$, but not the phaser terms $e^{-j\beta r_1}$ and $e^{-j\beta r}$, as there is a significant difference between the values of βr_1 and βr. So, let replace $r_1 = r - z \cos\theta$ in equation (10.16), and find out the total magnetic vector potential as follows;

$$A_z = \frac{\mu I_0}{4\pi r} \int_{-\lambda/4}^{\lambda.4} e^{-j\,\beta(r - z\cos\theta)} \cos \beta\, z.d$$

$$z = \frac{\mu I_0}{2\pi r} e^{-j\beta r} \int_0^{\pi/4} e^{j\beta z\cos\theta} \cos \beta z.d\, z$$

After simplification, this gives

$$A_z = \frac{\mu\, I_0\, e^{-jvr}}{(2\pi\, \beta r)} \frac{\cos\left(\frac{\pi}{2}\cos\theta\right)}{\sin^2\theta} \quad \text{for} \quad L = \lambda/2 \quad \text{...(10.17)}$$

Similar to Hertzian dipole case (Section 10.3),

Using $B = \mu\, H = \nabla \times A$ and Maxwell's equation $\nabla \times H = j\omega\varepsilon E$

The expressions for magnetic and electric fields come as follows;

$$H_\phi = \frac{JI_0\, e^{-j\,\beta\, r}}{22\pi\, r} \frac{\cos\left(\frac{\pi}{2}\cos\theta\right)}{\sin\theta} \quad \text{...(10.18a)}$$

and

$$E_\theta = \frac{j\eta I_0\, e^{-j\beta\, r}}{2\pi\, r} \frac{\cos\left(\frac{\pi}{2}\cos\theta\right)}{\sin\theta} \quad \text{...(10.18b)}$$

It is also seen that expressions of E_θ and H_ϕ indicate that they are in time phase and orthogonal. Also the ratio of E_θ and H_ϕ is constant and equal to η, known as characteristic impedance of medium. The term showing variation of field with angle θ. i.e. it determines the pattern of antenna is termed as *pattern factor*, and defined by

$$F(\theta) = \frac{\cos\left(\frac{\pi}{2}\cos\theta\right)}{\sin\theta} \quad \text{...(10.18c)}$$

This is normalized electric field pattern of a half wave dipole antenna.

With the similar procedure to Hertzian dipole, the time–average power density are found to be

$$P_{\text{rad}} = \frac{\eta I_0^2}{8\pi^2 r^2} \frac{\cos^2\left(\frac{\pi}{2}\cos\theta\right)}{\sin^2\theta} a_r$$

Hence the time–average radiated power can be determined as follows;

$$P_{\text{rad}} = \int P_{av}\, dx$$

$$= \int_0^{2\pi}\int_0^{\pi} \frac{\eta I_0^2 \cos^2\left(\frac{\pi}{2}\cos\theta\right)}{8\pi^2 r^2 \sin^2\theta} \times r^2 \sin^2\theta\, d\theta\, d\phi$$

After simplification, which gives;

$$P_{\text{rad}} = 36.56\, I_0^2$$

And since $P_{\text{rad}} = \frac{1}{2} I_{rms}^2 \times R_{rad}$, therefore,

R_{rad} is defined as $P_{\text{rad}} = \frac{2\, P_{rad}}{I_0^2} = 73\Omega$

Neglecting ohmic loss, that is $P_{\text{diss}} = 0$ and hence $R_l = 0$.

If length of Hertizian dipole is $\lambda/2$, we found $R_{\text{rad}} = 2\Omega$, which is very much less than 73Ω. Therefore $\lambda/2$ dipole antenna is capable of delivering greater amounts of power to space than the Hertzian dipole. As the total input impedance of an antenna is represented by $Z_{\text{in}} = R_{\text{in}} + J\, X_{in}$ (where R_{in} is resistance and X_{in} reactance), where $R_{\text{rad}} = R_{\text{in}}$ for a loss–less dipole antenna and reactance X_{in} is to be 42.5 Ω for a dipole of length $\lambda/2$, however X_{in} approaches zero, if the length of dipole reduced slightly. Typically, for $\lambda = 0.485\, \lambda$, X_{in} is zero and $Z_{\text{in}} = 3\ \Omega$. That why, in practice, dipole antenna designed maintaining its length lesser than the actual length. The reason behind the standard 75 Ω co–axial cable is only that the radiation resistance of $\lambda/2$ dipole antenna is 73 Ω.

A monopole antenna is half in the length of half wave–dipole antenna and generally fed by a co-axial cable connected to its base. The monopole antenna is perpendicular to the plane, which is usually assumed to be infinite and perfect conducting. Basic geometry and radiating characteristics of both the antennas are shown in (Fig. 10.4).Using the image theory one can replace monopole antenna as a half-wave dipole antenna. The simple structure of monopole coupled with unique properties such as a pure vertical polarization and horizontal Omni directional coverage, hence attract its extensive possible uses in verity of applications. Therefore, monopole antenna is an attractive option for broad band communications. Monopole antenna is one of the most widely used antennas throughout the RF spectrum ranging from VHF to UHF. Therefore the field produced by monopole antenna in the region above the ground plane with its image is same as the field due to $\lambda/2$ wave dipole i.e.

$$H_\phi = \frac{j\, I_0\, e^{-j\beta r}}{2\pi r} \frac{\cos\left(\frac{\pi}{2}\cos\theta\right)}{\sin\theta} \qquad \text{...(10.19}a\text{)}$$

$$E_\theta = \frac{\eta\, jI_0\, e^{-j\beta r}}{2\pi r} \frac{\cos\left(\frac{\pi}{2}\cos\theta\right)}{\sin\theta} \quad ...(10.19b)$$

However, monopole radiates only over the hemispherical surface above the ground plane i.e. $0 \le \theta \le \pi$. That is radiated power is half of power radiated from the dipole with same amount of current. Thus, for λ/4 monopole antenna the radiated power will be $P_{rad} = 18.28\, I_0^2$ and therefore $R_{rad} = \frac{73}{2} = 36.5\,\Omega$. The main difference is that λ/4 monopole radiates only in hemisphere surface whereas λ/2 dipole radiates more or less in all directions (Fig 10.4). The λ/2 dipole and λ/4 monopole antenna also called *Hertz antenna* and *Marconi antenna* respectively.

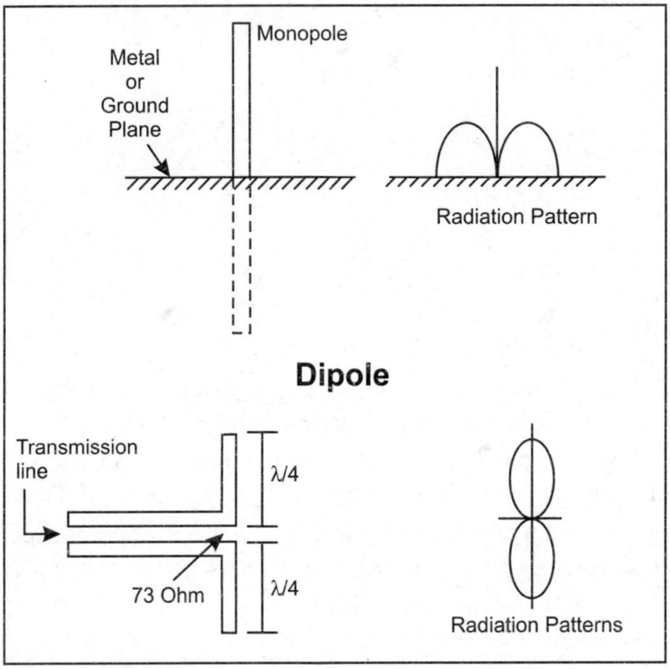

Fig 10.4. Radiation pattern of monopole and dipole antennas

In addition, a monopole antenna requires a physical ground plane while a dipole antenna uses an additionally radiator to generate a synthetic ground plane between the symmetric radiator elements. For a dipole antenna, the radiator elements are connected 180 degrees out-of-phase to each other (i.e. inner and outer conductors of a coaxial cable). For a monopole antenna, outer conductor of a coaxial cable is used as the ground plane. The size and design constraints for monopole antennas are often restrictive and the radiation pattern of the monopole depend on the orientation of the ground plane, however dipole antennas, can relatively easily be oriented in the direction of optimum reception/transmission.

10.5 HENTENNA

The Hentenna is a loop style antenna developed by Mr. Tadashi Okubo and Mr. Someya, in Japan in the 70's. In Japan, the word HEN means *'interesting, unusual, strange'* etc, so it is a strange antenna and geometrically it is an Asymmetrical Double Rectangle (ADR) Loop Antenna. It has two loop sections L_1 and L_2, as shown in Fig. 10.5*a*, the values of these sections are (λ/6) and (λ/12) respectively. The L_1 section is responsible for radiation while L_2 provides matching and helps to keep low angle radiation (10 to 13 degrees). In general both the ends of the feeds are connected with the 12 mm × 12 mm and 20 mm long aluminium solid blocks with screw fixing holes. A 1 mm brass road is used to connect the above block sections to the feed line. This antenna is not an Omni and it is a bidirectional type. The HEN Part (Strange or Interesting) is as follows; If this antenna is fixed horizontally then it will produce a vertical polarization. If it is fixed vertically the polarization will be in horizontal, "*The vertical slot is horizontally polarized and vice versa*". The radiation pattern is just like peanut shell (Fig. 10.5*c*).

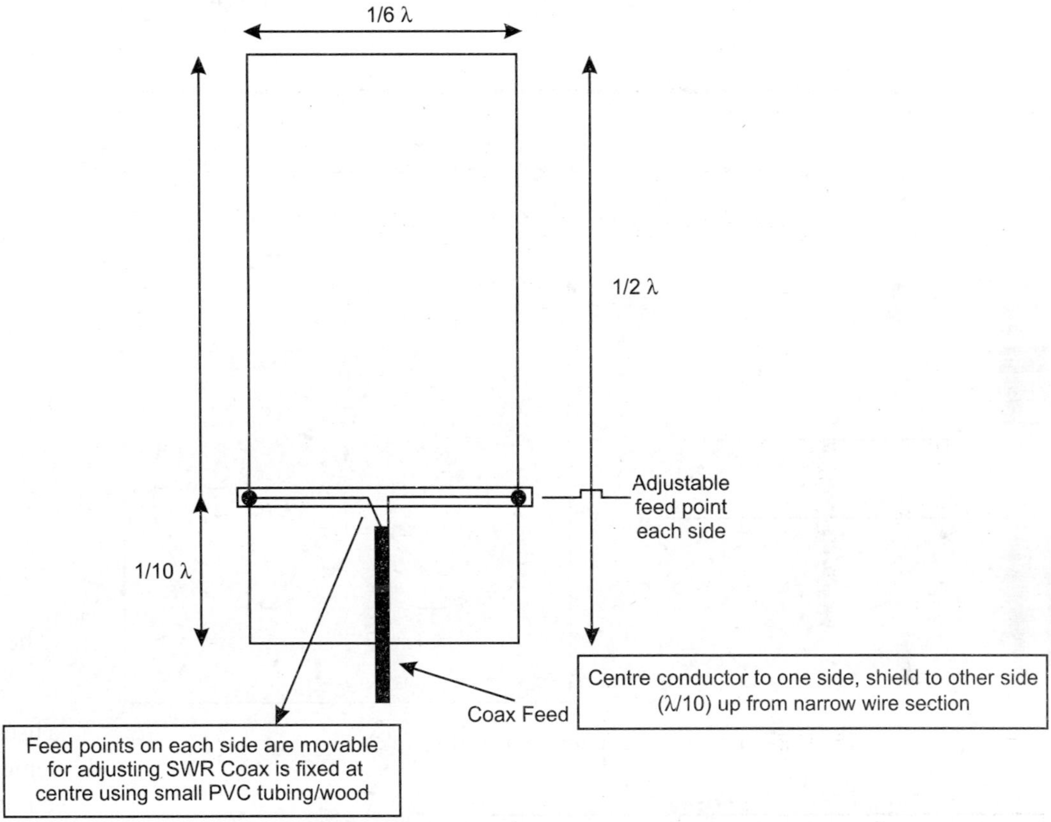

Fig 10.5. (*a*) Basic geometry of a hentenna

Advantages of the Hentenna

(*a*) Good performance and around 5.3 dBd gain, and Low angle radiation

(*b*) Total performance is equivalent to 2-3 element Yagi-uda antenna,

(*c*) Wide band width and easy to adjust impedance and SWR perfectly.

(*d*) It appears to be a vertical antenna but has mainly horizontal radiation, This is one of the reasons this antenna is "Hen", (STRANGE)

Hentenna system has more gain than an ordinary square loop and has the less impedance (nearby 70 ohms). Hentenna designed at 145 MHz (Fig 10.5 b), produces a 5.1 dBd gain and with a 1.3:1 SWR with an amazing band width of 6.5 MHz. However, this band width can be upto 10 MHz bandwidth with 2:1 SWR.

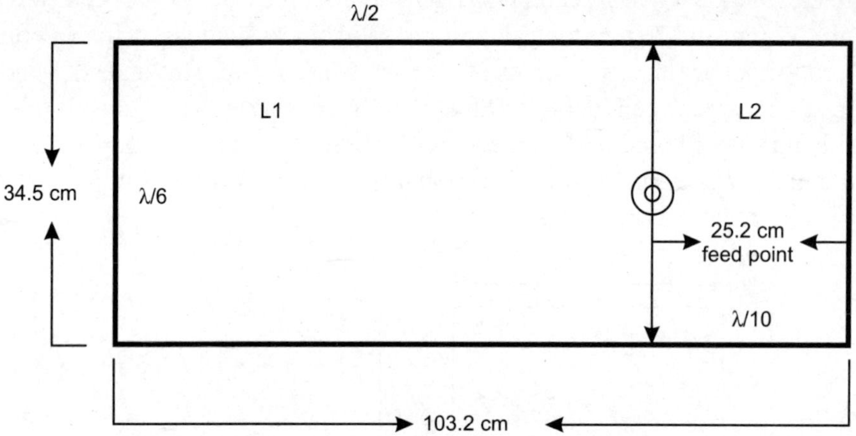

Fig 10.5. (*b*) Hentenna designed at 145 MHz

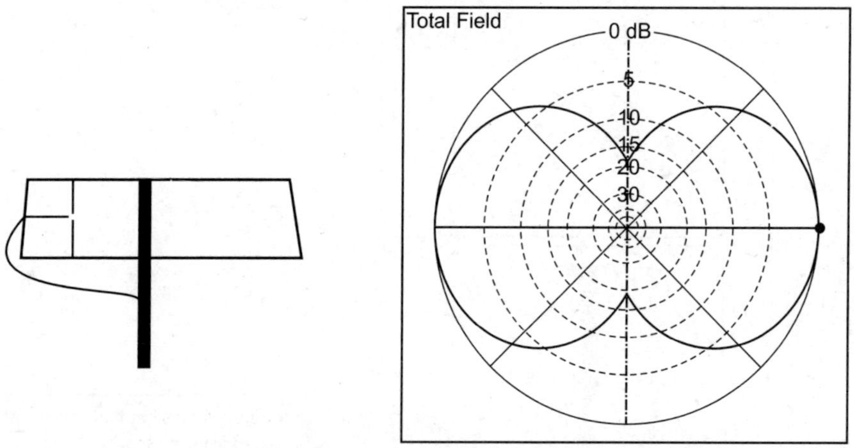

Fig 10.5. (*c*) A practical hentenna and its radiation pattern

10.6 BASIC ANTENNAS PARAMETERS

Isotropic radiator

If an isotropic radiator radiates power P_t, the power density P_d at a distance R from the source will be

$$P_d = \frac{P_t}{4\pi R^2} \, (W/m^2) \qquad ...(10.20)$$

Although the isotropic antenna is not practical, it is commonly used as a reference to compare other antennas performances. For a large value of R, the transmitted wave can be considered as uniform plane. The electric field (E) is perpendicular to direction of wave travelling and H is normal to E and $\bar{n}$ is unit rector along the direction of propagation. Both $\bar{E}$ and $\bar{H}$ lie in the constant-phase-plane and wave is a (TEM) wave and one can find the electric field at a distance R from isotropic antennas as follows;

$$E = \frac{\sqrt{60\,P_d}}{R} = \sqrt{2}\,E_{rms} \qquad \text{...(10.21)}$$

Where E is peak magnitude of electric field and E_{rms} is its *r.m.s* value.

Far-field region

Generally, it is assumed that the antenna operated properly in the far field region and radiation pattern are measured in this region only. In the far-field region, the transmitted wave of the transmitting antenna resembles a spherical wave from the point source that only locally resembles a uniform plane wave. And condition far field operation is found to be

$$R \geq \frac{2D^2}{\lambda_0} \qquad \text{...(10.22)}$$

Where λ_0 the free-space wavelength and D is maximum aperture / dimension of the antenna.

Bandwidth

The band width of antenna is usually defined as the frequency range within the performance of the antenna, with respect to certain characteristic, conforms to a specified standard particularly; the antenna gain and F/B ratio hold up. It is noticed that most of the antennas operate around there resonant frequency. This is because outside particular frequency range (BW) the levels of reactance goes high and deteriorates performance of the antenna. Bandwidth is expressed as the percentage of the difference between upper and lower frequency to the center frequency. In term of quality factor Q and VSWR, the bandwidth of an antenna also is defined as

$$\text{BW} = \frac{f_c}{Q}, \text{ and } BW = \frac{VSWR - 1}{Q\sqrt{VSWR}} \qquad \text{...(10.23)}$$

Radiation Resistance

Radiation resistance describes the relation between the total radiated energy from a transmitting antenna and the current flowing in it. It is fictitious parameter and represented by Radiation resistance acts as a load for the transmitter or for the radio-frequency Tx line connecting the transmitter and antenna. The total energy supplied to antenna is used in radiation as well as heating the antenna. Hence associated with each one there are resistances; *Radiation resistance* (R_r) *and ohmic resistance* (R_l).Therefore the total energy given to the antenna will be

$$P = P_r + P_l = I^2 R_r + I^2 R_l = I^2 (R_r + R_l) = I^2 R$$

Where $R = R_r + R_l$. Presence of ground significantly affects the radiation resistance, because the EM waves radiated from it are reflected from ground which induced current in the antenna while flowing through it. In general since reflected waves are weaker than incident waves, the fluctuation in radiation resistance decreases as height increases. Radiation resistance of wire antenna and rod-tubing antenna are found to be varied between 55 to 65 ohms, however for half-wave dipole it is 73 ohms.

Directivity, gain and efficiency

Directivity of an antenna in a given direction is defined as the ratio of the radiation intensity in that direction and the radiation intensity averaged over all directions. The maximum directivity D_{max} is defined as the value of the directive gain in the direction of its maximum values. The directive gain $D(\theta, \phi)$ over the power density radiated by an isotropic radiator fed by the same account of power is given by;

$$D(\theta, \phi) = \frac{P(\theta, \phi)}{P_T/4\pi} \quad ...(10.24a)$$

and

$$D_{max} = D_0 = \frac{P_{max}(\theta, \phi)}{P_T/4\pi} = 4\pi \frac{P_{max}(\theta, \phi)}{P_T} \quad ...(10.24b)$$

Where $P(\theta, \phi)$ = Poynting vector

and P_T = Total radiated power.

The directivity is a dimension less quantity. The maximum directivity is always ≥ 1. *Directivity of an isotropic source;* $D_0 = 1$. Thus, directivity other antenna will be greater than one.

The gain of an antenna is the directivity multiplied by the aperture or illumination efficiency of the antenna to radiate the energy presented at its terminal i.e.

$$G = \eta D_{max} \quad ...(10.25)$$

where η = Illumination efficiency.

The gain of an antenna also defined as the ratio of the radiation intensity P in a given direction and radiation intensity that would be obtained if the power fed to the antenna is radiated isotropically.

$$G(\theta, \phi) = 4\pi \frac{P(\theta, \phi)}{P_{in}} \quad ...(10.26)$$

Gain is calculated via the input power P_{in}, which is measurable quantity unlike directivity, which is calculated via the radiated power P_r. In general, the narrower the beam width the higher the gain of antenna.

The *efficiency* of antenna represents the fraction of total energy supplied to it, which is converted into free space EM waves. In general it is defined as the power radiated from antenna to total power supplied to the antenna i.e.

Antenna efficiency

$$(\eta) = \frac{\text{Radiated power}}{\text{Total in put power}} = \frac{P_{rad}}{P_{in}} = \frac{P_{rad}}{P_{rad} + P_{loss}} \quad ...(10.27a)$$

Where P_{rad} = Actual power radiated.

P_{in} = The power coupled to the antenna.

P_{loss} = Power lost in the antenna (conductor dielectric losses).

Sine, power is proportional to resistance ($P_o = I^2R$), Antenna efficiency also expressed as

$$\eta = \frac{R_r}{R_r + R_l} \quad \text{Where } R_r \text{ and } R_l \text{ are the radiation and loss resistances respectively} \qquad ...(10.27b)$$

The total efficiency of the antenna (η_T) represents the total loss of energy at the input terminals of the antenna and within the antenna structure.

FBR

FBR means forward to backward radiation ratio and defined as

$$\text{FBR} = \left(\frac{\text{Power radiated in frword direction}}{\text{Power radiated in Backword direction}}\right)$$

FBR is mostly observed in array/Yagi antenna. The FBR changes with frequency of operation and its value decreases with increasing the spacing between elements. Higher value of FBR is obtained at cost of gain of antenna; because diverting more radiation in opposite direction reduces the gain of antenna.

Radiation Pattern

The radiation pattern of antenna is the graphical representation of the radiation properties of the antenna as a function of space coordinates (Angles and distances). Radiation pattern generally describes the normalized field / power values with respect to the maximum values. The RP is measured at electrically large distance, where the spatial (Angular) distribution of the radiated power does not depend on the distance. At a large distance, the power density drops off as r^{-2} in any direction. The power pattern and field pattern are the same when they are computed and plotted in dB. However, both the patterns are related to each-other-*field pattern is square root of power pattern* i.e.

$$\frac{|\bar{E}(\theta, \phi)|}{\eta} \equiv \eta\,|H(\theta, \phi)| \propto \sqrt{|P(\theta, \phi)|} \qquad ...(10.28)$$

In the view of properties of EM waves, the antenna has E plane and H plane patterns each with co-and cross- polarizations. The E-plane pattern refers to the plane containing the electric field vector (E_θ) and the direction of maximum radiation with E_ϕ as cross polarization components. Similarly, the H-plane pattern contains the magnetic field vector (H_θ) and the direction of maximum radiation. The radiation patterns in 2D and 3 D, for an antenna to be located at the origin.

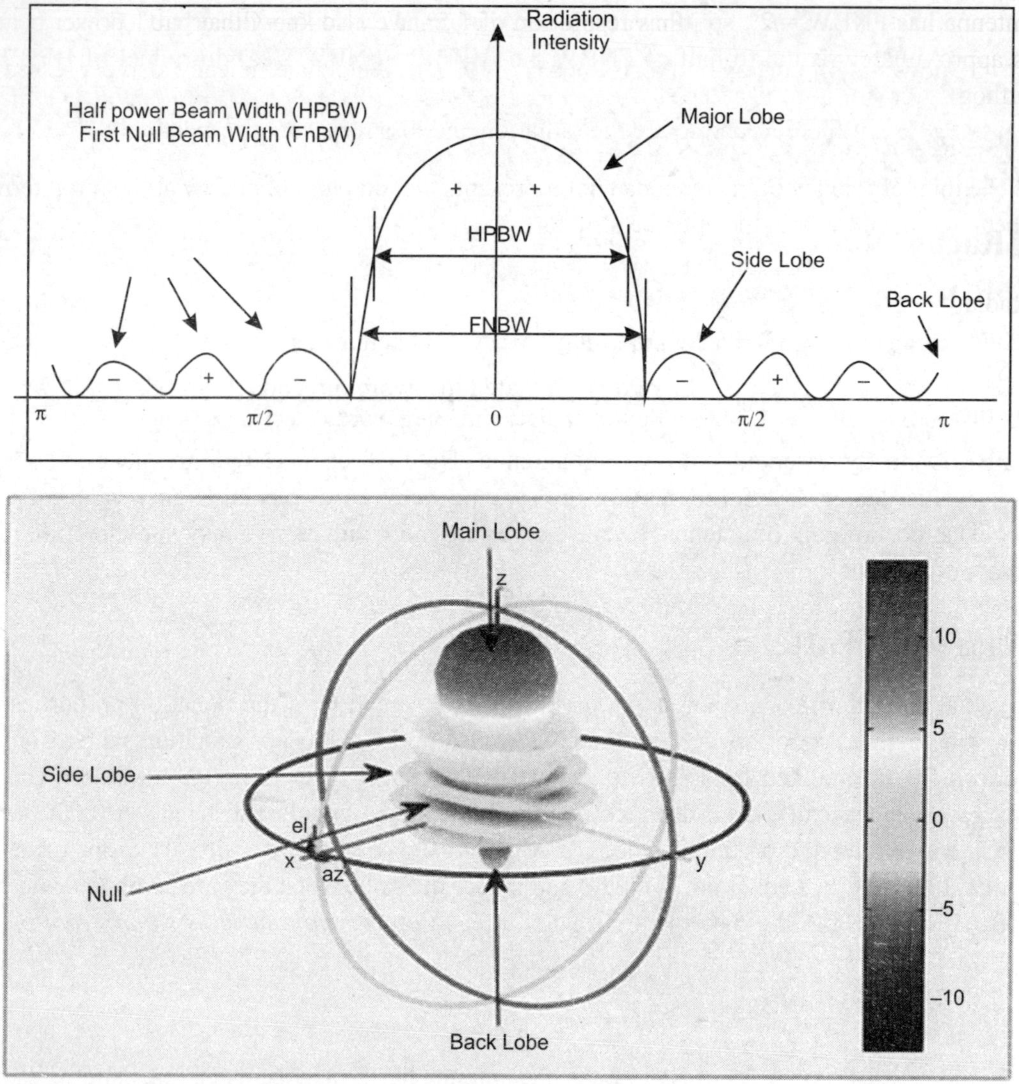

Fig. 10.6. 2 D and 3 D Radiation pattern of an antenna

HPBW, FNBW and Antenna resolution

In direction of the maximum beam, 'the angle between the two points at which the radiation intensity is half of that of maximum value' termed as HPBW. Usually it referred as 3-dB beam width and it is increases as side lobes decreases and vice-versa. The half-power beam width is approximately equal to half of FNBW. FNBW referred to first null beam width. The side lobes are power radiation peaks in addition to the main lobe. They are given as the number of decibels below the main lobe peak. In practice the test antenna is used as receiver and Tx antenna is placed in the far-field region of the test antenna and vice-versa, this is because radiated fields are plane waves in the vicinity of the AUT.

Resolution of an antenna is equal to half the first null beam width i.e. FNBW/2. That is if any antenna has FNBW = 2°, so it has resolution of 1.5°. We also know that half –power beam-width is approximately equal to half of FNBW. i.e. HPBW= FNBW/2. And product of HPBW in two orthogonal planes of the antenna pattern is referred as solid beam angle or antenna beam area i.e.

$$\Omega_A = (\theta_E, \theta_H)_{\text{HPBW}} = \left(\frac{FNBW}{2}\right)_E \left(\frac{FNBW}{2}\right)_H \quad ...(10.29)$$

Radiation Intensity

Radiation intensity is useful in calculating *gain* and *directivity* of an antenna. It is measured in $\frac{Watt}{S_r^2}$. Radiation Intensity in a given direction is defined as the power per unit solid angle radiated by the antenna in that direction i.e.

$$U = \frac{dP_r}{d\Omega}\left(\frac{W}{S_r}\right) \quad ...(10.30)$$

The radiation power density 'P'is the Poynting vector magnitude of the far-field i.e.

$$P = \frac{dP_r}{ds}\left(\frac{W}{m^3}\right) \quad ...(10.31)$$

There is radiation between radiation intensity and radiation power density.

$$U = r^2 \times P \quad ...(10.32)$$

The radiation intensity U is independent the distance from the source (r) and depends only on the direction (θ, ϕ) i.e. Angular positions. The radiation intensity can also be expressed in terms of the fields as;

$$U(\theta, \phi) = \frac{r^2}{2\eta}|\overline{E}|^2 = \frac{1}{2}r^2\eta|\overline{H}|^2 \quad ...(10.33)$$

Radiation intensity and radiation pattern of an isotropic radiator;

$$P(r, \theta.\phi) = \frac{P_r}{4\pi r^2}$$

$$U(\theta; \phi) = r^2 P = \frac{P_r}{4\pi} = cons\tan t$$

$$\Rightarrow \quad \overline{U}(\theta, \phi) = 1$$

Where $\overline{U}$ (θ, ϕ) is normalized radiation intensity.

Beam efficiency

It is defined as the ratio of the power radiated in a cone of angle $2\theta_1$ (say) and the total radiated power. The angle $2\theta_1$ can be generally any angle, but usually this is the first null bandwidth.

$$\text{Beam efficiency} = \frac{P_{\theta 1}}{P_T} = \frac{\int_0^{2\pi}\int_0^{\theta_1} U(\theta, \phi)\sin\theta.d\theta.d\phi}{\int_0^{2\pi}\int_0^{\pi} U(\theta, \phi)\sin\theta.d\theta.d\phi} \quad ...(10.34)$$

Equation (10.48) defines beam efficiency provided the antenna has its major lobe directed along the z axis ($\theta = 0$). If θ_1 is the angle where the FNBW's minima occurs in two orthogonal planes, beam efficiency will show only what part of the total radiated power is channeled through the main beam.

Effective aperture

It is the ratio of the available power at the terminals of the antenna to the power flux density of a plane wave incident upon the antenna, which is polarization matched to the antenna i.e.

$$A_e = \frac{P_A}{W_i} \qquad ...(10.35)$$

Where

P_A = The power delivered to the load from the antenna.

W_i = Power flux density of the incident wave.

If there is specific direction chosen, the direction of maximum radiation intensity is implied. For aperture types of antenna, the effective area is smaller than the physical aperture area. Aperture antennas (parabolic reflector and dish) with constant amplitude and phase distribution across the aperture have the maximum effective area, which is practically equal to the geometrical area. The effective aperture of the wire antenna is much larger than the surface area of the wire itself.

The effective area, beam solid angle and gain of an antenna is linked by the expression

$$A_e\Omega_A = \lambda^2 \quad \text{and} \quad G = \frac{4\pi A_e}{\lambda^2} \qquad ...(10.36)$$

Hence, Maximum effective area (A_{em}) can be given by

$A_{em} = \frac{D\lambda^2}{4\pi} = \frac{\lambda^2}{4\pi} = 0.077\,\lambda^2$, that is all the loss-less antenna must have $A_e \geq 0.0077\,\lambda^2$.

The aperture efficiency of an antenna is defined as ratio of the effective aperture area and its physical area i.e. $\eta_p = A_e/A_p$, where A_e effective and A_p physical areas of antenna.

Effective height

The effective height of antennas is another parameter, useful for transmitting tower type antennas. It is related to the aperture of antenna and applications for small antenna. It may be defined as the ratio of induced voltage to the incident field across the antenna i.e

$$h_e = \frac{V}{E} \text{ meter} \qquad ...(10.37)$$

In other way, the effective height of a transmitting antenna is to equate effective height/length multiplied by the average current. i.e.

$$h_e = \frac{1}{I}\int_0^h I(z)\,dz = \frac{I_{av}}{I_0} h \text{ meter} \qquad ...(10.38)$$

Where, h_e = effective height (m)

h = physical height (m)

I_{av} = Average current (A)

I_o = peak current (A)

Effective height can also be expressed as a vector quantity;

$$V = \overline{h_e}.\overline{E} = h_e.E \cos\theta \quad ...(10.39)$$

Where, θ = Angle between polarization angles of antenna and wave, *deg*

$\overline{h_e}$ = Effective height m

E = Field intensity of incident wave, V/m

Effective height of an antenna can also be linked with radiation resistance (R_r), efficiency (η) and effective aperture (A_e). We know, the power delivered to the load is

$$P = \frac{1}{4}\frac{V^2}{R_r} = \frac{h_e^2 E}{4R_r} \quad ...(10.40)$$

Which is also equal to

$$P = S.A_e = \frac{E^2}{\eta_0} A_e \quad ...(10.41)$$

Implies that

$$h_e = 2\sqrt{\frac{R_r A_e}{\eta}} \; (m) \quad ...(10.42)$$

Signal to Noise ratio

It is defined as the ratio of power received to system noise power i.e.

$$\text{SNR} = \frac{P_r}{P_N}$$

In which the system noise power 'P_n' (in watts) is related with system noise temperature 'T_s' as follows;

That is

$$\text{SNR} = \frac{P_r}{k.T_s.\Delta f} \quad ...(10.43a)$$

Both signal and noise power must be measured at the same /equivalent point in a system, and within the same bandwidth. If the signal and the noise are measured across the same impedances, then the SNR can also be calculated in terms of amplitudes;

$$\text{SNR} = \frac{P_r}{P_N} = \left(\frac{A_S}{A_N}\right) rms]$$

$$\Rightarrow \quad \text{SNR (dB)} = 10 \log\left(\frac{P_r}{P_N}\right) = 20 \log\left(\frac{A_S}{A_N}\right) \quad ...(10.43b)$$

Where A_s and A_N are the r.m.s values of amplitudes.

10.7 RECEIVING ANTENNA- *YAGI-UDA ANTENNA*

Yagi-Uda antenna is an array system of the closely spaced dipoles or similar elements. In this array system only one element is fed which is termed as driven element, while others are mutually coupled with the driven element. A reflector is placed back side and one/(more) directors are placed on the front side of the driven element. Yagi antennas are directional along the axis perpendicular to dipole in the plane of elements, from the reflector through the driven element and out via the director(s). In general, the length of elements of the antenna decreases 5 % from reflector to last director. The dual end fire beam characteristics of Yagi-antenna can be changed to more desirable single end fire beam by choosing suitable design parameters. The *E*-plane pattern of the antenna is equal to the *H*-plane pattern multiplied by array factor which is that of a half-wave dipole. Yagi-Uda Antenna with at least several directors is an end fire traveling wave antenna that support surface wave of the slow wave type $(c/v) > 1$. That is the driver-reflector pair launches a wave onto directors that shows the wave down such that the wave phase velocity v is less than that of velocity of light in free space (c). The phase delay per unit distance along the axis of the array in the forward direction is greater than that of the ordinary end fire condition. Simplest Yagi-Uda antenna with reflector length (0.49 λ), driver length (0.47 λ) and director length (0.45λ) each spaced only (0.04 λ apart) is shown in Fig. 11.4, where the radius of each wire is (0.001 λ).

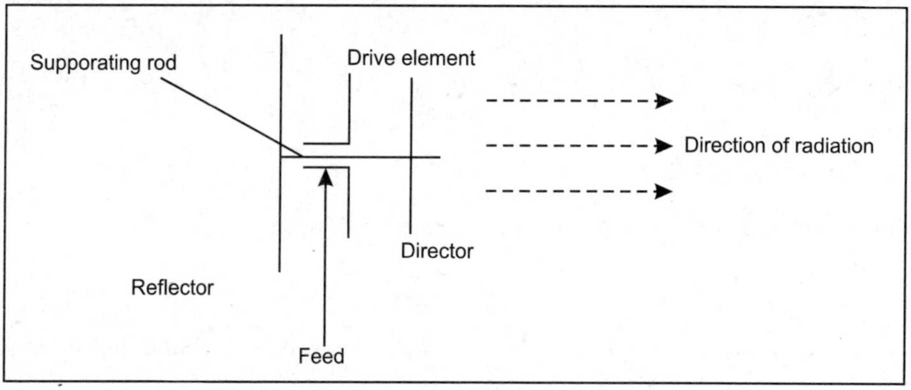

Fig. 10.7. Configuration of three elements Yagi-Uda Antenna

Design parameters

Basic Yagi-Uda antenna consists of three elements namely reflector, driver and director. Lengths of these elements are calculated as follows:

$$\text{Length of reflector} = \frac{500}{f_{MHz}} \text{ feet}$$

$$\text{Length driven element} = \frac{475}{f_{MHz}} \text{ feet}$$

$$\text{Length of director} = \frac{455}{f_{MHz}} \text{ feet}$$

Gain

Gain of Yagi antenna is related to its boom length, but for a parasitic array there is a smaller increase in gain per element as directors are added to the arrays. The addition of directors up to

5/6 only provides significant increase in gain. Adding one director in 3 elements (i.e. 4 elements Yagi) provides 1 dB gain enhancement particularly adding one director to increase from 9 to 11 elements in array increases 0.2 dB gain. However addition of reflectors results in fractional increase in gain and hence usually not preferred. Fig. 10.8 shows the gain versus number of elements N in the array (including one reflector and one driven element) with an inter element spacing for all the elements of 0.15 λ. The gain is mostly controlled by the director element, by changing their length and spacing in between, both are interrelated. The more sensitive parameter is the director length, which becomes more critical as the boom length increases. Yagi-Uda Antenna has relatively low input impedance and narrow band width. However both can be improved at the expense of others (such as gain, magnitudes of lobes,..etc.) One way to increase the input impedance without affecting other parameter is to use an impedance step up element as feed (two element folded dipole with a step-up ratio of about 4). Front to back ratio of about 30 (≈ 15dB) can be achieved at wider than optimum element spacing, and for optimum designs, the minor lobes are about 30% or less of maximum lobes.

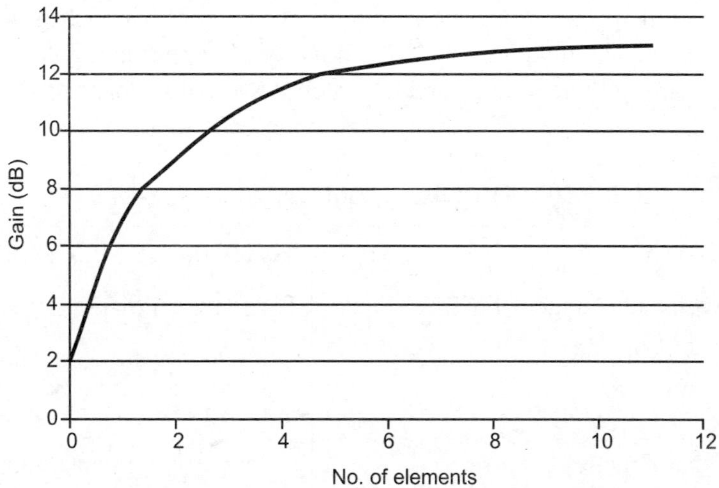

Fig. 10.8. Variation of gain of a typical Yagi-Uda antenna with total number of elements

Directivity

The maximum directivity of this antenna is found to be 9 dBi or 7 dBd for reflector spacing between 0.15 λ and 0.25 λ. Director to director spacing are typically 0.25 λ and 0.35 λ, with larger spacing being common for long arrays and closer spacing for shorter arrays. Typically, the reflector length is λ/2 and the director length always 11 to 20 % shorter than their resonant length

10.8 MICROSTRIP ANTENNAS

Microstrip antennas have been one of the most innovative topics in recent antenna technology, and increasingly finding applications in wide range of modern microwave systems. Microstrip antennas are one of the common printed antennas, in which only a portion of metallization provides radiation. A microstrip antenna (MSA) is a layered structure with two parallel conductors separated by a thin dielectric substrate. The upper portion is termed as patch that is responsible for radiation and lower portion acts as a ground plane. This is the reason microstrip antenna is also known as *patch antenna*. There are various shapes of patches; Rectangular, square, elliptical, circular and triangular/equilateral *etc*. The patch antennas are typically used for frequencies from 110 MHz to 110 GHz.

Generally, a MSA consists of rectangular radiating patch element, fed with a coaxial line in which length L is most critical dimension and slightly less than half-wave-length in the dielectric substrate. Variety of dielectric substrates having dielectric constant in the range of 2.2 to 12 can be used to design the microstrip antenna. However since fringing fields are the main sources of radiation, dielectric substrate with low ε_r is usually preferred because it allows the enhanced fringing fields; hence *enhanced bandwidth*. The choice of patch thickness also affects the bandwidth over which the antennas required to be operated. The MSA is designed in such a way that its radiation is maximum normal to the patch (broadside direction), which is accomplished by properly choosing the mode of excitation. The radiating patch is normally of copper and gold, and can assume any virtual shapes. Square, elliptical, and circular patches are commonly used due to simplicity of analysis and fabrication and also offered low cross polarized radiation.

Table 10.1 Advantages and Disadvantages of MSA

Advantages of MSA	**Disadvantages of MSA**
There are following advantages of MSA: • Compact, low profile and light weight • Cheap and easily developed • Thin and planar structure • Compatibility with feed and matching network • Mountable on missiles, rocket and satellites	There are following limitations of MSA: • Narrow band width and low efficiency • Low power handling capability • Limitation on maximum gain • Poor isolation between feed and antenna

Applications of MSA

Major applications of microstrip patch antennas are listed in Table 10.2.

Table 10.2 Applications of microstrip patch antennas in the various fields

S. No.	Platform	Systems
1.	Aircraft	Radars, Navigation, Communications, Altimeter, and Landing systems
2.	Missiles	Radars, Fuzing and Telemetry
3.	Satellites	Communications, Direct broadcasting TV, Remote sensing, Radars and Radio-meters
4.	Ships	Communications, Radars and Navigations

5.	Land vehicles	Mobile satellites and Mobile radios
6.	Others	As biomedical radiator in microwave therapy As a feed elements in large and complex antennas

10.9 POINCARE'S SPHERE

Herri Poincare: One of the greatest mathematical physicists at the end of 19th century. He made a series of profound innovations in *geometry, the theory of differential equations, Electromagnetism, topology, and the philosophy of mathematics.*

Poincare's Sphere

The Poincare Sphere, conceived by Herri, Poincare in 1892 provides a convenient way of representing polarized wave and predicting how any given yet order will change the polarization form. The upper and lower poles represent left and right circularly polarized waves. Points on the equator indicate linear polarization however other points on the Sphere represent elliptical polarization. An arbitrary point chosen on the equator designates horizontal polarization, and the diametrically opposite point designates vertical polarization, a general point *p* on the surface of the (Unit radius) Poincare Sphere (in terms of the longitude and the latitude) represent complete polarization of waves. Basic Poincare sphere is shown in Fig 10.10.

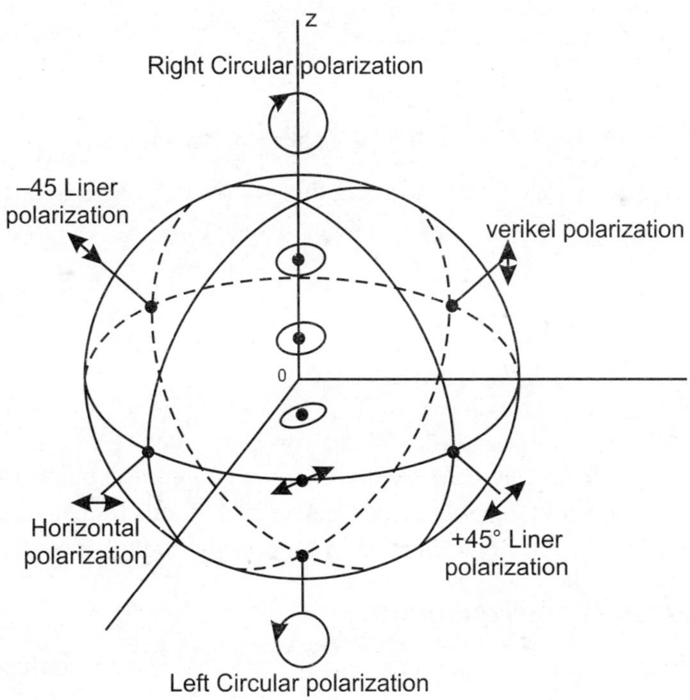

Fig. 10.10. The Poincare sphere

Every point on the surface of the Poincare Sphere represent a unique state of polarization and general pattern can be found. The Poincare Sphere provides a very useful way of characterizing polarization state. In addition, it is simple way understanding the de – polarization process and of seeing the comparative advantages of using different polarizations for meteorological measurements.

Any mixture of different kinds of the polarized wave may be reduced to an equivalent amount of one kind of polarized wave characterized by the orientation and length of axes of an ellipse. It may, of course happen that the ellipse degenerates to a straight line or circle. A 3-D Poincare sphere with Stoke's parameters and latitude and azimuth angles is shown in Fig 10.11.

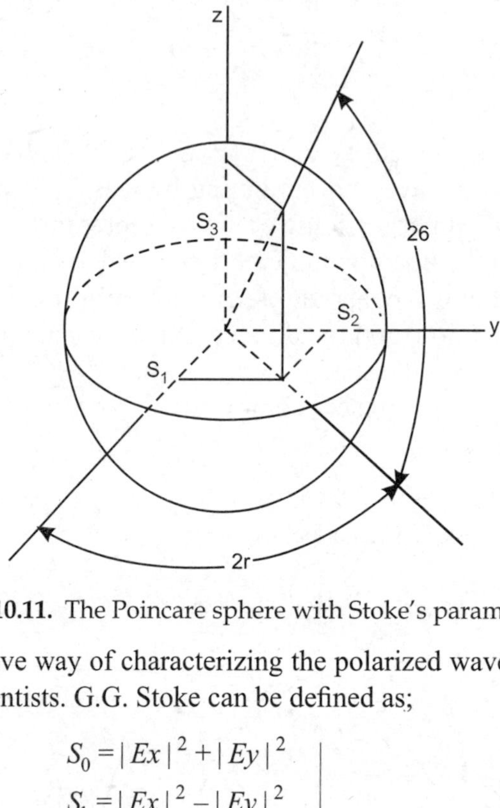

Fig 10.11. The Poincare sphere with Stoke's parameters

Using this, an alternative way of characterizing the polarized wave is by means of the Stokes parameter named after scientists. G.G. Stoke can be defined as;

$$\left.\begin{aligned} S_0 &= |Ex|^2 + |Ey|^2 \\ S_1 &= |Ex|^2 - |Ey|^2 \\ S_2 &= 2|Ex||Ey|\cos\phi \\ S_3 &= 2|Ex||Ey|\sin\phi \end{aligned}\right| \quad ...(10.44)$$

Stokes parameters (S_0, S_1, S_2, and S_3) constitute a matrix representation of polarized wave. This kind of representation makes available certain powerful methods for calculating how the state of a beam of polarized wave changes on reflection or on passing through a medium.

Conjugate and cross – polarized points;

If a wave with polarization ratio '*p*' is represented by S_1, S_2, and S_3 on the Poincare Sphere, then the points represent conjugate of '*p*' i.e. '*p**'

$$p = j\left|\frac{Ey}{Ex}\right|e^{j\phi} \qquad ...(10.45)$$

$$p^* = (-j)\left|\frac{Ey}{Ex}\right|e^{-j\phi} = (-1)\,j\left|\frac{Ey}{Ex}\right|e^{-j\phi}$$

$$p^* = j\left|\frac{Ey}{Ex}\right|e^{-j(-\pi+\phi)} \qquad ...(10.46)$$

As $\quad e^{-j\pi} = \cos\pi - j\sin\pi = -1 + 0 = (-1)$

Corresponding Stoke's parameter are defined as

$$S'_0 = |Ex|^2 + |Ey|^2 = S_0$$

$$S'_1 = |Ex|^2 - |Ey|^2 = S_1$$

$$S'_2 = 2\,|Ex||Ey|\cos(\lambda - \phi) = -S_2$$

$$S'_3 = 2\,|Ex||Ey|\sin(\lambda - \phi\_ = S_3$$

The polarization factor $p' = -\dfrac{1}{p}$ has a special significance and $p' = -\dfrac{1}{p} = j\left|\dfrac{Ex}{Ey}\right|e^{-j\phi}$ corresponding Stokes parameters are defined as

$$S'_{op} = S_0, \qquad S'_{1p} = -S_1$$

$$S'_{2p} = S_2, \quad \text{and} \quad S'_{3p} = -S_3$$

Stokes Parameters and the Poincare Sphere

To characterize the amplitude and polarization of a wave, Stock's introduced four quantities termed as Stocks parameters. They are defined in equation number (10.44), in which E_x and E_y are the field components of a wave representing polarization ellipse and ϕ is the phase difference between the wave and ϕ may be determined from either S_2 or S_3. The parameter S_o gives the amplitude, directly while $|E_x|$ and $|E_y|$ can be found from S_0 and S_1. From these equations it is found that:

$$S_1^2 = |Ex|^4 + |Ey|^2 - 2\,|Ex||Ey|$$

$$S_2^2 = 4\,|Ex|^2|Ey|^2\cos^2\phi$$

$$S_3^2 = 4\,|Ex|^2|Ey|^2\sin^2\phi$$

$$S_1^2 + S_2^2 + S_3^2 = |Ex|^4 + |Ey|^2 - 2\,|Ex|^2|Ey|^2 + 4\,|Ex|^2\{\cos^2\phi + \sin^2\phi\}$$

$$= |Ex|^4 + |Ey|^4 + 2\,|Ex|^2|Ey|^2 = S_0^2$$

$$\Rightarrow \quad S_0^2 = S_1^2 + S_2^2 + S_3^2 \qquad ...(10.47)$$

The description of polarization ellipse reveals that

$$\tan\alpha = \frac{|E_y|}{|E_x|} \quad 0 \le \alpha \le \frac{\pi}{2} \qquad ...(10.48)$$

$$\tan 2\tau = \tan 2\alpha .\, \cos\phi \qquad ...(10.49)$$

and $$\sin 2\delta = \frac{2|E_x||E_y|}{|E_x|^2+|E_y|^2}.\sin\phi \qquad ...(10.50)$$

In which, α = The auxiliary angle, τ = Ellipse tilt angle, and δ = Another auxiliary angle

Again, from (10.53) and (10.48) we get

$$\tan 2\tau = \left(\frac{2\tan\alpha}{1-\tan^2\alpha}\right)\cos\phi = \frac{2.\frac{|E_y|}{|E_x|}}{1-\frac{|E_y|^2}{|E_x|^2}}.\cos\phi\,\frac{2|E_x||E_y|}{|E_x|^2-|E_y|^2}$$

$$\Rightarrow \qquad \tan 2\tau = \frac{S_2}{S_1} \text{ (From 10.44)} \qquad ...(10.51)$$

From (10.50) and (10.44)

$$\sin 2\delta = \frac{S_3}{S_0} \Rightarrow S_3 = S_0 \sin 2\delta \qquad ...(10.52)$$

From (10.50) and (10.51) and (10.4) we get;

$$S_1 = S_0 \cos 2\delta.\cos 2\tau \qquad ...(10.53)$$

Again substituting (10.53) into (10.50) yields;

$$S_2 = S_0 \cos 2\delta.\sin 2\tau \qquad ...(10.54)$$

Recalling expressions for S_o S_1 S_2 and S_3;

$$S_0 = |E_x|^2+|E_y|^2$$

$$S_1 = S_0 \cos 2\delta \cos 2\tau$$

$$S_2 = S_0 \cos 2\delta \cos 2t$$

$$S_3 = S_0 \sin 2\delta$$

So, it is clear that these educations give a geometrical interpretation of the Stokes parameters. S_1, S_2 and S_3 are the Cartesian co-ordinates of a point on a Sphere of radius S_0. The angle 2δ and 2t are the latitude and azimuth angles measurement to the point. This interpretation was first introduced by Poincare hence the Sphere is called Poincare Sphere. Since we can describe the amplitude of a wave by S_0 and its polarization by, S_1, S_2 and S_3 hence any wave can be described by a point on the Sphere. To every state of polarization there corresponds one point on the Sphere and vice versa.

Special cases;

(*a*) For left circular polarization:

$$|E_x| = |E_y| \text{ and } f = \frac{\pi}{2}$$

Hence $$S_0 = 2|Ex|^2,\quad S_1 = S_2 = 0 \text{ and } S_3 = S_0$$

(*b*) For right circular polarization:

$$|E_x| = |E_y| \quad \text{and} \quad \phi = \pm\frac{\pi}{2}$$

Hence $$S_0 = 2|Ex|^2, \quad S_1 = S_2 = 0 \quad \text{and} \quad S_3 = S_0$$

(*c*) (*i*) *Left elliptical*

$$0 < \phi < \pi \text{ and therefore from equation (1.d), } S_3 > 0$$

All the points for this polarization are plotted on the upper hemisphere of *Poincare Sphere*.

(*ii*) *Right elliptical*

$$\pi < \phi < 2\pi \quad \text{and} \quad S_3 < 0.$$

i.e. all the points are in the lower hemisphere

(*iii*) *Linear polarization*

If E_x and $E_y \neq 0$, then $\phi = 0, \pi$ and $S_3 = 0$ and all linear polarizations are at the equator.

For the linear vertical polarization, the polarization point is at the (–) x-axis with the intersection of Sphere and for linear horizontal it at the (+) x-axis intersection. The (+) y axis intersection corresponds to linear polarization with a tilt angle $\left(\frac{\pi}{4}\right)$, and the (–) y axis intersection to a tilt angle of $\left(\frac{3\pi}{4}\right)$.

10.10 POLARIZATION RATIO

In general, the polarization ratio *P*, (which carries all the necessary information about polarization) is defined as

$$P = \frac{E_y}{E_x} = \frac{b}{a}e^{j\phi} \qquad ...(10.55)$$

Where '*a*' and '*b*' are real and related such as; $a^2 + b^2 = 1$.

Whereas for circular wave component polarization ratio is termed as *circular polarization ratio* and defined as

$$q = \frac{E_L}{E_R} = \frac{L}{R}e^{j\phi}$$

Which is also, $$q = \frac{a - jbe^{j\phi}}{a + jbe^{j\phi}}$$

Therefore relation between *P* and *q* will be

$$\text{q} = \frac{1 - jP}{1 + jP} \quad \text{or} \quad P = -\frac{1-q}{1+q}.$$

From these relations it is clear that P s multiplied by *j*, therefore we can re-write; $p = jP$. Hence we get $q = \frac{1-p}{1+p}$ or $p = \frac{1-q}{1+q}$. This new term 'q' is called modified polarization ratio and relates with electric field components as

$$q = j\frac{E_y}{E_x} = \frac{b}{a}e^{j\phi}$$

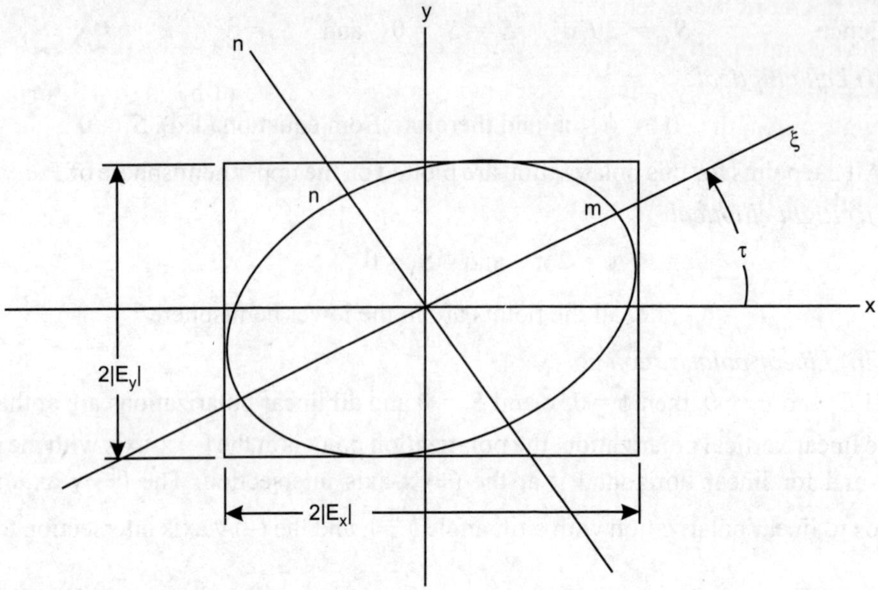

Fig. 10.12. Tilted polarization ellipse.

Where E_x, E_y, a, b and ϕ are defined in Fig.10.12.*

Some special values of polarization factor P is listed in Table 10.3.

Table 10.3. Polarization factor for different polarization

Waves	Characteristics	P	p	Q
Liner Horizontal	$E_y = b = 0$	0	0	+ 1
Liner Vertical	$E_x = a = 0$	∞	$j\,\infty$	– 1
Right circular	$a = b, \phi = \pi/2$	– j	+ 1	0
Left circular	$a = b, \phi = + \pi/2$	+ j	– 1	∞

SOLVED EXAMPLES

Example 1. Calculate the maximum radiation intensity and directivity of a $\lambda/2$ dipole antenna carrying current 2.5 A° r.m.s value at $f = 200$ MHz. Also, find the maximum effective area responsible for radiation.

Solution: We know that

$$I_{rms}^2 = \frac{I_0^2}{2} \Rightarrow I_0^2 = 2 \times 2.5 \times 2.5 = 12.50$$

Maximum radiation intensity,

And directivity is already defined as equal to 1.64 or 2.25 dB.

The operating wave-length, $\lambda = c/f = 150$ cm.

The maximum effective area, $A_{em} = 0.013 \times \lambda^3 = 0.13 \times 150^2 = 3.0 \times 10^3$ cm-square.

Example 2. Find directivity, gain, effective aperture area and solid beam angle for a centre-fed dipole of length ($\lambda/10$). Assume field distribution across antenna is $E(\theta) = \sin\theta$ aqnd $R_l = 0.4\ \Omega$

Solution: (*i*) The solid beam angle

$$\Omega_A = \int_{4}\int_{\pi} \sin^2\theta\, d\Omega = 2\pi \int_0^{\pi} \sin^3 d\theta = \frac{8\pi}{3} = 8.83\ S_r$$

(*ii*) The directivity

$$D = \frac{4\pi}{\Omega_A} = \frac{4\pi \times 3}{8\pi} = 1.5 = 1.76\ dB$$

(*iii*) The radiation resistance

$$R_r = 790\left(\frac{I_{av}^2}{I}\right)L_\lambda^2 = 790\left(\frac{1}{2}\right)\left(\frac{1}{10}\right)^2 = 1.975\ \Omega$$

(*iv*) The efficiency

$$\eta = \frac{R_r}{R_r + R_l} = \frac{1.975}{1.975 + 0.4} = 818.34\ \%$$

(*v*) The Gain

$$G = \eta.D = 0.833 \times 1.5 = 0.967 \text{ dBd}$$

(*vi*) The effective aperture

$$\text{Ae} = \eta\, A_m = 0.833\frac{\lambda^2}{\Omega_A} = 0.833\frac{3\lambda^2}{8\pi} = 0.099\lambda^2$$

in case l = 3.2 cm Hence $A_e = 0.099 \times 3.2^2 = 1.014$ cm^2

Example 3. The field pattern of an antenna is given by

$$F(\theta) = \cos^4\theta, \quad 0 < \theta < \frac{\pi}{2}, \text{ and } 0 < \phi < 2\pi$$

And zero elsewhere. Then find the directivity.

Solution: The directivity of antenna is given by

$$D = \frac{4\pi}{\int\int P(\theta)\sin\theta d\theta d\phi}$$

Where $P = (\theta)$ = power pattern = du, Then we get

If $\theta = u \Rightarrow -\sin\theta\, d\theta = du$, Then we get

Example 4. Show that the gain of a Hertz dipole is 1.5, provided normalized radiation pattern, $E(\theta) = \sin\theta$.

Solution: Assuming, a loss-less dipole i.e 100% efficiency.

Similar to above Question

$$D = \frac{4\pi}{\int\int_0^{2\pi} \sin^2\theta \sin\theta d\theta d\phi} = \frac{4\pi}{2\pi \int_0^{\pi} \sin^3\theta d\theta}$$

$$\Rightarrow \qquad D = 1.5$$

EXERCISES

1. What do mean by 'EM wave radiation'? Write essential condition for it.
2. Derive the expressions for Radiation resistance of dipole and monopole antennas.
3. Define Scalar and Vector potentials. How does they relate with electric and magnetic fields?
4. What do you mean by 'Retarded Potentials'? How do they play role in the radiation of EM waves?
5. What is Lorentz gauge condition? "Static source do not radiate" Justify the statement.
6. What are the major components of a complete antenna system?
7. What is Hertz dipole? What types of fields are generated from it?
8. What do you mean by Near and Far-field zone of an antenna? And how they are important in the radiation of EM waves?
9. At what distance all fields generated from a small dipole becomes equal?
10. Describe patch antenna, also mention it a applications.
11. Define the following parameters of an antenna;
    (*a*) Directivity (*b*) Bandwidth
    (*c*) Radiation Pattern (*d*) Effective aperture (e) SNR
12. Find the directivity of a one wave-length long dipole.
13. Write the expressions of radiation fields for a half-wave dipole antenna.
14. What are the similarities and differences between Marconi and Hertz antennas.
15. Find the radiation efficiency of an antenna having loss-resistance 0.5% that of radiation resistance.
16. What is FBR? How does it relate with gain of an antenna? Find FBR, if 12% of power radiated by antenna reflected in the backward direction.
17. What will be the bandwidth of an antenna having quality factor 30. Also, find the bandwidth assuming VSWR 1.65.
18. Calculate the maximum radiation intensity as well as directivity of a $\lambda/2$ dipole antenna carrying current 10.5 A° r.m.s value at $f = 250$ MHz. Also, find the maximum effective area responsible for radiation.
19. Find directivity, gain, effective aperture area and solid beam angle for a centre- fed dipole of length (λ/10). Assume field distribution across antenna is $E(\theta) = \sin\theta$ and $R_l = 0.5\ \Omega$
20. The field pattern of an antenna is defined by

$$F(\theta) = \cos^3\theta, \quad 0 \le \theta \le \frac{\pi}{2}. \text{ and } 0 \le \phi \le 2\pi$$

And zero elsewhere. Then find the directivity.

- Honesty is the first chapter in the book of wisdom.

Thomas Jefferson

Chapter 11

ELECTROMAGNETIC INTERFERENCE AND COMPACTIBILITY

11.1 INTRODUCTION

Every electronic devices radiate electromagnetic waves known as electromagnetic (EM) emissions. The world is highly saturated with such emissions, this is because of existance of variety of sources for such emissions. The effect of such FM radiating sources is spread throughout the electronic spectrum. Electromagnetic interference / Electromagnetic compatibly (EMI/EMC) is a unique and specialized subject to Electromagnetic Engineering. It came into prominence with the tremendous advances in modern technology and proliferation of a variety of electronic devices which produce electromagnetic emissions. Area surrounding above devices are known as Electromagnetic Environment. An Electromagnetic environment consists of various apparatus such as radio, TV, RADARs and other similar devices that radiate EM waves as they operate.

In general, EMI is nothing but the undesired signals which cause disturbances on the operation of a circuit or device. In other words, EMIs degrad performance of a device due to external fields making up electromagnetic environment. Every electronic device is susceptible to EMI and its influence are observed all ground us. The effect of EMI is seen like 'ghosts' in TV screen, power line transient interference with personal computors and self-oscillation of a radio receiver or transmitter devices.

Electromagnetic compatibility is defined as the ability of equipments/devices to be able to operate satisfactory in the presence of interference and do not cause interference to the nearby equipments. Electromagnetic compatibility (EMC) is achieved when a device functions satisfactorily without introducing intolerable disturbances to the electromagnetic environment or to other neighboring devices. EMI is like a problem that occurs when undesired voltages or currents appear and interrapt the performance of the devices, however EMC offers solution to this problem. The EMC ensure the compatibility of system and this is achieved by applying proven design technique, which ensures a system relatively free from EMI.

EMC is fast growing area, this is because of the ever increasing density of electronic devices in the modern behaviouraly. EMC is not only a concern to electrical, electronic and computer engineers, but also automotive engineers as well, and its main goal is to correct the operation of different equipments/devices in common electromagnetic environments. EMC deals with three main issues namely; Emission, susceptibility or immunity and coupling. Susceptibility is tendency of equipment to break down due to unwanted emission, (which is known as RFI). Susceptibility is also referred as victim. A coupling is defined as the mechanism by which emitted interferences approach to the victim. In sort, EMI is property of equipment, while EMC ensures the correct operation of equipment in that EMI environments.

Source of EMI

In general, the cause of EM interference may be either within the system or from the outside. They are refered as an intra and inter system problems. Thus;

Intra EMI : An EMI is said to be intra-EMI, if the functional characteristics of one module within an electronic equipment or system is disturbed due to EMI from another module.

Inter EMI : An EMI said to be inter EMI, if the functional characteristics of one equipment is affected due to EMI generated by same equipment/device itself.

EMI sources can also be grouped into (*i*) Narrow band EMI e.g. RADIOs, TV, and mobile sets (*ii*) Broad band EMI Electrical power Tx lines.

EMI refers to any frequency of electrical noise on the EM spectrum, while RFI represents specific sub group of electrical noise on EMI spectrum.

As well as controlling of EMI is concern both of the EMIs can be controlled by the system designers using some design guidelines and techniques.

Methods of Control EMI

To suppress or control EMI, several means are employed in the design process, and each techniques has got a distinct role in the system. These methods are :

(i) Grounding	(vii) Circuit impedance level control
(ii) Schielding	(viii) Cable design
(iii) Bonding	(ix) Cancellation techniques in frequency or time domain.
(iv) Filter	(x) Proper selection of cables and passive components.
(v) Isolation	(xi) Antenna polarization control
(vi) Separation and orientation	(xii) Balancing

However, the effective and popular methods are first, *i.e.*, Grounding, Shielding, Bondding and Filtering. The proper grounding may some times minimize the need for shielding and filtering, also proper shielding may minimize the need for filtering. The following sections describe the Grounding, Shielding, Bonding and Filtering.

(A) Grounding: Grounding is nothing but the establishing the concern between two points of electrical and electronic elements of a system. Between those two points, one may be the some reference termed as ground. An ideal ground is chareterized by Zero-potential and Zero-Imedance. It can be used as a reference for all signals associated in circuit. Basically, there are two types of grounding techniques : (*i*) Floating ground; (*ii*) Single-point groundings.

First one, isolates circuit or equipment electrically from a common ground plane, while second one is used to minimize the effects of facility ground currents. There is another grounding technique known as Multiple point grounding, which reduces ground lead lengths. In grounding, the ground plane may be a ground wire that is carried throughout the system or a large conductive body.

(B) Bonding : Similar to grounding, bonding is the establishment of a low-impedance path between two conductive metal surfaces. Bonding denotes the physical implementation of grounding. Bonding is helpful to protect against the effect of shocks and to protect circuits from current return paths, as they have ability to carry large fault current. They also provide protection from antenna ground plane connections and minimize the potential difference between the devices. Basically, there are two types of bonding: (i) Direct bonding (ii) Indirect bonding.

The direct bonding is metal to metal contact between the elements in circuit, however, Indirect bonding is made by contact using conductive jumpers.

The quality of bonding is represented in terms of three parameters : *AC* and *DC* resistances and bonding effectiveness.

$$DC \text{ Resistance} : (R_d) = \frac{l}{\sigma S} \qquad ...(11.1)$$

The dc resistance (R_d) is often used as an indication of bond quality. Where (l) is length of bond, S is its cross-sectional area and 'σ' is its conductivity. Bond resistance increases as frequency increases, this is due to skin-effect.

$$\text{The AC resistance } (R_A) = \frac{l}{\sigma \delta W}$$

where

$$W = \text{Width of bond}$$

$$\delta = \text{Skin depth}$$

$$= \left(\frac{1}{\pi f \mu \sigma}\right)^{1/2}$$

Bonding effectiveness : It is defined as the difference between induced voltage in the case of an equipment with a bond trap and that of without a bond trap. It is expressed as terms of (*dB*).

(C) Filtering : In this technique, the level of conducted interference are substantially reduced by using the electrical filter. Filter is a network of jumper/distributed elements; Resistors, inductors and Capacitors. They offer comparatively little opposition to certain frequencies, while blocking the passage of other frequencies. The most significant characteristics of filter is expressed in terms of its Insertion loss I_L which is defined as :

$$I_L = 20 \log \frac{V_{01}}{V_{02}} \text{ (dB)} \qquad ...(11.2)$$

where

$$V_{01} = \text{Output voltage of a signal source with filter}$$

$$V_{02} = \text{Output voltage of a signal source without filter.}$$

Generally, the Low-pass filters are used for filtering the EM interferences. The insertion loss for these filters are given as;

$$I_L = 10 \log_{10} (1 + K^2), (dB) \qquad ...(11.3)$$

where K is a constant and defined as

$$K = \pi fRC = \frac{\omega RC}{2} \text{ for capacitive filter} \qquad ...(11.4a)$$

$$K = \frac{\pi fL}{R} = \frac{\omega L}{2R} \text{ for inductive filter} \qquad ...(11.4b)$$

where $\omega = 2\pi f$, known as angular frequency.

(D) Shielding : In this case, EM interference is controlled by the way of confining radiated energy to a specific region or to prevent it from entering into a specific region. Shielding is done in the form of portions and baxes, as well as in the form of cable and connector shields. The main types of shield includes solid, screens and braids, and is used on cables. However, in all causes, the quality of shielding is characterized by its shielding effectiveness, which is defined as the ratio of incident power density to transmitted power density, *i.e.*,

$$S.E = 10 \log_{10} \frac{P_i}{P_t} \qquad ...(11.5a)$$

in which, the incident power density (P_i) is the power density measured before a shield is installed and the transmitted power density (P_t) is the power density at the same point after the shield is in place.

However, in terms of field strengths (E and H), the shielding effectiveness may also be defined as the ratios of incident to transmitted fields.

In particular, for electric field $S_e\,(dB) = 20 \log_{10} \left(\frac{E_i}{E_t}\right)$ ...(11.5b)

and for magnetic fields, $S_n\,(dB) = 20 \log_{10} \left(\frac{H_i}{H_t}\right)$ ...(11.5c)

Good conducting material used for shielding are : Silver, Copper, Aluminium, Gold, Brass and Bronze, however the good magnetic shielding materials are mumetal and steel.

The shielding effectiveness for electric fields in dB can also be written as the sum of the three terms;

$$S_e\,(dB) = R(dB) + M(dB) + A\,(dB) \qquad ...(11.6)$$

where $R(dB)$ = reflection loss and expressed as

$$= 168 + 10 \log_{10} \left(\frac{\sigma_r}{\mu_r \cdot f}\right) \qquad ...(11.7a)$$

where σ_r = Conductivity of a material relative to copper.

$A(dB)$ = The absorption loss for the material and is given by

$$= \frac{8.6859.t}{\delta}$$

in which, t = Thickness of the shield.

δ = Depth of pentration

$M(dB)$ = Multiple reflection loss and given by

$$= 20 \log_{10}\left[1-\left(\frac{\eta_0-\eta}{\eta_0+\eta}\right)^2 e^{-\frac{2t}{\delta}} . e^{-j2\beta.t}\right] \quad ...(11.7b)$$

$M(dB)$ is negligible for shields made of good conductors ($\eta << \eta_0$). If $t >> \delta$ (*i.e.*, if thickness of shield is far greater than skin-depth), $\alpha = \beta = \frac{1}{\delta}$ and therefore

M(dB) is modified to $20 \log_{10}\left[1-e^{-\frac{2t}{\delta}} . e^{-\frac{j2t}{\delta}}\right]$ ...(11.7*c*)

An alluminium sheet is considered a highly effective shield for a computor cabinet. A cabinet that effectively shields the circuits from external fields is also equally effective in preventing radiation from those circuits to the external world. However, radiated emission from the (effectively shielded computer system) are caused by openings in the cabinet such as cracks, holes from disc drives and also from wires that pentrate the cabinet such as power cord and cables to external circuits / devices.

Variation of shielding effectiveness (*dB*) with percentage of EMI and electric field intensity is listed in Table 11.1

Table 11.1

Shielding effectiveness (dB)	% EMI locked	*E* (*V*/*m*, no shield) : *E* (*V*/*m*, shield)
6	50	2 : 1
20	90	10 : 1
40	99	100 : 1
60	99.9	1000 : 1
80	99.99	10,000 : 1
100	99.9999	1,00,000 : 1

(Source : Haffman Apntair Company : EMC)

Basically, There are two additional EMI sources, namly:

(*i*) Natural (*ii*) Man-made /Artifical

The origin of EMI are basically conducted and radiated emissions. The conducted emissions are referred to voltags/current that are carried by metallic paths and placed on the common power network where they may cause interference with other connected devices. Radiated emissions refered to electric and/or magnetic fields, and they concern with radiated electric fields by the devices that may be received by other electronic devices causing interference in these devices.

Natural EMI sources are further devided into : **Terrestial** and **Extra-Terrestial EMI sources.** The initial terrestial sources are lightning discharges and atmospheric thundersforms, however extra-terrestial sources; sun-distrubed and quiet, cosmic noise and rodio stars.

EMI coupling mechanisms

In fact, there are many devices that can act as both receptors and emitters of electromagnetic interference. Communication systems like cell phones for example, utilize both transmitters and receivers, which can emit an EMI signal that has the potential to not only affect other systems, but can also couple back onto itself.

There are many ways where electromagnetic interference can be coupled from the source to the receiver. Understanding particular coupling method which brings the interference to the receiver is an important concern to use proper EMC. Simple electromagnetic interference coupling mechanisms are shown in Fig. 11.1

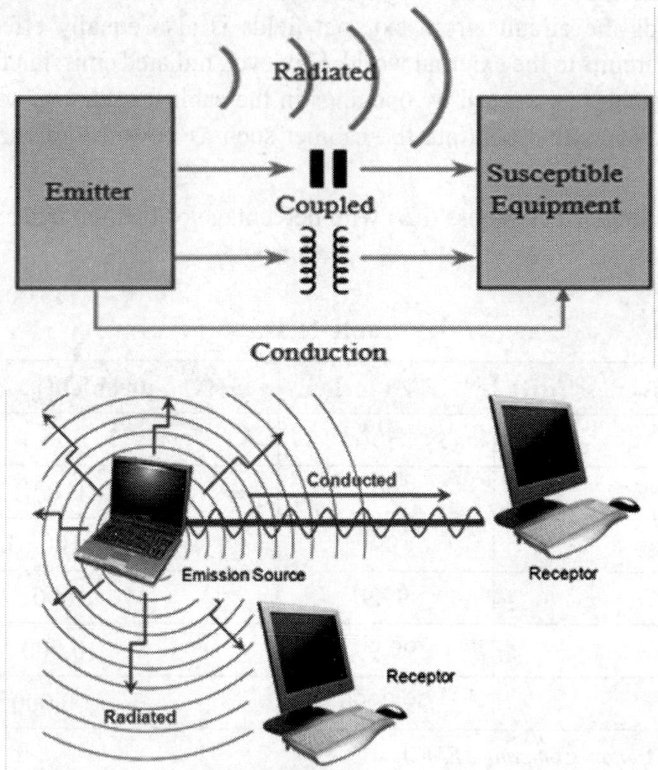

Fig 11.1. Electromagnetic interference coupling mechanisms

- **Radiated:** This type of EMI coupling is probably the most common one, where EMI coupling is normally experienced when the source and victim are at a large distance (i.e. at multiple wavelengths). The source radiates a signal irrespective of wanted or unwanted, and the receiver (or victim) receives same signal in a way that disrupts its performance. Radiated emissions are higher frequency signals coupled from one device to another one through non-conductive materials such as air, space, plastic and insulators.
- **Conducted:** As the name implies, conducted emissions occur when there is a conduction route along the path of the signal travelling. This may be along power cables or other interconnection cablings. EMI within this type is usually o be low frequency nature, 30 MHz generally considered as the upper limit.

- The conduction may be in one of two modes:
- **Common mode:** Such EMI coupling occurs if the noise appears is in the same phase on the two conductors, e.g. out and return for signals, or +ve and -ve for power cables.
- **Differential mode:** This EMI coupling occurs when the noise is out of phase on the two conductors. The filtering techniques required will vary according to the type of EMI coupling experienced.

  However, in the situations where both conducted and radiated emissions are at work together, signal, power and ground cables can behaves as the receiving or transmitting antennas.
- **Coupled:** Termed coupled EMI can be one of two forms, namely capacitive coupling and magnetic induction.
- **Capacitive coupling:** This coupling occurs when a changing voltage from the source capacitively transfers a charge to the receiver (victim) circuitry.
- **Magnetic coupling:** This type of EMI coupling occurs when a varying magnetic field exists between the source and victim - typically two conductors may run close together (less than λ apart), and induces a current in the victim circuitry, thereby transferring the signal from source to victim.

By knowing the form of coupling present and also the way in which it is reaching the victim, the most effective method of reducing the EMI is by putting measures in place to reduce the coupling as well as the level of interference to an acceptable level.

So, by understanding the source, the coupling methods and the susceptibility of the victim, the level of interference can be reduced to a value where the EMI causes no undue degradation in performance of device.

EMI Testing

EMI testing is essential because EMI predictions and analysis, are not sufficient enough to meet the EMI regulations. It is also necessary due to the complexity of designs for EMC. There are a number of methods and nomenclature used in measuring EMI emission and susceptibility characteristics of equipment and subsystems.

There are three levels of testing that exist and generally, as the level of testing increases, it becomes the more complicated and expensive.

- **Low-level testing** – It is component, equipment, and subsystem testing. The low-level testing ensures that there will be very few EMI problems when testing at the next higher level would be done.
- **Intermediate-level testing** – This is system and vehicle testing, and involves testing for EMI malfunction or degradation because of self-jamming.
- **High-level testing** – It is electromagnetic environment (EME) interaction with the testing items and necessary because low-level and intermediate level testing is inadequate for the products operating in its natural environment.

  EMI testing can also be divided into three categories: compliance, engineering, and audit testing.

- **Compliance tests** – The compliance testing is very precise and absolute measurements, so expensive equipments and experienced personnel is only preferred and it is done to prove that the design meets appropriate EMI requirements. This test is performed at the end of the design and before sale of the product. Testing can be performed in two types of laboratory: (*a*) An independent EMI test laboratory or (*b*) An in-house EMI test laboratory.
- **Engineering tests** – The main objectives of such tests are to uncover potential problems caused in design process. This is when one has the most time and flexibility in making design changes. The engineering testing does not require such high precision and accuracy like compliance testing to obtain good results and may be performed in-house. Some tests include emission prescreening, ESD prescreening, and power-disturbance and RFI prescreenings.
- **Audit tests** – Audit tests are basically associated with manufacturing and quality and it is done to ensure that the design stays intact through its product life.

Hot Spots and Limb Currents

The absorption of RF energy in the human body can sometimes result in a non-uniform distribution of heat causing hot spots. Hot spots (high local SARs) might observe in the human body over the frequency range: 30 MHz to 400 MHz, with 5 to 10 times the average whole body SAR. These hot spots will generally be evident at frequencies around body resonance where absorption is greatest and at sub-resonances in parts of the body.

It has also been approximated that dry cloths might act as an impedance transformer and increase the amount of energy coupled into the body. The thickness of clothing in this frequency band should be a significant fraction of the incident wavelength. At the frequency about 100 MHz, theoretical consideration of currents induced in the human body and especially the limbs, has some important concern. As a result, it shows that the current in the legs of an adult in an RF field provides large SAR values at the points where the effective conductive cross-sectional areas are comparatively small. The leg currents are proportional to operating frequency and also to the square of height of the person being exposed. The approximate formula for the electric field exposed vertically (valid up to 27 MHz) for the leg current of an erect barefoot person, is given as follows:

$$I\,(mA) = 0.108 \times h^2 \times f \times E \qquad ...(11.8)$$

Where

h = subject height (m)

f = frequency (MHz)

E = electric field (V/m)

For a field $E = 60\text{V/m};\quad f = 2\text{ MHz}\quad \text{and}\quad h = 1.75\text{ m}$

$$I\,(mA) = 0.108 \times (1.75)^2 \times 2 \times 60$$

$$= 39.70\text{ mA.}$$

However, for the frequency above 27 MHz, the measured currents peaked around 40 MHz, show the resonance of subject. An empirical expression for current above 27 MHz indicating an element representing the resonance frequency of the 'standard man', gives a sine wave shape for I/E versus frequency. The current when wearing shoes ranged from 0.8 to 0.82 times the barefoot current. However, previous work done at frequency 1 MHz showed that corresponding fractions vary between 0.62 to 0.64. The increase of current with frequency is due to the fall in the impedance to ground as the frequency increases.

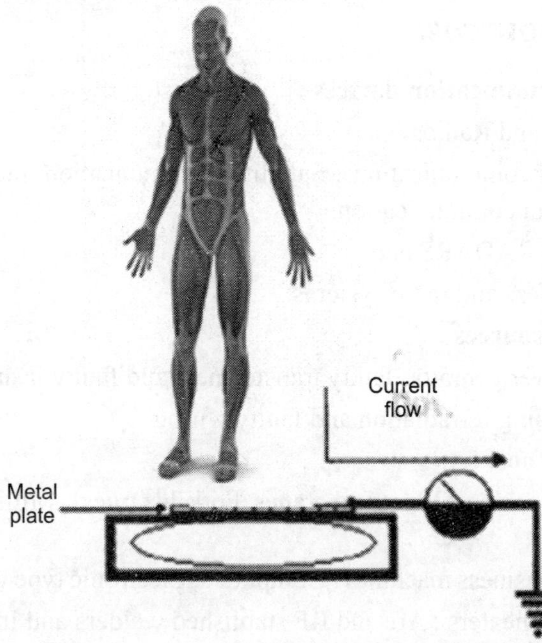

Fig. 11.2. Limb current measurement concept

The general principle of Limb current measurement is shown in Fig. 11.2, where the subject stands in the RF field on an insulated plinth and the feet are connected to ground through a current measurement device, particularly an RF milliammeter that would take readings directly. In practice it is important not to use any unshielded connections, wires and a RF meter having frequency limitations. A common method is to let the current flow through a known resistor, measure the voltage across it and convert the measurement to current.

Using HUMAN BODY MODEL, a human body can sometimes be approximated as follows;

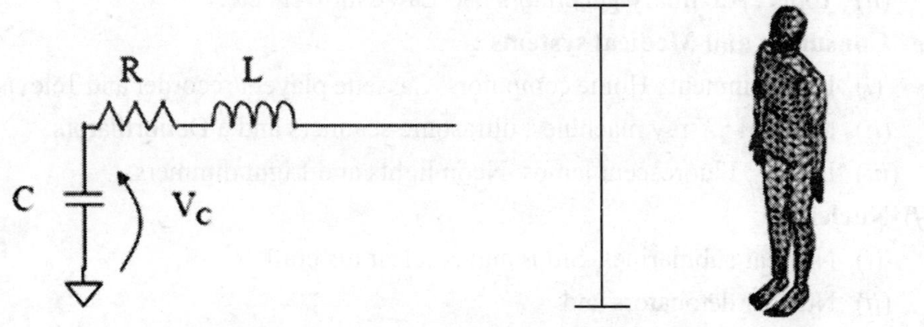

Fig. 11.3. Human Body Model.

Where the value of these parameters ranges;

$C = 50$ to 250 pF

$R = 500$ ohms to 10k ohms

$L = 50 - 200$nH

$|V_C| = 0$ to 30kV (above 30V will have corona effect)

Man-Made EMI sources:

(*a*) Electronic communication devices :

(*i*) Televisions and Radios

(*ii*) All types of communications; Satellite communication, mobile communication and point to point communication.

(*iii*) All types of RADARS and

(*iv*) Police wireless and radio systems.

(*b*) Electric power sources :

(*i*) Electric power genrator, faulty transformers and faulty insulators.

(*ii*) Poor grounding re-radiation and faulty wiring.

(*c*) Machines and Tools :

(*i*) Industrial machines; Electrical cranes, Fork-list trucks, Milling machines and printing presses.

(*ii*) Office and business machines : Computers, Electronic type writters and Photocopiers.

(*iii*) Welders and heaters : Arc and RF stablished welders and Induction heaters.

(*iv*) Transporters : Elevators, Escalators and Conveyer bults etc.

(*v*) Power Tools : Electric drills and Grinders and Mixer and Electric hand wash machines etc.

(*vi*) Appliances : Air-conditions
: Microwave-ovens
: Refregerators and
: Vacuum cleaners

(*d*) Ignition Systems :

(*i*) Vehicles : Automobiles, Air-craft, Tanks, Trucks and Tractors.

(*ii*) Tools : Auxilliary generators nad Lawn mowers etc.

(*e*) Consumer and Medical systems :

(*i*) Entertainment : Home computors, Cassette players/recorder and Television receivers.

(*ii*) Medical : *X*-ray machines, ultrasonic scanners and a Detibrillators.

(*iii*) Lights : Fluorescent lamps, Neon lights and Light dimmers.

(*f*) Nuclear :

(*i*) Nuclear submarines, Ships and Nuclear air-craft.

(*ii*) Nuclear detonators and

(*iii*) Nuclear power stations.

Electromagnetic Compatibility

EMC is very much related with frequency variation, therefore frequency allocation and assignments is done according to the constants established by international treaties. The International Telecommunication Union (ITU) publishes the radio regulations resulted from such treaties.

Standards for Measuring Shielding Effectiveness of Enclousers

1. **MIL-STD-285 :** Attention measurements for Enclosures, Electromagnetic shielding for electronic test purposes.
2. **IEEE-STD-299 :** Method of measuring the effectiveness of the Electromagnetic shielding of Enclosures.
3. **ASTM E1851 :** Method of measuring the Electromagnetic shielding Effectiveness of Durable rigid wall relocatable structures.
4. **ASTM D4935 :** Method of measuring of Electromagnetic shielding effectiveness of planar materials.
5. **VG 95373 :** Test method for measuring coupling and screening.

Cut-off frequency : In terms of EMC cut-off frequency is defined as the maximum possible frequency beyond which the wavelength will no longer shield EMI.

Electromagnetic Immunity : Defined as the ability of an electronic product to function satisfactorily EMI environment in the presence of Electromagnetic radiation.

Radio Frequency Interference : It is nothing but a EMI that extends over the 1 KHz-10 GHz, frequency band. RFI (EMI) electromagnetic radiation is made of both electric (E) and magnetic (H) fields. High-frequency radiation tends to have a large electric field (E) component, while low-frequency radiation tends to have a large magnetic field (H) component. High current devices produce magnetic field that could cause interference and affects the device performances.

Most high frequency wave energy is reflected off a conductive wall, however, in the low frequency range, the magnetic waves can penetrate the shield. As well as absorption of the wave is concern, it is related to the magnetic susceptibility and wall thickness of the shielding material.

EMC requirement sometimes can be met by using standard metal enclosures, because these enclosures can attenuate about 20 dB at 1 GHz, depending upon the type of cover overlap. However, specially designed enclosures can provide attenuation levels between 40 dB to 100 dB.

EMC Standards: These are two types of standards :

(*i*) **Military Standards :** These standards ensure system to system compatibility in a real time military environment. Equipments are classified based on their deployment environment. Most of these standards are broadly based on MIL-STD 461 and 462 Standards.

Military standard includes emission and susceptibility standards. First one specify emission limits in voltage/current and power/field strengths in particular frequency ranges, however, second one specify conducted spike or radiated field parameters.

(*ii*) **Civilian EMC Standards :** Civilian standards are applicable for equipments used for commercial, domestic and industrial applications. The emision standards are specified protect broadcast services from interference.

These standards also take into account the physiological interference effect experience by human beings.

Advantages of EMC standards

(*i*) Higher profits are possible.

(*ii*) Product life is increased.

(*iii*) The equipment operates satisfactorily in EMI scenario.

(*iv*) Design safety margin is provided.

(*v*) Compatibility, Maintainability and Reliability are increased.

EMC standards used in different conifers are given in Table 11.2.

Table 11.2 : EMC Standards in Different Countries

S.No.	Standard	Meaning	Country
1.	CISPR (IEC)	Committee international special perturations radioelectriques	Europe
2.	FCC	Federal communications council	USA
3.	VG	Military standard	Germany
4.	VDE	Verband deutscher electrotecknikev	Germany
5.	ISI	EMI measurements and measuring apparatus	India
6.	DEF STD	59–41 British MIL–STD	UK
8.	GAM-EG-13	France MIL–STD	France
9.	GENELEC	European committee for electrotechnical standardisation	Europe
10.	EN	European norms	Europe

SOLVED EXAMPLES

Example 1. Write the expressions for shielding effectiveness in terms of power and field intensity.

Solution : (a) In terms of power

$$\text{S.E.} = 10 \log_{10} \frac{P_t}{P_\tau}$$

(b) In terms of fields

$$S_e\,(\text{dB}) = 20 \log_{10} \left(\frac{E_t}{E_\tau}\right) = 20 \log_{10} \left(\frac{H_t}{H_\tau}\right)$$

Example 2. Find the leg current of a person of height 1.75 m, if a vertical electric field intensity of 65 V/m is generated across the body at frequency of 1.5 MHz.

Solution : For a field $E = 60$ V / m; $f = 1.5$ MHz and $h = 1.75$ m, the leg current is given by;

$$I\,(mA) = 0.108 \times h^2 \times f \times E$$

when h = The height of person/object.

$$I\,(mA) = 0.108 \times (1.75)^2 \times 1.5 \times 65 = 32.25 \text{ mA}$$

Example 3. The microwave exposure limits for the frequency range 60-300 MHz is 5 V/m. If the conductivity of the biological substance is 150 S/m and mass density is 1.3×10^3 kg/m^3. Calculate the power density of incident waves, then magnetic field intensity.

Solution : The power density is

$$P = \frac{E^2}{377} = \frac{5^2}{377} = 6.63 \times 10^{-2} \text{ W/m}^2$$

As $$P = \eta H^2$$

$\Rightarrow$ $$H = \sqrt{\frac{P}{\eta}} = \sqrt{\frac{6.63 \times 10^{-2}}{377}} = 68 \text{ mA/m.}$$

Example 4. Estimate the parameters of EMI bonding at $f = 20$ MHz, under following specifications;

The length of bonding = 4.2 cm

The width of bonding = 2.5 cm

Cross-sectional area = 9.5 cm^2

Conductivity = 1.8×10^8 S/m

Solution : DC resistance $(R_d) = \frac{1}{\sigma S} = \frac{4.2 \times 10^{-2}}{1.8 \times 10^8 \times 9.5 \times 10^{-4}}$

$$= 0.2248 \times 10^{-6} = 0.2448.\ \mu\Omega$$

Skin depth $$(\delta) = \left(\frac{1}{3.14 \times 20 \times 10^6 \times 4 \times 3.14 \times 10^{-7} \times 1.8 \times 10^8}\right)^{1/2} = 0.0083 \text{ mm}$$

Hence, the AC resistance (R_A) is

$$R_A = \frac{1}{\sigma \delta W} = \frac{4.2 \times 10^{-2}}{1.8 \times 10^8 \times 0.0083 \times 10^{-3} \times 2.5 \times 10^{-2}}$$

$$= 112.44 \times 10^{-5} = 1124.4\ \mu\Omega$$

Example 5. The low-pass filters are used for filtering the EM interference at $f = 145.6$ MHz in the capacitive arrangement. Find the insertion loss, if circuit components of the filters are as follows;

$$R = 115\ \Omega,\ L = 234\ \mu\text{H and } C = 145 \text{ pF}$$

Solution : The insertion loss for the filter is given as;

$$I_L = 10 \log_{10} (1 + K^2)\ (\text{dB})$$

where K is a constant and for capacitive filter it is defined as;

$$K = \pi fRC = 3.14 \times 145.6 \times 10^6 \times 115 \times 145 \times 10^{-11} = 0.922$$

Hence $$I_L = 10 \log_{10} [(1 + (9.022)^2] = 10 \log_{10} (82.396) = 19.16 \text{ (dB)}$$

Example 6. An electrostatic discharge (ESD) can be modeled as a capacitance of 115 pF charged 1600 V and discharges through a 2 kΩ resistor. Find the current wave form.

Solution : The circuit diagram can be drawn as follows (Fig 11.1).

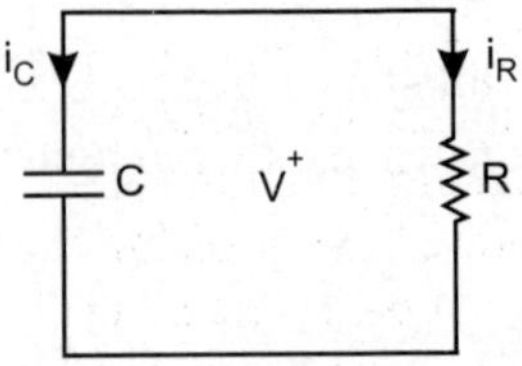

Fig. 11.1 Circuit diagram for Example 1.

As per circuit,

$$i_C + i_R = 0, \text{ hence } \frac{C\,dv}{dt} + \frac{v}{R} = 0$$

$$\frac{dv}{v} = -\frac{dt}{RC}$$

So that $$\ln v = -\frac{t}{\tau} + \ln v_0$$ v_0 is constant

and $$\tau = RC = 115 \times 10^{-11} \times 2 \times 10^3 = 0.230 \ \mu\text{.s}$$

Hence $$v = v_0\, e^{-(t/\tau)}, v_0 = 1600$$

Therefore, $$i_C = \frac{C\,dv}{dt} = C\left(-\frac{1}{\tau}\right) v_0\, e^{-(t/\tau)}$$

$$= \frac{-155 \times 10^{-11}}{0.230 \times 10^{-6}} \times 1600.e^{-(t/\tau)} = -0.8.\, e^{-(t/\tau)}$$

Example 7. Describe EMI receiver, EMI emitter and Respecter.

Solution:

EMI Receiver : EMI receiver is a tunable, sensitive voltmeter and measures E and H field intensity. It has a housing shielding effectiveness of a t least 90 dB, and operating frequency ranges from 30 Hz to 22 GHz, that is it can cover narrow band and broad band measurements accordingly.

EMI Emitter : In EMI application, Emitter is source of interference, and it applies to unintentional radiators. It is not like used emitter such transmitting antennas.

EMI–Receptor : It is advice which receives conducted or radiataed EM emission, because it has the potential for being susceptible to undesired interferences. Since it is susceptible to EMI from the emission received, it is considered to be a victim.

Example 8. Describe cross-talk, and susceptibility.

Solution:

Cross-talk : cross-talk causes due to the coupling of conducted emissions between two pairs of wire, such that one pair connected from source of emission, while other pair from the susceptible device. One that creating cross-talk called aggressor, while cross-talk receiver called avictim.

Susceptibility : It is often characterized as a lack immunity. Susceptibility represents the inability of device to function without degrading its performances in the influence of EM disturbances. The level of susceptibility at which performances of testing devices start degrading is termed as threshold susceptibility. This is often frequency dependent. The nominal level of conducted and radiated EM signal and noise present at a specific location is called ambient level. In general it is function of naturally existing atmospheric noise and interferences at that location.

Example 9. Describe different types of EMI

Solution: *Types of EMI - Electromagnetic Interference*

Since Electromagnetic Interferences can arise in many ways and from different sources. The EMI can be grouped into the following ways.

(*a*) The type of EMI based on it is created:

- **Man-made EMI:** This type of EMI generally arises from other electronics circuits, although some EMI can arise from switching of large currents, etc.
- **Naturally occurring EMI:** This type of EMI can arise from many sources - cosmic noise as well as lightning and other atmospheric types of noise all contribute.

(*b*) The type of EMI based on its duration:

- **Continuous interference:** This type of EMI generally arises from a source such as a circuit that is emitting a continuous signal. However background noise, which is continuous, may be created in a number of ways, either manmade or naturally occurring. Continuous noise hence interference is generally characterized as being low voltage in nature and common low frequency sources would be Switch Mode Power Supplies (SMPS), electric brush motors, ignition systems and fluorescent lighting. Radio Frequency Interference is a common term that generally defines a wide range of continuous higher frequency sources, ranging from high power radio transmitters to computer clock oscillators.

- **Impulse or Intermittent noise:** Again, this type of EMI may be man-made or naturally occurring. Lightning, ESD, and switching systems all contribute to impulse noise which is a form of EMI. It is generally distinguished from continuous noise as having duration of less than 60 Hertz. Intermittent signals are further classified as being either repeatable or random in nature. Repeatable one is internally generated within the circuit, and is predictable hence can be quantified in terms of amplitude, energy and duration. So the designer can safeguard the system by selecting a suitable protection device.
- Random noise, is however not predictable and cannot be quantified, examples are electromagnetic pulse in the form of lightning, solar flares, cosmic noise, or a nuclear reaction, or an electrostatic discharge. The selection of a suitable safeguard is the use of approximated statistical data and trend analysis.

(*c*) The types of EMI by their bandwidth.

- **Narrowband:** Such EMI is likely to be a single carrier source - possibly generated by an oscillator. Another narrowband EMI is the spurious signals and caused by inter-modulation and distortion in a transmitter such as a mobile phone of Wi-Fi router. These spurious signals appear at different points in the spectrum and may cause interference to another user of the spectrum. Thus such spurious signals must be maintained within tight limits.
- **Broadband:** These types of EMI can arise from a great variety of sources. Man-made broadband interference can arise from sources such as arc welders because a spark is continuously generated. Naturally occurring broadband noise can also be observed from the Sun, hence EMI.

Example 10. Describe Compliance Testing

Solution: *Compliance Testing*

Electromagnetic Compatibility or EMC is established through two testing:

Emissions testing: It identifies the frequency and amplitude of EMI generated by a device. Emissions are measured by connecting a Line Impedance Stabilization Network (LISN), a current probe, or an antenna to an EMI receiver, scanning the desired frequency range and measuring the amplitude of the signals detected. This approach would act much in the same way as the 'scan" button on an FM radio.

Immunity testing: It verifies the performance of a device or system, when subjected to known levels of EMI. Immunity / Susceptibility testing by comparison, connects the LISN, a current probe or antenna to an RF amplifier and injects a signal into the system being evaluated. The equipment still scans for EMI signals, but the intention is different, it is to monitor the performance of the testing device while subjected to the energy and response of the equipment is observed.

That is EMC = EME + EMI, where EME = Electromagnetic Emission (Interference) and EMI = Electromagnetic Immunity (Susceptibility).

Emissions and Immunity testing are further broken down into four basic EMC tests;

(*i*) Radiated Emissions, (*ii*) Conducted Emissions (*iii*) Radiated Immunity and iv. Conducted Immunity. Radiated tests utilize an antenna in the experimental setup, while conducted tests deal with wires and , and are easy to identify as no antennas are used.

Government agencies tasked with instituting EMC standards have established emission limits of 100 μV or 40 dBμV maximum and immunity levels of 10 V or 140 dBμV minimum as general guidelines for operation of unlicensed equipment.

UNSOLVED QUESTIONS

1. Determine the reflection loss for aluminium, brass, and stainless steel at 2.2 MHz.
2. Determine the absorption loss for 1/8-inch-thick aluminium, brass and stainless-steel shields at 2 MHz.
3. What are sources of EMI? Describe in detail.
4. Describe any two methods of controlling the EMI.

5. What do you mean by the terms; RFEMI and ESD.
6. If an electrostatic discharge (ESD) is modeled as a capacitance of 225 pF charged to 1600 V and discharging though a 2 kΩ resistor. Find the current wave form.
7. Find the leg current of a person of height 1.7m, if a vertical electric field intensity of 60 V/m is generated across the body at frequency of 1.5 MHz.
8. Estimate the parameters of EMI bonding at $f = 30$ MHz, under following specifications;

   The length of bonding = 3.45 cm

   The width of bonding = 2.546 cm

   Cross-sectional area = 9.56 cm^2

   Conductivity = 18 × 107 S/m
9. The low-pass filters are used for filtering the EM interferences at $f = 110$ MHz in th capacitive and inductive arrangements. Find the insertion loss, if circuit components of the filters are as follows :

$$R = 115\ \Omega, L = 235\ \mu\text{H and } C = 145\ \text{pF}$$

10. Find the reflection loss and absorption loss in the shielding, used to reduce radiated EMI reflection or absorption such that relative conductivity of material is 3 and thickness of the shield is 0.65 cm, at operating frequency 2.5 MHz.

- The journey of a thousand miles begins with one step.

Lao Tzu

SOLVED EXAMPLES

Examples 1. Good conductor has very high value of permitivity, so the corresponding value of electrostatic force is :

(*a*) Smaller (*b*) Higher

(*c*) Zero (*d*) None

Solution : (*a*) Smaller

Example 2. Permitivity of glass is greater than that air :

(*a*) 5 times (*b*) 10 times

(*c*) 2 times (*d*) None

Example 3. The permitivity of vacuum is 8.854×10^{-12} F-m^{-1}. So, it is:

(*a*) Lowest value (*b*) Highest value

(*c*) Not known (*d*) None

Solution : (*a*) Lowest value.

Example 4. What will be flux (ψ) radiating from a positive point charge of magnitude 200 pc.

(*a*) 50 pc (*b*) 100 mc

(*d*) 1500 pc (*d*) 200 pc

Solution : (d) 200 pc.

Example 5. Determine the flux that flows through an area open on the surface of 10m radius Gaussian sphere carrying a charge of 100 pc. Also, find total flowing through the area of 200 cm^3.

Solution : We known that, at the surface of sphere :

$$D = \frac{q}{4\pi r^2}\bar{r}$$

$$= \frac{100 \times 10^{-12}}{4 \times 3.14 \times (1 \times 10)^2} = 7.967 \times 10^{-12} \text{ C/m}$$

Therefore, total flux density flowing through the 200cm^3 surface, will be

$$\psi = D \times \text{area} = 7.96 \times 200 \times 10^{-12} \times 10^{-6}$$

$$= 1.6 \times 10^{-15} \text{ C}$$

Example 6. Determine the flux density and electric field intensity at distance of 0.5 m from an isolated point charge of 20 μC. If an identical charge of 10 mC is placed at this points, what force will it experience.

Solution :

$$D = \frac{q}{4\pi r^2} = \frac{20 \times 10^{-6}}{4 \times 3.14 \times (0.5)^2} = 6.4 \times 10^{-6} \text{ Cm}^{-2}$$

$$E = \frac{D}{\varepsilon_0} = \frac{6.4 \times 10^{-6}}{8.854 \times 10^{-12}} = 7.2 \times 10^5 \frac{V}{m}$$

Hence,

$$F = q.E = 7.2 \times 10^5 \times 10 \times 10^{-6} = 7.2 \text{ N}$$

Example 7. Repeat the Q. 6, if the amount of charge is double, however distance between them is same.

Solution :

$$D = 12.7 \times 10^{-6} \text{ C-m}^2$$

$$F_t = 1.44 \times 10^6 \text{ V/m}$$

$$F = 14.4 \text{ N}$$

Example 8. Determine the absolute potential at distance of 0.2 m from an isolated point charge of 10mc. Also, find the potential difference between this point and another point at 10 m from the charge.

Solution : The potential will be total workdone in moving the charge from infinity to 0.2 m, from the to the charge.

That,

$$dV_1 = -\int_{r_1}^{r_2} \frac{q}{\pi\varepsilon_0 r^2}.dV$$

$$V_1 = -\frac{q}{4\pi\varepsilon_0}\left[-\frac{1}{r}\right]_0^{0.2} = \frac{10 \times 10^{-6} \times 5}{4 \times 3.14 \times 8.854 \times 10^{-12}}$$

$$= 4.5 \times 10^5 \text{ volt}$$

Similarly

$$V_2 = \frac{10 \times 10^6}{4 \times 3.14 \times 8.854 \times 10^{-12}} \times \frac{1}{10} = 9.0 \times 10^3 \text{ Volt}$$

$$V_1 - V_2 = \Delta V = 4.41 \times 10^5 \text{ Volt}$$

Example 9. Two metal plates of area 15 cm^2 are placed at 5 mm part. If a dielectric material having $\varepsilon_r = 2.5$ and thickness 3 mm is placed at voltage mid way between the plates and of 200 volt is applied across Determine:

(*a*) equivalent capacitance

(*b*) surface charge densities

(*c*) flux density at all points

(*d*) '*E*' at all points

(*e*) bound charge densities

Solution :

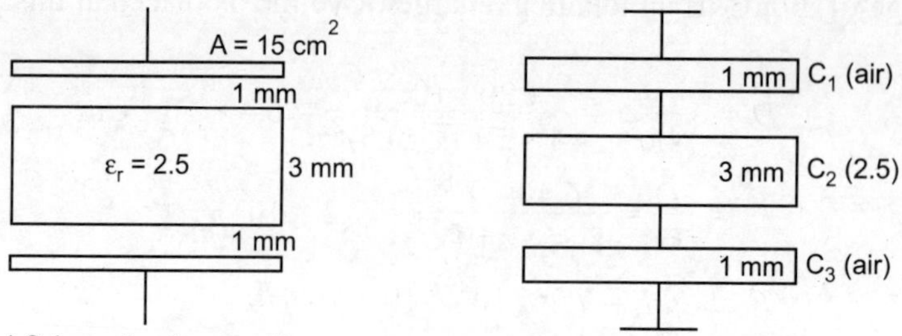

(a) Schematic of composite dielectric capacitor

(b) Equivalent circuit of (a)

Fig. 1. Geometry for example 9.

(*a*) We know that

$$C = \frac{E.A}{d}$$

and if more then one captions are arranged in parallel, the equivalent capacitance (C_T) will be.

$$\frac{1}{C_T} = \frac{1}{C_1} + \frac{1}{C_2} + \frac{1}{C_3}$$

$$= \frac{d_1}{\varepsilon.A_1} + \frac{d_2}{\varepsilon.A_2} + \frac{d_3}{\varepsilon.A_3}$$

$$= \frac{1 \times 10^{-3}}{\varepsilon_0 \times 15 \times 10^{-4}} + \frac{3 \times 10^{-3}}{\varepsilon_0 \times 2.5 \times 15 \times 10^{-4}} + \frac{1 \times 10^{-3}}{\varepsilon_0 \times 15 \times 10^{-4}}$$

$$= \frac{10^{-3}}{\varepsilon_0 \times 15 \times 10^{-5}}\left[1 + \frac{3}{2.5} + 1\right]$$

$$= \frac{2}{3\varepsilon_0} \times 3.2 = \frac{6.4}{3\varepsilon_0}$$

$$C_T = \frac{3 \times 8.854 \times 10^{-12}}{6.4} = 4.15 \text{ pF}$$

Since, voltage applied across the plates is 200 V, so the total stored charge is

$$Q_T = C \times V = 4.15 \times 10^{-12} \times 200 = 830 \text{ nc}$$

(*b*) Surface charge density,

$$r = \frac{Q_T}{A} = \frac{830 \times 10^{-9}}{10 \times 10^{-4}} = 830 \text{ nc-m}^{-2}$$

(*c*) Since, each unit charge will radiate a unit of flux, so the flux density in the air gap will be:

$$D = 830 \text{ nc} - \text{m}^{-2}$$

This flux will remain constant through the capacitor.

(*d*) Therefore, the electric field strength in each air gap is;

$$E = \frac{D}{\varepsilon_0} = \frac{830 \times 10^{-9}}{8.854 \times 10^{-12}} = 93.75 \text{KV m}^{-1}$$

Again, the flow-density at surface of dielectric:

$$D = \varepsilon_0 \, \varepsilon_r \, . \, E$$

$$\Rightarrow \quad E = \frac{D}{\varepsilon_0 \varepsilon_r} = \frac{93.15}{2.5} = 37.5 \text{ KV}$$

So, the difference between field strength in air and dielectric is

$$DE = (93.75 - 37.5) = 56.25 \text{ KV-m}^{-1}$$

And this difference is due to bound charge on the costal of dielectric. Hence, the surface/bound charge density will be

$$P_d = \varepsilon_0 \times \Delta E = 8.854 \times 10^{-12} \times 56.25 \times 10^3$$

$$= 498 \text{ nC-m}^{-2}$$

Example 10. A electric field make an angle of 30° to the horizontal boundary between air and a dielectric. If the relative permittivity of dielectric (ε_r) is 6, determine the angle of refraction. Assume that there are no bound charges.

Solution : As per boundary condition,

$$E_t = E \cos 30° = \frac{\sqrt{3}}{2} \text{E}$$

Since, there are no bound charges at the surface, the normal components of the D is continuous across the boundary. Thus,

$$D_n = D \sin 30° = \varepsilon_0 E \sin 30° = \frac{\varepsilon_0 E}{2}$$

Since the dielectric has, $\varepsilon_r = 6$, hence,

$$6 \times \varepsilon_0 \times E_n = \varepsilon_0 \, . \, E \sin 30°$$

$$\Rightarrow \quad E_n = \frac{E}{6} \times \frac{1}{2} = \frac{E}{12}$$

Therefore, the angle of refraction will be;

$$\tan \alpha = \frac{E_n}{E_t} = \frac{E/12}{E\sqrt{3}/2} = \frac{1}{12} \times \frac{2}{\sqrt{3}}$$

$$\tan \alpha = \frac{1}{6\sqrt{3}} = 0.096$$

$$\alpha = 5.486°$$

Hence, reflected E-field makes an angle of 5.486° from the horizontal surface.

Example 11. A 500 m length of co-axial cable has an inner conductor of radius 2 mm and outer conductor of radius 1 cm. Dielectric constant of dielectric material between the conductor is 5. Determine

(*i*) Capacitance per unit length.

(*ii*) If the inner conductor is at potential of 2 kV above the conductor, what will be maximum value of 'E' in the dielectric.

Solution : We known that

$$C = \frac{2\pi\varepsilon_0 \, \varepsilon_r.}{\text{In } (b/a)}$$

$$= \frac{2 \times 3.14 \times 8.854 \times 10^{-12} \times 5}{\ln\left(\dfrac{10^{-12}}{2 \times 10^{-3}}\right)}$$

$$= \frac{278.0156 \times 10^{-12}}{1.609} = 172.78 \text{ pc}$$

Therefore, the line charge density, $\rho = C.V$

$$\rho = 172.78 \times 10^{-12} \times 2 \times 10^3 = 345.57 \text{ nC-m}^{-1}$$

Since, E is inversely proportional to the distance (r). E_{max} will occur at the surface of conductor.

i.e.,

$$E_{max} = \frac{\rho}{2\pi\varepsilon_0\, \varepsilon_r .r}$$

$$= \frac{345.57 \times 10^{-9}}{2 \times 3.14 \times 8.854 \times 10{-}12 \times 5 \times 10^{-3}}$$

$$= 622 \text{ kV-m}^{-1}$$

Example 12. Given the potential $V = \frac{5}{r^2}$ sin θ. cos ϕ, Find the diuretic flux density at point $(1, \frac{\pi}{2}, 0)$.

Solution:

$$D = \varepsilon E$$

$$E = -\nabla V = \left[\frac{\partial V}{\partial r} a_r + \frac{1}{V}\frac{\partial V}{\partial a} a_\theta + \frac{1}{r \sin\theta}\frac{\partial V}{\partial \phi} a_\phi\right]$$

$$= q\frac{10}{r^3} \sin\theta \cos\phi .\, a_r - \frac{5}{r^3} \cos\theta \cos\phi\, a_\theta + \frac{5}{r^3} \sin\phi .\, a_\phi$$

D at $(1, \frac{\pi}{2}, 0)$

$$= E_0\,(10\, a_r - 0\, a_\theta + 0\, a_\phi)$$

$$= 8.854 \times 10^{-11} \frac{C}{m^2}$$

Example 13. A high voltage power line consists of 1 cm radius copper wire placed 25 m above the ground. Determine the capacitance that the has to the ground.

Solution :

$$C = \frac{2\pi\varepsilon_0}{In\,(d/2a)} \text{F m}^{-1}$$

$$= \frac{2 \times 3.14 \times 8.854 \times 10^{-12}}{In\,(25/2 \times 10^{-2})} = 7.8 \text{ pF-m}^{-1}$$

Example 14. A 3 mm wide track is etched on one side of some double-sided printed circuit board. If the thickness of the board is 2.5 mm and dielectric constants is 5. What will be capacitance?

Solution :

$$C = \frac{\varepsilon_0\, \varepsilon_r\, d}{h} = \text{Fm}^{-1}$$

$$= \frac{8.854 \times 10^{-12} \times 5 \times 3 \times 10^{-3}}{2.5 \times 10^{-3}}$$

$$= 53.12 \text{ pF-m}^{-1}$$

Example 15. A 10 μF capacitor has a potential of 200 d.c. across its terminals. If the capacitance increases to 60 μF, find the change in flux.

Solution :
$$\psi_1 = \Delta CV = (60 - 10)\text{m}F \times 200$$
$$= 50 \times 200 \times 10^{-6} = 10 \text{ mC}$$

Example 16. A single 10 μWb magnetic monopole is situated in air. Find the flux density and the force on an identical monopole placed at distance of 0.6 meter.

Solution : The magnetic strength :
$$|H| = \frac{\rho_1}{4\pi\mu r^2} = \frac{10 \times 10^{-6}}{4\pi \times 4\pi \times 10^{-7} \times (0.6)^2} = \frac{100}{56.79} = 1.76$$
$$|H| = 1.76 \text{ NWb}^{-1}$$
$$\therefore \quad |B| = \mu|H| = 1.76 \times 4\pi \times 10^{-7}$$
$$= 2.210 \times 10^{-6} = 2.21 \ \mu\text{WB m}^{-2}$$

The force on identical monopole :
$$F = p_2 |\text{H}|$$
$$= 10 \times 10^{-6} \times 1.76 = 17.6 \ \mu\text{N}_0$$

Example 17. A toroidal coil is wound on a ferrite frame with inner radius 4 cm and outer radius 6 cm. If coil has 1000 turn and the core has a relative permeability 200. Determine the current required to produce a mean flux of 1.5 m Wb in the core.

Solution : Average radius $\frac{4+6}{2} = 5 \text{ cm} = 0.05$

Thus, the average length of the magnetic path
$$l = 2\pi r = 2 \times 3.14 \times 0.05 = 0.314 \text{ m}$$

The frame has a circular cross-sectional area of radius :
$$r = \frac{6-4}{2} = 1 \text{ cm} = 0.01$$

Therefore, the reluctance of the core :
$$S = \frac{l}{\mu \times area} = \frac{0.314}{200 \times 4\pi \times 10^{-7} \times \pi r^2}$$
$$= 4 \times 10^6 \text{ At Wb}^{-1} \qquad \text{(Amper–turn per Wb)}$$

Thus,
$$V_m = 1.5 \times 10^{-3} \times 4 \times 10^6$$
$$= 6 \times 10^3 \text{ At} \qquad \text{(Amper–turn)}$$

As the coil has 1000 terms, the coil current is :
$$I = \frac{V_m}{N} = \frac{6 \times 10^{+3}}{1000} = 6 \text{ Amp.}$$

Example 18. An infinitesimal length 10^{-3} m of wire located at a point (1, 0, 0) and carrying current 3 A in the direction of the unit-vector a_x. Find the magnetic flux density due to the current element at the point (0, 2, 2).

Solution : We know that the current element is given by

$$I.\, dl = (3)\,(10^{-3}).\, a_x = 0.003\, \overline{a}_x$$

Radial vector between point (1, 0, 0) of the current element to the point (0, 0, 2) will be:

$$R = (0-1)a_x + (2-0)a_y + (2-0)a_z = -a_x + 2a_y + 2a_z$$

$$|R| = (4+4+1)^{1/2} = 3$$

Hence,

$$|B|_{(0,\,2,\,2)} = \frac{\mu_0}{4\pi}\frac{I.dl \times a_R}{R^2}$$

$$= \frac{\mu_0}{4\pi}\frac{I.dl \times \overline{R}}{R^3}$$

$$= \frac{\mu_0}{4\pi} \times \frac{0.003a_x \times (-a_x + 2a_y + 2a_z)}{3^3}$$

$$= \frac{0.001\,\mu_0}{18\pi}(-a_y + a_z)\, Wb/m^2 = 1.567 \times 10^{-4}\,(-a_y + a_z)\,\frac{Wb}{m^2}.$$

Example 19. If volume charge density 'ρ' is defined by

$$\rho = (x + y + z)\ \text{C/m}^2$$

Then, find the total charge contained by a cubic volume bounded by the planes $x = 0$, $x = 1$, $y = 0$, $y = 1$ and $z = 0$, $z = 1$.

Solution : From the Gauss's low,

$$Q_T = \int_V \rho dv$$

$$\Rightarrow \qquad Q_T = \int_{x=0}^{x=1}\int_{y=0}^{1}\int_{z=0}^{1}(x+y+z).\,dx.\,dy.\,dz$$

$$= \int_{x=0}^{1}\int_{y=0}^{1}\left[xz + yz + \frac{z^2}{2}\right]_0^1 dx.\,dy$$

$$= \int_0^1\int_0^1\left[x + y + \frac{1}{2}\right]dx.\,dy$$

$$= \int_0^1\left[xy + \frac{y^2}{2} + \frac{y}{2}\right]_0^1 .dx$$

$$= \int_0^1\left[xy + \frac{y^2}{2} + \frac{y}{2}\right]_0^1 .dx$$

$$\int_{x=0}^{1}(x+1)\,dx = \left[\frac{x^2}{2} + x\right]_0^1 = \frac{3}{2}C$$

Example 20. Find the directivity of antenna corresponding to the power density pattern function;

$$f(\theta.\phi) = \sin^2\theta.\,\cos^2\theta$$

Solution :

$$D = 4\pi\,\frac{[f(\theta,\phi)]_{\max}}{\int_{\theta=0}^{\pi}\int_{\phi=0}^{2\pi} f(\theta.\,\phi)\sin\theta.d\theta.\,d\phi}$$

$$= 4\pi \frac{[\sin^2\theta\cos^2\theta]_{\max}}{\int_{\theta=0}^{\pi}\int_{\phi=0}^{2\pi} \sin^3\theta.\cos^2\theta.\,d\theta.d\phi}$$

$$= 4\pi \frac{\left[\frac{1}{4}\sin^2 2\theta\right]}{2\pi\int_{\theta=0}^{\pi}(\sin^3\theta - \sin^5\theta)d\theta}$$

$$= \frac{1}{2}\frac{1}{\left[\left(\frac{4}{3}\right)-\left(\frac{16}{15}\right)\right]} = 1\frac{7}{8}$$

Example 21. The dominant modes TE_{10} is propagated in rectangular guide of 8 × 5 cm^2. Find the single frequency, if the distance between maximum and minimum points is 4.47 cm.

Solution : We know that

$$f_c = \frac{c}{2}\left[\left(\frac{m}{a}\right)^2 + \left(\frac{n}{b}\right)^2\right]^{1/2}$$

$$(f_c)_{10} = \frac{c}{2}\left[\frac{1}{a}\right]$$

$$= 1.5 \times 10^8 \left[\frac{1}{8\times10^{-2}}\right] = \frac{15}{8}\times10^9$$

$$= 1.875 \text{ GHz}$$

$$\lambda_g = 2 \times 4.47 = 8.94 = 0.0894 \text{ m}$$

$$\lambda_0 = \frac{c}{fc} = \frac{3\times10^8}{1.875\times10^9} = 0.16 \text{ m}$$

As $$\lambda_g = \frac{\lambda_0}{\left[1-\left(\frac{fc}{f}\right)^2\right]^{1/2}}$$

$$1-\left(\frac{fc}{f}\right)^2 = \left(\frac{\lambda_0}{\lambda_g}\right)^2 = \left(\frac{0.16}{0.0894}\right)^2 = 3.2$$

$$\left(\frac{fc}{f}\right)^2 = 2.2$$

⇒ $$f = \frac{fc^2}{2.2} = \frac{1.875}{2.2} = 1.6 \text{ GHz}$$

Example 22. An RF extension having Z_1 60 Ω is terminated with Z_1 = (400 + *j*200)W. Find VSWR.

Solution : $$\rho = \frac{Z_L - Z_0}{Z_L + Z_0} = \frac{400 + j200 - 600}{400 + j200 + 600} = \frac{-200 + j200}{100 + j300}$$

$$\rho = \frac{282 < 45°}{1019 < 11.31°}$$

$$\rho = 0.277 < -50.31°$$

$$\text{SWR} = \frac{1+|\rho|}{1-|\rho|} = \frac{1+0.277}{1-0.277} = 1.766$$

Example 23. Show that $\overline{H}_y = E_0 \sin(\omega t - \beta z)$ and $\overline{H}_z = -\dfrac{\beta E_0}{\mu_0 \omega} \sin(\omega t - \beta z)$ travels with velocity of light in air. Also obtain $\left(\dfrac{E}{H}\right)$ ratio.

Solution :
$$\left|\frac{E_y}{H_x}\right| = \frac{E_0 \sin(\omega t - \beta z)}{\dfrac{\beta E_0}{\mu_0 \omega} \sin(\omega t - \beta z)} = \frac{\mu_0 \omega}{\beta} = \frac{\mu_0 \times 2\pi f}{2\pi/\lambda}$$

$$\Rightarrow \quad \left|\frac{E}{H}\right| = \mu_0 \times f_\lambda \times \mu_0 \times c = \mu_0 \frac{1}{\sqrt{\mu_0 \varepsilon_0}}$$

$$= \sqrt{\frac{\mu_0}{\varepsilon_0}} = \eta = 120\,\pi = 377\ \Omega.$$

which is characteristic impedance of free-space.

Therefore,
$$c = \frac{1}{\sqrt{\mu_0 \varepsilon_0}} = \frac{\omega}{\beta}$$

$$= \frac{1}{\sqrt{4 \times 3.14 \times 10^{-7} \times 8.8854 \times 10^{-12}}} = 3 \times 10^8 \text{ m/sec}$$

is velocity of light in free-space.

Example 24. Short-vertical antenna situated on the surface of a perfect conducting earth, produces effective field strength given by, $E_\theta = 100 \sin\theta \left(\dfrac{mV}{m}\right)$ at a distance of 1 m from the antenna. Find the total power radiated.

Solution : Given $\quad E_\theta = 100 \sin\theta \left(\dfrac{mV}{m}\right)$

Power flow in the radial direction]

$$P_r = \int_0^{2\pi} \int_0^{\pi} .P_\theta \sin\theta.d\theta.d\phi \qquad \text{As, } P = \frac{E^2}{R}$$

$$= \frac{10000}{377} \times 2\pi \int_0^{\pi} \sin^3\theta.d\theta.d\phi$$

$$= \frac{166.57}{4} \times \int_0^{\pi} [3 \sin\theta - \sin 3\theta]$$

$$= \frac{166.57}{4}\left[3 \cos\theta - \frac{(-\cos 30)}{3}\right]_0^{\pi}$$

$$P_r = \frac{166.6}{4}\left[3(\cos\pi - \cos 0) + \frac{1}{3}(\cos 3\pi - \cos 0)\right]$$

$$= \frac{166.6}{4}\left[3(-1-1) + \frac{1}{3}(\cos 3\pi - 1)\right]$$

$$= \frac{166.6}{4}\left[+6 + \frac{1}{3}(-2)\right]$$

$$= \frac{166.6}{4} \times \frac{16}{3} = 222.13 \text{ watts}$$

Examples 25. In a loss-less T_x line, the velocity of propagation is 2.5×10^8 m/sec. If capacitance of the line is 30 pF, find :

(*i*) Inductance of the line

(*ii*) Phase-constant at 100 MHz

(*iii*) Characteristic impedance of the line.

Solution : Phase-constant $(\beta) = \frac{\omega}{V_p} = \frac{2 \times 3.14 \times 100 \times 10^6}{2.5 \times 10^8} = 2.51$ rad/m

Phase-velocity, $V_p = \frac{1}{\sqrt{LC}}$

$$\Rightarrow \quad L = \frac{1}{V_p^2.C} = \frac{1}{30 \times 10^{-12} \times (2.5 \times 10^8)^2}$$

$$L = 5.33 \times 10^{-7} \text{ H/m}$$

Characteristic impedance, $Z_0 = \sqrt{\frac{L}{C}}$

$$= \frac{\sqrt{5.33 \times 10^{-7}}}{3 \times 10^{-11}} = 133.3\Omega$$

Example 26. A loss-less *Tx* line having, $Z_0 = 75\Omega$ is terminated with unknown load impedance 'Z_L' and VSWR is 3. If first voltage minima is occured from the load at 20 cm. Find, Z_1 at operating frequency 150 MHz.

Solution : $Z_0 = 75\Omega$ and $\lambda = \frac{c}{f} = \frac{3 \times 10^8}{150 \times 10^6} = 2$ m

Since minima occurs at 20 cm

$$2\beta d - \theta = \pi$$

$$\Rightarrow \quad 2 \times \frac{2\pi}{2} \times 20 - \theta = \pi$$

$$\Rightarrow \quad \theta = -0.6\,\pi = -108°$$

$$\text{SWR} = \frac{1+|\rho|}{1-|\rho|}, \text{ or } |\rho| = \frac{SWR-1}{SWR+1} = \frac{3-1}{3+1} = \frac{1}{2} = 0.5$$

$$\rho = |\rho| e^{j\theta} = \frac{Z_L - Z_0}{Z_L + Z_0}$$

$$\Rightarrow \quad 0.5e^{-j108} = \frac{Z_L - Z_0}{Z_L + Z_0}$$

$$\Rightarrow 0.5\,[(\cos 108) - j\,(108°)] = \frac{Z_L - Z_0}{Z_L + Z_0}$$

$$\Rightarrow 0.5\,[-0.309 - j \times 0.951] = \left(\frac{Z_L - Z_O}{Z_L + Z_O}\right)$$

$$\Rightarrow \quad [0.1545 + 0.4755\,j] = \left(\frac{Z_L - Z_0}{Z_L + Z_0}\right)$$

$$\Rightarrow (Z_L + Z_0)\,(0.155 + j\,0.48) = Z_0 - Z_L$$

$$\Rightarrow 0.155\,Z_L + 0.155\,Z_0 + j\,0.48\,Z_L + j\,0.48\,Z_0 - Z_L = Z_0 - Z_L$$

$$\Rightarrow (1.155 Z_L + j\,0.48 Z_L) = 0.845\,Z_0 - j\,0.48\,Z_0$$

$$\Rightarrow \qquad Z_L = \frac{(0.845 - j\,0.48)\,Z_0}{(1.155 + j\,0.48)} = \left[\frac{63.38 - j\,36}{1.155 + j\,0.48}\right]$$

$$= \frac{(63.38 - j36) \times (1.55 + j0.48)}{(1.155)^2. + (0.48)^2\,(58 - j.7.16)} = (58 - \text{j}\,7.16)\,\Omega$$

Example 27. A 2 × 3 cm² wave-guide is filled with a dielectric of $\varepsilon_r = 4$. If the wave-guide operates at 20 GHz in TM_{11} mode, find cut-off frequency, phase-constant and phase-velocity.

Solution : (*a*) The f_c (TM$_{11}$ mode) $= \dfrac{c}{2\sqrt{\varepsilon_r}}\left[\left(\dfrac{1}{a}\right)^2 + \left(\dfrac{1}{b}\right)^2\right]^{1/2}$

$$= \frac{3 \times 10^8}{2\sqrt{4}}\left[\left(\frac{1}{0.02}\right)^2 + \left(\frac{1}{0.03}\right)^2\right]$$

$$= \frac{9.014}{2}\ \text{GHz} = 4.57\ \text{GHz}$$

(*b*)

$$\beta = \beta_0\left[1 - \left(\frac{f_c}{f}\right)^2\right]^{1/2}$$

$$= \frac{2 \times 3.14}{1.5}\left[1 - \left(\frac{4.57}{20}\right)^2\right]^{1/2}$$

$$= 4.2 \times [1 - 0.0522]^{1/2}$$

$$= 4.2 \times [0.95]^{1/2} = 4.093\ \text{rad/cm}$$

$$= 409.3\ \text{rad/m}$$

(*c*) Phase-velocity (v_p)

$$= \frac{c/\sqrt{\varepsilon_r}}{\left[1 - \left(\frac{f_c}{f}\right)^2\right]^{1/2}}$$

$$= \frac{3 \times 10^8\sqrt{4}}{[0.95]^{1/2}} = \frac{1.5 \times 10^8}{0.975}$$

$$V_p = 1.538 \times 10^8\ \text{m/sec}$$

Example 28. An electric field of strength 10 μV/m is to be measured at a point, $\theta = \pi/2$, and 500 km from a half-wave dipole antenna in air at 50 MHz. Find

(*i*) Length of the dipole.

(*ii*) Current that is fed to the antenna.

(*iii*) Average power radiated by the antenna.

Solution : (*i*)

$$\lambda = \frac{c}{f} = \frac{3 \times 10^8}{50 \times 10^6} = 6\text{m}$$

$$l = \frac{\lambda}{2} = 3m$$

(*ii*)

$$E_\phi = \frac{\eta I_0 \cos\left(\frac{\pi}{2}\cos\theta\right)}{2\pi\,r\sin\theta} = \frac{120\pi \times I_0}{2\pi \times 500 \times 10^3 \times 1}$$

$$\Rightarrow \qquad 10 \times 10^{-6} = \frac{60.I_0}{500 \times 10^3}$$

$$\Rightarrow \qquad I_0 = \frac{10 \times 10^{-6}\ 500 \times 10^3}{60}$$

$$= 83.3 \times 10^{-3}\ A = 83 \text{ mA}$$

(*iii*) Since, for $\lambda/2$ dipole antenna, Radiation resistance $(R_r) = 73\Omega$

$$P_{av} = \frac{1}{2}.I_0^2 R_r$$

$$= \frac{1}{2} \times (83)^2 \times 73\,(10^{-3})^2$$

$$P_{av} = 253.6\ \mu\text{W}$$

Example 29. A uniform plane wave at 1 GHz, is propagating through a non-magnetic dielectric medium with $\varepsilon_r = 4$ and $\tan\delta = 0.02$. The RMS value of the forward travelling wave at $z = 10$ cm is 100 V/m. Calculate the loss in power density between $z = 10$ cm and $z = 70$ cm.

Solution : Given $f = 1$ GHz $= 10^9$ Hz

Hence, velocity $$v = \frac{3 \times 10^8}{\sqrt{\varepsilon_r}} = \frac{3 \times 10^8}{\sqrt{4}} = 1.5 \times 10^8 \text{ m/sec}$$

$$\beta = \frac{2\pi}{\lambda} = \frac{2\pi \times f}{v} = \frac{2 \times \pi \times 10^9}{1.5 \times 10^8}$$

$$\Rightarrow \qquad \beta = \frac{2\pi}{0.15}$$

$$\text{Attenuation factor } (\alpha) = \frac{\beta}{2}\tan\delta = \frac{3.14}{0.15} \times 0.02 = 0.42 \text{ Np/m}$$

$$\text{Hence} \qquad \left(\frac{E\ at\ 70\ cm}{E\ at\ 10}\right) = -\alpha\,(z_2 - z_1) = e^{-0.42 \times 0.6} = 0.78$$

$$\Rightarrow \qquad E \text{ at } 70 \text{ cm} = E_{10} \times 0.78 = 100 \times 0.78 = 78 \text{ V/m}$$

$$\frac{Power\ at\ 70\ cm}{Power\ at\ 10\ cm} = \left(\frac{78}{100}\right)^2 = 0.6084$$

Therefore, loss of power density is

$$(1 - 0.6084) \times 100\% = 39.2\% \text{ of the lost of power in the dielectric}$$

Example 30. Plane $z = 0$ and $z = 4$, carry current $K_0 = -15\ a_x$ A/m and $K_z = 20\ a_x$ A/m, respectively Find H at (1, 1, 1).

Solution : Let the parallel sheet is shown as follows :

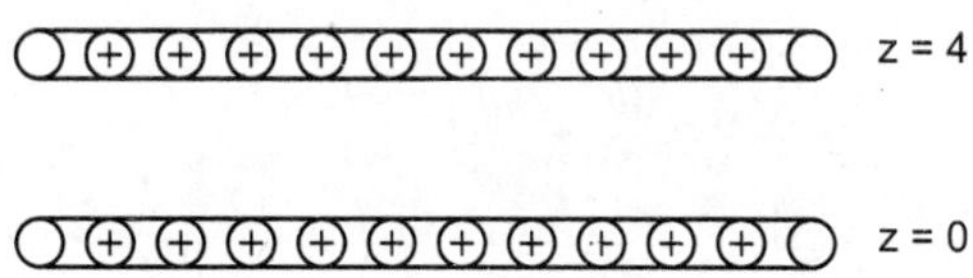

Fig. 2. Geometry for example 30.

Also, let $H = H_{z=0} + H_{z=4}$

We know that, magnetic field intensity, H' due to an infinite sheet of current density K (A/m) is given by

$$H = \frac{1}{2} K \times a_x$$

where, a_x = unit normal vector directed from the current sheet to the point of observation.

$$H_{z=0} = \frac{1}{2} K \times a_n = \frac{1}{2}(-15a_x) \times a_z = 7.5a_y \text{ A/m}$$

$$H_{z=4} = \frac{1}{4} K \times a_n = (20a_x) \times (-a_z) = 7.5\, a_y \text{ A/m}$$

$$H_T = (7.5\, a_y + 7.5\, a_y) = 15a_y \text{ A/m}$$

Example 31. A small current loop 'l_1' with magnetic moment 5 a_z A-m² is located at the origin while another loop l_2 with magnetic moment 3 a_y A-m² is located at (4, –3, 10). Determine the torque on l_2.

Solution : The torque T_2 on the loop L_2 will be will be due to the field B_1 produced by loop L_1, Hence

$$T = m_0 \times B_1$$

Since, m_1 is magnetic moment for loop (L_1) and is along a_z we can find the B_1 using

$$B_1 = \frac{\mu_0 . m_1}{4\pi . r^3}(2\cos\theta . a_r + \sin\theta\, a_\theta)$$

Since, $m_2 = 3a_y = 3(\sin\theta\sin\phi . a_r + \cos\theta\sin\phi . a_\theta + \cos\phi a_z)$

At given point (4, –3, 10)

$$r = [4^2 + 3^2\ 10^2]^{1/2} = 5\sqrt{5}$$

$$\tan\theta = \frac{\rho}{z} = \frac{5}{10} = \frac{1}{2}$$

which gives $\sin\theta = \frac{1}{\sqrt{5}}$ and $\cos\theta = \frac{2}{\sqrt{5}}$

Hence, $\tan\phi = \frac{y}{x} = \frac{-3}{4} \Rightarrow \sin\phi = \frac{-3}{5}, \cos\phi = \frac{4}{5}$

$$\therefore \quad B_1 = \frac{4\times 3.14\times 10^{-7}}{4\times 3.14\times(5\sqrt{5})^3}\left(2.\frac{2}{\sqrt{5}}a_\theta + \frac{1}{\sqrt{5}}a_\theta\right)$$

$$= \frac{10^{-7}}{625}(4a_r + a_\theta)$$

and

$$m_2 = 3\left[-\frac{3}{5}.\frac{1}{\sqrt{5}}a_r + \frac{2}{\sqrt{5}}\left(-\frac{3}{\sqrt{5}}\right)a_\theta + \frac{4}{5}a_\phi\right]$$

$$= \left(\frac{-9.a_r}{5\sqrt{5}} - \frac{18a_\theta}{5\sqrt{5}} + \frac{12a_\phi}{5}\right)$$

$$T = 10^{-7}625.\,(5\sqrt{5})\,[-9a_r - 18a_\theta + 12\sqrt{5}\,a_\phi] \times (4a_r + a_\phi)$$

$$= 4.3 \times 10^{-11} (-6a_r + 38.78a_\theta + 24\, a_\phi)$$

$$= (-0.258a_r + 1.665a_\theta + 1.83a_\phi) \text{ N-m}$$

Example 32. The magnetic vector potential, $A = \frac{\rho^2}{4} a_z$. Calculate the total magnetic flux crossing the surface $\phi = \frac{\pi}{2}$, $1 < \rho \leq 2\text{m}$, $0 \leq z \leq 5$ m.

Solution:
$$B = \nabla \times A = \begin{vmatrix} a_\rho & a_\phi & a_z \\ \frac{\partial}{\partial \rho} & \frac{\partial}{\partial \phi} & \frac{\partial}{\partial z} \\ A_x & A_y & A_z \end{vmatrix}$$

$$= \begin{vmatrix} a_f & a_\phi & a_z \\ \frac{\partial}{\partial \rho} & \frac{\partial}{\partial \phi} & \frac{\partial}{2z} \\ 0 & 0 & -\frac{\rho^2}{4} \end{vmatrix} = -\frac{\partial}{\partial \rho}\left(-\frac{\rho^2}{4}\right) = \frac{\rho}{2} a_\phi$$

$$B = \frac{\rho}{2} \text{ and } d_S = d\rho .\, dz .\, a_\phi$$

$$\phi = \int B.ds = \frac{1}{2}\int_{z=0}^{5}\int_{\rho=1}^{2} \rho^2.d\rho.dz$$

$$= \frac{1}{4}[\rho^2]_2^1 = \frac{15}{4} = 3.75 \text{ Wb.}$$

Example 33. If the region $0 \leq z \leq 2$ m is occupied infinite slab of permeable material ($\mu_r = 3.0$) and $B = (10y\, a_x - 5x\, a_y)$ mW/m², within the slab, then determine (*a*) J (*b*) J_b (*c*) M and (*d*) K_b on $z = 0$.

Solution : (*a*) We known that

$$J = \Delta \times H = \frac{1}{\mu_0 \mu_r} \Delta \times B = \frac{1}{4p \times 10^{-1} \times 3.0}\left(\frac{\partial B_y}{\partial x} . \frac{\partial B_x}{\partial y}\right) a_z$$

$$= \frac{10^7}{12\pi}[-5 - 10] \times 10^{-3}\, a_z = -3.98 \text{ KA/m}^2$$

(*b*) $$J_b = X_m J = (\mu_r - 1)\, J = 2 \times -3.98 \text{ KA/m}^2$$

$$= -7.96 \text{ KA/m}^2$$

(*c*) $$M = X_m H = \frac{(\mu_r - 1)B}{4\pi \times \mu_r} = \frac{2 \times (10y\, a_x - 5x\, a_y) \times 10^{-3}}{4 \times 3.14 \times 10^{-7} \times 3.0}$$

$$= \frac{20 \times (10y\, a_x - 5x\, a_y).10^3}{37.60}$$

$$= (5.32y\, a_x - 2.65x\, a_y) \text{ KA/m}$$

(*d*) $K_b = M \times a_z$, since $z = 0$ is the lower side of the slab occupying, $0 < z < 2$, $a_n = -a_z$, hence,

$$K_b = (5.32y\, a_x - 2.65x\, a_y)(-a_z)$$

$$= (2.6x\, a_x - 5.32\, a_y) \text{ KA/m}$$

Example 34. A microwave exposure limits for the frequency range 60–320 MHz is $5\frac{V}{m}$. If the conductivity of the biological substance is 160 $\frac{S}{m}$ and mass-density is 1.3×10^3 $\frac{kg}{m^3}$. Find the power density of the incident waves and SAR.

Solution : The power density

$$P = \frac{E^2}{377} = \frac{5^2}{377} = 6.63 \times 10^{-2} \frac{W}{m^2}$$

and

$$\text{SAR} = \frac{\sigma . E^2}{m_d} = \frac{160 \times 25}{1.3 \times 10^3} = 3.076 \frac{W}{kg}$$

Example 35. Estimate the parameters of EMI bonding at $f = 20$ MHz under the following specifications :

Bonding length = 3.5 cm

Width of bonding = 2.5 cm

Cross-section area = 9.5 cm

Conductivity = $1.8 \times 10^8 \frac{S}{m}$

Solution : DC resistance $(R_d) = \frac{l}{\sigma . s} = \frac{3.5 \times 10^{-2}}{1.8 \times 10^8 \times 9.5 \times 10^{-4}} = 0.204$ mΩ

$$\text{Skin-depth } (\delta) = \left(\frac{1}{\pi f \mu \rho}\right)^{1/2}$$

$$= \frac{1}{(3.14 \times 20 \times 10^{-6} \times 4 \times 3.14 \times 10^{-7} \times 1.8 \times 10^8)^{\frac{1}{2}}} = 0.0083 \text{ mm}$$

Hence, AC resistance (R_A) is

$$= \frac{l}{\sigma . S . W} = \frac{3.5 \times 10^{-2}}{1.8 \times 10^8 \times 0.0083 \times 10^3 \times 2.5 \times 10^{-2}} = 937 \ \mu\Omega$$

Example 36. The low-pass filter are used for filtering the EM interference at $f = 122$ MHz, in two arrangement (*i*) capactive (*ii*) inductive. Find the insertion loss, if the circuit components of the filter are $R = 125\Omega$, $L = 234$ μH and $C = 145$ μF.

Solution : For the capacitive management

$$K = \pi f RC = 3.14 \times 122 \times 10^6 \times 125 \times 145 \times 10^{-12} = 6.94$$

Hence, the insertion loss,

$$I_l = 10 \log (1 + K^2) = 10 \log [1 + (6.94)^2 \} = 16.92 \text{ (dB)}$$

For inductive arrangement :

$$K = \frac{\pi f L}{R} = \frac{3.14 \times 122 \times 10^6 \times 234 \times 10^{-6}}{125} = 717.12$$

Hence,

$$I_L = 10 \log 10 [1 + (717.12)^2] = 57.12 \text{ dB}$$

Example 37. Find the reflection loss and absorption loss in the shielding used to refuse EMI reflection and absorption, such that the relative permittivity of material is 3 and thickness of the shield is 0.55 cm, at operating frequency 2.22 MHz.

Solution : The reflection loss

$$R\,(\text{dB}) = 168 + 10\log\left(\frac{\sigma_r}{\mu_r.\,f_r}\right)$$

$$= 168 + 10\log\left(\frac{3}{1\times 2.2\times 10^6}\right)$$

$$= 109.34 \text{ dB}$$

Skin-depth

$$(\delta) = \left(\frac{1}{\pi f \mu\sigma}\right)^{1/2} - \frac{1}{(3.14\times 2.2\times 10^6\times 4\times 3.14\times 10^{-7}\times 1.8\times 10^8)^{1/2}}$$

$$= 0.025 \text{ mm}$$

Hence, the absorption loss

$$A\,(\text{dB}) = \frac{8.6859\times 0.55\times 10^{-2}}{0.025\times 10^{-3}} = 19.11 \text{ dB}$$

Example 38. Write the expressions for shielding effectiveness in terms of power and field intensity.

Solution : (*a*) In terms of power. S.E, $= 10\log\left(\frac{P_i}{P_t}\right)$

(*b*) In terms of field intensity

$$\text{S.E.} = 20\log_{10}\left(\frac{E_i}{E_t}\right) = 20\log 10\left(\frac{H_i}{H_t}\right)$$

Example 39. If r is position vector of the point (x, y, z) and A is a constant vector, using vector analogy, show that

(*i*) $(r - A).\, r = 0$, is the equation of a sphere

(*ii*) $(\text{r} - A).\, \text{A} = 0$, is the equation of a plane

(*iii*) $(A - r)\times r \neq 0$

Solution : (*i*) Let $A = (A_x, A_y, A_z)$ and $r = (x, y, z)$

$$(r - A).\, r = (x - A_x)\,x + (y - A_y)\,y + (z - A_z)z$$

$$= x^2 - x.\,dx + y^2 - A_y\,.\,y + z^2 - A_z\,.\,z$$

So, if $(r - A).\, r = 0$

$$\Rightarrow \quad x^2 + y^2 + z^2 = x\,A_x + y\,A_y + z\,A_z = r^2 \qquad \text{(say)}$$

which is the equation of sphere whose surface touches the origin and radius $\sqrt{x - A_x + y.A_y + z.A_z}$

(*ii*) $(r - A).\, A = (x - A_x).\, A_x + (y - A_y)\, A_y + (z - A_z)\, A_z$

$$= x.\,A_x + y\,A_y + z\,A_z - A_y^2 - A_z^2 - A_x^2$$

$$= x\,.\,A_x + y.\,A_y + z\,.\,A_z + D \text{ (say)}$$

where $D = -\,(A_x^2\;\; A_y^2 + A_z^2)$

So, if $\quad (r - A).\,\text{A} = 0$

$\Rightarrow x.\,A_y + y.\,A_y + z.\,A_z + D = 0$

Which is the equation of plane.

$\Rightarrow \quad \{(A_y - y)\,z - (A_z - z)\,y\}i + [(A_z - z)\,x - (A_x - x)z]\,j + [(A_x - x)y - (A_y - y)k]$

$= [(A_y.\,z - yz - A_z - A_z\,.\,y + zy)\,i + (A_z\,.\,x - zx - A_x\,.\,z + x_z)\,j + (A_x\,.\,y - xy - A_y\,.\,x + xy)k]$

$\Rightarrow \quad A_x\,(-zj + yk) + A_y\,(zi - xk) + A_z\,(xj - y_i)\text{k}$

Example 40. (a) Given $\rho_x = 2x^2 + xy$ then calculate $\int_S \rho_x dS$ over the region $y \le x^2, 0 < x < 1$.

Solution : Hence $\int_S \rho_x dS = \int_{x=0}^{x=1} \int_{y=0}^{y=x^2} (2x^2 + xy)\,dx\,dy$

$$= \int_{x=0}^{x=1} \left(2x^2 y + \frac{xy^2}{2}\right)_0^{x^2} dx = \int_0^1 \left(2x^4 + \frac{x^5}{2}\right) dx$$

$$= \left[\frac{2x^5}{5} + \frac{x^6}{12}\right]_0^1 = \frac{2}{5} + \frac{1}{12} = \frac{29}{60} = 0.48$$

In this question integral is evaluated w. r. to y first, because, the region is enclosed by 'y' first, and then second by 'x'

Example 40. (*b*) The electric flux density of a region is defined by

$$D = 2z\rho\cos^2\phi\,a_z\left(\frac{C}{m^2}\right)$$

Calculate the change density at $\left(1, \frac{\pi}{4}, 3\right)$ and using Gauss's law also find the total charge enclosed by the cylinder of radius 1 m with in vegion $-2 \le z \le 2$ m.

Solution : The charge density,

$$\rho_v = \Delta.\,D = \frac{\partial D_z}{\partial z} = 2\rho\cos^2\phi$$

$$\rho_v\left(1, \frac{\pi}{4}, z\right) = 2 \times 1 \times \cos^2\frac{\pi}{4}$$

$$= 2 \times \frac{1}{4} \times 1 = 1.0\,\frac{C}{m^3}$$

From the Gauss's law

$$Q = \psi = \int D.dS = \left[\int_S + \int_t + \int_b\right] D.ds$$

$$= \psi_s + \psi_t + \psi_b$$

where ψ_s, ψ_t and ψ_b represents, flux through the side, the top surface and bottom surface of the cylinder respectively. Since D does not have component along a_ρ, $\psi_s = 0$.

Therefore, for ψ_t, $\quad dS = \rho.\,d\phi.\,d\rho.\,a_z$

$$\psi_t = \int_{\rho=0}^{1} \int_{\phi=0}^{2\pi} 2z.\,\rho\cos^2\phi.\rho\,d\phi.d\rho\big|_{z=2}$$

$$= 4\int_0^1 \rho^2\,d\rho \int_0^{2\pi} \cos^2\phi.\,d\phi$$

$$= 4\left(\frac{\rho^3}{2}\right)_0^1.\rho = \frac{4\pi}{3}$$

and for ψ_b. $\quad dS = -\rho\,d\phi.\,d\rho\,a_z$

So, $$\psi_b = -\int_{\rho=0}^{1}\int_{\phi=0}^{2\pi} 2z\pi\cos^2\phi d\phi d\rho\Big|_{z=-2}$$

$$= 4\left(\frac{1}{3}\right).p = \frac{4\pi}{3}$$

Thus, $$Q = 0 + \frac{4\pi}{3} + \frac{4\pi}{3} = \frac{8\pi}{3}C$$

Example 41. The total electrostatic energy due to 'n' point charges is

$$W_E = \frac{1}{2}\sum_{k=1}^{n} Q_k.V_k \ Joules \qquad ...(i)$$

Then, show that for a continuous volume charge distribution ;

$$W_E = \frac{1}{2}\int_v D.\,E dv = \frac{1}{2}\int \varepsilon_0 |E|^2 dv \qquad ...(ii)$$

Solution : We known that, for continuous charge distribution, the W_E may be expressed as

$$W_E = \frac{1}{2}\int \rho_v V\, dv = \frac{1}{2}\int_v (\Delta.\,D)\, V.\, dv \qquad ...(iii)$$

Also, for any vector 'A' and scalar 'V' the identity

$$\Delta.\, VA = A.\,\Delta V + V(\Delta.V)$$

$$\Rightarrow \qquad (\Delta.\,A)\, V = \Delta.\, VA - A.\,\Delta V \qquad ...(iv)$$

Therefore, from (*iii*) and (*iv*), we get

$$W_E = \frac{1}{2}\int (D.\, VD)\, dv - \frac{1}{2}\int (D.\Delta V) dv \qquad ...(v)$$

Using Divergence Theorem, on the first term of equation gives :

$$W_E = \frac{1}{2}\int (VD).dS = \frac{1}{2}\int_v (D.\Delta V) dv \qquad ...(vi)$$

It is also well know that, V varies as $\left(\frac{1}{r}\right)$ and D as $\frac{1}{r^2}$ for point charges and

V varies as $\frac{2}{r^2}$ and D as $\frac{1}{r^2}$ for dipoles and so on.

Therefore, (VD) together must be atleast as $\frac{1}{r^3}$ and dS varies as r^2, consequently, the first integral of equation (*vi*) must tend to zero, as the surface S becomes large. That is

$$W_E = -\frac{1}{2}\int_V (D.\,\Delta V)\, dv = \frac{1}{2}\int (D.\,E)\, dv$$

and as $E = -\Delta V$ and $D = \varepsilon_0 E$;

$$W_E = \frac{1}{2}\int D.E\, dv = \frac{1}{2}\int \varepsilon_0 E^2 dv$$

Example 42. Derive the expression for Ohm's law, $J = \sigma E$.

Solution : Let a conductor is placed in a region of electric field of intensity E, then the force on an electron with charge ($-e$) is

$$F = -eE \quad ...(i)$$

Due to this force an electron suffers constant collision with the atomic lattice and drive from one atom to another. And if 'm' is mass of a electron and 'u' average drift velocity due to electric field E, then according to Newton's law, average change in momentum must be equal to applied force, that is

$$\frac{mu}{\tau} = -eE$$

$$u = -\frac{e\tau}{m}E \quad ...(ii)$$

where 'τ' is average time interval between collisions. Therefore, the drift velocity of electron is directly proportional to the applied field. If there are n electron per unit volume, the electron charge density may be given as :

$$\rho_v = -n.e$$

So, the conduction current density will be

$$J = \rho_u \, . \, u = (-ne)\left(\frac{e\tau}{m}\right)E$$

$$= ne^2\left(\frac{\tau}{m}\right). \; E = \sigma E \quad ...(iii)$$

$$J = \sigma E$$

where $\sigma = \frac{me^2\tau}{m}$ is termed as the conductivity of the material.

Example 43. (*a*) Define magnetization. Show that, the relative permeability (μ_r) and magnetic susceptibility (X_m) for a magnetic material can be given by ;

$$\mu_r = 1 \times X_m$$

Solution : The magnetic dipole moment per unit volume, is defined as magnetization (M).

The value of M is zero for free-space, whereas for any other material medium $\mu \neq 0$ and as a result, the magnetic flux density 'B' change to

$$\Delta \times \left(\frac{B}{\mu_0}\right) = I_f + J_b = \Delta \times H + \Delta \times M \quad ...(i)$$

or $$B = \mu\,(H + M),$$

this relationship is true for all materials whether they are linear or not. However, for linear materials, Magnetization depends linearly on H such that ;

$$M = X_m H \quad ...(ii)$$

where X_m is known as magnetic susceptibility of the medium, and measure sensitivity of the material to magnetic field. hence, from (*i*) and (*ii*), we get

$$B = \mu_0\,(1 + X_m)\,H = \mu H$$

Or $$B = \mu_r . \, \mu_0 H$$

Where $\mu_r = 1 + X_m = \frac{\mu}{\mu_0}$

Example 43. (*b*) A rod of length '*l*' is rotated about *z*-axis, with angular velocity of ω. If $B = B_0 \cdot a_z (y)$, derive the expression for the voltage induction of the conductor. Also, find panicking volume if $l = 4.4$ cm $f = 2$MHz and $H_0 = 5.2$A/m.

Solution : We know that, $V_{caf} = \int (u \times \overline{B}) dl$ ...(*i*)

where in this question,

$$dl = d\rho.\, a_\rho \text{ and } \mu = \rho \frac{d\phi}{dt}$$

$$= \rho\omega a_\phi,$$

Hence $$u \times \overline{B} = \rho\omega_a \phi \times B_0 a_z = B_0 . \rho\omega (\overline{a}_\phi \times \overline{a}_z)$$

$$= B_0 \rho\omega a_\rho \quad ...(ii)$$

as, $$\overline{a}_\phi \times \overline{a}_z = \alpha_\rho$$

Therefore from (*iv*) and (*ii*), we get

$$V_{emf} = \int_{\rho=0}^{1} B\omega\rho\overline{a}\rho.\, d\rho\, \overline{a}\rho$$

$$= B_0\omega \left[\frac{\rho^2}{2}\right]_0^l = \frac{1}{2} B_0 \omega l^2$$

where $B_0 = \mu H_0 = 4 \times 3.14 \times 10^{-7} \times 5.2 = 65.312 \times 10^{-7} \frac{Wb}{m^2}$

Hence, $$V_{emf} = \frac{1}{2} \times 65.312 \times 10^{-7} \times 3.14 \times 2 \times 10^6 \times (4.4)^2 \times 10^{-4}$$

$$= 3.97 \times 10^{-5} = 39.701\ \mu V$$

Example 44. A distorted less Tx line is characterized by :

$Z_0 = 55\Omega.\ \alpha = 20\text{m}\left(\frac{N_p}{m}\right)$ and $u = 0.6\, u_0$ in which u_0 is the velocity of light in free-space. Find., R, L and *C* at 100 MHz.

Solution : $R = \alpha.\, Z_0 = 20 \times 10^{-3} \times 55 = 1.1\ \Omega$

$$L = \frac{Z_0}{u} = \frac{55}{0.6 \times 3 \times 10^8} = 30.56 \times 10^{-8} \frac{H}{m}$$

$$C = \frac{1}{uZ_0} = \frac{1}{55 \times 1.8 \times 10^8} = \frac{1}{99} \times 10^{-8}$$

$$= 1.010 \times 10^{10} \frac{F}{m}$$

Example 45. "For a square wave-guide, attenuation constant α_c is minimum for TE_{10} mode, Provided $f = 2962 f_c$." Prove it.

Solution : In a rectangular wave-guide, for TE_{10} mode,

$$\alpha_c = \frac{2Rs}{b.\eta\sqrt{1-\left(\frac{fc}{f}\right)^2}}\left[\frac{l}{2} + \frac{b}{2}\left(\frac{fc}{f}\right)^2\right] \quad ...(i)$$

where R_s = Surface exactness = $\frac{l}{\sigma_c \delta} = \sqrt{\frac{\pi f \mu}{\sigma_c}}$

Hor square guide, $a = bl$ equation (i) reduces to :

$$\alpha_c = \frac{2\sqrt{\frac{\pi f \mu}{\sigma_c}}}{a\eta\sqrt{\left[1-\left(\frac{f_c}{f}\right)^2\right]}}\left[\frac{l}{2}+\left(\frac{f_c}{f}\right)^2\right]$$

$$= \frac{2\sqrt{\left(\frac{\pi\mu}{\sigma_c}\right)}}{a.\eta} \times \frac{\sqrt{f}\left[\frac{l}{2}+\left(\frac{f_c}{f}\right)^2\right]}{\sqrt{1-\left(\frac{f_c}{f}\right)^2}}$$

where, let $K = \frac{2\sqrt{\frac{\pi\mu}{\sigma_c}}}{a.\eta}$, is a constant

Now for minimum value of a_c, $\frac{d\alpha_c}{df}$ should be zero.

That is

$$\frac{d\alpha_c}{df} = K\left[1-\left(\frac{f_c}{f}\right)^2\right]^{\frac{1}{2}}\left[\frac{1}{2}\frac{1}{2}f^{-1/2} - \frac{3}{2}f_c^2 . f^{-5/2}\right]\left[1-\left(\frac{f_c}{f}\right)^2\right] = 0$$

$$\Rightarrow \quad \left[\frac{1}{4\sqrt{f}} - \frac{3}{2}f_c^2 f^{-5/2}\right] - \left[\frac{1\sqrt{f}}{4\sqrt{f}} + f_c^2\, f^{-3/2}\right]\frac{(2f_c^2\, f^{-3/2})}{\sqrt{1-\left(\frac{f_c}{f}\right)^2}} = 0$$

$$\left[1-\left(\frac{f_c}{f}\right)^2\right]^{-1/2}\left[\frac{1}{4\sqrt{f}} - \frac{3}{2}\frac{f_c^2}{f^{5/2}}\right] - \left[\frac{l\sqrt{f}}{4\sqrt{f}} - \frac{f_c^2}{f^{3/2}}\right]\frac{2f_c^2}{f^3\sqrt{1-\left(\frac{f_c}{f}\right)^2}} = 0$$

As $\quad \left[1-\left(\frac{f_c}{f}\right)^2\right]^{1/2} \neq 0$

$$\Rightarrow \quad \left[\frac{l}{4\sqrt{f}} - \frac{3}{2}\frac{f_c^2}{f^{5/2}}\right] = \left[\frac{l\sqrt{f}}{4\sqrt{f}} - \frac{f_c^2}{f^{3/2}}\right]\frac{2f_c^2}{f^3\left[1-\left(\frac{f_c}{f}\right)^2\right]}$$

$\Rightarrow$ Which gives $f = 2.962\, f_c$

Example 46. For a medium, having $\varepsilon_r = 2.5$, An antenna has far-zone field given by

$$E_S = \frac{5\sin 2\theta}{r}e^{-j\beta r} . a_\theta\left(\frac{V}{m}\right)$$

Where $\beta = \omega\sqrt{\mu_0 . \varepsilon_0}$, Find the average radiated power.

Solution : We know that

$$P_r = \frac{|E_s|^2}{2\eta}.a_r = \frac{25\sin^2 2\theta}{2\eta r^2}a_r$$

Hence
$$P_{ral} = \frac{25}{2\eta}\int\int(2\sin\theta\cos\theta)^2\sin\theta\, d\theta\text{-}d\phi$$

$$= \frac{25}{2\times 120\pi}\sqrt{\varepsilon_r}\times 2\pi\int 4\sin^2\theta.\cos^2\theta d(-\cos\theta)d\theta$$

$$= \frac{15}{120}\sqrt{2.5}\int(\cos^4\theta - \cos^2\theta)\, d(-\cos\theta)d\theta$$

$$= \frac{25\sqrt{2.5}}{120}\left(\frac{\cos^5\theta}{5} - \frac{\cos^3\theta}{3}\right)_\theta^\pi$$

$$= \frac{25}{120}\sqrt{2.5}\times\left(-\frac{2}{5}+\frac{2}{3}\right)$$

$$= \frac{25\sqrt{2.5}}{120}\times\frac{-6+10}{15}$$

$$= \frac{25\sqrt{2.5}}{120}\times\frac{4}{15} = 87.84 \text{ mW}$$

Example 47. A parallel plate capacitor is filled with a dielectric of power factor 0.01 and ε_r = 9.6 The plates are 300 mm square and 20 mm distance apart. If 2000 V (r.m.s) is applied at 1.5 MHz, find the power dissipated as heat.

Solution : We know that for the parallel plate capacitor.

$$C = \frac{\varepsilon_0.\varepsilon_r.A}{d} = \frac{9.6\times 8.854\times 10^{-12}\times 900\times 10^{-6}}{20\times 10^{-3}}$$

$$= 0.382\times 10^{-9} \text{ Farad}$$

$$= 382 \text{ pF}$$

Power dissipated in the lossy dielectric is

$$P = V.I_C\cos\phi = VI_C\tan\delta$$

$$= V(V\omega C)\tan\delta = V^2\omega_C\tan\delta$$

Hence $P = 2000\times 2000\times 3.14\times 1.5\times 10^6\times 0.01\times 382\times 10^{-12} = 143.94$ W

Example 48. Design a cylinder cavity resonator for the following specification :

Resonant at TH_{011} mode frequency 9.40 GHz

Resonant at TE_{111} mode frequency 8.2 GHz.

Solution : We known that for circular cylinder resonator:

For TE_{111} mode.

$$\lambda_\omega = \frac{2l}{\left[1+\left(\frac{2l}{3.4\, la}\right)^2\right]} = \frac{3\times 10^{10}}{8.2\times 10^9} = 3.66 \text{ m} \qquad ...(i)$$

For $TE_{0.01}$ mode

$$\lambda_\omega = \frac{2l}{1+\left(\frac{2l}{1.64a}\right)} = \frac{3\times 10^{10}}{9.40\times 10^3} = 3.19 \text{ m} \qquad ...(ii)$$

Therefore form (*i*) and (*ii*)

$$\frac{3.66}{3.19} = \sqrt{\frac{1+\left(\frac{2l}{1.64a}\right)^2}{1+\left(\frac{2l}{3.4la}\right)^2}}$$

$$1.316 = \frac{1+1.488\left(\frac{l}{a}\right)^2}{1+0.344\left(\frac{l}{a}\right)^2}$$

$$\Rightarrow \qquad \left(\frac{l}{a}\right) = 0.553 \qquad \text{...(iii)}$$

Again, from equation (*i*)

$$2l = 3.66\ 3.66\sqrt{1+\left(\frac{2}{3.41}\right)^2 \times 0.3052}$$

$$\Rightarrow \qquad l = 1.93 \text{ cm}$$

Hence $$a = \frac{l}{0.533} = \frac{1.93}{0.533} = 3.49 \text{ cm}$$

Example 49. Design a single section quarter wave transformer at 8 GHz from 1 × 3 cm^2 guide to 1 × 2cm^2 guide, if the guide is filled wit air.

Solution : If we choose, TE_{10} mode;

$$\lambda_c = 2a = 2\text{cm}$$

For which, $f_c = \frac{3 \times 10^{10}}{2}$ Hz = 15GHz, which is not possible.

For TE_{10} mode, λ_c for the first guide is

$$2 \times 3 = 6 \text{ cm}$$

and corresponding cut-off frequency will be;

$$f_c = \frac{3 \times 10^{10}}{6} \text{ Hz} = 5 \text{ GHz}$$

For second wave-guide

$$\lambda_c = 2 \times 2 = 4\text{cm}$$

$$f_c = \frac{3 \times 10^{10}}{4} = 7.5 \text{ cm}$$

For 8 GHz frequency

$$\lambda = \frac{v}{f} = \left(\frac{3 \times 10^8}{8 \times 10^9}\right) = 3.75 \text{ cm}$$

$$\lambda_g = \frac{3.75}{\left[1-\left(\frac{3.75}{6}\right)^2\right]} = \frac{375}{0.609} = 6 \text{ cm}$$

For second wave-guide

$$\lambda_g = \frac{3.75}{\left[1-\left(\frac{3.75}{4}\right)^2\right]} = \frac{3.75}{0.721} = 30.99 \text{ cm}$$

For first wave-guide

$$Z_{01} = 377 \frac{\lambda_a}{\lambda}(b/a) = 377 \times \frac{6.16}{3.75} \times \left(\frac{1}{3}\right)$$
$$= 206.428\ \Omega$$

For second wave-guide

$$Z_{02} = 377 \times \frac{30.99}{3.75} \times \left(\frac{1}{2}\right) = 1557.764\text{W}$$
$$Z\,(\text{trasfer}) = \sqrt{Z_{01} \times Z_{02}} = \sqrt{206.428 \times 1557.764}$$
$$= 567.067 \text{ W}$$
$$\lambda_c = 2b$$

and
$$567.067 = 377 \times \frac{1}{b} \frac{1}{\sqrt{1-\left(\frac{3.75}{2b}\right)^2}} = \frac{377}{\sqrt{b^2-\left(\frac{3.75}{2}\right)^2}}$$

$$\Rightarrow \quad 2.262 = \frac{1}{b^2 - 3.515}$$

$$\Rightarrow \quad b^2 - 3.515 = \frac{1}{2.262}\ 0.442$$

$$\Rightarrow \quad b = 1.989 \text{ cm}$$

For matching section :

$$\lambda_g = \frac{3.75}{\sqrt{1-\left(\frac{3.75}{3.978}\right)^2}} = \frac{3.75}{0.3368} = 11.25\text{cm}$$

Hence, the length of matching section

$$\frac{\lambda_g}{4} = \frac{11.25}{4} = 2.8 \text{ cm}$$

Example 50. A 50 Ω loss-less Tx line connects a signal of 100 KHz to a local line of 100Ω The local power is 100mW. Calculate (*i*) VSWR (*ii*) positions of the first V_{min} and V_{max} (*iii*) impedance at V_{min} and V_{max} and values of V_{max} and V_{min}.

Solution: $\rho_L = \frac{Z_L - Z_0}{Z_L + Z_0} = \frac{100-50}{100+50} = \frac{1}{3}$

$$\lambda = \frac{c}{f} = \frac{30 \times 10^8}{100 \times 10^3}\ 3 = 3000$$

(*i*)
$$\text{VSWR} = \frac{1+|\rho_L|}{1-|\rho_L|} = \frac{1+\frac{1}{3}}{1-\frac{1}{3}} = 2$$

(*ii*) For real $Z_L > Z_0$, first V_{max} is located at the local and V_{min} is located at a distance $\left(\frac{\lambda}{4}\right) = \frac{3000}{4}$ = 750 m from the local.

(*iii*) Local power, $P_L = \dfrac{V_{max}^2}{Z_2} = 100$ min

$$V_{max} = [100 \times 100 \times 10^{-3}] = \sqrt{10} \text{ volts}$$

$$V_{max} = 3.16 \text{ volts}$$

$$V_{min} = \frac{V_{max}}{S} = \frac{3.16}{2} = 1.58 \text{ V}$$

at V_{max}, $Z_{in} = S.Z_0 = 2 \times 50 = 100\Omega$

at V_{min}, $Z_{in} = \dfrac{Z_0}{S} = \dfrac{50}{2} = 25\Omega.$

- There is nothing on this earth more to be prized than true friendship.

Thomas Aquinas

OBJECTIVE QUESTION BANK

1. The electric flux density (D) is also known as

(*a*) Displacement vector (*b*) Electric induction

(*c*) Both (a) and (b) (*d*) None of these

2. The equation, $\nabla^2\phi = 0$, (where ϕ is scalar function) is called

(*a*) Lenz equation (*b*) Laplace's equation

(*c*) Poisson's equation (*d*) None of these

3. The equation, $J = \sigma E$, is known as

(*a*) Ohm's law (*b*) Helmholtz law

(*c*) Faaday law (*d*) None of these

4. An audio frequency transformer has a core made of iron with $\sigma = 0.5 \times 10^7$ and $\mu = 1000\ \mu_0$, at $f = 15$ kHz. What will be its skin-depth

(*a*) 1.07 mm (b) 0.058 mm

(*c*) 5.87 mm (*d*) None of these

5. Which is correct statement

(*a*) Attenuation constant is inversely proportion to the skin-depth

(*b*) Attenuation is directly proportional to the skin-depth

(*c*) Skin-depth is frequency independent

(*d*) None of these

6. In general, the characteristic impedance of a co-axial line lies between

(*a*) 100-120 Ω (*b*) 60-100 Ω

(*c*) 10-50 Ω (*d*) 50 to 80 Ω

7. The reflection coefficient across a T_X line of $Z_0 = 50\ \Omega$, is $\left(\frac{1}{3}\right)$, then what will be the input impedance

(*a*) 70 Ω (*b*) 150 Ω

(*c*) 65 Ω (*d*) 100 Ω

8. The propagation constant (γ) of a T_X line is expressed as

(*a*) $\gamma = \alpha + j\beta$ (*b*) $\gamma = \sqrt{2Y}$

(*c*) $\sqrt{\frac{Z}{Y}}$ (*d*) $\gamma = j\omega\sqrt{LC}$ for ideal T_X line

9. The intrinsic impedance (η) of the free-space is approximately equal to 120 $\pi\Omega$. What is the exact value

(*a*) 377 Ω (*b*) 376.73 Ω

(*c*) 377.12 (*d*) None of these

10. Which is not true for a uniform plane wave

(*a*) Velocity is $\frac{1}{\sqrt{\mu\varepsilon}}$ (*b*) E is normal to H

(*c*) It is TEM wave (*d*) None of these

11. If for a material, $\eta = (1 + j)R_S$, where η is intrinsic impedance and R_S is surface resistivity. Then material is

(*a*) Semi-Conductor (*b*) Good conductor

(*c*) Good dielectric (*d*) None of these

12. A wave incident at angle θ_1 = 60° from air to fused quartz with ε_r = 3.78, find the transmitted angle

(*a*) 25° (*b*) 26.4°

(*c*) 33° (*d*) 50°

13. If for TE component of an EM wave impedaces are 756 Ω and 217 Ω. Then, transmission coefficient will be

(*a*) – 0.478 (*b*) 0.446

(*c*) 0.96 (*d*) 0.87

14. If the quality factor of a resonant cavity at 10 GHz is 10^4. Then the bandwidth between half-power points will be

(*a*) 1 MHz (*b*) 2 MHz

(*c*) 5 MHz (*d*) None of these

15. The directivity of a Hertzian-dipole is found to be

(*a*) 1.5 (*b*) 1.64

(*c*) 2.15 (*d*) None of these

16. The ratio of magnetic momentum to angular momentum is termed as

(*a*) Momentum ratio (*b*) Magnetic ratio

(*c*) Angular ratio (*d*) Gyromagnetic ratio

17. Which is not a ferrite device

(*a*) Gyrator (*b*) Circular

(*c*) Isolater (*d*) None of these

18. Which is incorrect for the hybrid mode

(*a*) HE_{11} mode (*b*) No cut-off frequency

(*c*) Dominant mode (*d*) None of these

19. A material in which the response to the magnetic field leads to losses, is called the

(*a*) Ferrite (*b*) Ferrimagnetic magnetic

(*c*) Para magnetic material (*d*) Both (a) and (b)

20. The field (*s*) that do not pass straight from one plate to other plate is known as
(*a*) Magnetic field (*b*) Fringing field
(*c*) Diverging field (*d*) Electric field

21. The product [$A \cdot (B \times C)$] is referred as the
(*a*) Dot and cross product (*b*) Triple vector product
(*c*) Scalar triple product (*d*) None of these

22. The product [A × (B × C)] is referred as
(*a*) Vector triple product (*b*) Scalar triple product
(*c*) Both (a) and (b) (*d*) None of these

23. Four fundamental equation of electromagnetic are grouped into
(*a*) Ohm's law (*b*) Faraday law
(*c*) Maxwell's equation (*d*) Lenz law

24. Maxwell's equations of electromagnetic are valid under all condition except one and that is
(*a*) Do not apply to isotropic media
(*b*) Do not apply to non-homogeneous media
(*c*) Do not apply to mediums which move with respect to system of co-ordinates
(*d*) Do not apply to non-linear media

25. For a plane EM wave travelling in a linear, isotropic, homogeneous and stationary medium
(*a*) E and H are in the same plane (*b*) E and H are the mutually perpendicular
(*c*) $E \times H \neq 0$ (*d*) Above all these

26. For a loss-less T_X line the following is correct;
(*a*) $R = 0, G = 0$ (*b*) $R \neq 0, G = 0$
(*c*) $R \neq 0, G \neq 0$ (*d*) None of these

27. If the return loss for a load is 18 dB, what will be VSWR
(*a*) 2.7 (*b*) 1.7
(*c*) 1.56 (*d*) 1.29

28. TM_{01} and TE_{01} modes are
(*a*) Symmetrical modes (*b*) Asymmetrical modes
(*c*) Not known (*d*) None of these

29. If velocity factor of a T_X line is 0.8, then what will be the velocity of EM waves
(*a*) 3.5×10^{10} cm/sec (*b*) 3×10^{8} cm/ec
(*c*) 2.4×10^{8} m/sec (*d*) None

30. SWR of a T_X line, unity means
(*a*) The line is terminated Z_0 (*b*) The line ($\lambda/2$) long
(*c*) The line is short circuited (*d*) None of these

31. The SWR on a line in infinity, when the line is open circuited or short-circuited or when it is terminated with:

(*a*) Its Z_0 (*b*) A reactive load
(*c*) A resistive load (*d*) None of these

32. A co-axial cable belongs to
(*a*) TE types of T_X line (*b*) TEM types of T_X line
(*c*) Quasi-TEM types of T_X line (*d*) None of these

33. The characteristics impedance of a T_X line is
(*a*) Independent of its length (*b*) Directly proportional to its length
(*c*) Inversely proportional to its length (*d*) None of these

34. A short-circuited $\left(\frac{\lambda}{4}\right)$ line has an input impedance of
(*a*) Infinity (*b*) Zero
(*c*) Not known (*d*) 50 Ω

35. The distance between successive maximum of a standing wave on a T_X line is
(*a*) $\frac{\lambda}{2}$ (*b*) λ
(*c*) $\frac{\lambda}{4}$ (*d*) $\frac{3\lambda}{4}$

36. If the operative frequency is considered to be equal to cut-off frequency. Then phase and group velocity will be
(*a*) The V_g equal V_p (*b*) V_g becomes zero
(*c*) V_p becomes zero (*d*) None of these

37. The most dominant mode (TE) in an evacuated circular wave-guide is
(*a*) TE_{10} mode (*b*) TE_{01} mode
(*c*) TE_{20} mode (*d*) None of these

38. Which is incorrect statement
(*a*) Pencil beam is highly directive beam
(*b*) A fan beam is narrow in one direction and wide in other
(*c*) Polarization is orientation of electric field vector
(*d*) None of these

39. Directivity of an antenna is
(*a*) Same as its directive gain (*b*) Maximum value of its directive gain
(*c*) Same as its efficiency (*d*) None of these

40. Radar reciprocity is put to use in
(*a*) Radar system (*b*) Television broadcast
(*c*) Radio broadcast (*d*) Microwave heating

41. The quasi-fields with the distance as
(*a*) $\frac{1}{r}$ (*b*) $\frac{1}{r^2}$
(*c*) r^2 (*d*) None of these

42. Ohmic losses on an antenna are

(*a*) Antenna Temp (*b*) Noise figure
(*c*) Noise Temp (*d*) None of these

43. Maxwell's-Faraday equation is

(*a*) $\nabla \times E = -\frac{\partial B}{\partial t}$ (*b*) $\iint B . dS = 0$
(*c*) Both (a) and (b) (*d*) None of these

44. Choose the incorrect option for the plans waves

(*a*) Plane waves are linearly polarised only
(*b*) *E* and *H* fields are always in time phase
(*c*) The magnitude of the two fields is always variable
(*d*) The stored energy is equally divided between the fields

45. Choose incorrect one

(*a*) Dielectric loss increase with frequency
(*b*) Dielectric loss decrease with frequency
(*c*) Dielectric losses are negligible for air
(*d*) None

46. In case of TEM wave in parallel plane wave guide which is incorrect

(*a*) Cut-off frequency always greater than zero
(*b*) The phase-velocity is equal to velocity of light
(*c*) All frequencies can propagate along the wave guide
(*d*) None

47. Which is incorrect

(*a*) Curl $H = 0$ (*b*) $\nabla \times H = 0$
(*c*) $H = -\text{grad}\ \phi$ (*d*) None of these

48. The ratio of magnetization (*M*) to magnetic field intensity is

(*a*) Magnetic susceptibility (*b*) Relative permeability
(*c*) Torque (*d*) None of these

49. Which is not similar

(*a*) MnO (*b*) FeO
(*c*) $Fecl_3$ (*d*) None

50. Unit of power density

(*a*) $\left(\frac{W}{m^3}\right)$ (*b*) $\left(\frac{W}{m}\right)$
(*c*) $\frac{W}{m^2}$ (*d*) None of these

51. The specific resistance of Ferrite is
(*a*) Very high (*b*) Very low
(*c*) Around 100 (*d*) None of these

52. The permeability of the Ferrite is
(*a*) Negative number (*b*) Unit vector
(*c*) Asymmetric tensor (*d*) Simple tensor

53. The depth of penetration does not depend, upon
(*a*) Angle of incidence (*b*) Frequency
(*c*) Permeability (*d*) Conductivity

54. If two vectors are defined by

$$A = 2i - 5j - 4k \text{ and } B = 3i + 5j + 2k$$

What is angle between these vector
(*a*) 0° (*b*) 90°
(*c*) 150.7° (*d*) 130.77°

55. The unit of current density (J) is
(*a*) $\frac{A}{m^2}$ (*b*) $\frac{A}{m^3}$
(*c*) $\frac{V}{m^2}$ (*d*) $\frac{H}{m^2}$

56. The radiation hazards to electro-explosive devices are termed as
(*a*) Man-made hazards (*b*) Natural hazard
(*c*) Radhaz (*d*) None of these

ANSWERS

1.	(*c*)	**2.**	(*b*)	**3.**	(*a*)	**4.**	(*b*)	**5.**	(*a*)	**6.**	(*d*)	**7.**	(*d*)	**8.**	(*c*)
9.	(*b*)	**10.**	(*b*)	**9.**	(*b*)	**10.**	(*b*)	**11.**	(*b*)	**12.**	(*b*)	**13.**	(*b*)	**14.**	(*a*)
15.	(*d*)	**16.**	(*d*)	**17.**	(*d*)	**18.**	(*d*)	**19.**	(*d*)	**20.**	(*b*)	**21.**	(*c*)	**22.**	(*a*)
23.	(*c*)	**24.**	(*c*)	**25.**	(*b*)	**26.**	(*a*)	**27.**	(*d*)	**28.**	(*a*)	**29.**	(*c*)	**30.**	(*a*)
31.	(*b*)	**32.**	(*b*)	**33.**	(*a*)	**34.**	(*a*)	**35.**	(*a*)	**36.**	(*b*)	**37.**	(*d*)	**38.**	(*d*)
39.	(*b*)	**40.**	(*a*)	**41.**	(*b*)	**42.**	(*d*)	**43.**	(*c*)	**44.**	(*a*)	**45.**	(*b*)	**46.**	(*a*)
47.	(*d*)	**48.**	(*a*)	**49.**	(*d*)	**50.**	(*c*)	**51.**	(*a*)	**52.**	(*c*)	**53.**	(*a*)	**54.**	(*d*)
55.	(*a*)	**56.**	(*c*)												

- A leader is one who knows the way, goes the way, and shows the way.

John C. Maxwell

OBJECTIVE QUESTIONS TEST SERIES

Chapter 1

1. Who predicted the existence of the electromagnetic waves?
(*a*) Faraday
(*b*) H.C. Oersted
(*c*) William Gilbert
(*d*) J.C. Maxwell

3. Find the simplet atom in the following
(*a*) Nitrogen
(*b*) Hydrogen
(*c*) Oxygen
(*d*) Helium

3. Who invented the antenna?
(*a*) Faraday
(*b*) Maxwell
(*c*) Yagi Uda
(*d*) None

4. Optical waveguide based on
(*a*) total internal reflection
(*b*) total internal refraction
(*c*) Brewester angle
(*d*) Snell's law

5. Optical waveguide has two layers
(*a*) Core-cladding
(*b*) Core-1, and Core-2
(*c*) Cladding and Cladding-2
(*d*) None

6. Energy of the atom E_n related with band levels as
(*a*) $E_n \propto \frac{1}{n}$
(*b*) $E_n \propto \frac{1}{n^2}$
(*c*) $E_n \propto -n^2$
(*d*) $E_n \propto 1 - n^2$

7. Which one is correct for optical fiber?
(*a*) $n_1 \geq n_2$
(*b*) $n_1 \leq n_2$
(*c*) $n_1 \approx n_2$
(*d*) $n_1 = n_2$

8. J.C. Maxwell's given
(*a*) Three equation
(*b*) Four equation
(*c*) one equation which relates electric field and magnetic fields
(*c*) $\frac{1}{2}$ (E × H*) = Pav

9. No. of electron in one coulomb's

(*a*) $\frac{1}{1.6 \times 10^{-19}}$ (*b*) 1.6×10^{-9}

(*c*) 9.1×10^{-31} (*d*) $\frac{1}{9.1 \times 10^{-31}}$

10. Value of ε_0 is

(*a*) $4\pi \times 10^{-7}$ (*b*) $\frac{1}{36\pi \times 10^{-9}}$

(*c*) 7.745×10^{-7} (*d*) None

Chapter 2

1. The vector A × (–1) is equal to:

(*a*) – A (*b*) A–1

(*c*) A (*d*) None

2. The electric potential in a region is a kind of :

(*a*) Vector fields (*b*) Electric fields

(*c*) Scalar fields (*d*) None

3. The $a_x . a_x = a_y . a_y$ is equal to :

(*a*) $a_z . z$ (*b*) $a_z . a_z$

(*c*) 1 (*d*) both (b) and (c)

4. The $a_x . a_x = a_y . a_y = a_z . a_z$ is equal to :

(*a*) 0 (*b*) 2

(*c*) $a_x . a_z$ (*d*) none

5. The $a_x \times a_y$ is equal to

(*a*) a_y (*b*) a_z

(*c*) $a_x . a_z$ (*d*) 1

6. The vector, $A = 2a_x + a_y + 4a_z$ is characterized as

(*a*) Unit vector (*b*) Uniform vector

(*c*) Module vector (*d*) None

7. The vector $A = 2xa_z x + ya_y + za_z$

(*a*) Unit vector (*b*) Uniform vector

(*c*) None-uniform vector (*d*) Both (*a*) and (*b*)

8. The gradient of a scalar *V* is represented by :

(*a*) ∇ V (*b*) ∇. V

(*c*) ∇ V (*d*) both (*a*) and (*b*)

9. If $A = a_x + a_y$, $B = a_x + 2a_y - 2a_z$ and $C = a_x + a_z$. Then A.B × C is

(*a*) 4 (*b*) 3

(*c*) 5 (*d*) 9

10. If there are two points (1, –2,2) and (3,1,0), what will be straight line distance between them :
    (*a*) 4 (*b*) 3
    (*c*) 5 (*d*) 9
11. If a vector, $A = 3a_z + 4a_y + 5a_z$ is converted into spherical co-ordinate, the value of 'φ' will be:
    (*a*) 4 (*b*) 3.5
    (*c*) 5.3 (*d*) 9
12. If ar and ax, are unit vectors then, $(a_r \cdot a_x)$, is equal to :
    (*a*) 1 (*b*) sin θ
    (*c*) cos φ (*d*) sin 0. cos φ
13. The curl of a vector is :
    (*a*) zero (*b*) constant
    (*c*) vector (*d*) a scalar
14. If A = ∇ V, V is said to be the :
    (*a*) Divergence of A (*b*) Curl of A
    (*c*) Vector potential of A (*d*) Scalar potential of A
15. The divergence of a vector is :
    (*a*) a scalar (*b*) a vector
    (*c*) not define (*d*) both

Chapter 3

1. Two point charges Q1 =2nC and Q2 =5nC are at distance apart. Which of the following statement are incorrect?
    (*a*) The force on Q1 is repulsive
    (*b*) The force on Q2 is the same in magnitude as that on Q1
    (*c*) The force on Q2 is along the line joining them
    (*d*) A point charge Q3 = –5nC located at the midpoint between Q1 and Q2 will also experiences force
2. The electric field in free space is
    (*a*) $\frac{D}{\varepsilon_0}$ (*b*) $\frac{D}{\mu_0}$
    (*c*) D.ε0 (*d*) $\frac{\sigma}{\varepsilon_0}$
3. The work done by the force $F = 4ax - 3ay + 2az$ N in given a 1 Nc charge and a displacement of $10ax + 2ay - 10az$ m is
    (*a*) 103 nJ (*b*) 60 nJ
    (*c*) 64 nJ (*d*) 20 nj

4. Equation $\nabla.(-\varepsilon\nabla V) = -\rho_v$ may be regarded as Poisson's equation for an in homogeneous medium.
   (*a*) True (*b*) False
   (*c*) None (*d*) Partially true

5. In a cylindrical coordinate system, the equation, $-\frac{\partial^2 \psi}{\partial r^2} + \frac{1}{r}\frac{\partial \psi}{\partial r} + \frac{\partial^2 \psi}{\partial z^2} + 10 = 0$ is called
   (*a*) Maxwell's equation (*b*) Laplace equation
   (*c*) Poisson's equation (*d*) Helmholtz equation

6. Which of the following equation does not satisfy the Laplace equation?
   (*a*) $V = 3.x + 6$ (*b*) $V = 11\, x.y$
   (*c*) $V = r \cos \varphi$ (*d*) $V = \frac{10}{r}$

7. A parallel plate capacitor connected to a battery stores twice as much as charge with a given dielectric as it does with air as dielectric, then susceptibility of the dielectric is,
   (*a*) 0 (*b*) 1
   (*c*) 2 (*d*) 3

8. If E is a vector, then $\nabla . \nabla \times E$ is
   (*a*) 0 (*b*) 1
   (*c*) Does not exist (*d*) None of these

9. If E = cos (6 × 107 .t – βz). *az* then β is,
   (*a*) 0.1 rad/sec (*b*) 0.2 rad/sec
   (*c*) 0.3 rad/sec (*d*) 0.4 rad/sec

10. Which is not true for electric flax elensity '*D*'
   (*a*) It is related with E (*b*) also called displacement vector
   (*c*) does not depends upon the medium (*d*) None

Chapter 4

1. Which one is not a source of magnetostatic field:
   (*a*) A DC current in wire
   (*b*) A permanent magnet
   (*c*) An accelerated charge
   (*d*) An electric fields linearly changing with time

2. The first Maxwell's equation in free space
   (*a*) $\nabla \times H = \frac{\partial D}{\partial t} + J$ (*b*) $\nabla \times D = 0$
   (*c*) $\nabla \times H = 0$ (*d*) $\nabla \times H = J$

3. One of these equations is not Maxwell's equation for static electromagnetic fields in a linear homogenous medium.

(*a*) $\nabla . B = 0$ (*b*) $\nabla \times D = 0$

(*c*) $\oint B .dl = \mu_0 I$ (*d*) $\nabla^2 A = \mu_0 I$

4. The electric field just above a conductor is always,

(*a*) Normal to the surface (*b*) Tangential to surface

(*c*) Zero (*d*) ∞

5. If the flux density is 10 Wb/m^2 and the area of the coils is 2 m^2, the flux is

(*a*) 5 (*b*) 10

(*c*) 15 (*d*) 20

6. The unit of the magnetic susceptibility is;

(*a*) Nil (*b*) Amp

(*c*) H/m (*d*) Wb

7. $\oint B\, ds$ is equal to

(*a*) Zero (*b*) Q

(*c*) H (*d*) J

8. $\nabla \times A$ is equal to

(*a*) H (*b*) B

(*c*) J (*d*) 0

9. $\nabla \times \nabla V_m$ is

(*a*) Zero (*b*) $\nabla^2 V_m$

(*c*) J (*d*) $\nabla . \nabla V_m$

10. The ratio of E/H is

(*a*) $\sqrt{\frac{\mu}{\varepsilon}}$ (*b*) $\sqrt{\mu\varepsilon}$

(*c*) $\sqrt{\frac{\varepsilon_0}{\varepsilon}}$ (*d*) $\sqrt{\frac{\mu}{\mu_0}}$

Chapter 5

1. Electromagnetic wave produced by

(*a*) time varying voltage (*b*) time varying current

(*c*) static charges (*d*) None of these

2. Which one is not Maxwell's equation

(*a*) $\nabla \times H = j + \frac{\partial D}{\partial t}$ (*b*) $\nabla \times E = -\frac{\partial B}{\partial t}$

(*c*) $\nabla . B = 0$ (*d*) $\nabla . D = \frac{\partial \rho_u}{\partial t}$

3. Which is correct expression?

(*a*) $D = \frac{E}{\varepsilon}$ (*b*) $H = \mu B$

(*c*) $J = \sigma E$ (*d*) $\nabla . B = 1$

4. Which one is not integral form of Maxwell's equation

(*a*) $\int H \cdot dl = \int_S (D + J) ds$ (*b*) $\int dl = -\int_S B \cdot ds$

(*c*) $\int B \cdot ds = \int_\upsilon P_\upsilon d\upsilon$ (*d*) $\int_S B \cdot ds = \rho\upsilon$

5. Which one is Maxwell's equation in phaser form

(*a*) $\nabla \times H = J + j\omega\varepsilon E$ (*b*) $\nabla \times H + j\omega\varepsilon = 0$

(*c*) $\nabla \times H = J$ (*d*) None of these

6. Boundary condition is

(*a*) $D_{n1} + D_{n2} = \rho_s$ (*b*) $D_{n1} = D_{n2}$

(*c*) $D_{n1} - D_{n2} = \rho_s$ (*d*) None of .these.

7. Free space impedance is

(*a*) $\frac{H}{E} = 120\,\pi$ (*b*) $\frac{E}{H} = 120\pi$

(*c*) $\frac{H}{E} = 3772$ (*d*) None

8. The wavelength of a wave with propagation constant $\gamma = 0.1\pi + 0.2\pi j$, is

(*a*) 10 m (*b*) 30 m

(*c*) 25 m (*d*) 20 m

9. The electric field just above a conductor is always

(*a*) Normal to the surface (*b*) tangential to source

(*c*) zero (*d*) ∞

10. The intrinsic impedance of a median whose $\sigma = ,0$, $\varepsilon, = 9$, $\mu_r = 1$ is

(*a*) $40\,\pi\,\Omega$ (*b*) $9\,\Omega$

(*c*) $120\,\pi\,\Omega$ (*d*) $60\,\pi\,\Omega$

11. If E is a vector, then $\nabla . \nabla \times E$ is

(*a*) 0 (*b*) 1

(*c*) Does not exist (*d*) None of this

12. Which is correct

(*a*) $8 = \frac{1}{\alpha}$ (*b*) $\delta = \frac{1}{\beta}$

(*c*) $\delta = \alpha$ (*d*) both (*a*) & (*b*)

Chapter 6

1. Poynting vector is given by

(*a*) $E \times H$ (*b*) $E . H$

(*c*) $H \times E$ (*d*) $H . E$

2. Poynting vector gives

(*a*) rate of energy flow (*b*) direction of polarization

(*c*) electric fields (*d*) magnetic fields

3. Absolute permeability of free space is

(*a*) $4\pi \times 10^{-7}$ A/m (*b*) $4\pi \times 10^{-7}$ H/m

(*c*) $4\pi \times 10^{-7}$ F/m (*d*) $4\pi \times 10^{-7}$ H/m^2

4. For uniform plane wave in x-direction

(*a*) $E_x = 0$ (*b*) $H_x = 0$

(*d*) $E_x = 0$ and $H_x = 0$ (*d*) None

5. Skin depth is given by

(*a*) α (*b*) $\frac{1}{\alpha}$

(*e*) α (*d*) 0

6. Poynting vector in complex form is

(*a*) $P = E \times H$ (*b*) $\overline{P} = \overline{E} \times \overline{H}*$

(*c*) $P = \frac{1}{2}(\overline{E} \times \overline{H}*)$ (*d*) None

7. Uniform plane wave is

(*a*) longitudinal in nature (*b*) transverse in nature

(*c*) neither transverse nor longitudinal (*d*) vertically directed

8. A field can exist if it satisfies

(*a*) Gauss's law (*b*) Faraday's law

(*c*) Coulomb's law (*d*) all Maxwell's equation

9. Displacement current density is

(*a*) D (*b*) J

(*c*) $\frac{\partial D}{\partial t}$ (*d*) $\frac{\partial J}{\partial t}$

10. If E is 2V/m, of a wave is free space, then H is

(*a*) $\frac{1}{60\pi} A/m$ (*b*) 60π A/m

(*c*) 120π A/m (*d*) 240π A/m

(**Hint :** Ans. $V_P = 1.9782 \times 10^8$ m/s, $\sigma = 318,03 \times 10^{-4}$ neper/m.)

11. A conducting medium is,

(*a*) $\sigma = 0$ (*b*) $\sigma \neq 0$

(*c*) $\sigma = \infty$ (*d*) None

12. A conductor is said to be good conductor if,

(a) $\frac{\sigma}{\omega\varepsilon} \approx 1$ (b) $\frac{\sigma}{\omega\varepsilon} >> 1$

(c) $\frac{\sigma}{\omega\varepsilon} << 1$ (c) $\frac{\sigma}{\omega\varepsilon} = 1$

13. If a medium is a good dielectric then which one is correct?

(a) $\frac{\sigma}{\omega\varepsilon} = 0$ (b) $\frac{\sigma}{\omega\varepsilon} \approx 1$

(c) $\frac{\sigma}{\omega\varepsilon} << 1$ (d) $\frac{\sigma}{\omega\varepsilon} >> 1$

14. The loss tangent (tanθ) is equal to

(a) $\tan\theta = \frac{\mu}{\varepsilon}$ (b) $\tan\theta = \frac{\sigma}{\omega\varepsilon}$

(c) $\tan\theta = \frac{\omega\varepsilon}{\sigma}$ (d) $\tan\theta = \mu\varepsilon$

15. Attenuation constant for a good dielectric is

(a) $\alpha = \frac{\sigma}{2}\sqrt{\frac{\mu}{\varepsilon}}$ (b) $\alpha = \omega\sqrt{\frac{\omega\varepsilon}{2}\left[\left(1+\frac{\sigma^2}{2\omega^2\varepsilon^2}\right)-1\right]}$

(c) $\alpha = \beta$ (d) None

16. For a good conductor

(a) $\alpha = \beta = \sqrt{\frac{\omega\mu\sigma}{2}}$ (b) $\alpha \neq \beta \neq \sqrt{\frac{\omega\mu\sigma}{2}}$

(c) $\alpha = \frac{\alpha}{2}\sqrt{\frac{\mu}{2}}$ (d) None

17. Surface resistance for good conductor is

(a) $R_s = \sqrt{\frac{j\omega\mu}{2\sigma}}$ (b) $R_s = \sqrt{\frac{\omega\mu}{2\sigma}}$

(c) $R_s = \sqrt{\frac{j\omega\mu}{\sigma}}$ (d) $R_s = \eta = \sqrt{\frac{\mu}{\varepsilon}}$

18. Which one is correct for depth of penetration for good conductor

(a) $\delta \propto \frac{1}{\sqrt{f}}$ (b) $\delta \propto = \frac{1}{f}$

(c) $\delta \propto \sqrt{f}$ (d) $\delta \propto \sqrt{\alpha}$

19. The Brester angle θ_b is

(a) $\theta_b = \tan^{-1}$ (b) $\theta_b = \tan^{-1}\sqrt{\frac{\varepsilon_2}{\varepsilon_1}}$

(c) $\theta_b = \tan^{-1}$ (d) $\theta_b = \tan^{-1}\left(\frac{\sigma}{\varepsilon}\right)$

20. The boundary condition on $\overline{E}$ is

(a) $a_n \times (\overline{E}_1 - \overline{E}_2)$ (b) $a_n \cdot (\overline{E}_1 - \overline{E}_2)$

(c) $\overline{E}_1 - \overline{E}_2$ (d) None of these

21. The velocity of EM wave is
(*a*) inversely to β (*b*) direction proportional to β
(*c*) inversely proportional to α (*d*) direction proportional to α

22. Reflection coefficient on normal incidence is
(*a*) $\rho = \frac{\eta_2 - \eta_1}{\eta_2 + \eta_1}$ (*b*) $\rho = \frac{\eta_1}{\eta_2}$
(*c*) $\rho = \frac{\eta_2 + \eta_1}{\eta_2 - \eta_1}$ (*d*) $\rho = \frac{2\eta_2}{\eta_1 + \eta_2}$

Chapter 7

1. Transmission line carries
(*a*) TE mode (*b*) TM mode
(*c*) H-E mode (*d*) TEM mode
2. The general expression for propagation constant in transmission line is
(*a*) $\gamma = \sqrt{\frac{R + j\omega L}{(G + j\omega C)}}$ (*b*) $\gamma^2 = \left(\frac{R + j\omega L}{G + j\omega C}\right)^2$
(*c*) $\gamma = \sqrt{(R + j\omega L)(G + j\omega C)}$ (*d*) $\gamma^2 = (R + j\omega L)^2 (G + j\omega C)^2$
3. Which one is not correct for transmission line equations?
(*a*) $\frac{dI}{dZ} = -(G + j\omega C)V$ (*b*) $\frac{dV}{dZ} = -(R + j\omega L)\, I$
(*c*) $\frac{d^2 I}{dZ^2} = \gamma^2 I$ (*d*) $\frac{d^2 V}{dZ^2} = (R + j\omega L)\, V$
4. Characteristic impedance at microwave frequencies when $\omega L >> R$ and $\omega C >> G$ is.
(*a*) $Z_0 = \sqrt{\frac{R + j\omega L}{G + j\omega C}}$ (*b*) $Z_0 = \sqrt{\frac{L}{C}}$
(*c*) $Z_0 = 120\pi$ (*d*) None of these.
5. For the short circuited load the reflection coefficients of the transmission line is
(*a*) $\rho = 0$ (*b*) $\rho = -1$
(*c*) $\rho = 1$ (*d*) None
6. Which one is correct choice.?
(*a*) $Z_0 = Z_{in} Z_L$ (*b*) $Z_0^2 = Z_{in} Z_L$
(*c*) $Z_0 = 120\pi$ (*d*) $Z0 = \infty$
7. Which one is correct for VSWR?
(*a*) $1 \le \text{VSWR} \le \infty$ (*b*) $1 < \text{VSWR} < \infty$
(*c*) $\text{VSWR} = 1$ (*d*) $0 \le \text{VSWR} \le 1$

8. Baluns are used in transmission line for

(*a*) Impedance matching (*b*) as amplifier
(*c*) as voltage follower (*d*) None of these

9. Quarter wave transmission line with open and output side is equivalent to;

(*a*) parallel *R-L* circuit (*b*) series R-L circuit
(*c*) series LC circuit (*d*) parallel RLC circuit

10. Which one is better for variable frequency?

(*a*) single stub (*b*) double stub
(*c*) both (*d*) None

Chapter 8

1. Which is not true for source supplying energy in circuit?

(*a*) It has its own internal resistance.
(*b*) Maximum power takes place when temp is 2 °C.
(*c*) Load resistance should be same as the source resistance.
(*d*) Both (*a*) and (*d*)

2. The *hot spots* is resultant of;

(*a*) High VSWR (*b*) Low VSWR
(*c*) NO relation with VSWR (*d*) None

3. It is a quarter wave length long Tx line of characteristic impedance $\sqrt{Z_0 R_A}$.

(*a*) Quarter weave length transformer (*b*) Half wave length transformer
(*c*) Single stub transformer (*d*) None

4. What is main disadvantage with matching devices?

(*a*) High volume (*b*) Narrow BW
(*c*) Flat Bw (*d*) Frequency independent.

5. Transmission lines used in antenna system have characteristic impedance that is;

(*a*) Purely resistive (*b*) Purely inductive
(*c*) Purely capacitive (*d*) None

6. Direct and reflected waves propagating in opposite directions creates;

(*a*) Flat waves (*b*) Sine waves
(*c*) Cosine waves (*d*) Standing waves

7. If there is impedance matching between impedances of 100 ohms and 50, reflection calefacient is;

(*a*) 1.5 (*b*) 0.33
(*c*) 3.0 (*c*) 2.5

8. Theoretical values of VSWR lie between;

(*a*) 1.0 – 3.0 (*b*) 0.33 – 4.5
(*c*) 0 – 1 (*c*) 1 – ∞

9. The bandwidth of Chebyshev transformer is ---- a binomial multi– section matching transformer

(*a*) Larger than (*b*) Lesser than

(*c*) Equal to (*d*) None

10. Balun stands for;

(*a*) Balance transformer.

(*b*) Unbalance transformer to balance matching transformer

(*c*) Balance to balance matching transformer

(*d*) Balance to unbalance matching transformer

11. The baluns are used for home television antennas where impedance ration is ;

(*a*) 4:1 (*b*) 2:3

(*c*) 4.2 (*c*) 3.5

12. The ETC 41-2-3 is an example of;

(*a*) Quarter wave transformer (*b*) Half wave transformer

(*c*) Flux coupled balun transformer (*d*) Tx line balun transformer

13. The ETC 41-2 is an example of;

(*a*) Quarter wave transformer (*b*) Half wave transformer

(*c*) Flux coupled balun transformer (*d*) Tx line balun transformer

Chapter 9

1. Width of a waveguide must be around .... at the frequency of the wave to be propagated.

(*a*) a half wavelength (*b*) a quarter wavelength

(*c*) a wavelength (*d*) none

2. The wave-propagation of EM wave is represented by:

(*a*) $c = \frac{1}{\sqrt{\mu^2 \varepsilon}}$ (*b*) $c = \frac{1}{\sqrt{\mu \varepsilon}}$

(*c*) $c = \frac{1}{\sqrt{E}}$ (*d*) none

3. In the modes the electric or magnetic field lines nay one is entirely transverse to the direction of propagation (e.g. $E_z = 0$ and $H_z \neq 0$).

(*a*) TM modes (*b*) TE modes

(*c*) TEM modes (*d*) HE modes

4. TE_{10} mode has the highest cut-off wavelength (2*a*) in the rectangular waveguide. This mode is known as:

(*a*) odd mode (*b*) Even mode

(*c*) degenerate mode (*d*) dominant mode

5. The greater power-handling capability is directly related to the .... surface area.
   (*a*) decrease (*b*) increase
   (*c*) independent (*d*) none
6. A mode is dominant mode (among *TE* modes) in R-wave guide:
   (*a*) TE_{01} (*b*) TE_{11}
   (*c*) TE_{10} (*d*) none
7. Velocity factor of a transmission line is 0.75, then EM waves travel down the line with a velocity of
   (*a*) velocity of light (*b*) less than the velocity of light
   (*c*) cannot be calculated (*d*) none
8. TE_{01} mode exists in an evacuated circular waveguide of radius 2 cm, find its cut-off wavelength
   (*a*) 2 cm (*b*) 3 cm
   (*c*) 4 cm (*d*) 5 cm

Chapter 10

1. The basic requirements of transmitting antennas are:
   (*a*) High gain and high efficiency (*b*) High gain and Low side lobes
   (*c*) Large signal to noise ratio (*d*) Lone of the mentioned
2. The requirements of receiving antennas
   (*a*) Low side lobes and large SNR. (*b*) Low side lobes and small SNR
   (*c*) Low gain and Low side lobes (*d*) None
3. The beam width of the antenna pattern measured at half power points is called:
   (*a*) Half power beam width (*b*) Full null beam width
   (*c*) Beam width (*d*) None
4. An antenna has a field pattern of $E(\theta) = \cos^2 \theta, \theta$ varies between 0 and 90°. Half power beam width of the antenna is:
   (*a*) 33° (*b*) 66°
   (*c*) 120° (*d*) None
5. As the beam area of an antenna Increases, the directivity of the antenna:
   (*a*) Increases (*b*) Decreases
   (*c*) Remains unchanged (*d*) Depends on antenna
6. If an antenna radiates over half a sphere, directivity of the antenna is:
   (*a*) One (*b*) Four
   (*c*) Three (*d*) Two
7. Ideally, the number of point sources an antenna can resolve is numerically equal to:
   (*a*) Gain of the antenna (*b*) Directivity
   (*c*) Beam efficiency (*d*) Beam area

8. The solid area through which all the power radiated by the antenna is:
   (*a*) Beam area
   (*b*) Effective area
   (*c*) Aperture area
   (*d*) Beam efficiency
9. Power radiated from an antenna per unit solid angle is called radiation intensity.
   (*a*) True
   (*b*) False
   (*c*) Not known
   (*d*) None
10. Patch antennas are the antennas of small size and are made of:
   (*a*) Strip line
   (*b*) Microstrip lines
   (*c*) Coaxial cables
   (*d*) Rectangular waveguide.

Chapter 11

1. The HERP stands for
   (*a*) Hazards of Electromagnetic Radiation to Personnel
   (*b*) Hazards of Energy Radiation to Product
   (*c*) Harmful Effects of Radiation and Power
   (*d*) None of these
2. The HERO stands for
   (*a*) Hazards of Energy Radiation to Object
   (*b*) Hazards of Electromagnetic Radiation to Ordinance
   (*c*) High Effects of Radiation to Ordinance
   (*d*) None of these
3. Which is/are methods of controlling EMI
   (*a*) Grounding
   (*b*) Shielding
   (*c*) Bonding
   (*d*) All of these
4. In Shielding to reduce the radiated EMI reflection or absorption, what types of material is used?
   (*a*) conductive materials
   (*b*) semi-conductor materials
   (*c*) insulator material
   (*d*) None of these
5. The EMISM stands for
   (*a*) EMI shielding margin
   (*b*) EMI safety margin
   (*c*) EM induction nad safety mode
   (*d*) None of these
6. Radio frequency interference is an EMI only that occurs between frequency range
   (*a*) 1 kHz-10 GHz
   (*b*) 10 GHz-40 GHz
   (*c*) LF range only
   (*d*) None of these
7. The ability of equipment/device to be able to operate satisfactorily in presence of interference too, is defined as
   (*a*) Radio frequency interference
   (*b*) Microwave interference
   (*c*) Electromagnetic compatibility
   (*d*) None of these

8. The appearance of 'ghosts' in TV picture reception is due to

(*a*) EMI (*b*) EMC

(*c*) EM fluctuation (*d*) All of these

9. Select the incorrect statement in regards to noise and interference

(*a*) Noise can be caused by natural disturbances

(*b*) Noise cannot be eliminated, but interference can

(*c*) Interference is the undesirable effect of noise

(*d*) None of these

10. Which is/are not the electromagnetic interference (EMI) coupling mode(s)?

(*a*) Conductive coupling (*b*) Inductive coupling

(*c*) Resistive coupling (*d*) None of these

11. Which is not a source of EMI?

(*a*) Optical fiber (*b*) Police radar

(*c*) Aircraft (*d*) None of these

12. Test method for measuring coupling and screening it erm as

(*a*) VG 95373 (*b*) VG 96123

(*c*) MIL–STD–85 (*d*) None

- If opportunity doesn't knock, build a door.

Milton Berle

REFERENCES

1. H. Hayt. "Engineering electromagnetic", Tata McGraw-Hils, New Delhi,7th edition 2005, pp. 1-20.
2. M.N.O Sadkiku, "Elements of electromagnetic", Oxford Press, 3/e, 2002.
3. A. Panday "Electromagnetism", Prentice Hall of India, New Delhi, 2004.
4. C.A. Cullwick, "Fundamental of electromagnetism", Macmillan, NY, 1939.
5. Jordan E.C. and K.G, Balmin "Electromagnetic waves and radiating systems", Prentice Hall of India, New Delhi, 1993.
6. Y. Ramat-Samii, "Useful co-ordinate transformations for antenna application", IEEE, Trans, vol, AP-27, No. 4, July 1997.
7. S. Ramo *et.al*, "Fields and waves in communication electronics". Wiley-India, 3/e New Delhi, 2007.
8. M.N. Rao, "Elements of engineering electromagnetics", Prentice Hall of India 6/e, New Delhi, 2007.
9. S.Y. Liao, "Microwave devices and circuits", Prentic Hall, 3 edition, India, 2002.
10. Kracy and Flesich, "Electromagnetic with applications", Mc-Grew, Hill, 5th, Edition, 2001.
11. Gangadhar and Ramanathan, "Field theory, "Khanna Publisher" 5th edition, New Delhi, 2003.
12. Schaum's series, "Theory and problems of electromagnetics", Tata, Mc-Graw Hill, 2/*e* 2001.
13. K. Cheng, "Field and wave electromagnetic", Addison Wesley, 2nd Edition, 1998.
14. J.M. Crowley, "Fundametals of applied electrostatics", John Wiley & Sons, New York, 1998.
15. M.R. Spigel, "Mathematical handbook of formulas and tables", Mc-Graw Hill, New York, pp. 124-130, 1968.
16. A.D. Moore, ed. "Electrostatics and its applications" John Wiley & Sons, New York, 1973.
17. B. Boltan, "Electromagnetism and its applications" Van Nostrand, London, 1980.
18. D.T. Paris and F.K. Hurd, "Basic electromagnetic theory", Mc-Graw Hill, New York, 1969.
19. J.K. Waston, "Applications of magnetism", John Wiley & Sons, New York, 1980.

20. M.A.R. Guston, "Microwave transmission impedence data", Van Nostrand Reinhold, London 1972.
21. J.A. Seeger, "Microwave theory, component and devices", Engilwood Cliffs, NJ, Prance-Hall, 1986.
22. C.R.Paul *et, al*, "Introduction to electromagnetic fields, "*Web, Mc-Graw Hill*, USA.
23. R.L. Yadava, "Antenna and wave propagation", Prentic Hall of India, May 2011, India.
24. J.Van Bladel, "Lorentz," *IEEE, Ant. & Prop., Magazine*, vol. 22, pp. 67.April 1991.
25. J.D. Kraus, "Antennas-For all applications, "2/e, *TATA Mc-Graw Hill*, New Delhi, 2007.
26. C.A. Balanis, "Antennas theory; Analysis and design, "2/e *John Wiley*, India 2007
27. R.C. Johnson, "Antenna engineering handbook, "*Mc-Graw Hill*, NY 2/*e* 1992.
28. C. Tai and C.Pereira, "An approximate formula for calculating the directivity of an antenna," *IEEE , Trans, Antenna Propagate*, AP, vol 24, no 2, pp. 225-226, March 1976.
29. Harold Hott, "Polarization in antennas and radars," *John W & S* 1986.
30. R.E. Collin and F.J. Zucker, "Antenna theory, "*Part I, Mc-Graw Hill*, New Delhi.
31. K.D. Prasad and M.Prasad, "Antenna and wave propagation, "*Satya Praksham*, New Delhi, 2000.
32. A Das and S.K. Das, "Microwave engineering, "*Tata, Mc-Graw Hill*, New Delhi, 2000.
33. A. Guy, " Biological effects and medical applications of microwave energy, " *IEEE Transactions on Microwave Theory and Techniques*, vol. 32, no 9 September 1984.
34. Umesh Sinha, "Transmission lines and network," *Tech India Publications*, New Delhi, 2003
35. H.J. Riblet, "General synthesis of quarter wave impedance transformers," *IRE Trans, Microwave Theory Tech*, vol. MTT-5, pp. 36-43, January 1957.
36. J.R. Whinnery and H.W. Jamieson, "Equivalent circuits for discontinuities in transmission lines," *IRE proc*; vol. 32, pp.98-116, Feb. 1994.
37. P.V. Bernnan, "Impedance matching method using passive loss – less feedback, " *IEE proc*, H, vol. 139, no 5. Oct. 1992.
38. R.C. Hansen "Super conducting antennas," International conference on antenna and propagation 1991, Conference Publication, 333, pp.55-558, 1991.
39. R.C. Johnson, & H. Jasik "Antenna engineering hand book," *Tata Mc-Graw Hill*, 1961.
40. C.R. Paul, "Introduction to electromagnetic compatibility", John Wiley & Sons, New York, 1992.
41. P.M. Morse and H. Feshbach, "Methods of theoretical physics", Mc. Graw Hill Book Co., Inc., NY, 1953.
42. V.H. Ruinsey, "A new way of solving Maxwell's equations", IRE Transactions on antenna and Propagation, pp. 461-465, Sept. 1961.
43. G.S.N. Raju, "Electromagnetic field theory and transmission lines, "Pearson Education, 2006.

44. K.S. Yee, "Numerical solution of initial boundary value problem involving Maxwell's equation in isotropic media, "IEEE Trans. Antenna and Propagate, vol. 17, pp. 585-589, 1996.

45. A. Taflove and M. Brodwin, "Numerical solution of steady state EM scattering problems using the time-dependent Maxwell's equations," IEEE Microwave theory Tech., vol. 23, pp. 623-730, 1975.

46. D. M. Sullivan, "Electromagnetic simulation using the FDTD method," IEEE Press, series on RF and Microwave Technology, NY. 2003.

47. J. D. Kraus, "Electromagnetic with applications" Mc. Graw-Hill, NY. 2002.

48. R.K. Yadav and A. Goswami, "Microwaves and radar Engineering", Pragati Prakashan, Meerut (UP), 2005.

49. K.G. Gupta, "Microwaves" New Age International publishers, New Delhi, 2003.

50. R. Gowy, (*i.e.* Gowm) "Electromagnetic fields and waves", Kataria & Sons, New Delhi, 2008.

51. L. Hatkio, "Elliptically polarized waves," Proc. IRE, vol. 38, pp. 1455, Dec 1950.

- The secret of getting ahead is getting started.

Mark Twain

APPENDIX

APPENDIX -1 : MISCELLENEOUS TOPICS

1. The origin of 377 ohms

There is a wondering question, that who was the first to give the concept of '377 ohms'. For several years, we take this value for granted, however it reflects an evolutionary process of thinking in our system of units in the field of electromagnetism. The available literatures reveal that Dr. George A. Compell was the first man to say 377 Ω. He translated a report of Maxwell and found the value of 377 Ω, and dramatized this value as the impedance of an idealized parallel strip-line of square cross-section. It appeared in 1932.

After years, in 1938, Dr. Schelkunoft described, 377 ohms as the instrinsic impedance of medium in his article. Dr. Stratton was the first to state 377 ohms in a text book in year 1941. He dignify in a list of fundamental constants, along with the speed of light.

2. Poynting Theorem for moving medium

As well known, the Poynting Theorem's is a most commonly accepted concept of energy conservation for electromagnetic waves. Poynting theorem was generalized by Compton and Tai, based on Minkowski's Theory for homogeneous, isotropic, linear, non-dispersive media moving with velocity v, when $\frac{v}{c} << 1$.

In general, according to Minkowski Theory, Maxwell's equations are as follows;

$$\nabla \times E = -\frac{\partial B}{\partial t} \quad \text{...(1)}$$

$$\nabla \times H = \frac{\partial D}{\partial t} \quad \text{...(2)}$$

This is true for electromagnetic fields in a medium in uniform motion with constant velocity provided the current density is assumed to be zero.

The constitutive relations for this situation given by Tai in its useful form is expressed as;

$$D = \varepsilon . \overline{\alpha} . E - \Omega \times H \quad \text{...(3)}$$

$$B = \mu\, \overline{\alpha}, H - \Omega \times E \quad \text{...(4)}$$

in which
$$\Omega = \frac{(n^2-1).\beta}{(1-n^2\beta^2)C_0}$$

and
$$\beta = \frac{v}{c_0}$$

$$\bar{\alpha} = \begin{bmatrix} \alpha & 0 & 0 \\ 0 & \alpha & 0 \\ 0 & 0 & 1 \end{bmatrix}, \alpha_r = \frac{1-\beta^2}{1-n^2\beta^2}$$

$$n = \frac{c_0}{c}$$

where,

c = velocity of the stationary medium.

c_0 = velocity of light the in vacuum.

$\bar{v}$ = va_z, $a_z \rightarrow$ is unit vector along the z-axis.

Therefore, from (1) and (3), we get

$$\nabla \times E = -\frac{\partial}{\partial t}(\mu . \bar{\alpha}\, H - \Omega \times E)$$

$$= -\frac{\partial}{\partial t}(\mu\ \bar{\alpha}\, H) + \frac{\partial}{\partial t}(\Omega \times E)$$

$$H.(\nabla \times E) = -\frac{\partial}{\partial t}\left(\frac{\mu}{2} H . \bar{\alpha}\, H\right) + H . \frac{\partial}{\partial t}(\Omega \times E)$$

or
$$-\overline{H}.(\nabla \times \overline{E}) = -\frac{\partial}{\partial t}\left(\frac{\mu}{2}\overline{H} . \alpha . \overline{H}\right) - \Omega . \frac{\partial}{t}(\overline{E} \times \overline{H}) \quad ...(5)$$

Again from (2) and (3), we get

$$\overline{H}.(\nabla \times \overline{E}) = \frac{\partial}{\partial t}\left(\frac{\varepsilon \overline{E} \alpha \overline{E}}{2}\right) - \Omega . \frac{\partial}{\partial t}(\overline{E} \times \overline{H}) \quad ...(6)$$

Using the Vector Identity

$$-\nabla . (A \times B) = -B .(\nabla \times A) + A . (\nabla \times B) \quad ...(7)$$

It is found that the generalized Poynting Theorem to be

$$-\nabla . S = \frac{1}{2}\frac{\partial}{\partial t}[\mu\, \overline{H} . \alpha . \overline{H} + \varepsilon \overrightarrow{\overline{H}} . \bar{\alpha} . \overline{E}] - \Omega \frac{\partial}{\partial t}(S) \quad ...(8)$$

Where $S = \vec{E} \times \vec{H}$ and defined as Poynting Vector or Radiated power density.

The first term on the R.H.S. of (8) is total stored energy density, whereas the second term is a power density due to the change of electromagnetic momentum.

Again, combining (3), (4) and (8), we get

$$-\nabla . S = \frac{\partial}{\partial t}\left[\frac{1}{2}\vec{H} . (B + \Omega \times E) + \frac{1}{2}\overline{E}\,(D + \Omega \times E)\right] - \Omega . \frac{\partial}{\partial t} S$$

$$= \frac{\partial}{\partial t}\left[\frac{1}{2}\vec{H}.B\right]+\frac{\Omega}{2}.\frac{\partial S}{\partial t}+\frac{\partial}{\partial t}\left[\frac{1}{2}\overline{ED}\right]+\frac{\Omega}{2}\frac{\partial S}{\partial t}-\Omega.\frac{\partial S}{\partial t}$$

$$= \frac{\partial}{\partial t}\left(\frac{1}{2}\overline{H}.\overline{B}+\frac{1}{2}\overline{E}.\overline{D}\right)+\Omega\frac{\partial S}{\partial t}-\Omega\frac{\partial S}{\partial t}$$

Hence,

$$-\nabla.S = \frac{\partial}{\partial t}\left[\frac{1}{2}\overline{E}.\overline{D}+\frac{1}{2}\overline{H}.\overline{B}\right] \quad ...(9)$$

$$= \frac{\partial}{\partial t}\left(\frac{1}{2}\varepsilon E^2\right)+\frac{\partial}{\partial t}\left(\frac{1}{2}\mu H^2\right)$$

The right hand side of (9) is represented as U and termed as the total stored energy density.

3. Radiation resistance of antennas in lossy media

As we know, in a dissipative medium, the fields due to radiation consists of near-field and far-field components, like an antenna has three components in a loss-less medium. In particular for an electrically small dipole, the far-field is given by

$$E = \frac{\omega\mu}{4\pi r}p_e.\sin\theta\, e^{-jkr} \quad ...(10)$$

where p_e = The effective dipole moment = $(I_0 . h)$

r = The distance from the dipole centre to a point at an angle 'θ' from the dipole axis.

$k = (\beta - j\alpha)$, is the plane-wave complex propagation constant of the medium.

As, the intrinsic impedance of the medium, $\eta = \left(\frac{\omega\mu}{k}\right)$, the corresponding power density (far-field) is given as;

$$S = Re\left[\frac{E^2}{\eta}\right] = Re\left[\frac{E^2.k}{\omega\mu}\right] = \left(\frac{|E|^2.\beta}{\omega\mu}\right) \quad ...(11)$$

(Taking only real parts)

$$S_m = Se^{2\alpha r} \quad ...(12)$$

The total radiated power can be determined by integrating S_m over a sphere,

$$P'_{\text{rad}} = \iint S_m \sin\theta . r^2 . d\theta . d\phi$$

$$= \int_0^{2\pi}\int_0^{\pi}\frac{|E|^2\beta}{\omega\mu}.e^{\alpha r}.\sin\theta\, r^2.\ d\theta.d\phi 0$$

$$= 2\pi\int_0^{\pi}\frac{|E|^2.\beta}{\omega\mu}e^{2\alpha r}.r^2\sin\theta.d\theta$$

$$= 2\pi\int_0^{\pi}\frac{\omega\mu\beta(p_e)^2}{(4\pi r)^2}.e^{2\alpha r}.\sin^3\theta.r^2\, d\theta$$

$$= \frac{\omega\mu\beta(p_c)^2}{8\pi}\int\sin^3\theta.d\theta$$

$$P'_{\text{rad}} = \frac{\omega\mu\beta\,(p_c)^2}{8\pi} \times \frac{4}{3}$$

$$= \frac{\omega\mu\beta}{6\pi}\,(I_0 h)^2 \qquad ...(13)$$

If like in the case of antenna in free space, there is radiation resistance corresponding to input current (I), then

$$R'_{\text{rod}} = \frac{P'_{rod}}{I^2} = 20\,\frac{\beta_0}{\beta}\,(\beta h)^2$$

where,
β_0 = Phase-constant in air
β = Phase-constant in the medium

So, if $\beta_0 \to \beta$

$$R'_{\text{rad}} = 20\,(\beta_0 h)^2$$

$$= 20 \times \left(\frac{2\pi}{\lambda}\,h\right)^2 = 80\,\pi^2\left(\frac{h}{\lambda}\right)^2 \qquad ...(14)$$

λ = Operating wave-length
h = Height/length of the dipole

Similarly, the efficiency of the small antenna is defined as

$$\eta_r = \frac{R'_{rad}}{R_{in}} \qquad ...(15)$$

where R_{in} is the input resistance and it may differ considerably from R_{rad}, in lossy medium. Therefore, the modified power gain of the antenna is

$$G_m = \eta_r \,.\, D.$$

4. Solution of Maxwell's equation

As we know, that the general solution of Maxwell's equations for a fixed frequency is expressed as the combination of two types of solutions. Where each of them is characterized by the electric vector (E), and magnetic vector (H) as follows:

$$E = \eta_0 H < 45° = j\eta_0 H \qquad ...(1)$$

That is E and H are 45° out of phase.

Simplest way of solving Maxwell's equations is the decomposition of the field into Transverse Electric (TE) and Transverse Magnetic (TM) parts. The importance of this techneque rests on the fact that it is practically the way of getting a complete set of solutions. In this section, an alternative method of solving Maxwell's equation proposed by RUMSEY is described, where the field is decomposed into right handled and left-handed circularly polarized parts.

Let us consider, an *EM* wave is propagating along z-axis, so that $E_z = 0$, for the *TE* part and $H_z = 0$ for the *TM* part. That is, the *TM* part can be considered as due to electric dipoles parallel to the z-axis and the *TE* part due to magnetic dipoles parallel to the z-axis. Therefore, the circularly polarized parts can be considered as due to right and left handed circularly polarized (CP) dipoles parallel to the z-axis. The CP dipoles are nothing but the combination of electric and magnetic dipoles. In general, both the methods are found equivalent, however the C.P. method has the advantages in case propagation takes place over a sheet made of perfect conducting filaments, so that, at any point on this sheet, the conductivity is infinite in one direction but zero is the direction at right angles.

According to the Hertz Potential Method, the Maxwell's equation can be expressed as;

$$\nabla \times H = \frac{\partial E}{\partial t} = j\omega\varepsilon E \text{ and } \nabla \times E = -\frac{\partial H}{\partial t} = j\omega H \qquad ...(2)$$

where $\quad E = Ee^{j\omega t}$ and $H = He^{j\omega t} = j\omega\varepsilon E$

are solved by substituting :

$$E = E_e + E_m \text{ and } H = H_e + H_m \qquad ...(3)$$

In which $\quad E_e = \nabla \times a_x U$ and $H_m = \nabla \times \hat{a}_z\, V \qquad ...(4)$

where $\hat{a}_z$ = Unit-vector parallel to $\hat{z}$ axis.

and U and V satisfy the scalar wave equation

$$\nabla^2 f + \beta^2 f = 0 \qquad ...(5)$$

in which 'β' is a constant and defined as $\beta^2 = \omega^2\mu\,\varepsilon$.

If U and V are general solutions of (5), then Hertz Potential Method gives a general solution of (2). Where

$$E = \pm j\eta H$$

Let us consider, an alternative solutions of (2), for which $E = CH$, where C is being a new scalar constant. Therefore comparison with (2), show that, it is possible only if

$$C = \pm j\eta$$

where, $\quad \eta = \sqrt{\frac{\mu}{\varepsilon}} \qquad ...(6)$

That is, we will have two types of solutions of (2) corresponding to (+) and (–) signs, and they might be

$$E_1 = +j\eta H_1 \text{ and } E_2 = -j\eta H_2 \qquad ...(7)$$

Which is obtained by integrating the instantenous Poynting vector over the period and dividing by one period. The average poynting vector

$$P = R_e(E \times H^{\oplus}) \qquad ...(8)$$

Also, the instantaneous Poynting vector is expressed as :

$$P = Re\,(Ee^{j\omega t}) \times R_e\,(He^{j\omega t}) \qquad ...(9)$$

where $\quad Re\,(Ee^{j\omega t}) = E \cos \omega t$

$Re\,(He^{j\omega t}) = H \cos \omega t$

The quantities H and its complex conjugate $H^{\oplus}$ have the same space direction, but they differ in sign in their phase factors. That is if

$$H = |H|e^{j\omega t}, H^{\oplus} = He^{-j\omega t}$$

In present case, Eq, (8) gives, two types of Poynting vectors say P_1 and P_2.

$$P_1 = E_r \times H_i = \eta H_r \times H_i \quad \text{...(10a)}$$

$$P_2 = -E_r \times H_i = -\eta H_r \times H_i \quad \text{...(10b)}$$

where H_r and H_i one the real and imaginary parts of H_1 (or) H_2 respectively (acording to P_1 and P_2).

Equations (10a and 10b) reveal that, the instantaneous Poynting vector is independent of time.

Since, for a lossy medium with conductivity (σ), the Instinsic impedance (η) is complex quantity and defined by

$$\eta = \left(\frac{j\omega\mu}{j\omega\mu + \sigma}\right)^{1/2}$$

Equation (10) modified as follows;

$$P_1 = Re(\eta)\, H_r \times H_i \text{ and } P_2 = -Re(\eta)\, H_r \times H_i \quad \text{...(11)}$$

Eq. (7) reveals that, H_1 can be obtained, if we know E_1, but it is unknown that how the three components of E are related. Therefore, it will be much better to express the whole field by means of only one scalar function instead of three scalar functions which constitute three components.

This can be done by applying the Hertz Potential Technique to E_1, which results as follows :

$$E_1 = \nabla \times \nabla \times \hat{z}\, u_1 - \beta\nabla \times \hat{z}\, u_1 \quad \text{...(12)}$$

$$= \nabla\left(\frac{\partial u_1}{\partial z}\right) - \beta\nabla \times \hat{z}\, u_1 + \beta^2\, \hat{z}\, u_1 \quad \text{...(13)}$$

Similarly,

$$E_2 = \nabla \times \nabla \times \hat{z}\, u_2 + \beta\nabla \times \hat{z}\, u_2 \quad \text{...(14)}$$

$$= \left(\frac{\partial u_2}{\partial z}\right) + \beta\nabla \times \hat{z}\, u_2 \quad \text{...(15)}$$

Which are obtained by simply reversing the sign of β. The function u_1 is solution of (4) as well as source of E_1.

Let us consider

$$E = E_1 + E_2$$

where E_1 and E_2 are the components of E, corresponding to function u_1 and u_2. Then substitution of E_1 and E_2 from (12) and (14) gives;

$$E_1 + E_2 = E$$

$$= \nabla \times \nabla \times \hat{z}\,(u_2 + u_1) + \beta\nabla \times \hat{z}\,(u_2 - u_1) \quad \text{...(16)}$$

$$= \nabla \times \nabla \times \hat{z}\, u_+ + \beta\nabla \times (u) \quad \text{...(17)}$$

where $u_+ = u_2 + u_1$

and $u_- = u_2 - u_1$

And they are termed as the TM and TE Potential Functions for Electric field E.

That is, we can always find u_1 and u_2 to fit any such pair of Hertz Potential Functions and because of the generality of Hertz Potential Functions, we can also represent any solution of (2) as a linear combination of solutions of types one and two. That is, the representation of (14) is complete.

The orthogonality of the field E_1 and E_2 can be demonstrated with the help of the Reaction Method. Let us consider the reaction (coupling) between two sources '*a*' and '*b*'. The source '*a*' radiates a field or type E_1 and source '*b*' radiates a field or type E_2. The reaction between fields is represented by <*ab*> and is expressed by

$$<ab> = \oint\oint (E_a \times H_b - E_b \times H_a) \,.\, n\, dS \qquad ...(18)$$

where '*S*' is a closed surface separating sources '*a*' and '*b*' and '*n*' is vector pointing from '*a*' to '*b*'.

Simplication of (18), uing $E_a = j\eta H_a$ and $E_b = -j\eta H_b$, gives $<ab> = 0$. which implies that there is no reaction/coupling between two sources.

For examples, if sources '*a*' is a unit current generator connected to some antenna which radiates a field of type E_1, then <*ab*> is the open circuit voltage, this antenna receives away from source '*b*'. That $<ab> = 0$ means that the antenna which radiates field of type 'E_1' is blind to an incident field of types E_2.

APPENDIX - 2 : FINITE-DIFFERENCE TIME-DOMAIN (FTDT) TECHNIQUE

Finite-Difference Time-domain (FTDT) is a popular electromagnetic moduling technique. It is easy to understand and implement in software. Since, it is a time-domain technique, it can cover a wide frequency range with a single simulation run.

The FDTD method belongs to differential time domain numerical modeling methods. In this method Maxwell's (differential form) equations are simply modified to control-differencial equations, disenritized and implemnetal in software. The equations are solved in a leap-frog manner, that is the electric field is solved at a given instant in time, then the magnetic fields are calculated at the next instant in time, and the same process is repeated over and over again.

How Does FDTD Work

When differential form of Maxwell's equation is examined, $\left(i.e. \frac{\partial D}{\partial t} = \frac{\varepsilon \partial E}{\partial t} = \nabla \times H\right)$, it is found that the time -derivative of the electric field is dependent on the curl of the magnetic field.

This reveals that the any change in '*E*'-field is dependent on the change in the *H*-field across space. This is the basic of FDTD equations, *i.e.*, the new value of the E-field is dependent on the old value of the E-field (hence the difference in time) and the difference in the old value of H-field on either side of the E-field point in space. In this description, all the constant has been omitted, however the overall effect is same as described.

The 'H-field' is found in similar manner. The new value of the H-field is dependent on the old value of the H-field (hence the difference in time), and also dependent on the difference in the E-field on either side of the H-field point. Based on co-ordinate system, there are 1-D, 2-D and 3-D, FDTD techniques, and above description holds true for all the techniques.

Strengths and Weakness of FDTD : In general, every Model/Techniques has some strengths and some weakness. FDTD is very versatile and intisitive technique, so that users can easily understand how to use it, and know what do expect from a given model.

The wide frequency range quality of the technique is extremely useful in applications, where resonant frequencies are not known exactly, on any time that a broad band result is desired. Since FDTD is time-doamin technique, it lends itself to providing animation display of E/H fields movement throughout the model. FDTD allows the users to specify the material at all points within the computation domain.

FDTD allows the effects of a apertures and shielding to be determined directly. The fields both inside and outside a structure can also be found directly. FDTD also provides the E and H fields directly. Since FDTD requires that, the entire computation domain be grided, and grids must be small compare to the smallest wavelength and smallar than the smallest feature in the model, very large computational domain can be developed, which may in very long solution times. Since FDTD provides the E/H fields directly, every where in the computational domain, the fields value at some distance are difficult to find.

Maxwell's Equation and the Equivalent set of Finite Difference Equations : We know that the time-dependent Maxwell's Equations in free space are :

$$\frac{\partial E}{\partial t} = \frac{1}{\varepsilon_0} \nabla \times H \qquad ...(1)$$

$$\frac{\partial H}{\partial t} = \frac{1}{\mu_0} \nabla \times E \qquad ...(2)$$

and $$B = \mu_0 H \text{ and } D = \varepsilon_0 H \qquad ...(3)$$

where, E, H, B and D are assumed to be given functions of space and time. If we start with a simple one dimensional co-ordinate system, assuming only E_x and H_y components (z-is direction of propagation), above equation reduces to

$$\frac{\partial E_x}{\partial t} = + \frac{1}{\varepsilon_0} \frac{\partial H_y}{\partial z} \qquad ...(4a)$$

$$\frac{\partial H_y}{\partial t} = - \frac{1}{\mu_0} \frac{\partial E_x}{\partial z} \qquad ...(4b)$$

Taking a grid point of the space as

$(i, j, k) = (i\,\Delta x, j\,\Delta y, k\,\Delta z)$ and for any function of space and time, we put

$F(i\,\Delta x, j\,\Delta y, k\,\Delta z, n\Delta t) = F^n(i, j, k)$

Taking the central difference approximations for both the temporal and spatial derivatives yields;

$$\frac{E_x^{n+\frac{1}{2}}(k) - E_x^{n-\frac{1}{2}}(k)}{\Delta t} = + \frac{H_y^n\left(k+\frac{1}{2}\right) - H_y^n\left(k-\frac{1}{2}\right)}{\Delta x} \qquad ...(5)$$

$$\text{and}\quad \frac{H_y^{n+1}\left(k+\frac{1}{2}\right)-H_y^{n}\left(k+\frac{1}{2}\right)}{\Delta t} = -\frac{1}{\mu_0}\,\frac{E_x^{n+\frac{1}{2}}(k+1)-E_x^{n+\frac{1}{2}}(k)}{\Delta x} \qquad \text{...(6)}$$

where superscript, *i.e.*, '*n*' actually means a time $t = \Delta t \,.\, n$ and term '$n + 1$' means one time step later. The term 'K' actually means the distance, $z = \Delta x \,.\, K$.

In the formulation of Eqs. (5) and (6), it is considered that the $\vec{\text{E}}$ and $\vec{\text{H}}$ fields are inter leaved in both space and time. The magnetic field 'H' uses the argument $\left(k+\frac{1}{2}\right)$ and $\left(k-\frac{1}{2}\right)$ to indicate that the H-field values are assumed to be located between the E-field values. Similarly, the $\left(n+\frac{1}{2}\right)$ or $\left(n-\frac{1}{2}\right)$ superscript indicates that it occurs slightly after or before n, respectively. Fig (A2.1) illustrated the interleaving of the 'E' and H fields in space and time in FDTD formulation. To calculate $H_y\left(k+\frac{1}{2}\right)$, for instance, the neighbouring values of E_x at K and $(K + 1)$ are needed, and to calculate $E_x\,(K + 1)$, the value of H_y are needed at $\left(K+\frac{1}{2}\right)$ and $\left(K+1\frac{1}{2}\right)$.

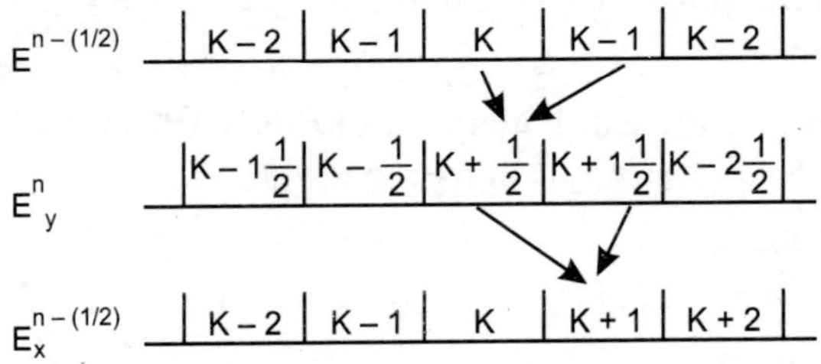

Fig. A2.1. Interleaving of the 'E' and 'H' fields

The equations (5) and (6) can be rearranged in an iterative algorithm as follows;

$$E_x^{n+\frac{1}{2}}(K) = E_x^{n-\frac{1}{2}}(K) - \frac{\Delta t}{\varepsilon_0\,.\,\Delta x}\left[H_y^n\left(K+\frac{1}{2}\right)-H_y^n\left(K-\frac{1}{2}\right)\right] \qquad \text{...(7)}$$

$$H_y^{n+1}\left(K+\frac{1}{2}\right) = H_y^n\left(K+\frac{1}{2}\right) - \frac{\Delta t}{\mu_0\,.\,\Delta x}\,[E_x^{n+\frac{1}{2}}(K+1)-E_x^{n+\frac{1}{2}}(K)] \qquad \text{...(8)}$$

These equations reveal that the calculation of E and H are interleaved in both space and time. For example, the new value of E_x is calculated from the previous value of E_x and the most recent value of H_y. This is the fundamental paradigm of the finite-difference time domain (FDTD) method.

It is also found that, both the equations are very similar, however as we know the values of ε_0 and μ_0 differs by several orders of magnitude, hence E_x and H_y will also differ by several orders of magnitude. This is circumvented by making the following change of variables.

$$\vec{E} = \sqrt{\frac{\varepsilon_0}{\mu_0}}\,E \qquad \text{...(9)}$$

Substituting, the value of $\vec{E}$ from (Eq. 9), into (7) and (8), gives :

$$\overline{E}_x^{\left(n+\frac{1}{2}\right)(K)} = \overline{E}_x^{n-\frac{1}{2}}(K) - \frac{1}{\sqrt{\mu_0\,\varepsilon_0}}\frac{\Delta t}{\Delta x}\left[H_y^x\left(K+\frac{1}{2}\right) - H_y^n\left(K-\frac{1}{2}\right)\right] \qquad ...(10)$$

and

$$H_y^{n+1}\left(K+\frac{1}{2}\right) = H_y^n\left(K+\frac{1}{2}\right) - \frac{1}{\sqrt{\varepsilon_0\mu_0}}\frac{\Delta t}{\Delta x}[\overline{E}^{\left(n+\frac{1}{2}\right)}(K+1) - \overline{E}_x^{n+\frac{1}{2}}(K)] \qquad ...(11)$$

Once, the cell size, Δx is chosen, then the time step 'Δt' is determined by

$$\Delta t = \frac{\Delta x}{2\varepsilon_0} \qquad ...(12)$$

in which 'c_0' is the speed of light in free-space.

Therefore, $$\frac{1}{\sqrt{\mu_0\varepsilon_0}}\frac{\Delta t}{\Delta x} = c_0 \cdot \frac{\Delta x / 2.c_0}{\Delta x} = \frac{1}{2} \qquad ...(13)$$

In order to determine time step, let us consider the hypothesis that an EM wave propagating in free space can not be faster than the speed of light. To propagate a distance to one cell the minimum time (Δt) will be $\frac{\Delta x}{c_0}$, however, for two dimensional case, it is needed to allow for the propagation in the diagonal direction, which brings the time requirement to be $\Delta t = \frac{\Delta x}{(\sqrt{2c_0})}$.

Similarly, for three dimension simulation, the minimum time required to be $\Delta t = \frac{\Delta x}{(\sqrt{3}\,c_0)}$. In general, requirement of minimum time may be summarized by the well-known "Courant condition"

$$t \leq \frac{\Delta x}{\sqrt{n}\,.\,c_0} \qquad ...(14)$$

where n is any value from to 1, 2, 3, ..., n.

Determine the cell size

Choosing the cell size to be used in a FDTD formulation, is similar to an approximation; enough sampling points must be taken into account to ensure that an adequate representation is made. The number of points per wave-length is dependent on many factors, however a good rule of thumb is 10 points per wave-length (λ). For example we are running simulation at 400 MHz, therefore in free space, EM energy will propagate at the wavelength

$$\lambda_0 = \frac{c}{400\,MHz} = \frac{3\times10^8}{4\times10^8\ \text{sec}^{-1}} = 0.75\ \text{m} = 75\ \text{cm}$$

Therefore, $$\Delta x = \frac{\lambda_0}{10} = 7.5\ \text{cm}$$

In another case, if EM simulation is taken in biological tissues, having a relative dielectric constant of 50 at 499 MHz, so

$$\lambda_m = \frac{c_0/\sqrt{50}}{400\,MHz}$$

$$\Delta x = \frac{\lambda_m}{10} = 1.06\ \text{cm}$$

Therefore, we would probability select a cell size of one centimeter.

Therefore, the time-increment is

$$\Delta t < \frac{\Delta r}{v} = \frac{1.06 \times 10^{-2}}{3 \times 10^{8} \frac{m}{s}} = 0.35 \times 10^{-11} \text{ sec.}$$

The solution grid for FD-TD technique is known as (Yee's Mesh) and it is shown in Fig. A2.2. Careful examination of the grid shows that the solution points for the electric field are spatially offset from the magnetic field solution points. This allows the electric or magnetic field at each node to be computed using the surrounding four magnetic or electric field values.

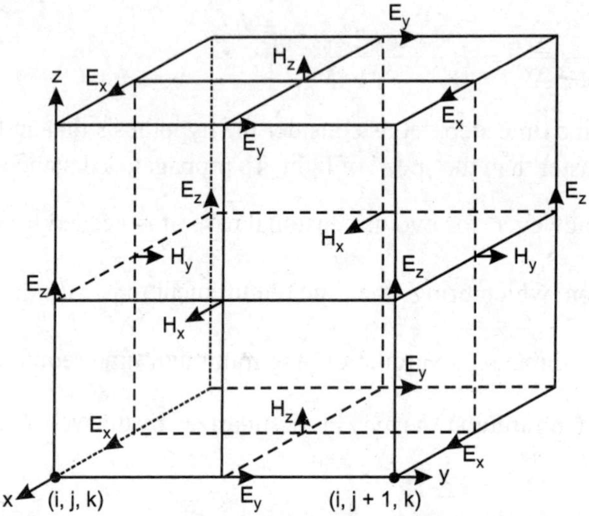

Fig. A. 2.2. Solution grid for FD-TD (Yee's mesh).

In addition to discretizsing $\vec{E}$ and $\vec{H}$ in space, in this method, the temporal changes in the field are also computed at descrite time-intervals. For the spatial grid sepations of Δx, Δy, and Δz and time-increment of Δt, the field may be written as

$$F(x, y, z, t) = F(i\,\Delta x, j\,\Delta y, k\,\Delta z, n\,\Delta t) = F^n(i, j, k) \qquad ...(15)$$

Grid size and Stability Criterion

As well as the size of space grid is concern, it must be such that over one increment, the EM field does not change significantly. That is, to have meaningful results, the linear dimension of the grid must be only a fraction of wave-length. Preferably, we shall choose, $\Delta x = \Delta y = \Delta z$. For computational stability, it is necessary to satisfy a relation between the space increments (Δs) and time increment Δt. In general a rigorous stability criterion is difficult to obtain for variables, ε and μ. However, for constant ε and μ it is required that

$$[(\Delta x)^2 + (\Delta y)^2 + (\Delta z)^2]^{1/2} > c\,\Delta t = \frac{1}{\sqrt{\mu\varepsilon}} . \Delta t \qquad ...(16)$$

where 'c' is the velocity of light. If c_{max} is maximum value of c, then we must choose

$$[(\Delta x)^2 + (\Delta y)^2 + (\Delta z)^2]^{1/2} > c_{max}\,\Delta t \qquad ...(17)$$

This requirement puts a restriction on Δt for chosen Δx, Δy and Δz.

FDTD Equations for TE and TEM waves :

As we know, the two common modes of Electromagnetic Wave Propagation are TE and TM modes. The Transverse Electric (TE) and Transverse Magnetic(TM) modes are characterized as follows:

1. Transverse Electric Wave (TE) : $H_z = 0, H_y = 0$ and $E_z \neq 0$ and the finite difference equations for the TE waves are :

$$H_z^{n+\frac{1}{2}}\left(i+\frac{1}{2}, j+\frac{1}{2}\right) = H_n^{n-\frac{1}{2}}\left(i+\frac{1}{2}, j+\frac{1}{2}\right) - \frac{\Delta r}{\Delta x}\left[E_y^n\left(i+1, j+\frac{1}{2}\right) - E_y^n\left(i, j+\frac{1}{2}\right)\right] + \frac{1}{z}\frac{\Delta r}{\Delta y}\left[E_x^n\left(i+\frac{1}{2}, j+1\right) - E_x^n\left(i+\frac{1}{2}, j\right)\right] \qquad ...(18a)$$

$$E_x^{n+1}\left(i+\frac{1}{2}, j\right) = E_x^n\left(i+\frac{1}{2}, j\right) + z\frac{\Delta r}{\Delta y}\left[H_z^{n+\frac{1}{2}}\left(i+\frac{1}{2}, j+\frac{1}{2}\right) - H_z^{n+\frac{1}{2}}\left(i+\frac{1}{2}, j-\frac{1}{2}\right)\right] \qquad ...(18b)$$

$$E_y^{n+1}\left(i+j+\frac{1}{2}\right) = -z\frac{\Delta r}{\Delta}\left[H_z^{n+\frac{1}{2}}\left(i+\frac{1}{2}, j+\frac{1}{2}\right) - H_z^{n+\frac{1}{2}}\left(i-\frac{1}{2}, j+\frac{1}{2}\right)\right] - H_z^{n+1}\left(i-\frac{1}{2}, j+\frac{1}{2}\right)\Big] \qquad ...(18c)$$

2. Transverse Magnetic (TM) Waces : $E_x = 0$, $E_y = 0$ and $H_z \neq 0$ and the finite difference equations for the TM waves are :

$$E_z^{n+1}(i, j) = E_n^z(i, j) + z\frac{\Delta r}{\Delta x}\left[H_y^{n+\frac{1}{2}}\left(i+\frac{1}{2}, j\right) - H_y^{n+\frac{1}{2}}\left(i-\frac{1}{2}, j\right)\right] - z\frac{\Delta r}{\Delta y}\left[H_x^{n+\frac{1}{2}}\left(i, j+\frac{1}{2}\right) - H_x^{n+\frac{1}{2}}\left(i, j-\frac{1}{2}\right)\right] \qquad ...(19a)$$

$$H_x^{n+\frac{1}{2}}\left(i, j+\frac{1}{2}\right) = H_x^{n-\frac{1}{2}}\left(i, j+\frac{1}{2}\right) - \frac{1}{z}\frac{\Delta r}{\Delta y}[E_z^n(i+1, j) - E_z^n(i, j)] \qquad ...(19b)$$

and
$$H_y^{n+1}\left(i+\frac{1}{2}, j\right) = H_y^{n-\frac{1}{2}}\left(i+\frac{1}{2}, j\right) + \frac{1}{z}\frac{\Delta r}{\Delta x}[E_z^n(i+1, j) - E_z^n(i, j)] \qquad ...(19c)$$

These are equations, (18*a*–18*c* and 19*a*–19*c*) with reference to wave propagation in the z-direction (z-axis) provided Electric field (E) and Magnetic field (H) are directed along x and y-axis respectively.

Despite, various computational burden of this method, the FDTD approach is being used to solve a wide range of dynamic electromagnetic problems, such as the design of microwave and millimeter-wave integrated circuit (MMICS). Design of sub picosecond photonic devics, as well as inter connections between high speed digital circuits in a multichip modules (MCMs) is also done using this method.

APPENDIX - 3: MATHEMATICAL FOR MULE

(a) Trigonometric Formulae

$$\sin^2\theta + \cos^2\theta = 1,$$

$$1 + \tan^2\theta = \sec^2\theta$$

$$1 + \cot^2\theta = \operatorname{cosec}^2\theta$$

$$\cos 2\theta = 2\cos^2\theta - 1 = 1 - 2\sin^2\theta$$

$$\sin(\theta \pm \phi) = \sin\theta\cos\phi \pm \cos\theta\sin\phi$$

$$\cos(\theta \pm \phi) = \cos\theta\cos\phi \mp \sin\theta\sin\phi$$

$$2\sin\theta\sin\phi = \cos(\theta - \phi) - \cos(\theta + \phi)$$

$$2\sin\theta\cos\phi = \sin(\theta + \phi) - \sin(\theta - \phi)$$

$$2\cos\theta\cos\phi = \cos(\theta + \phi) + \cos(\theta - \phi)$$

$$\tan 2\theta = \frac{2\tan\theta}{1 - \tan^2\theta},\ \tan(\theta + \phi) = \frac{\tan\theta \pm \tan\phi}{1 \mp \tan\theta\tan\phi}$$

$$\sin\theta = \frac{e^{j\theta} - e^{-j\theta}}{2j},\ \cos\theta = \frac{e^{j\theta} + e^{-j\theta}}{2}$$

$$e^{j\theta} = \sin\theta + j\cos\theta,\ 1 \text{ rad} = 57.296°$$

(b) Complex Variables

$$z = x + jy = r < \theta\ re^{j\theta} = r(\cos\theta + j\sin\theta)$$

$$j = \sqrt{(1)},\ \frac{1}{j} = -j, j^2 = -1$$

$$z^* = x - jy < -\theta = re^{-j\theta}$$

$$= r(\cos\theta - j\sin\theta)$$

$$z_1 \pm z_2 = (x_1 + x_2) \pm j(y_1 + y_2)$$

$$z_1 . z_2 = r_1 . r_2\, e^{j(\theta_1 + \theta_2)} = r_1 . r_2 < (\theta_1 + \theta_2)$$

$$= r_1 . r_2 [\cos(\theta_1 + \theta_2) + j\sin(\theta_1 + \theta_2)]$$

$$\frac{z_1}{z_2} = \frac{r_1}{r_2} e^{j(\theta_1 - \theta_2)} = \frac{r_1}{r_2} < (\theta_1 + \theta_2)$$

$$= \frac{r_1}{r_2} [\cos(\theta_1 + \theta_2) + j\sin(\theta_1 + \theta_2)]$$

$$\sqrt{z} = \sqrt{(x + jy)} = \sqrt{r}\, e^{j\theta/2} = \sqrt{r} < \frac{\theta}{2}$$

$$= \sqrt{r}\left[\cos\frac{\theta}{2} + j\sin\frac{\theta}{2}\right]$$

$$z^n = (x+jy)^n = r^n < n\theta = r^n\, e^{jn\theta}$$

$$= r^n(\cos n\theta + j\sin n\theta)$$

$$z^{\frac{1}{n}} = (x+jy)^{\frac{1}{n}} = r^{(1/n)}\,.\, e^{j\theta/n} = r^{\frac{1}{n}}\,.< \frac{\theta}{n} + \frac{2\pi k}{n}$$

(where $k = 0, 1, 2, 3, ..., n-1$)

$$ln\,(re^{j\theta}) = ln\, r + ln\; e^{j\theta}$$

$$= ln\, r + j\theta + j2k\pi$$

(*c*) Hybolic Functions

$$\sinh x = \left(\frac{e^x - e^{-x}}{2}\right)$$

$$\cosh x = \frac{e^x - e^{-x}}{2}$$

$$\sin jx = j\sinh x \quad \text{and} \quad \cos jx = \cosh x$$

$$\sinh jx = j\sin x \quad \text{and} \quad \cosh jx = \cos x$$

$$\sinh(x \pm y) = \sinh x\cosh y \pm \cosh x\sinh y$$

$$\cosh(x \pm y) = \cosh x\cosh y \pm \sinh x\sinh y$$

$$\sinh(x \pm jy) = \sinh x\cos y \pm j\cos x\sin y$$

$$\cosh(x \pm jy) = \cosh x\cos y \pm j\sinh x\sin y$$

$$\tanh(x \pm jy) = \frac{\sinh 2x}{\cosh 2x + \cos 2y} \pm \frac{j\sin 2y}{\cosh 2x + \cos 2y}$$

$$\cosh^2 x - \sinh^2 x = 1$$

$$\text{sech}^2 x + \tanh^2 x = 1$$

$$\sin(x \pm jy) = \sin x\cosh y \pm j\cos x\sinh y$$

$$\cos(x \pm jy) = \cos x\cosh y \pm j\sin x\sinh y$$

(*d*) Some Common Identities

$$e^x = 1 + x + \frac{x^2}{2!} + \frac{x^3}{3!} + ...$$

$$e = 2.7182$$

$$\text{If } |x| \leq 1,\ ln\,(1+x) = x$$

$$e^x = 1 + x$$

$$(1 \pm x)^n = 1 \pm nx$$

$$\sin x \approx x \text{ or } \lim_{x \to 0} \frac{\sin x}{x} = 1$$

$$\cos x \approx 1, \tan x \approx x$$

APPENDIX - 4 : DERIVATIVES

If $U = U(x)$, $V = V(x)$ and a = constant

$$\frac{d}{dx}(aU) = a\frac{dU}{dx}$$

$$\frac{d}{dx}\left(\frac{U}{V}\right) = \frac{U\dfrac{dV}{dx} + V\dfrac{dU}{dx}}{V^2}$$

$$\frac{d}{dx}\log_a U = \frac{\log_a e}{U}\cdot\frac{dU}{dx}$$

$$\frac{d}{dx}a^U = a^U \ln a\frac{dU}{dx}$$

$$\frac{d}{dx}e^U = e^U\frac{dU}{dx}$$

$$\frac{d}{dx}\sin U = \cos U\cdot\frac{dU}{dx}$$

$$\frac{d}{dx}\sinh U = \cosh U\frac{dU}{dx}$$

$$\frac{d}{dx}\cosh U = \sinh U\frac{dU}{dx}$$

$$\frac{d}{dx}\tanh U = \text{sech}^2 U\frac{dU}{dx}$$

APPENDIX - 5 : INDEFINITE INTEGRALS

If $U = U(x)$, $V = V(x)$, and a = constant

$$\int a\,dx = ax + C$$

$$\int U\,dV = UV - \int V\,dU \qquad \text{(integration by parts)}$$

$$\int U^n\,dU = \frac{U^{n+1}}{n+1} + C,\ n \neq -1$$

$$\int \frac{dU}{U} = \ln U + C$$

$$\int a^U\,dU = \frac{a^U}{\ln a} + C, a > 0,\ a \neq 1$$

$$\int e^U\,dU = e^U + C$$

$$\int e^{ax}\,dx = \frac{1}{a}e^{ax} + C$$

$$\int xe^{ax}\,dx = \frac{e^{ax}}{a^2}(ax-1)+C$$

$$\int x^2 e^{ax}\,dx = \frac{e^{ax}}{a^3}(a^2x^2-2ax+2)+C$$

$$\int \ln x\,dx = x\ln x - x + C$$

$$\int \sin ax\,dx = \frac{1}{a}\cos ax + C$$

$$\int \cos ax\,dx = \frac{1}{a}\sin ax + C$$

$$\int \tan ax\,dx = \frac{1}{a}\ln(\sec ax) + C = -\frac{1}{a}\ln(\cos ax) + C$$

$$\int \sec ax\,dx = \frac{1}{a}\ln(\sec ax + \tan ax) + C$$

$$\int \sin^2 ax\,dx = \frac{x}{2} - \frac{\sin 2ax}{4a} + C$$

$$\int \cos^2 ax\,dx = \frac{x}{2} + \frac{\sin 2ax}{4a} + C$$

$$\int x\sin ax\,dx = \frac{1}{a^2}(\sin ax - ax\cos ax) + C$$

$$\int x\cos ax\,dx = \frac{1}{a^2}(\cos ax + ax\sin ax) + C$$

$$\int e^{ax}\sin bx\,dx = \frac{e^{ax}}{a^2+b^2}(a\sin bx - b\cos bx) + C$$

$$\int e^{x}\cos bx\,dx = \frac{e^{ax}}{a^2+b^2}(a\cos bx + b\sin bx) + C$$

$$\int \sin ax\sin bx\,dx = \frac{\sin(a-b)x}{2(a-b)} - \frac{\sin(a+b)x}{2(a+b)} + C,\ a^2 \neq b^2$$

$$\int \sin ax\cos bx\,dx = -\frac{\cos(a-b)x}{2(a-b)} - \frac{\cos(a+b)x}{2(a-b)} + C,\ a^2 \neq b^2$$

$$\int \sin ax\cos bx\,dx = \frac{\cos(a-b)x}{2(a-b)} - \frac{\cos(a+b)x}{a(a+b)} + C,\ a^2 \neq b^2$$

$$\int \cos ax\cos bx\,dx = \frac{\sin(a-b)x}{2(a-b)} + \frac{\sin(a+b)x}{2(a+b)} + C,\ a^2 \neq b^2$$

$$\int \sinh ax\,dx = \frac{1}{a}\cosh ax + C$$

$$\int \cosh ax\,dx = \frac{1}{a}(\sinh ax) + C$$

$$\int \tanh ax\,dx = \frac{1}{a}\ln(\cosh ax) + C$$

$$\int \frac{dx}{x^2 + a^2} = \frac{1}{a}\tan^{-1}\left(\frac{x}{a}\right) + C$$

$$\frac{x\,dx}{x^2 + a^2} = \frac{1}{2}\ln(x^2 + a^2) + C$$

$$\frac{x^2\,dx}{x^2 + a^2} = x - a\tan^{-1}\left(\frac{x}{a}\right) + C$$

$$\int \frac{dx}{x^2 - a^2} = \begin{cases} \dfrac{2}{2a}\ln\left(\dfrac{x-a}{x+a}\right) + C, & x^2 > a^2 \\ \dfrac{1}{2a}\ln\left(\dfrac{a-x}{a+x}\right) + C, & x^2 < a^2 \end{cases}$$

$$\int \frac{dx}{\sqrt{a^2 - x^2}} = \sin^{-1}\left(\frac{x}{a}\right) + C$$

$$\int \frac{dx}{\sqrt{x^2 \pm a^2}} = \ln(x + \sqrt{x^2 \pm a^2}) + C$$

$$\int \frac{x\,dx}{\sqrt{x^2 + a^2}} = \sqrt{x^2 + a^2} + C$$

$$\int \frac{dx}{(x^2 + a^2)^{3/2}} = \frac{x/a^2}{\sqrt{x^2 + a^2}} + C$$

$$\int \frac{x\,dx}{(x^2 + a^2)^{3/2}} = -\frac{1}{\sqrt{x^2 + a^2}} + C$$

$$\int \frac{x^2\,dx}{(x^2 + a^2)^{3/2}} = \ln\left(\frac{\sqrt{x^2 + a^2}}{a} + \frac{x}{a}\right) - \frac{x}{\sqrt{a^2 + b^2}} + C$$

$$\int \frac{dx}{(x^2 + a^2)^2} = \frac{1}{2a^2}\left(\frac{x}{x^2 + a^2} + \frac{1}{a}\tan^{-1}\left(\frac{x}{a}\right)\right) + C$$

APPENDIX - 6 : DEFINITE INTEGRALS

$$\int_0^{\pi} \sin mx \sin nx\,dx = \int_0^{\pi} \cos mx \cos nx\,dx = \begin{cases} 0, & m \neq n \\ \pi/2 & m = n \end{cases}$$

$$\int_0^{\pi} \sin mx \cos nx\,dx = \begin{cases} 0, & m + n = even \\ \left(\dfrac{2m}{m^2 - n^2}\right), & m + n = odd \end{cases}$$

$$\int_0^{2\pi} \sin mx \sin nx\,dx = \int_{-\pi}^{\pi} \sin mx \sin nx\,dx = \begin{cases} 0, & m \neq n \\ \pi, & m \neq n \end{cases}$$

$$\int_0^{\infty} \frac{\sin ax}{x}\,dx = \begin{cases} \pi/2, & a > 0 \\ 0, & a = 0 \\ -\pi/2, & a < 0 \end{cases}$$

$$\int_0^\infty \frac{\sin^2 x}{x}\,dx = \frac{\pi}{2}$$

$$\int_0^\infty \frac{\sin^2 ax}{x^2}\,dx = |a|\frac{\pi}{2}$$

$$\int_0^\infty x^n e^{-ax}\,dx = \frac{n!}{a^{n+1}}$$

$$\int_0^\infty e^{-ax^2}\,dx = \frac{1}{2}\sqrt{\frac{\pi}{a}}$$

$$\int_{-\infty}^\infty e^{-ax^2}\,dx = \sqrt{\frac{\pi}{a}}$$

$$\int_{-\infty}^\infty e^{-(ax^2+bx+c)}\,dx = \sqrt{\frac{\pi}{a}}\,e^{(b^2-4ac)/4a}$$

$$\int_0^\infty e^{-ax}\cos bx\,dx = \frac{a}{a^2+b^2}$$

$$\int_0^\infty e^{-ax}\sin bx\,dx = \frac{b}{a^2+b^2}$$

APPENDIX - 7 : VECTOR IDENTITIES

If A and B are vector fields while U and V are scalar fields, then through vector identities they are related as

$$\nabla(U+V) = \nabla U + \nabla V$$

$$\nabla(UV) = U\nabla V + V\nabla U$$

$$\nabla V^n = nV^{n-1}\nabla V, \text{ where '}n\text{' is integer.}$$

$$\nabla(A\,.\,B) = (\mathrm{A}\,.\,\nabla)B + (B\,.\,\nabla)A + A\times(\nabla\times B) + B\times(\nabla\times A)$$

$$\nabla\,.\,(A+B) = \nabla\,.\,A + \nabla\,.\,B$$

$$\nabla\,.\,(A\times B) = \mathrm{B}\,.\,(\nabla\times A) - A\,.\,(\nabla\times B)$$

$$\nabla\,.\,(VA) = V\nabla\,.\,A + A\,.\,\nabla V$$

$$\nabla\,.\,(\nabla V) = \nabla^2 V$$

$$\nabla\,.\,(\nabla\times A) = 0$$

$$\nabla\times(A+B) = \nabla\times A + \nabla\times B$$

$$\nabla\times(A\times B) = A(\nabla\,.\,B) - B(\nabla\,.\,A) + (B\,.\,\nabla)A - (A\,.\,\nabla)B$$

$$\nabla \times (VA) = \nabla V \times A + V(\nabla \times A)$$

$$\nabla \times (\nabla V) = 0$$

$$\nabla \times (\nabla \times A) = (\nabla . A) - \nabla^2 A$$

$$\int_L A . dl = \int_S \nabla \times A . dS$$

$$\int_L V\, dl = -\int_S \nabla V \times dS$$

$$\int_S A . dS = \int_v \nabla . A\, dv$$

$$\int_S V\, dS = \int_v \nabla V\, dv$$

$$\int_S A \times dS = -\int_v \nabla \times A\, dv$$

APPENDIX - 8 : MATERIAL CONSTANTS

Table 9.1 Conductivity of some common materials at $t = 20°C$.

Material		Conductivity (siemens/meter)
(*a*)	*Conductors*	
	Silver	6.1×10^7
	Copper (standard annealed)	5.8×10^7
	Gold	4.1×10^7
	Aluminium	3.5×10^7
	Tungsten	1.6×10^7
	Zinc	1.7×10^7
	Brass	1.1×10^7
	Iron (pure)	10^7
	Lead	5×10^6
	Mercury	1×10^6
	Carbon	3×104
	Water (sea)	4
(*b*)	*Semiconductors*	

	Germanium (pure)	2.2
	Silicon (pure)	4.4×10^{-4}
(*c*)	*Insulators*	
	Water (distilled)	10^{-4}
	Earth (dry)	10^{-5}
	Bakelite	10^{-10}
	Paper	10^{-11}
	Glass	10^{-12}
	Porcelain	10^{-12}
	Mica	10^{-15}
	Paraffin	10^{-15}
	Rubber (hard)	10^{-15}
	Quartz (fused)	10^{-17}
	Wax	10^{-17}

Table 9.2 : Dielectric constant or relative permittivity (ε_r) and Strength of some common materials.

Material	Dielectric Constant ε_r (Dimensionless)	Dielectric Strength E (V/m)
Barium titanate	1200	7.5×10^6
Water (sea)	80	–
Water (distilled)	81	–
Nylon	8	–
Paper	7	12×10^6
Glass	5 – 10	35×10^6
Mica	6	70×10^6
Porcelain	6	–
Bakelite	5	20×10^6
Quartz (fused)	5	30×10^6
Rubber (hard)	3.1	25×10^6

Wood	2.5 – 8.0	–
Polystyrene	2.55	–
Polypropyolene	2.25	–
Paraffin	2.2	30×10^6
Petroleum oil	2.1	12×10^6
Air (1 atm.)	1	3×10^6

Table 9.3 : Relative permeability (μ_r) of some materials.

	Material	Relative permeability (μ_r)
(*a*)	*Diamagnetic*	
	Bismuth	0.999833
	Mercury	0.999968
	Silver	0.9999736
	Lead	0.9999831
	Copper	0.9999906
	Water	0.9999912
	Hydrogen (s.t.p) ≃	1.0
(*b*)	Paramagnetic	
	Oxygen (s.t.p.)	0.999998
	Air	1.000000037
	Aluminium	1.000021
	Tungsten	1.00008
	Platinum	1.0003
	Manganese	1.001
(*c*)	Ferromagnetic	
	Cobalt	250
	Nickel	600
	Soft	5000
	Silicon-iron	7000

Table 9.4 : Depth of typical metals at different (T).

Metal	Conductivity σ(S/m)	Skin-depth δ(m)	Surface resistivity Rs (Ω)
Copper (77°K)	18×10^7	$0.37 f^{-1/2}$	$1.5 \times 10^{-7} f^{1/2}$
Copper (300°K)	5.80×10^7	$0.066 f^{-1/2}$	$2.61 \times 10^{-7} f^{1/2}$
Brass (300°K)	1.57×10^7	$0.127 f^{-1/2}$	$5.01 \times 10^{-7} f^{1/2}$
Silver (300°K)	6.17×10^7	$0.0642 f^{-1/2}$	$2.52 \times 10^{-7} f^{1/2}$

(Where 'f' is operating frequency)

Index

ELECTROMAGNETIC FIELDS AND WAVES

R.L. YADAVA

FEEDBACK FORM

1. Which topics from your examination point of view are inadequately or not covered in this book?

____

____

2. Is there any misprints/ mistakes/ factual inaccuracy in this book? If so, please specify.

____

____

3. What is your assessment of this book as regards the presentation of the subject-matter, expression, precision and price in relation to the other recommended books available on this subject?

____

____

4. Which competing books you regard as better than this? Please mention their author and publishers.

____

____

5. Any further suggestions/comments you would like to make for the improvement of the book?

____

____

PS: If needed, please attach a separate sheet.

Name : ______________________

College : ______________________

Address : ______________________

Email : ______________________

Mobile : ______________________

Please mail the feedback form at:

KHANNA BOOK PUBLISHING CO. (P) LTD.

4C/4344, Ansari Road, Daryaganj, New Delhi-110002

You can also email this feedback form at contact@khannabooks.com or Whatsapp at +91 – 99109 09320

BEST FEEDBACK WILL BE REWARDED SUITABLY